THIRD EDITION

Fundamentals of Complex Analysis
with Applications to Engineering and Science

E. B. Saff
Professor of Mathematics
Vanderbilt University

A. D. Snider
Professor of Electrical Engineering
University of South Florida

With an appendix by
Lloyd N. Trefethen
Professor of Numerical Analysis
Oxford University
and
Tobin Driscoll
Professor of Mathematics
University of Delaware

Pearson Education, Inc.
Upper Saddle River, New Jersey 07458

Library of Congress Cataloging-in-Publication Data

Saff, E. B.
Fundamentals of complex analysis with applications for engineering and science / E.B.
Saff, A.D. Snider ; with an appendix by Lloyd N. Trefethen and Tobin Driscoll.
    p. cm.
Rev. ed. of: Fundamentals of complex analysis for mathematics, science, and engineering.
2nd ed. c1993.
Includes bibliographical references.
ISBN 0-13-907874-6
1. Mathematical analysis. 2. Functions of complex variables. I. Title: Fundmentals of
complex analysis. II. Snider, Arthur David. III. Saff, E. B., Fundamentals of complex
analysis for mathematics, science, and engineering. IV. Title.

QA300.S18 2003
515–dc21   2002037079

Editor-in-Chief: *Sally Yagan*
Acquisition Editor: *George Lobell*
Vice President/Director of Production and Manufacturing: *David W. Riccardi*
Executive Managing Editor: *Kathleen Schiaparelli*
Senior Managing Editor: *Linda Mihatov Behrens*
Production Editor: *Bob Walters*
Manufacturing Buyer: *Michael Bell*
Manufacturing Manager: *Trudy Pisciotti*
Marketing Manager: *Halee Dinsey*
Marketing Assistant: *Rachel Beckman*
Editorial Assistant: *Jennifer Brady*
Art Director: *Jayne Conte*
Creative Director: *Carole Anson*
Cover Design: *Bruce Kenselaar*
Cover Photo: *Eric Curry/COBIS/Tecmap Corporation*
Art Studio: *MacroTEX Services*

 © 2003, 1993, 1976 Pearson Education, Inc.
Pearson Education, Inc.
Upper Saddle River, New Jersey 07458

Printed in the United States of America
10 9 8 7 6 5 4 3 2 1

ISBN 0-13-907874-6

Pearson Education LTD., *London*
Pearson Education Australia PTY, Limited, *Sydney*
Pearson Education Singapore, Pte. Ltd
Pearson Education North Asia Ltd, *Hong Kong*
Pearson Education Canada, Ltd., *Toronto*
Pearson Educacion de Mexico, S.A. de C.V.
Pearson Education - Japan, *Tokyo*
Pearson Education Malaysia, Pte. Ltd

To Loretta and Dawn,
    who have added real joy to our complex lives

Contents

Preface

The raison d'existence for *Fundamentals of Complex Analysis with Applications to Engineering and Science, 3/e* is our conviction that engineering, science, and mathematics undergraduates who have completed the calculus sequence are capable of understanding the basics of complex analysis and applying its methods to solve engineering problems. Accordingly, we address ourselves to this audience in our attempt to make the fundamentals of the subject more easily accessible to readers who have little inclination to wade through the rigors of the axiomatic approach. To accomplish this goal we have modeled the text after standard calculus books, both in level of exposition and layout, and have incorporated engineering applications throughout the text so that the mathematical methodology will appear less sterile to the reader.

To be more specific about our mode of exposition, we begin by addressing the question most instructors ask first: To what extent is the book self contained, i.e., which results are proved and which are merely stated? Let us say that we have elected to include all the proofs that reflect the spirit of analytic function theory and to omit most of those that involve deeper results from real analysis (such as the convergence of Riemann sums for complex integrals, the Cauchy criterion for convergence, Goursat's generalization of Cauchy's theorem, or the Riemann mapping theorem). Moreover, in keeping with our philosophy of avoiding pedantics, we have shunned the ordered pairs interpretation of complex numbers and retained the more intuitive approach (grounded in algebraic field extensions).

Cauchy's theorem is given two alternative presentations in Chapter 4. The first is based on the deformation of contours, or what is known to topologists as homotopy. We have taken some pains to make this approach understandable and transparent to the novice because it is easy to visualize and to apply in specific situations. The second treatment interprets contour integrals in terms of line integrals and invokes Green's theorem to complete the argument. These parallel developments constitute the two parts of Section 4 in Chapter 4; either one may be read, and the other omitted, without disrupting the exposition (although it should not be difficult to discern our preference, from this paragraph).

Steady state temperature patterns in two dimensions are, in our opinion, the most familiar instances of harmonic functions, so we have principally chosen this interpretation for visualization of the theorems of analytic function theory. This application receives attention throughout the book, with special emphasis in Chapter 7 in the context of conformal mapping. There we draw the distinction between direct methods, wherein a mapping must be constructed to solve a specific problem, and indirect methods that postulate a mapping and then investigate which problems it solves. In doing so we hope to dispel the impression, given in many older books, that all applications of the technique fall in the latter category.

In this third edition L. N. Trefethen and T. Driscoll have updated an appendix that reflects the progress made in recent years on the numerical construction of conformal mappings. A second appendix compiles a listing of some useful mappings having closed form expressions.

Linear systems analysis is another application that recurs in the text. The basic ideas of frequency analysis are introduced in Chapter 3 following the study of the transcendental functions; Smith charts, circuit synthesis, and stability criteria are addressed at appropriate times; and the development culminates in Chapter 8 with the exposition of the analytic-function aspects of Fourier, Mellin, Laplace, Hilbert, and z transforms, including new applications in signal processing and communications. We hope thereby that our book will continue to serve the reader as a reference resource for subsequent coursework in these areas.

Features of the Third Edition

Novel features of the third edition are a discussion of the Riemann sphere, adding substance to the pragmatic concept of the "point at infinity" in complex analysis; an introduction to functional iteration and the picturesque Julia sets that thereby manifest themselves in the complex plane; an early exploration of the enrichment that the complex viewpoint provides in the analysis of polynomials and rational functions; and an introductory survey of harmonic function methods for calculating equilibrium temperatures for simple geometries. Optional sections are indicated with an asterisk so that readers can select topics of special interest. Summaries and suggested readings appear at the end of each chapter. As in previous editions, the text is distinguished by its wealth of worked-out examples that illustrate the theorems, techniques, and applications of complex analysis.

Instructors (and curious students) may benefit from a MATLAB toolbox developed by Francisco Carreras, available by Internet download from the web site

`http://ee.eng.usf.edu/people/snider2.html`

(click on complextools.zip). Instructions for its use are detailed in the file comp-man.doc. The toolbox provides graphic onscreen visualizations and animations of the algebraic manipulations of complex numbers and the common conformal maps, as well as a introductory guide for designing Joukowski airfoils.

A downloadable .pdf file of the inevitable errata that our helpful readers report to us is also available at this site.

The authors wish to acknowlege our mentors, Joseph L. Walsh and Paul Garabedian, who have inspired our careers, and to express their gratitude to Samuel Garrett, our longtime colleague at the University of South Florida; to acquisitions editor

George Lobell for encouraging this project; to Adam Lewenberg for providing the art work and technical support; to our production editor Bob Walters for his guidance in converting this work from manuscript to book; and to the following mathematicians, whose critical commentary contributed enormously to the development of the text:

Carlos Berenstein, University of Maryland
Keith Kearnes, University of Colorado
Dmitry Khavinson, University of Arkansas
Donald Marshall, University of Washington (Chapters 1-4, only)
Mihai Putinar, University of California at Santa Barbara
Sergei Suslov, Arizona State University
Rebecca Wahl, Butler University
G. Brock Williams, Texas Tech University

E. B. Saff
esaff@math.vanderbilt.edu

A. D. Snider
snider@eng.usf.edu

Chapter 1

Complex Numbers

1.1 The Algebra of Complex Numbers

To achieve a proper perspective for studying the system of complex numbers, let us begin by briefly reviewing the construction of the various numbers used in computation.

We start with the rational numbers. These are ratios of integers and are written in the form m/n, $n \neq 0$, with the stipulation that all rationals of the form n/n are equal to 1 (so we can cancel common factors). The arithmetic operations of addition, subtraction, multiplication, and division with these numbers can always be performed in a finite number of steps, and the results are, again, rational numbers. Furthermore, there are certain simple rules concerning the order in which the computations can proceed. These are the familiar commutative, associative, and distributive laws:

Commutative Law of Addition

$$a + b = b + a$$

Commutative Law of Multiplication

$$ab = ba$$

Associative Law of Addition

$$a + (b + c) = (a + b) + c$$

Associative Law of Multiplication

$$a(bc) = (ab)c$$

Distributive Law

$$(a + b)c = ac + bc,$$

for any rationals a, b, and c.

1

Notice that the rationals are the only numbers we would ever need, to solve equations of the form

$$ax + b = 0.$$

The solution, for nonzero a, is $x = -b/a$, and since this is the ratio of two rationals, it is itself rational.

However, if we try to solve quadratic equations in the rational system, we find that some of them have no solution; for example, the simple equation

$$x^2 = 2 \tag{1}$$

cannot be satisfied by any rational number (see Prob. 29 at the end of this section). Therefore, to get a more satisfactory number system, we extend the concept of "number" by appending to the rationals a new symbol, mnemonically written as $\sqrt{2}$, which is defined to be a solution of Eq. (1)). Our revised concept of a number is now an expression in the standard form

$$a + b\sqrt{2}, \tag{2}$$

where a and b are rationals. Addition and subtraction are performed according to

$$(a + b\sqrt{2}) \pm (c + d\sqrt{2}) = (a \pm c) + (b \pm d)\sqrt{2}. \tag{3}$$

Multiplication is defined via the distributive law with the proviso that the square of the symbol $\sqrt{2}$ can always be replaced by the rational number 2. Thus we have

$$(a + b\sqrt{2})(c + d\sqrt{2}) = (ac + 2bd) + (bc + ad)\sqrt{2}. \tag{4}$$

Finally, using the well-known process of *rationalizing the denominator*, we can put the quotient of any two of these new numbers into the standard form

$$\frac{a + b\sqrt{2}}{c + d\sqrt{2}} = \frac{a + b\sqrt{2}}{c + d\sqrt{2}} \frac{c - d\sqrt{2}}{c - d\sqrt{2}} = \frac{ac - 2bd}{c^2 - 2d^2} + \frac{bc - ad}{c^2 - 2d^2}\sqrt{2}. \tag{5}$$

This procedure of "calculating with radicals" should be very familiar to the reader, and the resulting arithmetic system can easily be shown to satisfy the commutative, associative, and distributive laws. However, observe that the symbol $\sqrt{2}$ has not been absorbed by the rational numbers painlessly. Indeed, in the standard form (2) and in the algorithms (3), (4), and (5) its presence stands out like a sore thumb. Actually, we are only using the symbol $\sqrt{2}$ to "hold a place" while we compute around it using the rational components, except for those occasional opportunities when it occurs squared and we are temporarily relieved of having to carry it. So the inclusion of $\sqrt{2}$ as a number is a somewhat artificial process, devised solely so that we might have a richer system in which we can solve the equation $x^2 = 2$.

With this in mind, let us jump to the stage where we have appended all the real numbers to our system. Some of them, such as $\sqrt[4]{17}$, arise as solutions of more complicated equations, while others, such as π and e, come from certain limit processes.

Each irrational is absorbed in a somewhat artificial manner, but once again the resulting conglomerate of numbers and arithmetic operations satisfies the commutative, associative, and distributive laws.[†]

At this point we observe that we still cannot solve the equation

$$x^2 = -1. \tag{6}$$

But now our experience suggests that we can expand our number system once again by appending a symbol for a solution to Eq. (6); instead of $\sqrt{-1}$, it is customary to use the symbol i. (Engineers often use the letter j.) Next we imitate the model of expressions (2) through (5) (pertaining to $\sqrt{2}$) and thereby generalize our concept of number as follows:[‡]

Definition 1. A **complex number** is an expression of the form $a + bi$, where a and b are real numbers. Two complex numbers $a + bi$ and $c + di$ are said to be equal ($a + bi = c + di$) if and only if $a = c$ and $b = d$.

The operations of addition and subtraction of complex numbers are given by

$$(a + bi) \pm (c + di) := (a \pm c) + (b \pm d)i,$$

where the symbol := means "is defined to be."

In accordance with the distributive law and the proviso that $i^2 = -1$, we postulate the following:

The multiplication of two complex numbers is defined by

$$(a + bi)(c + di) := (ac - bd) + (bc + ad)i.$$

To compute the quotient of two complex numbers, we again "rationalize the denominator":

$$\frac{a + bi}{c + di} = \frac{a + bi}{c + di} \frac{c - di}{c - di} = \frac{ac + bd}{c^2 + d^2} + \frac{bc - ad}{c^2 + d^2} i.$$

Thus we formally postulate the following:

The division of complex numbers is given by

$$\frac{a + bi}{c + di} := \frac{ac + bd}{c^2 + d^2} + \frac{bc - ad}{c^2 + d^2} i \qquad (\text{if } c^2 + d^2 \neq 0).$$

These are rules for computing in the complex number system. The usual algebraic properties (commutativity, associativity, etc.) are easy to verify and appear as exercises.

[†]The algebraic aspects of extending a number field are discussed in Ref. 5 at the end of this chapter.

[‡]Karl Friedrich Gauss (1777–1855) was the first mathematician to use complex numbers freely and give them full acceptance as genuine mathematical objects.

Example 1

Find the quotient

$$\frac{(6 + 2i) - (1 + 3i)}{(-1 + i) - 2}.$$

Solution.

$$
\begin{aligned}
\frac{(6 + 2i) - (1 + 3i)}{(-1 + i) - 2} &= \frac{5 - i}{-3 + i} = \frac{(5 - i)}{(-3 + i)} \frac{(-3 - i)}{(-3 - i)} \\
&= \frac{-15 - 1 - 5i + 3i}{9 + 1} \\
&= -\frac{8}{5} - \frac{1}{5}i. \quad \blacksquare
\end{aligned}
\tag{7}
$$

(A slug marks the end of solutions or proofs throughout the text.)

Historically, i was considered as an "imaginary" number because of the blatant impossibility of solving Eq. (6) with any of the numbers at hand. With the perspective we have developed, we can see that this label could also be applied to the numbers $\sqrt{2}$ or $\sqrt[4]{17}$; like them, i is simply one more symbol appended to a given number system to create a richer system. Nonetheless, tradition dictates the following designations:[†]

Definition 2. The **real part** of the complex number $a + bi$ is the (real) number a; its **imaginary part** is the (real) number b. If a is zero, the number is said to be a **pure imaginary number**.

For convenience we customarily use a single letter, usually z, to denote a complex number. Its real and imaginary parts are then written Re z and Im z, respectively. With this notation we have $z = \text{Re } z + i \text{ Im } z$.

Observe that the equation $z_1 = z_2$ holds if and only if Re $z_1 = $ Re z_2 and Im $z_1 = $ Im z_2. Thus any equation involving complex numbers can be interpreted as a pair of real equations.

The set of all complex numbers is sometimes denoted as **C**. Unlike the real number system, there is no natural ordering for the elements of **C**; it is meaningless, for example, to ask whether $2 + 3i$ is greater than or less than $3 + 2i$. (See Prob. 30.)

EXERCISES 1.1

1. Verify that $-i$ is also a root of Eq. (6).

2. Verify the commutative, associative, and distributive laws for complex numbers.

[†]René Descartes introduced the terminology "real" and "imaginary" in 1637. W. R. Hamilton referred to a number's "imaginary part" in 1843.

3. Notice that 0 and 1 retain their "identity" properties as complex numbers; that is, $0 + z = z$ and $1 \cdot z = z$ when z is complex.

   (a) Verify that complex subtraction is the inverse of complex addition (that is, $z_3 = z_2 - z_1$ if and only if $z_3 + z_1 = z_2$).

   (b) Verify that complex division, as given in the text, is the inverse of complex multiplication (that is, if $z_2 \neq 0$, then $z_3 = z_1/z_2$ if and only if $z_3 z_2 = z_1$).

4. Prove that if $z_1 z_2 = 0$, then $z_1 = 0$ or $z_2 = 0$.

In Problems 5–13, write the number in the form $a + bi$.

5. **(a)** $-3\left(\dfrac{i}{2}\right)$ **(b)** $(8 + i) - (5 + i)$ **(c)** $\dfrac{2}{i}$

6. **(a)** $(-1 + i)^2$ **(b)** $\dfrac{2 - i}{\frac{1}{3}}$ **(c)** $i(\pi - 4i)$

7. **(a)** $\dfrac{8i - 1}{i}$ **(b)** $\dfrac{-1 + 5i}{2 + 3i}$ **(c)** $\dfrac{3}{i} + \dfrac{i}{3}$

8. $\dfrac{(8 + 2i) - (1 - i)}{(2 + i)^2}$

9. $\dfrac{2 + 3i}{1 + 2i} - \dfrac{8 + i}{6 - i}$

10. $\left[\dfrac{2 + i}{6i - (1 - 2i)}\right]^2$

11. $i^3(i + 1)^2$

12. $(2 + i)(-1 - i)(3 - 2i)$

13. $((3 - i)^2 - 3)i$

14. Show that $\operatorname{Re}(iz) = -\operatorname{Im} z$ for every complex number z.

15. Let k be an integer. Show that

$$i^{4k} = 1, \qquad i^{4k+1} = i, \qquad i^{4k+2} = -1, \qquad i^{4k+3} = -i.$$

16. Use the result of Problem 15 to find

   (a) i^7 **(b)** i^{62} **(c)** i^{-202} **(d)** i^{-4321}

17. Use the result of Problem 15 to evaluate

$$3i^{11} + 6i^3 + \frac{8}{i^{20}} + i^{-1}.$$

18. Show that the complex number $z = -1 + i$ satisfies the equation

$$z^2 + 2z + 2 = 0.$$

19. Write the complex equation $z^3 + 5z^2 = z + 3i$ as two real equations.

20. Solve each of the following equations for z.

(a) $iz = 4 - zi$

(b) $\dfrac{z}{1-z} = 1 - 5i$

(c) $(2-i)z + 8z^2 = 0$

(d) $z^2 + 16 = 0$

21. The complex numbers z_1, z_2 satisfy the system of equations

$$(1-i)z_1 + 3z_2 = 2 - 3i,$$
$$iz_1 + (1+2i)z_2 = 1.$$

Find z_1, z_2.

22. Find all solutions to the equation $z^4 - 16 = 0$.

23. Let z be a complex number such that Re $z > 0$. Prove that $\text{Re}(1/z) > 0$.

24. Let z be a complex number such that Im $z > 0$. Prove that $\text{Im}(1/z) < 0$.

25. Let z_1, z_2 be two complex numbers such that $z_1 + z_2$ and $z_1 z_2$ are each negative real numbers. Prove that z_1 and z_2 must be real numbers.

26. Verify that

$$\text{Re}\left(\sum_{j=1}^{n} z_j\right) = \sum_{j=1}^{n} \text{Re } z_j$$

and that

$$\text{Im}\left(\sum_{j=1}^{n} z_j\right) = \sum_{j=1}^{n} \text{Im } z_j.$$

[The real (imaginary) part of the sum is the sum of the real (imaginary) parts.] Formulate, and then *disprove*, the corresponding conjectures for multiplication.

27. Prove the *binomial formula* for complex numbers:

$$(z_1 + z_2)^n = z_1^n + \binom{n}{1}z_1^{n-1}z_2 + \cdots + \binom{n}{k}z_1^{n-k}z_2^k + \cdots + z_2^n,$$

where n is a positive integer, and the *binomial coefficients* are given by

$$\binom{n}{k} := \frac{n!}{k!(n-k)!}.$$

28. Use the binomial formula (Prob. 27) to compute $(2-i)^5$.

29. Prove that there is no rational number x that satisfies $x^2 = 2$. [HINT: Show that if p/q were a solution, where p and q are integers, then 2 would have to divide both p and q. This contradicts the fact that such a ratio can always be written without common divisors.]

30. The definition of the order relation denoted by $>$ in the real number system is based upon the existence of a subset $\mathcal{P}$ (the positive reals) having the following properties:

   (i) For any number $\alpha \neq 0$, either α or $-\alpha$ (but not both) belongs to $\mathcal{P}$.

   (ii) If α and β belong to $\mathcal{P}$, so does $\alpha + \beta$.

   (iii) If α and β belong to $\mathcal{P}$, so does $\alpha \cdot \beta$.

When such a set $\mathcal{P}$ exists we write $\alpha > \beta$ if and only if $\alpha - \beta$ belongs to $\mathcal{P}$.[†] Prove that the *complex* number system does not possess a nonempty subset $\mathcal{P}$ having properties (i), (ii), and (iii). [HINT: Argue that neither i nor $-i$ could belong to such a set $\mathcal{P}$.]

31. Write a computer program for calculating sums, differences, products, and quotients of complex numbers. The input and output parameters should be the corresponding real and imaginary parts.

32. The straightforward method of computing the product $(a + bi)(c + di) = (ac - bd) + i(bc + ad)$ requires four (real) multiplications (and two signed additions). On most computers multiplication is far more time-consuming than addition. Devise an algorithm for computing $(a + bi)(c + di)$ with only three multiplications (at the cost of extra additions). [HINT: Start with $(a + b)(c + d)$.]

1.2   Point Representation of Complex Numbers

It is presumed that the reader is familiar with the Cartesian coordinate system (Fig. 1.1) which establishes a one-to-one correspondence between points in the xy-plane and ordered pairs of real numbers. The ordered pair $(-2, 3)$, for example, corresponds to that point P that lies two units to the left of the y-axis and three units above the x-axis.

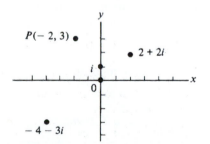

Figure 1.1 Cartesian coordinate system.

[†]On computers this is, in fact, the method by which the statement $\alpha > \beta$ is tested.

The Cartesian coordinate system suggests a convenient way to represent complex numbers as points in the xy-plane; namely, to each complex number $a + bi$ we associate that point in the xy-plane that has the coordinates (a, b). The complex number $-2 + 3i$ is therefore represented by the point P in Fig. 1.1. Also shown in Fig. 1.1 are the points that represent the complex numbers $0, i, 2 + 2i$, and $-4 - 3i$.

When the xy-plane is used for the purpose of describing complex numbers it is referred to as the *complex plane* or z-plane. (The term *Argand diagram* is sometimes used; the representation of complex numbers in the plane was proposed independently by Caspar Wessel in 1797 and Jean Pierre Argand in 1806.) Since each point on the x-axis represents a real number, this axis is called the *real axis*. Analogously, the y-axis is called the *imaginary axis* for it represents the pure imaginary numbers.

Hereafter, we shall refer to the point that represents the complex number z as simply *the point z*; that is, the point $z = a + bi$ is the point with coordinates (a, b).

Example 1

Suppose that n particles with masses $m_1, m_2, \ldots, m_n$ are located at the respective points $z_1, z_2, \ldots, z_n$ in the complex plane. Show that the center of mass of the system is the point

$$\widehat{z} = \frac{m_1 z_1 + m_2 z_2 + \cdots + m_n z_n}{m_1 + m_2 + \cdots + m_n}.$$

Solution.   Write $z_1 = x_1 + y_1 i, z_2 = x_2 + y_2 i, \ldots, z_n = x_n + y_n i$, and let M be the total mass $\sum_{k=1}^{n} m_k$. Presumably the reader will recall that the center of mass of the given system is the point with coordinates $(\widehat{x}, \widehat{y})$, where

$$\widehat{x} = \frac{\sum\limits_{k=1}^{n} m_k x_k}{M}, \qquad \widehat{y} = \frac{\sum\limits_{k=1}^{n} m_k y_k}{M}.$$

But clearly $\widehat{x}$ and $\widehat{y}$ are, respectively, the real and imaginary parts of the complex number $(\sum_{k=1}^{n} m_k z_k)/M = \widehat{z}$. ∎

Absolute Value. By the Pythagorean theorem, the distance from the point $z = a + bi$ to the origin is given by $\sqrt{a^2 + b^2}$. Special notation for this distance is given in

---

Definition 3. The **absolute value** or **modulus** of the number $z = a + bi$ is denoted by $|z|$ and is given by

$$|z| := \sqrt{a^2 + b^2}.$$

In particular,

$$|0| = 0, \qquad \left|\frac{i}{2}\right| = \frac{1}{2}, \qquad |3 - 4i| = \sqrt{9 + 16} = 5.$$

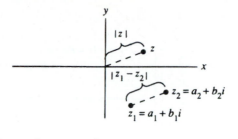

Figure 1.2 Distance between points.

The reader should note that $|z|$ is always a nonnegative real number and that the *only* complex number whose modulus is zero is the number 0.

Let $z_1 = a_1 + b_1 i$ and $z_2 = a_2 + b_2 i$. Then

$$|z_1 - z_2| = |(a_1 - a_2) + (b_1 - b_2)i| = \sqrt{(a_1 - a_2)^2 + (b_1 - b_2)^2},$$

which is the distance between the points with coordinates (a_1, b_1) and (a_2, b_2) (see Fig. 1.2). Hence *the distance between the points z_1 and z_2 is given by $|z_1 - z_2|$*. This fact is useful in describing certain curves in the plane. Consider, for example, the set of all numbers z that satisfy the equation

$$|z - z_0| = r, \tag{1}$$

where z_0 is a fixed complex number and r is a fixed positive real number. This set consists of all points z whose distance from z_0 is r. Consequently Eq. (1) is the equation of a circle.

Example 2
Describe the set of points z that satisfy the equations

    (a) $|z + 2| = |z - 1|$,   **(b)** $|z - 1| = \mathrm{Re}\, z + 1$.

    Solution.   **(a)** A point z satisfies Eq. (a) if and only if it is equidistant from the points -2 and 1. Hence Eq. (a) is the equation of the perpendicular bisector of the line segment joining -2 and 1; that is, Eq. (a) describes the line $x = -\frac{1}{2}$.

A more routine method for solving Eq. (a) is to set $z = x + iy$ in the equation and perform the algebra:

$$|z + 2| = |z - 1|,$$
$$|x + iy + 2| = |x + iy - 1|,$$
$$(x + 2)^2 + y^2 = (x - 1)^2 + y^2,$$
$$4x + 4 = -2x + 1,$$
$$x = -\frac{1}{2}.$$

    (b) The geometric interpretation of Eq. (b) is less obvious, so we proceed directly with the mechanical approach and derive $\sqrt{(x - 1)^2 + y^2} = x + 1$, or $y^2 = 4x$, which describes a parabola (see Fig. 1.3). ■

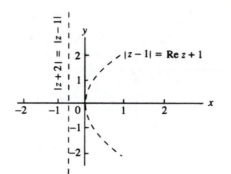

Figure 1.3 Graphs for Example 2.

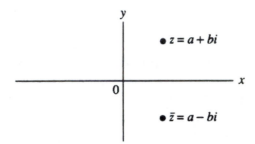

Figure 1.4 Complex conjugates.

Complex Conjugates. The reflection of the point $z = a + bi$ in the real axis is the point $a - bi$ (see Fig. 1.4). As we shall see, the relationship between $a + bi$ and $a - bi$ will play a significant role in the theory of complex variables. We introduce special notation for this concept in the next definition.

Definition 4. The **complex conjugate** of the number $z = a + bi$ is denoted by $\bar{z}$ and is given by
$$\bar{z} := a - bi.$$

Thus,
$$\overline{-1 + 5i} = -1 - 5i, \qquad \overline{\pi - i} = \pi + i, \qquad \overline{8} = 8.$$

Some authors use the asterisk, z^*, to denote the complex conjugate.

It follows from Definition 4 that $z = \bar{z}$ if and only if z is a real number. Also it is clear that the conjugate of the sum (difference) of two complex numbers is equal to the sum (difference) of their conjugates; that is,

$$\overline{z_1 + z_2} = \overline{z_1} + \overline{z_2}, \quad \overline{z_1 - z_2} = \overline{z_1} - \overline{z_2}.$$

Perhaps not so obvious is the analogous property for multiplication.

Example 3

Prove that the conjugate of the product of two complex numbers is equal to the product of the conjugates of these numbers.

   Solution.   It is required to verify that

$$\overline{(z_1 z_2)} = \overline{z_1}\, \overline{z_2}. \tag{2}$$

Write $z_1 = a_1 + b_1 i$, $z_2 = a_2 + b_2 i$. Then

$$\overline{(z_1 z_2)} = \overline{a_1 a_2 - b_1 b_2 + (a_1 b_2 + a_2 b_1)i}$$
$$= a_1 a_2 - b_1 b_2 - (a_1 b_2 + a_2 b_1)i.$$

On the other hand,

$$\overline{z_1}\, \overline{z_2} = (a_1 - b_1 i)(a_2 - b_2 i) = a_1 a_2 - b_1 b_2 - a_1 b_2 i - a_2 b_1 i$$
$$= a_1 a_2 - b_1 b_2 - (a_1 b_2 + a_2 b_1)i.$$

Thus Eq. (2) holds.   ∎

   In addition to Eq. (2) the following properties can be seen:

$$\overline{\left(\frac{z_1}{z_2} \right)} = \frac{\overline{z_1}}{\overline{z_2}} \quad (z_2 \ne 0); \tag{3}$$

$$\operatorname{Re} z = \frac{z + \bar{z}}{2}; \tag{4}$$

$$\operatorname{Im} z = \frac{z - \bar{z}}{2i}; \tag{5}$$

Property (4) demonstrates that the sum of a complex number and its conjugate is real, whereas (5) shows that the difference is (pure) imaginary. The conjugate of the conjugate of a complex number is, of course, the number itself:

$$\overline{(\bar{z})} = z. \tag{6}$$

   It is clear from Definition 4 that

$$|z| = |\bar{z}|;$$

that is, the points z and $\bar{z}$ are equidistant from the origin. Furthermore, since

$$z\bar{z} = (a + bi)(a - bi) = a^2 + b^2,$$

we have

$$z\bar{z} = |z|^2. \tag{7}$$

This is a useful fact to remember: *The square of the modulus of a complex number equals the number times its conjugate.*

Actually we have already employed complex conjugates in Sec. 1.1, in the process of rationalizing the denominator for the division algorithm. Thus, for instance, if z_1 and z_2 are complex numbers, then we rewrite z_1/z_2 as a ratio with a real denominator by using $\overline{z_2}$:

$$\frac{z_1}{z_2} = \frac{z_1\overline{z_2}}{z_2\overline{z_2}} = \frac{z_1\overline{z_2}}{|z_2|^2}. \tag{8}$$

In particular,

$$\frac{1}{z} = \frac{\overline{z}}{|z|^2}. \tag{9}$$

In closing we would like to mention that there is another, possibly more enlightening, way to see Eq. (2). Notice that when we represent a complex number in terms of two real numbers and the symbol i, as in $z = a + bi$, then the action of conjugation is equivalent to changing the sign of the i term. Now recall the role that i plays in computations; it merely holds a place while we compute around it, replacing its square by -1 whenever it arises. Except for these occurrences i is never really absorbed into the computations; we could just as well call it j, λ, $\sqrt{-1}$, or any other symbol whose square we agree to replace by -1. *In fact, without affecting the validity of the calculation, we could replace it throughout by the symbol* $(-i)$, *since the square of the latter is also* -1. Thus, for instance, if in the expression $(a_1 + b_1i)(a_2 + b_2i)$ we replace i by $-i$ and then multiply, the only thing different about the product will be the appearance of $-i$ instead of i. But expressed in terms of conjugation, this is precisely the statement of Example 3.[†]

EXERCISES 1.2

1. Show that the point $(z_1 + z_2)/2$ is the midpoint of the line segment joining z_1 and z_2.

2. Given four particles of masses 2, 1, 3, and 5 located at the respective points $1 + i$, $-3i$, $1 - 2i$, and -6, find the center of mass of this system.

3. Which of the points i, $2 - i$, and -3 is farthest from the origin?

4. Let $z = 3 - 2i$. Plot the points z, $-z$, $\overline{z}$, $-\overline{z}$, and $1/z$ in the complex plane. Do the same for $z = 2 + 3i$ and $z = -2i$.

5. Show that the points 1, $-1/2 + i\sqrt{3}/2$, and $-1/2 - i\sqrt{3}/2$ are the vertices of an equilateral triangle.

6. Show that the points $3 + i$, 6, and $4 + 4i$ are the vertices of a right triangle.

[†]By the same token we should be able to replace $\sqrt{2}$ by $-\sqrt{2}$ in $(3 + 2\sqrt{2})(4 - 3\sqrt{2})$ *either before or after multiplying* and obtain the same result. (Try it.)

7. Describe the set of points z in the complex plane that satisfies each of the following.

   (a) Im $z = -2$ (b) $|z - 1 + i| = 3$

   (c) $|2z - i| = 4$ (d) $|z - 1| = |z + i|$

   (e) $|z| = $ Re $z + 2$ (f) $|z - 1| + |z + 1| = 7$

   (g) $|z| = 3|z - 1|$ (h) Re $z \geq 4$

   (i) $|z - i| < 2$ (j) $|z| > 6$

8. Show, both analytically and graphically, that $|z - 1| = |\bar{z} - 1|$.

9. Show that if r is a nonnegative real number, then $|rz| = r|z|$.

10. Prove that $|$ Re $z| \leq |z|$ and $|$ Im $z| \leq |z|$.

11. Prove that if $|z| = $ Re z, then z is a nonnegative real number.

12. Verify properties (3), (4), and (5).

13. Prove that if $(\bar{z})^2 = z^2$, then z is either real or pure imaginary.

14. Prove that $|z_1 z_2| = |z_1||z_2|$. [HINT: Use Eqs. (7) and (2) to show that $|z_1 z_2|^2 = |z_1|^2 |z_2|^2$.]

15. Prove that $(\bar{z})^k = \overline{(z^k)}$ for every integer k (provided $z \neq 0$ when k is negative).

16. Prove that if $|z| = 1$ ($z \neq 1$), then Re$[1/(1 - z)] = \frac{1}{2}$.

17. Let $a_1, a_2, \ldots, a_n$ be real constants. Show that if z_0 is a root of the polynomial equation $z^n + a_1 z^{n-1} + a_2 z^{n-2} + \cdots + a_n = 0$, then so is $\bar{z_0}$.

18. Use the familiar formula for the roots of a quadratic polynomial to give another proof of the statement in Prob. 17 for the case $n = 2$.

19. We have noted that the conjugate $(\bar{z})$ is the reflection of the point z in the real axis (the horizontal line $y = 0$). Show that the reflection of z in the line $ax + by = c$ (a, b, c real) is given by
$$\frac{2ic + (b - ai)\bar{z}}{b + ai}.$$

20. (*Matrices with Complex Entries*) Let **B** be an m by n matrix whose entries are complex numbers. Then by $\mathbf{B}^{\dagger}$ we denote the n by m matrix that is obtained by forming the transpose (interchanging rows and columns) of **B** followed by taking the conjugate of each entry. In other words, if $\mathbf{B} = [b_{ij}]$, then $\mathbf{B}^{\dagger} = [\bar{b}_{ji}]$. For example:

$$\begin{bmatrix} i & 3 \\ 4-i & -2i \end{bmatrix}^{\dagger} = \begin{bmatrix} -i & 4+i \\ 3 & 2i \end{bmatrix}, \qquad \begin{bmatrix} 1+i \\ 3 \end{bmatrix}^{\dagger} = \begin{bmatrix} 1-i & 3 \end{bmatrix}.$$

For an n by n matrix $\mathbf{A} = \begin{bmatrix} a_{ij} \end{bmatrix}$ with complex entries, prove the following:

(a) If $\mathbf{u}^\dagger \mathbf{A} \mathbf{u} = \mathbf{0}$ for all n by 1 column vectors $\mathbf{u}$ with complex entries, then $\mathbf{A}$ is the zero matrix (that is, $a_{ij} = 0$ for all i, j). [HINT: To show $a_{ij} = 0$, take $\mathbf{u}$ to be a column vector with all zeros except for its ith and j^{th} entries.]

(b) Show by example that the conclusion ("$\mathbf{A}$ is the zero matrix") can fail if the hypothesis for part (a) only holds for vectors $\mathbf{u}$ with *real* number entries. [HINT: Try to find a 2 by 2 real *nonzero* matrix $\mathbf{A}$ such that $\mathbf{u}^\dagger \mathbf{A} \mathbf{u} = 0$ for all real 2 by 1 vectors $\mathbf{u}$.]

21. Let $\mathbf{A}$ be an n by n matrix with complex entries. We say that $\mathbf{A}$ is *Hermitean* if $\mathbf{A}^\dagger = \mathbf{A}$ (see Prob. 20).

(a) Show that if $\mathbf{A}$ is Hermitean, then $\mathbf{u}^\dagger \mathbf{A} \mathbf{u}$ is real for any n by 1 column vector $\mathbf{u}$ with complex entries.

(b) Show that if $\mathbf{B}$ is any m by n matrix with complex entries, then $\mathbf{B}^\dagger \mathbf{B}$ is Hermitean.

(c) Show that if $\mathbf{B}$ is any n by n matrix and $\mathbf{u}$ is any n by 1 column vector (each with complex entries), then $\mathbf{u}^\dagger \mathbf{B}^\dagger \mathbf{B} \mathbf{u}$ must be a nonnegative real number.

1.3   Vectors and Polar Forms

With each point z in the complex plane we can associate a *vector*, namely, the directed line segment from the origin to the point z. Recall that vectors are characterized by length and direction, and that a given vector remains unchanged under translation. Thus the vector determined by $z = 1 + i$ is the same as the vector from the point $2 + i$ to the point $3 + 2i$ (see Fig. 1.5). Note that every vector parallel to the real axis corresponds to a real number, while those parallel to the imaginary axis represent pure imaginary numbers. Observe, also, that the length of the vector associated with z is $|z|$.

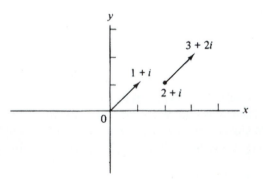

Figure 1.5 Complex numbers as vectors.

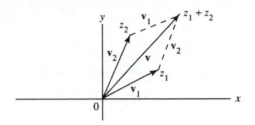

Figure 1.6 Vector addition.

Let $\mathbf{v}_1$ and $\mathbf{v}_2$ denote the vectors determined by the points z_1 and z_2, respectively. The vector sum $\mathbf{v} = \mathbf{v}_1 + \mathbf{v}_2$ is given by the parallelogram law, which is illustrated in Fig. 1.6. If $z_1 = x_1 + iy_1$ and $z_2 = x_2 + iy_2$, then the terminal point of the vector $\mathbf{v}$ in Fig. 1.6 has the coordinates $(x_1 + x_2, y_1 + y_2)$; that is, it corresponds to the point $z_1 + z_2$. Thus we see that *the correspondence between complex numbers and planar vectors carries over to the operation of addition.*

Hereafter, the vector determined by the point z will be simply called *the vector z.*

Recall the geometric fact that the length of any side of a triangle is less than or equal to the sum of the lengths of the other two sides. If we apply this theorem to the triangle in Fig. 1.6 with vertices 0, z_1, and $z_1 + z_2$, we deduce a very important law relating the magnitudes of complex numbers and their sum:

Triangle Inequality. For any two complex numbers z_1 and z_2, we have

$$|z_1 + z_2| \le |z_1| + |z_2|. \tag{1}$$

The triangle inequality can easily be extended to more than two complex numbers, as requested in Prob. 22.

The vector $z_2 - z_1$, when added to the vector z_1, obviously yields the vector z_2. Thus $z_2 - z_1$ can be represented as the directed line segment from z_1 to z_2 (see Fig. 1.7). Applying the geometric theorem to the triangle in Fig. 1.7, we deduce another form of the triangle inequality:

$$|z_2| \le |z_1| + |z_2 - z_1|$$

or

$$|z_2| - |z_1| \le |z_2 - z_1|. \tag{2}$$

Inequality (2) states that the difference in the lengths of any two sides of a triangle is no greater than the length of the third side.

Example 1

Prove that the three distinct points z_1, z_2, and z_3 lie on the same straight line if and only if $z_3 - z_2 = c(z_2 - z_1)$ for some real number c.

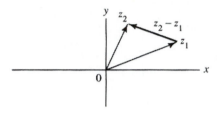

Figure 1.7 Vector subtraction.

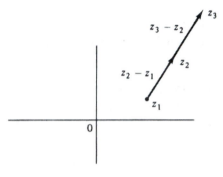

Figure 1.8 Collinear points.

Solution. Recall that two vectors are parallel if and only if one is a (real) scalar multiple of the other. In the language of complex numbers, this says that z is parallel to w if and only if $z = cw$, where c is real. From Fig. 1.8 we see that the condition that the points z_1, z_2, and z_3 be collinear is equivalent to the statement that the vector $z_3 - z_2$ is parallel to the vector $z_2 - z_1$. Using our characterization of parallelism, the conclusion follows immediately. ■

There is another set of parameters that characterize the vector from the origin to the point z (other, that is, than the real and imaginary parts of z), which more intimately reflects its interpretation as an object with magnitude and direction. These are the polar coordinates, r and θ, of the point z. The coordinate r is the distance from the origin to z, and θ is the angle of inclination of the vector z, measured positively in a counterclockwise sense from the positive real axis (and thus negative when measured clockwise) (see Fig. 1.9). *We shall always measure angles in radians in this book*; the use of degree measure is fine for visualization purposes, but it becomes quite treacherous in any discipline where calculus is involved. Notice that r is the modulus, or absolute value, of z and is never negative: $r = |z|$.

From Fig. 1.9 we readily derive the equations expressing the rectangular (or *Cartesian*) coordinates (x, y) in terms of the polar coordinates (r, θ):

$$x = r\cos\theta, \qquad y = r\sin\theta. \tag{3}$$

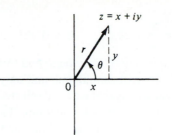

Figure 1.9 Polar coordinates.

On the other hand, the expressions for (r, θ) in terms of (x, y) contain some minor but troublesome complications. Indeed the coordinate r is given, unambiguously, by

$$r = \sqrt{x^2 + y^2} = |z|. \tag{4}$$

However, observe that although it is certainly true that $\tan \theta = y/x$, the natural conclusion

$$\theta = \tan^{-1} \left(\frac{y}{x} \right)$$

is *invalid* for points z in the second and third quadrants (since the standard interpretation of the arctangent function places its range in the first and fourth quadrants). Since an angle *is* fixed by its sine *and* cosine, θ is uniquely determined by the pair of equations

$$\cos \theta = \frac{x}{|z|}, \qquad \sin \theta = \frac{y}{|z|}, \tag{5}$$

but in practice we usually compute $\tan^{-1}(y/x)$ and adjust for the quadrant problem by adding or subtracting π (radians) when appropriate (see Prob. 14).

The nuisance aspects of θ do not end here, however. Even using Eqs. (5) one can, because of its identification as an angle, determine θ only up to an integer multiple of 2π. To accommodate this feature we shall call the value of any of these angles an *argument*, or *phase*, of z, denoted

$$\arg z.$$

Thus if θ_0 qualifies as a value of $\arg z$, then so do

$$\theta_0 \pm 2\pi, \ \theta_0 \pm 4\pi, \ \theta_0 \pm 6\pi, \dots,$$

and every value of $\arg z$ must be one of these.[†] In particular, the values of $\arg i$ are

$$\frac{\pi}{2}, \frac{\pi}{2} \pm 2\pi, \frac{\pi}{2} \pm 4\pi, \dots$$

[†] An alternative way to express $\arg z$ is to write it as the *set*

$$\arg z = \{\theta_0 + 2k\pi : k = 0, \pm 1, \pm 2, \dots\}.$$

and we write

$$\arg i = \frac{\pi}{2} + 2k\pi \qquad (k = 0, \pm 1, \pm 2, \ldots).$$

It is convenient to have a notation for some *definite* value of arg z. Notice that any half-open interval of length 2π will contain one and only one value of the argument. By specifying such an interval we say that we have selected a particular *branch* of arg z. Figure 1.10 illustrates three possible branch selections. The first diagram (Fig. 1.10(a)) depicts the branch that selects the value of arg z from the interval $(-\pi, \pi]$; it is known as the *principal value of the argument* and is denoted Arg z (with capital A). The principal value is most commonly used in complex arithmetic computer codes; it is inherently discontinous, jumping by 2π as z crosses the negative real axis. This line of discontinuities is known as the *branch cut*.

Of course, *any* branch of arg z must have a jump of 2π somewhere. The branch depicted in Fig. 1.10(b) is discontinuous on the *positive* real axis, taking values from the interval $(0, 2\pi]$. The branch in Fig. 1.10(c) has the same branch cut but selects values from the interval $(2\pi, 4\pi]$.

The notation $\arg_\tau z$ is used for the branch of arg z taking values from the interval $(\tau, \tau + 2\pi]$. Thus $\arg_{-\pi} z$ is the principal value Arg z, and the branches depicted in Fig. 1.10(b) and 1.10(c), respectively, are $\arg_0 z$ and $\arg_{2\pi} z$. Note that arg 0 cannot be sensibly defined for any branch.

With these conventions in hand, one can now write $z = x + iy$ in the *polar form* [recall Eq. (3)]

$$z = x + iy = r(\cos\theta + i\sin\theta) = r\operatorname{cis}\theta, \tag{6}$$

where we abbreviate the "cosine plus i sine" operator as cis.

Example 2

Find $\arg(1 + \sqrt{3}i)$ and write $1 + \sqrt{3}i$ in polar form.

Solution. Note that $r = |1 + \sqrt{3}i| = 2$ and that the equations $\cos\theta = 1/2$, $\sin\theta = \sqrt{3}/2$ are satisfied by $\theta = \pi/3$. Hence $\arg(1 + \sqrt{3}i) = \pi/3 + 2k\pi$, $k = 0, \pm 1, \pm 2, \ldots$ [in particular, $\operatorname{Arg}(1 + \sqrt{3}i) = \pi/3$]. The polar form of $1 + \sqrt{3}i$ is $2(\cos\pi/3 + i\sin\pi/3) = 2\operatorname{cis}\pi/3$.   ∎

In many circumstances one of the forms $x + iy$ or $r\operatorname{cis}\theta$ may be more suitable than the other. The rectangular form, for example, is very convenient for addition or subtraction, whereas the polar form can be a monstrosity (see Prob. 21). On the other hand, the polar form lends a very interesting geometric interpretation to the process of multiplication. If we let

$$z_1 = r_1\left(\cos\theta_1 + i\sin\theta_1\right), \qquad z_2 = r_2\left(\cos\theta_2 + i\sin\theta_2\right),$$

then we compute

$$z_1 z_2 = r_1 r_2\left[(\cos\theta_1\cos\theta_2 - \sin\theta_1\sin\theta_2) + i\left(\sin\theta_1\cos\theta_2 + \cos\theta_1\sin\theta_2\right)\right],$$

and so

$$z_1 z_2 = r_1 r_2\left[\cos\left(\theta_1 + \theta_2\right) + i\sin\left(\theta_1 + \theta_2\right)\right]. \tag{7}$$

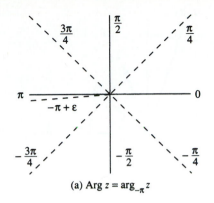

(a) Arg $z = \arg_{-\pi} z$

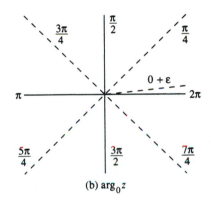

(b) $\arg_0 z$

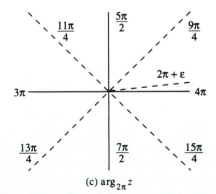

(c) $\arg_{2\pi} z$

Figure 1.10 Branches of arg z.

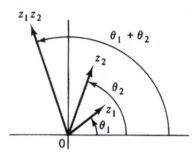

Figure 1.11 Geometric interpretation of the product.

The abbreviated version of Eq. (7) reads as follows:

$$z_1 z_2 = (r_1 \operatorname{cis} \theta_1)(r_2 \operatorname{cis} \theta_2) = (r_1 r_2) \operatorname{cis}(\theta_1 + \theta_2)$$

and we see that

The modulus of the product is the product of the moduli:

$$|z_1 z_2| = |z_1| \, |z_2| \ (= r_1 r_2); \tag{8}$$

The argument of the product is the sum of the arguments:

$$\arg z_1 z_2 = \arg z_1 + \arg z_2 \ (= \theta_1 + \theta_2). \tag{9}$$

(To be precise, the ambiguous Eq. (9) is to be interpreted as saying that if particular values are assigned to any pair of terms therein, then one can find a value for the third term that satisfies the identity.)

Geometrically, the vector $z_1 z_2$ has length equal to the product of the lengths of the vectors z_1 and z_2 and has angle equal to the sum of the angles of the vectors z_1 and z_2 (see Fig. 1.11). For instance, since the vector i has length 1 and angle $\pi/2$, it follows that the vector iz can be obtained by rotating the vector z through a right angle in the counterclockwise direction.

Observing that division is the inverse operation to multiplication, we are led to the following equations:

$$\frac{z_1}{z_2} = \frac{r_1}{r_2} \left[\cos(\theta_1 - \theta_2) + i \sin(\theta_1 - \theta_2)\right] = \frac{r_1}{r_2} \operatorname{cis}(\theta_1 - \theta_2), \tag{10}$$

$$\arg\left(\frac{z_1}{z_2}\right) = \arg z_1 - \arg z_2, \tag{11}$$

and

$$\left|\frac{z_1}{z_2}\right| = \frac{|z_1|}{|z_2|}. \tag{12}$$

Equation (10) can be proved in a manner similar to Eq. (7), and Eqs. (11) and (12) follow immediately. Geometrically, the vector z_1/z_2 has length equal to the quotient of the lengths of the vectors z_1 and z_2 and has angle equal to the difference of the angles of the vectors z_1 and z_2.

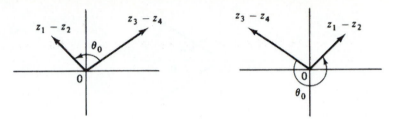

Figure 1.12 Perpendicular vectors.

Example 3

Write the quotient $(1 + i)/(\sqrt{3} - i)$ in polar form.

Solution.   The polar forms for $(1 + i)$ and $(\sqrt{3} - i)$ are

$$1 + i = |1 + i| \operatorname{cis}(\arg(1 + i)) = \sqrt{2} \operatorname{cis}(\pi/4),$$

$$\sqrt{3} - i = 2 \operatorname{cis}(-\pi/6).$$

Hence, from Eq. (10), we have

$$\frac{1 + i}{\sqrt{3} - i} = \frac{\sqrt{2}}{2} \operatorname{cis}\left[\frac{\pi}{4} - \left(-\frac{\pi}{6}\right)\right] = \frac{\sqrt{2}}{2} \operatorname{cis}\frac{5\pi}{12}. \quad ■$$

Example 4

Prove that the line l through the points z_1 and z_2 is perpendicular to the line L through the points z_3 and z_4 if and only if

$$\operatorname{Arg}\frac{z_1 - z_2}{z_3 - z_4} = \pm\frac{\pi}{2}. \tag{13}$$

Solution.   Note that the lines l and L are perpendicular if and only if the vectors $z_1 - z_2$ and $z_3 - z_4$ are perpendicular (see Fig. 1.12). Since

$$\arg\frac{z_1 - z_2}{z_3 - z_4} = \arg(z_1 - z_2) - \arg(z_3 - z_4)$$

gives the angle from $z_3 - z_4$ to $z_1 - z_2$, orthogonality holds precisely when this angle (up to an integer multiple of 2π) is equal to $\pi/2$ or $-\pi/2$. But this is the same as saying that (13) holds.   ■

Recall that, geometrically, the vector $\bar{z}$ is the reflection in the real axis of the vector z (see Fig. 1.13). Hence we see that *the argument of the conjugate of a complex number is the negative of the argument of the number*; that is,

$$\arg \bar{z} = -\arg z. \tag{14}$$

In fact, as a special case of Eq. (11) we also have

$$\arg \frac{1}{z} = -\arg z.$$

Thus $\bar{z}$ and z^{-1} have the same argument and represent parallel vectors (see Fig. 1.13).

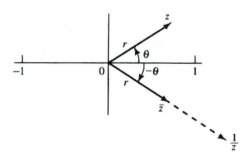

Figure 1.13 The argument of the conjugate and the reciprocal.

EXERCISES 1.3

1. Let $z_1 = 2 - i$ and $z_2 = 1 + i$. Use the parallelogram law to construct each of the following vectors.

   (a) $z_1 + z_2$ (b) $z_1 - z_2$ (c) $2z_1 - 3z_2$

2. Show that $|z_1 z_2 z_3| = |z_1| |z_2| |z_3|$.

3. Translate the following geometric theorem into the language of complex numbers: The sum of the squares of the lengths of the diagonals of a parallelogram is equal to the sum of the squares of its sides. (See Fig. 1.6.)

4. Show that for any integer k, $|z^k| = |z|^k$ (provided $z \neq 0$ when k is negative).

5. Find the following.

   (a) $\left| \dfrac{1 + 2i}{-2 - i} \right|$ (b) $\left| \overline{(1 + i)}(2 - 3i)(4i - 3) \right|$ (c) $\left| \dfrac{i(2 + i)^3}{(1 - i)^2} \right|$ (d) $\left| \dfrac{(\pi + i)^{100}}{(\pi - i)^{100}} \right|$

6. Draw each of the following vectors.

   (a) $7 \operatorname{cis}(3\pi/4)$ (b) $4 \operatorname{cis}(-\pi/6)$ (c) $\operatorname{cis}(3\pi/4)$ (d) $3 \operatorname{cis}(27\pi/4)$

7. Find the argument of each of the following complex numbers and write each in polar form.

   (a) $-1/2$ (b) $-3 + 3i$ (c) $-\pi i$
   (d) $-2\sqrt{3} - 2i$ (e) $(1 - i)(-\sqrt{3} + i)$ (f) $(\sqrt{3} - i)^2$
   (g) $\dfrac{-1 + \sqrt{3}i}{2 + 2i}$ (h) $\dfrac{-\sqrt{7}(1 + i)}{\sqrt{3} + i}$

8. Show geometrically that the nonzero complex numbers z_1 and z_2 satisfy $|z_1 + z_2| = |z_1| + |z_2|$ if and only if they have the same argument.

9. Given the vector z, interpret geometrically the vector $(\cos \phi + i \sin \phi)z$.

10. Show the following:

    (a) $\arg z_1 z_2 z_3 = \arg z_1 + \arg z_2 + \arg z_3$

    (b) $\arg z_1 \overline{z_2} = \arg z_1 - \arg z_2$.

11. Using the complex product $(1 + i)(5 - i)^4$, derive

$$\pi/4 = 4 \tan^{-1}(1/5) - \tan^{-1}(1/239).$$

12. Find the following.

    (a) $\text{Arg}(-6 - 6i)$ (b) $\text{Arg}(-\pi)$ (c) $\text{Arg}(10i)$ (d) $\text{Arg}(\sqrt{3} - i)$

13. Decide which of the following statements are always true.

    (a) $\text{Arg } z_1 z_2 = \text{Arg } z_1 + \text{Arg } z_2$ if $z_1 \neq 0$, $z_2 \neq 0$.

    (b) $\text{Arg } \overline{z} = - \text{Arg } z$ if z is not a real number.

    (c) $\text{Arg}(z_1/z_2) = \text{Arg } z_1 - \text{Arg } z_2$ if $z_1 \neq 0$, $z_2 \neq 0$.

    (d) $\arg z = \text{Arg } z + 2\pi k, k = 0, \pm 1, \pm 2, \ldots$, if $z \neq 0$.

14. Show that a correct formula for $\arg(x + iy)$ can be computed using the form

$$\arg(x + iy) = \begin{cases} \tan^{-1}(y/x) + (\pi/2)[1 - \text{sgn}(x)] & \text{if } x \neq 0, \\ (\pi/2)\,\text{sgn}(y) & \text{if } x = 0 \text{ and } y \neq 0, \\ \text{undefined} & \text{if } x = y = 0, \end{cases}$$

    where the "signum" function is specified by

$$\text{sgn}(t) := \begin{cases} +1 & \text{if } t > 0, \\ 0 & \text{if } t = 0, \\ -1 & \text{if } t < 0. \end{cases}$$

    Show also that the expression $\text{sgn}(y) \cos^{-1}(x/\sqrt{x^2 + y^2})$, at its points of continuity, equals $\text{Arg}(x + iy)$.

15. Prove that $|z_1 - z_2| \leq |z_1| + |z_2|$.

16. Prove that $||z_1| - |z_2|| \leq |z_1 - z_2|$.

17. Show that the vector z_1 is parallel to the vector z_2 if and only if $\text{Im}(z_1 \overline{z_2}) = 0$.

18. Show that every point z on the line through the distinct points z_1 and z_2 is of the form $z = z_1 + c(z_2 - z_1)$, where c is a real number. What can be said about the value of c if z also lies strictly between z_1 and z_2?

19. Prove that $\arg z_1 = \arg z_2$ if and only if $z_1 = c z_2$, where c is a positive real number.

20. Let z_1, z_2, and z_3 be distinct points and let ϕ be a particular value of $\arg[(z_3 - z_1)/(z_2 - z_1)]$. Prove that

$$|z_3 - z_2|^2 = |z_3 - z_1|^2 + |z_2 - z_1|^2 - 2|z_3 - z_1||z_2 - z_1| \cos \phi.$$

    [HINT: Consider the triangle with vertices z_1, z_2, z_3.]

21. If $r \operatorname{cis} \theta = r_1 \operatorname{cis} \theta_1 + r_2 \operatorname{cis} \theta_2$, determine r and θ in terms of r_1, r_2, θ_1, and θ_2. Check your answer by applying the law of cosines.

22. Use mathematical induction to prove the *generalized triangle inequality*:

$$\left| \sum_{k=1}^{n} z_k \right| \leq \sum_{k=1}^{n} | z_k |.$$

23. Let m_1, m_2, and m_3 be three positive real numbers and let z_1, z_2, and z_3 be three complex numbers, each of modulus less than or equal to 1. Use the generalized triangle inequality (Prob. 22) to prove that

$$\left| \frac{m_1 z_1 + m_2 z_2 + m_3 z_3}{m_1 + m_2 + m_3} \right| \leq 1,$$

and give a physical interpretation of the inequality.

24. Write computer programs for converting between rectangular and polar coordinates (using the principal value of the argument).

25. Recall that the dot (scalar) product of two planar vectors $\mathbf{v}_1 = (x_1, y_1)$ and $\mathbf{v}_2 = (x_2, y_2)$ is given by

$$\mathbf{v}_1 \cdot \mathbf{v}_2 = x_1 x_2 + y_1 y_2 .$$

Show that the dot product of the vectors represented by the complex numbers z_1 and z_2 is given by

$$z_1 \cdot z_2 = \operatorname{Re} (\bar{z}_1 z_2) .$$

26. Use the formula for the dot product in Prob. 25 to show that the vectors represented by the (nonzero) complex numbers z_1 and z_2 are orthogonal if and only if $z_1 \cdot z_2 = 0$. [HINT: Recall from the discussion following Eq. (9) that orthogonality holds precisely when $z_1 = i c z_2$ for some real c.]

27. Recall that in three dimensions the cross (vector) product of two vectors $\mathbf{v}_1 = (x_1, y_1, 0)$ and $\mathbf{v}_2 = (x_2, y_2, 0)$ in the xy-plane is given by

$$\mathbf{v}_1 \times \mathbf{v}_2 = (0, 0, x_1 y_2 - x_2 y_1) .$$

    (a) Show that the third component of the cross product of vectors in the xy-plane represented by the complex numbers z_1 and z_2 is given by $\operatorname{Im} (\bar{z}_1 z_2)$.

    (b) Show that the vectors represented by the (nonzero) complex numbers z_1 and z_2 are parallel if and only if $\operatorname{Im} (\bar{z}_1 z_2) = 0$. [HINT: Observe that these vectors are parallel precisely when $z_1 = c z_2$ for some real c.]

28. This problem demonstrates how complex notation can simplify the kinematic analysis of planar mechanisms.

Consider the crank-and-piston linkage depicted in Fig. 1.14. The crank arm a rotates about the fixed point O while the piston arm c executes horizontal motion. (If this were a gasoline engine, combustion forces would drive the piston and the connecting arm b would transform this energy into a rotation of the crankshaft.)

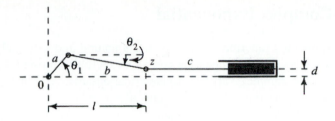

Figure 1.14 Crank-and-piston linkage.

For engineering analysis it is important to be able to relate the crankshaft's angular coordinates—position, velocity, and acceleration—to the corresponding linear coordinates for the piston. Although this calculation can be carried out using vector analysis, the following complex variable technique is more "automatic."

Let the crankshaft pivot O lie at the origin of the coordinate system, and let z be the complex number giving the location of the base of the piston rod, as depicted in Fig. 1.14,

$$z = l + id,$$

where l gives the piston's (linear) excursion and d is a fixed offset. The crank arm is described by $A = a(\cos\theta_1 + i\sin\theta_1)$ and the connecting arm by $B = b(\cos\theta_2 + i\sin\theta_2)$ (θ_2 is negative in Fig. 1.14). Exploit the obvious identity $A + B = z = l + id$ to derive the expression relating the piston position to the crankshaft angle:

$$l = a\cos\theta_1 + b\cos\left[\sin^{-1}\left(\frac{d - a\sin\theta_1}{b}\right)\right].$$

29. Suppose the mechanism in Prob. 28 has the dimensions

$$a = 0.1 \text{ m}, \qquad b = 0.2 \text{ m}, \qquad d = 0.1 \text{ m}$$

and the crankshaft rotates at a uniform velocity of 2 rad/s. Compute the position and velocity of the piston when $\theta_1 = \pi$.

30. For the linkage illustrated in Fig. 1.15, use complex variables to outline a scheme for expressing the angular position, velocity, and acceleration of arm c in terms of those of arm a. (You needn't work out the equations.)

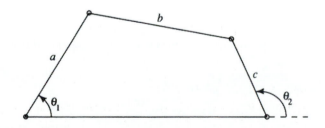

Figure 1.15 Linkage in Prob. 30.

1.4   The Complex Exponential

The familiar exponential function $f(x) = e^x$ has a natural and extremely useful extension to the complex plane. Indeed the complex function e^z provides a basic tool for the application of complex variables to electrical circuits, control systems, wave propagation, and time-invariant physical systems in general.

To find a suitable definition for e^z when $z = x + iy$, we want to preserve the basic identities satisfied by the real function e^x. So, first of all, we postulate that the multiplicative property should persist:

$$e^{z_1} e^{z_2} = e^{z_1 + z_2}. \tag{1}$$

This simplifies matters considerably, since Eq. (1) enables the decomposition

$$e^z = e^{x+iy} = e^x e^{iy} \tag{2}$$

and we see that to define e^z, we need only specify e^{iy} (in other words we will be able to exponentiate complex numbers once we discover how to exponentiate pure imaginary ones).

Next we propose that the differentiation law

$$\frac{de^z}{dz} = e^z \tag{3}$$

be preserved. Differentiation with respect to a complex variable $z = x + iy$ is a very profound and, at this stage, ambiguous operation; indeed Chapter 2 is devoted to a painstaking study of this concept (and the rest of the book is dedicated to exploring its consequences). But thanks to the factorization displayed in Eq. (2) we need only consider (for the moment) a special case of Eq. (3)—namely,

$$\frac{de^{iy}}{d(iy)} = e^{iy}$$

or, equivalently (by the chain rule),

$$\frac{de^{iy}}{dy} = ie^{iy}. \tag{4}$$

The consequences of postulating Eq. (4) become more apparent if we differentiate again:

$$\begin{aligned}
\frac{d^2 e^{iy}}{dy^2} &= \frac{d}{dy}(ie^{iy}) \\
&= i^2 e^{iy} \\
&= -e^{iy};
\end{aligned}$$

in other words, the function $g(y) := e^{iy}$ satisfies the differential equation

$$\frac{d^2 g}{dy^2} = -g. \tag{5}$$

Now observe that any function of the form

$$A \cos y + B \sin y \qquad (A, B \text{ constants})$$

satisfies Eq. (5). In fact, from the theory of differential equations it is known that every solution of Eq. (5) must have this form. Hence we can write

$$g(y) = A \cos y + B \sin y. \tag{6}$$

To evaluate A and B we use the conditions that

$$g(0) = e^{i0} = e^0 = 1 = A \cos 0 + B \sin 0$$

and

$$\frac{dg}{dy}(0) = ig(0) = i = -A \sin 0 + B \cos 0.$$

Thus $A = 1$ and $B = i$, leading us to the identification

$$\boxed{e^{iy} = \cos y + i \sin y.} \tag{7}$$

Equation (7) is known as *Euler's equation*.[†] Combining Eqs. (7) and (2) we formulate the following.

Definition 5. If $z = x + iy$, then e^z is defined to be the complex number

$$e^z := e^x (\cos y + i \sin y). \tag{8}$$

It is not difficult to verify directly that e^z, as defined above, satisfies the usual algebraic properties of the exponential function—in particular, the multiplicative identity (1) and the associated division rule

$$\frac{e^{z_1}}{e^{z_2}} = e^{z_1 - z_2} \tag{9}$$

(see Prob. 15a). In Sec. 2.5 we will obtain further confirmation that we have made the "right choice" by showing that Definition 5 produces a function that has the extremely desirable property of *analyticity*. Another confirmation is exhibited in the following example.

[†]Leonhard Euler (1707–1783).

Example 1

Show that Euler's equation is formally consistent with the usual Taylor series expansions

$$e^x = 1 + x + \frac{x^2}{2!} + \frac{x^3}{3!} + \frac{x^4}{4!} + \frac{x^5}{5!} + \cdots,$$

$$\cos x = 1 - \frac{x^2}{2!} + \frac{x^4}{4!} - \cdots,$$

$$\sin x = x - \frac{x^3}{3!} + \frac{x^5}{5!} - \cdots.$$

Solution. We shall study series representations of complex functions in full detail in Chapter 5. For now we ignore questions of convergence, etc., and simply substitute $x = iy$ into the exponential series:

$$e^{iy} = 1 + iy + \frac{(iy)^2}{2!} + \frac{(iy)^3}{3!} + \frac{(iy)^4}{4!} + \frac{(iy)^5}{5!} + \cdots$$

$$= \left(1 - \frac{y^2}{2!} + \frac{y^4}{4!} - \cdots\right) + i\left(y - \frac{y^3}{3!} + \frac{y^5}{5!} - \cdots\right)$$

$$= \cos y + i \sin y. \quad \blacksquare$$

Euler's equation (7) enables us to write the polar form (Sec. 1.3) of a complex number as

$$z = r \operatorname{cis} \theta = r(\cos \theta + i \sin \theta) = re^{i\theta}.$$

Thus we can (and do) drop the awkward "cis" artifice and use, as the standard polar representation,

$$z = re^{i\theta} = |z|e^{i \arg z}. \tag{10}$$

In particular, notice the following identities:

$$e^{i0} = e^{2\pi i} = e^{-2\pi i} = e^{4\pi i} = e^{-4\pi i} = \cdots = 1,$$

$$e^{(\pi/2)i} = i, \qquad e^{(-\pi/2)i} = -i, \qquad e^{\pi i} = -1.$$

(Students of mathematics, including Euler himself, have often marveled at the last identity. The constant e comes from calculus, π comes from geometry, and i comes from algebra—and the combination $e^{\pi i}$ gives -1, the basic unit for generating the arithmetic system from the counting numbers, or cardinals!)

Observe also that $|e^{i \arg z}| = 1$ and that Euler's equation leads to the following representations of the customary trigonometric functions:

$$\cos \theta = \operatorname{Re} e^{i\theta} = \frac{e^{i\theta} + e^{-i\theta}}{2}, \tag{11}$$

$$\sin \theta = \operatorname{Im} e^{i\theta} = \frac{e^{i\theta} - e^{-i\theta}}{2i}. \tag{12}$$

The rules derived in Sec. 1.3 for multiplying and dividing complex numbers in polar form now find very natural expressions:

$$z_1 z_2 = \left(r_1 e^{i\theta_1} \right) \left(r_2 e^{i\theta_2} \right) = (r_1 r_2)\, e^{i(\theta_1 + \theta_2)}, \tag{13}$$

$$\frac{z_1}{z_2} = \frac{r_1 e^{i\theta_1}}{r_2 e^{i\theta_2}} = \left(\frac{r_1}{r_2} \right) e^{i(\theta_1 - \theta_2)}, \tag{14}$$

and complex conjugation of $z = r e^{i\theta}$ is accomplished by changing the sign of i in the exponent:

$$\bar{z} = r e^{-i\theta}. \tag{15}$$

Example 2

Compute (a) $(1 + i)/(\sqrt{3} - i)$ and (b) $(1 + i)^{24}$.

Solution. (a) This quotient was evaluated using the cis operator in Example 1.11 of Sec. 1.3; using the exponential the calculations take the form

$$1 + i = \sqrt{2}\,\text{cis}(\pi/4) = \sqrt{2} e^{i\pi/4}, \qquad \sqrt{3} - i = 2\,\text{cis}(-\pi/6) = 2 e^{-i\pi/6},$$

and, therefore,

$$\frac{1+i}{\sqrt{3}-i} = \frac{\sqrt{2} e^{i\pi/4}}{2 e^{-i\pi/6}} = \frac{\sqrt{2}}{2} e^{i5\pi/12}.$$

(b) The exponential forms become

$$(1 + i)^{24} = (\sqrt{2} e^{i\pi/4})^{24} = (\sqrt{2})^{24} e^{i24\pi/4} = 2^{12} e^{i6\pi} = 2^{12}. \quad \blacksquare$$

In the solution to part (b) above we glossed over the justification for the identity $(e^{i\pi/4})^{24} = e^{i24\pi/4}$. Actually, a careful scrutiny yields much more—a powerful formula involving trigonometric functions, which we describe in the next example.

Example 3

Prove *De Moivre's formula:*[†]

$$\boxed{(\cos\theta + i \sin\theta)^n = \cos n\theta + i \sin n\theta, \qquad n = 1, 2, 3, \ldots} \tag{16}$$

Solution. By the multiplicative property, Eq. (1),

$$(e^{i\theta})^n = \underbrace{e^{i\theta} e^{i\theta} \cdots e^{i\theta}}_{(n \text{ times})} = e^{i\theta + i\theta + \cdots + i\theta} = e^{in\theta}.$$

Now applying Euler's formula (7) to the first and last members of this equation string, we deduce (16). $\blacksquare$

De Moivre's formula can be a convenient tool for deducing multiple-angle trigonometric identities, as is illustrated by the following example. (See also Probs. 12 and 20.)

[†]Published by Abraham De Moivre in 1707.

Example 4

Express $\cos 3\theta$ in terms of $\cos\theta$ and $\sin\theta$.

Solution. By Eq. (16) (with $n = 3$) we have

$$\cos 3\theta = \operatorname{Re}(\cos 3\theta + i \sin 3\theta) = \operatorname{Re}(\cos\theta + i \sin\theta)^3. \tag{17}$$

According to the binomial formula,

$$(a + b)^3 = a^3 + 3a^2 b + 3ab^2 + b^3.$$

Thus, making the obvious identifications $a = \cos\theta$, $b = i \sin\theta$ in (17), we deduce

$$\cos 3\theta = \operatorname{Re}\left[\cos^3\theta + 3\cos^2\theta\,(i \sin\theta) + 3\cos\theta\,(-\sin^2\theta) - i \sin^3\theta\right]$$
$$= \cos^3\theta - 3\cos\theta \sin^2\theta. \quad\blacksquare \tag{18}$$

Example 5

Compute the integral

$$\int_0^{2\pi} \cos^4\theta\, d\theta$$

by using the representation (11) together with the binomial formula (see Exercises 1.1, Prob. 27).

Solution. We can express the integrand as

$$\cos^4\theta = \left(\frac{e^{i\theta} + e^{-i\theta}}{2}\right)^4 = \frac{1}{2^4}\left(e^{i\theta} + e^{-i\theta}\right)^4,$$

and expanding via the binomial formula gives

$$\cos^4\theta = \frac{1}{2^4}\left(e^{4i\theta} + 4e^{3i\theta}e^{-i\theta} + 6e^{2i\theta}e^{-2i\theta} + 4e^{i\theta}e^{-3i\theta} + e^{-4i\theta}\right)$$
$$= \frac{1}{2^4}\left(e^{4i\theta} + 4e^{2i\theta} + 6 + 4e^{-2i\theta} + e^{-4i\theta}\right)$$
$$= \frac{1}{2^4}(6 + 8\cos 2\theta + 2\cos 4\theta).$$

Thus

$$\int_0^{2\pi} \cos^4\theta\, d\theta = \int_0^{2\pi} \frac{1}{2^4}(6 + 8\cos 2\theta + 2\cos 4\theta)\, d\theta$$
$$= \frac{1}{2^4}[6\theta + 4\sin 2\theta + \frac{1}{2}\sin 4\theta]\Big|_0^{2\pi} = \frac{6}{2^4}2\pi = \frac{3}{4}\pi. \quad\blacksquare$$

EXERCISES 1.4

In Problems 1 and 2 write each of the given numbers in the form $a + bi$.

1. (a) $e^{-i\pi/4}$ **(b)** $\dfrac{e^{1+i3\pi}}{e^{-1+i\pi/2}}$ **(c)** e^{e^i}

2. (a) $\dfrac{e^{3i} - e^{-3i}}{2i}$ **(b)** $2e^{3+i\pi/6}$ **(c)** e^z, where $z = 4e^{i\pi/3}$

In Problems 3 and 4 write each of the given numbers in the polar form $re^{i\theta}$.

3. (a) $\dfrac{1-i}{3}$ **(b)** $-8\pi(1 + \sqrt{3}i)$ **(c)** $(1 + i)^6$

4. (a) $\left(\cos\dfrac{2\pi}{9} + i\sin\dfrac{2\pi}{9}\right)^3$ **(b)** $\dfrac{2+2i}{-\sqrt{3}+i}$ **(c)** $\dfrac{2i}{3e^{4+i}}$

5. Show that $|e^{x+iy}| = e^x$ and $\arg e^{x+iy} = y + 2k\pi$ $(k = 0, \pm1, \pm2, \ldots)$.

6. Show that, for real θ,

   (a) $\tan\theta = \dfrac{e^{i\theta} - e^{-i\theta}}{i(e^{i\theta} + e^{-i\theta})}$ **(b)** $\csc\theta = \dfrac{2}{e^{i(\theta-\pi/2)} - e^{-i(\theta+\pi/2)}}$

7. Show that $e^z = e^{z+2\pi i}$ for all z. (*The exponential function is periodic with period $2\pi i$.*)

8. Show that, for all z,

   (a) $e^{z+\pi i} = -e^z$ **(b)** $\overline{e^z} = e^{\bar{z}}$

9. Show that $(e^z)^n = e^{nz}$ for any integer n.

10. Show that $|e^z| \le 1$ if Re $z \le 0$.

11. Determine which of the following properties of the real exponential function remain true for the complex exponential function (that is, for x replaced by z).

   (a) e^x is never zero.   **(b)** e^x is a one-to-one function.
   (c) e^x is defined for all x.   **(d)** $e^{-x} = 1/e^x$.

12. Use De Moivre's formula together with the binomial formula to derive the following identities.

   (a) $\sin 3\theta = 3\cos^2\theta\sin\theta - \sin^3\theta$

   (b) $\sin 4\theta = 4\cos^3\theta\sin\theta - 4\cos\theta\sin^3\theta$

13. Show how the following trigonometric identities follow from Eqs. (11) and (12).

   (a) $\sin^2\theta + \cos^2\theta = 1$

   (b) $\cos(\theta_1 + \theta_2) = \cos\theta_1\cos\theta_2 - \sin\theta_1\sin\theta_2$

14. Does De Moivre's formula hold for negative integers n?

15. **(a)** Show that the multiplicative law (1) follows from Definition 5.

    (b) Show that the division rule (9) follows from Definition 5.

16. Let $z = re^{i\theta}$, $(z \neq 0)$. Show that $\exp(\ln r + i\theta) = z$.†

17. Show that the function $z(t) = e^{it}$, $0 \leq t \leq 2\pi$, describes the unit circle $|z| = 1$ traversed in the counterclockwise direction (as t increases from 0 to 2π). Then describe each of the following curves.

    (a) $z(t) = 3e^{it}$, $0 \leq t \leq 2\pi$ **(b)** $z(t) = 2e^{it} + i$, $0 \leq t \leq 2\pi$
    (c) $z(t) = 2e^{i2\pi t}$, $0 \leq t \leq 1/2$ **(d)** $z(t) = 3e^{-it} + 2 - i$, $0 \leq t \leq 2\pi$

18. Sketch the curves that are given for $0 \leq t \leq 2\pi$ by

    (a) $z(t) = e^{(1+i)t}$ **(b)** $z(t) = e^{(1-i)t}$
    (c) $z(t) = e^{(-1+i)t}$ **(d)** $z(t) = e^{(-1-i)t}$

19. Let n be a positive integer greater than 2. Show that the points $e^{2\pi i k/n}$, $k = 0, 1, \ldots, n - 1$, form the vertices of a regular polygon.

20. Prove that if $z \neq 1$, then

$$1 + z + z^2 + \cdots + z^n = \frac{z^{n+1} - 1}{z - 1}.$$

    Use this result and De Moivre's formula to establish the following identities.

    (a) $1 + \cos\theta + \cos 2\theta + \cdots + \cos n\theta = \dfrac{1}{2} + \dfrac{\sin[(n + \frac{1}{2})\theta]}{2\sin(\theta/2)}$

    (b) $\sin\theta + \sin 2\theta + \cdots + \sin n\theta = \dfrac{\sin(n\theta/2)\sin((n + 1)\theta/2)}{\sin(\theta/2)}$, where

    $0 < \theta < 2\pi$.

21. Prove that if n is a positive integer, then

$$\left|\frac{\sin(n\theta/2)}{\sin(\theta/2)}\right| \leq n \qquad (\theta \neq 0, \pm 2\pi, \pm 4\pi, \ldots).$$

    [HINT: Argue first that if $z = e^{i\theta}$, then the left-hand side equals $|(1 - z^n)/(1 - z)|$.]

22. Show that if n is an integer, then

$$\int_0^{2\pi} e^{in\theta}\, d\theta = \int_0^{2\pi} \cos(n\theta)\, d\theta + i \int_0^{2\pi} \sin(n\theta)\, d\theta = \begin{cases} 2\pi & \text{if } n = 0, \\ 0 & \text{if } n \neq 0. \end{cases}$$

23. Compute the following integrals by using the representations (11) or (12) together with the binomial formula.

    (a) $\int_0^{2\pi} \cos^8\theta\, d\theta$ **(b)** $\int_0^{2\pi} \sin^6(2\theta)\, d\theta$.

---

†As a convenience in printing we sometimes write $\exp(z)$ instead of e^z.

1.5 Powers and Roots

In this section we shall derive formulas for the nth power and the mth roots of a complex number.

Let $z = re^{i\theta} = r(\cos\theta + i\sin\theta)$ be the polar form of the complex number z. By taking $z_1 = z_2 = z$ in Eq. (13) of Sec. 1.4, we obtain the formula

$$z^2 = r^2 e^{i2\theta}.$$

Since $z^3 = zz^2$, we can apply the identity a second time to deduce that

$$z^3 = r^3 e^{i3\theta}.$$

Continuing in this manner we arrive at the formula for the nth power of z:

$$z^n = r^n e^{in\theta} = r^n(\cos n\theta + i\sin n\theta). \tag{1}$$

Clearly this is just an extension of De Moivre's formula, discussed in Example 3 of Sec. 1.4.

Equation (1) is an appealing formula for raising a complex number to a positive integer power. It is easy to see that the identity is also valid for negative integers n (see Prob. 2). The question arises whether the formula will work for $n = 1/m$, so that $\zeta = z^{1/m}$ is an mth root of z satisfying

$$\zeta^m = z. \tag{2}$$

Certainly if we define

$$\zeta = \sqrt[m]{r}\, e^{i\theta/m} \tag{3}$$

(where $\sqrt[m]{r}$ denotes the customary, positive, mth root), we compute a complex number ζ satisfying Eq. (2) [as is easily seen by applying Eq. (1)]. But the matter is more complicated than this; the number 1, for instance, has *two* square roots: 1 and -1. And each of these has, in turn, two square roots—generating *four* fourth roots of 1, namely, 1, -1, i, and $-i$.

To see how the additional roots fit into the scheme of things, let's work out the polar description of the equation $\zeta^4 = 1$ for each of these numbers:

$$1^4 = (1e^{i0})^4 = 1^4 e^{i0} = 1,$$
$$i^4 = (1e^{i\pi/2})^4 = 1^4 e^{i2\pi} = 1,$$
$$(-1)^4 = (1e^{i\pi})^4 = 1^4 e^{i4\pi} = 1,$$
$$(-i)^4 = (1e^{i3\pi/2})^4 = 1^4 e^{i6\pi} = 1.$$

It is instructive to trace the consecutive powers of these roots in the Argand diagram. Thus Fig. 1.16 shows that i, i^2, i^3, and i^4 complete one revolution before landing on 1; (-1), $(-1)^2$, $(-1)^3$, and $(-1)^4$ go around twice; the powers of $(-i)$ go around three times counterclockwise, and of course 1, 1^2, 1^3, and 1^4 never move.

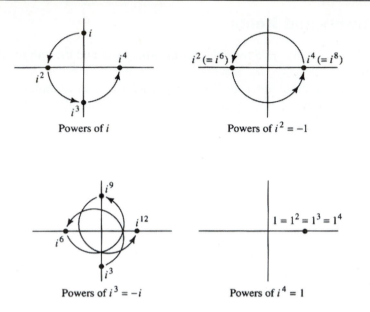

Figure 1.16 Successive powers of the fourth roots of unity.

Clearly, the multiplicity of roots is tied to the ambiguity in representing 1, in polar form, as e^{i0}, $e^{i2\pi}$, $e^{i4\pi}$, etc. Thus to compute *all* the mth roots of a number z, we must apply formula (3) to *every* polar representation of z. For the cube roots of unity, for example, we would compute as shown in the table opposite. Obviously the roots recur in sets of three, since $e^{i2\pi m_1/3} = e^{i2\pi m_2/3}$ whenever $m_1 - m_2 = 3$.

Generalizing, we can see that *there are exactly m distinct mth roots of unity, denoted by $1^{1/m}$, and they are given by*

$$\boxed{1^{1/m} = e^{i2k\pi/m} = \cos\frac{2k\pi}{m} + i\sin\frac{2k\pi}{m} \quad (k = 0, 1, 2, \ldots, m - 1).} \qquad (4)$$

The arguments of these roots are $2\pi/m$ radians apart, and the roots themselves form the vertices of a regular polygon (Fig. 1.17).

Taking $k = 1$ in Eq. (4) we obtain the root[†]

$$\omega_m := e^{i2\pi/m} = \cos\frac{2\pi}{m} + i\sin\frac{2\pi}{m},$$

and it is easy to see that the complete set of roots can be displayed as

$$1, \omega_m, \omega_m^2, \ldots, \omega_m^{m-1}.$$

[†]A number w is said to be a *primitive mth root of unity* if w^m equals 1 but $w^k \neq 1$ for $k = 1, 2, \ldots, m - 1$. Clearly, ω_m is a primitive root.

Cube Roots of Unity

Polar representation of 1	Application of (3)
$\vdots$	$\vdots$
$1 = e^{-i6\pi}$	$1^{1/3} = e^{-i6\pi/3} = 1$
$1 = e^{-i4\pi}$	$1^{1/3} = e^{-i4\pi/3} = -\frac{1}{2} + i\frac{\sqrt{3}}{2}$
$1 = e^{-i2\pi}$	$1^{1/3} = e^{-i2\pi/3} = -\frac{1}{2} - i\frac{\sqrt{3}}{2}$
$1 = e^{i0}$	$1^{1/3} = e^{i0/3} = 1$
$1 = e^{i2\pi}$	$1^{1/3} = e^{i2\pi/3} = -\frac{1}{2} + i\frac{\sqrt{3}}{2}$
$1 = e^{i4\pi}$	$1^{1/3} = e^{i4\pi/3} = -\frac{1}{2} - i\frac{\sqrt{3}}{2}$
$1 = e^{i6\pi}$	$1^{1/3} = e^{i6\pi/3} = 1$
$\vdots$	$\vdots$

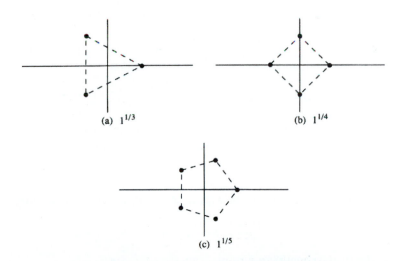

Figure 1.17 Regular polygons formed by the roots of unity.

Example 1

Prove that

$$1 + \omega_m + \omega_m^2 + \cdots + \omega_m^{m-1} = 0. \tag{5}$$

Solution. This result is obvious from a physical point of view, since, by symmetry, the center of mass $(1 + \omega_m + \omega_m^2 + \cdots + \omega_m^{m-1})/m$ of the system of m unit masses located at the mth roots of unity must be at the origin (see Fig. 1.17).

To give an algebraic proof we simply note that

$$(\omega_m - 1)(1 + \omega_m + \omega_m^2 + \cdots + \omega_m^{m-1}) = \omega_m^m - 1 = 0.$$

Since $\omega_m \neq 1$, Eq. (5) follows. ∎

To obtain the mth roots of an *arbitrary* (nonzero) complex number $z = re^{i\theta}$, we generalize the idea displayed by Eq. (4) and, reasoning similarly, conclude that *the m distinct mth roots of z are given by*

$$z^{1/m} = \sqrt[m]{|z|}\, e^{i(\theta + 2k\pi)/m} \qquad (k = 0, 1, 2, \ldots, m - 1). \tag{6}$$

Equivalently, we can form these roots by taking any single one such as given in (3) and multiplying by the mth roots of unity.

Example 2

Find all the cube roots of $\sqrt{2} + i\sqrt{2}$.

Solution. The polar form for $\sqrt{2} + i\sqrt{2}$ is

$$\sqrt{2} + i\sqrt{2} = 2e^{i\pi/4}.$$

Putting $|z| = 2$, $\theta = \pi/4$, and $m = 3$ into Eq. (6), we obtain

$$(\sqrt{2} + i\sqrt{2})^{1/3} = \sqrt[3]{2}\, e^{i(\pi/12 + 2k\pi/3)} \qquad (k = 0, 1, 2).$$

Therefore, the three cube roots of $\sqrt{2} + i\sqrt{2}$ are $\sqrt[3]{2}(\cos \pi/12 + i \sin \pi/12)$, $\sqrt[3]{2}(\cos 3\pi/4 + i \sin 3\pi/4)$, and $\sqrt[3]{2}(\cos 17\pi/12 + i \sin 17\pi/12)$. ∎

Example 3

Let a, b, and c be *complex* constants with $a \neq 0$. Prove that the solutions of the equation

$$az^2 + bz + c = 0 \tag{7}$$

are given by the (usual) quadratic formula

$$z = \frac{-b \pm \sqrt{b^2 - 4ac}}{2a}, \tag{8}$$

where $\sqrt{b^2 - 4ac}$ denotes one of the values of $(b^2 - 4ac)^{1/2}$.

Solution.   After multiplying Eq. (7) by $4a$, one can manipulate it into the form

$$4a^2z^2 + 4abz + b^2 = b^2 - 4ac.$$

The left hand side is $(2az + b)^2$, so

$$2az + b = (b^2 - 4ac)^{1/2} = \pm\sqrt{b^2 - 4ac},$$

which is equivalent to Eq. (8).   ■

EXERCISES 1.5

1. Prove identity (1) by using induction.

2. Show that formula (1) also holds for negative integers n.

3. Let n be a positive integer. Prove that $\arg z^n = n \operatorname{Arg} z + 2k\pi$, $k = 0, \pm 1, \pm 2, \ldots,$ for $z \neq 0$.

4. Use the identity (1) to show that

   (a) $(\sqrt{3} - i)^7 = -64\sqrt{3} + i64$ (b) $(1 + i)^{95} = 2^{47}(1 - i)$

5. Find all the values of the following.

   (a) $(-16)^{1/4}$ (b) $1^{1/5}$ (c) $i^{1/4}$

   (d) $(1 - \sqrt{3}i)^{1/3}$ (e) $(i - 1)^{1/2}$ (f) $\left(\dfrac{2i}{1 + i}\right)^{1/6}$

6. Describe how to construct geometrically the fifth roots of z_0 if

   (a) $z_0 = -1$ (b) $z_0 = i$ (c) $z_0 = 1 + i$

7. Solve each of the following equations.

   (a) $2z^2 + z + 3 = 0$

   (b) $z^2 - (3 - 2i)z + 1 - 3i = 0$

   (c) $z^2 - 2z + i = 0$

8. Let a, b, and c be real numbers and let $a \neq 0$. Show that the equation $az^2 + bz + c = 0$ has

   (a) two real solutions if $b^2 - 4ac > 0$.

   (b) two nonreal conjugate solutions if $b^2 - 4ac < 0$.

9. Solve the equation $z^3 - 3z^2 + 6z - 4 = 0$.

10. Find all four roots of the equation $z^4 + 1 = 0$ and use them to deduce the factorization $z^4 + 1 = (z^2 - \sqrt{2}z + 1)(z^2 + \sqrt{2}z + 1)$.

11. Solve the equation $(z + 1)^5 = z^5$.

12. Show that the n points $z_0^{1/n}$ form the vertices of a regular n-sided polygon inscribed in the circle of radius $\sqrt[n]{|z_0|}$ about the origin.

13. Show that $\omega_3 = (-1 + \sqrt{3}i)/2$ and that $\omega_4 = i$. Use these values to verify identity (5) for the special cases $n = 3$ and $n = 4$.

14. Let m and n be positive integers that have no common factor. Prove that the set of numbers $(z^{1/n})^m$ is the same as the set of numbers $(z^m)^{1/n}$. We denote this common set of numbers by $z^{m/n}$. Show that

$$z^{m/n} = \sqrt[n]{|z|^m} \left[\cos \frac{m}{n}(\theta + 2k\pi) + i \sin \frac{m}{n}(\theta + 2k\pi) \right] \tag{9}$$

for $k = 0, 1, 2, \ldots, n - 1$.

15. Use the result of Prob. 14 to find all the values of $(1 - i)^{3/2}$.

16. Show that the real part of any solution of $(z + 1)^{100} = (z - 1)^{100}$ must be zero.

17. Let m be a fixed positive integer and let l be an integer that is not divisible by m. Prove the following generalization of Eq. (5):

$$1 + \omega_m^l + \omega_m^{2l} + \cdots + \omega_m^{(m-1)l} = 0.$$

18. Show that if α and β are nth and mth roots of unity, respectively, then the product $\alpha\beta$ is a kth root of unity for some integer k.

19. (*Electric Field*) A uniformly charged infinite rod, standing perpendicular to the z-plane at the point z_0, generates an electric field at every point in the plane. The intensity of this field varies inversely as the distance from z_0 to the point and is directed along the line from z_0 to the point.

(a) Show that the (vector) field at the point z is given by the function $F(z) = 1/(\bar{z} - \bar{z_0})$, in appropriate units. (Recall Fig. 1.13, Sec 1.3.)

(b) If three such rods are located at the points $1 + i$, $-1 + i$, and 0, find the positions of equilibrium (that is, the points where the vector sum of the fields is zero).

20. Write a computer program for solving the quadratic equation

$$az^2 + bz + c = 0, \qquad a \neq 0.$$

Use as inputs the real and imaginary parts of a, b, c and print the solutions in both rectangular and polar form.

21. Some complex integer square roots can be obtained by a modification of the polynomial factoring strategy. For example, if $3 + 4i$ equals $(a + bi)^2$, then $4 = 2ab$ and $3 = a^2 - b^2$. A little mental experimentation yields the answer $a = 2$, $b = 1$; of course $-2 - i$ is the other square root. Use this strategy to find the square roots of the following numbers:

(a) $8 + 6i$ (b) $5 + 12i$ (c) $24 + 10i$

(d) $3 - 4i$ (e) $-8 + 6i$ (f) $8 - 6i$

1.6 Planar Sets

In the calculus of functions of a real variable, the main theorems are typically stated for functions defined on an *interval* (open or closed). For functions of a complex variable the basic results are formulated for functions defined on sets that are 2-dimensional "domains" or "closed regions." In this section we give the precise definition of these point sets. We begin with the meaning of a "neighborhood" in the complex plane.

The set of all points that satisfy the inequality

$$|z - z_0| < \rho,$$

where ρ is a positive real number, is called an *open disk* or *circular neighborhood of* z_0. This set consists of all the points that lie inside the circle of radius ρ about z_0. In particular, the solution sets of the inequalities

$$|z - 2| < 3, \qquad |z + i| < \frac{1}{2}, \qquad |z| < 8$$

are circular neighborhoods of the respective points 2, $-i$, and 0. We shall make frequent reference to the neighborhood $|z| < 1$, which is called the *open unit disk*.

A point z_0 which lies in a set S is called an *interior point of* S if there is some circular neighborhood of z_0 that is completely contained in S. For example, if S is the right half-plane Re $z > 0$ and $z_0 = 0.01$, then z_0 is an interior point of S because S contains the neighborhood $|z - z_0| < 0.01$ (see Fig. 1.18).

If every point of a set S is an interior point of S, we say that S is an *open set*. Any open disk is an open set (Prob. 1). Each of the following inequalities also describes an open set: (a) $\rho_1 < |z - z_0| < \rho_2$, (b) $|z - 3| > 2$, (c) Im $z > 0$, and (d) $1 < $ Re $z < 2$. These sets are sketched in Fig. 1.19. Note that the solution set T of the inequality $|z - 3| \geq 2$ is *not* an open set since no point on the circle $|z - 3| = 2$ is an interior point of T. Note also that an open interval of the real axis is *not* an open set since it contains no open disk.

Let $w_1, w_2, \ldots, w_{n+1}$ be $n + 1$ points in the plane. For each $k = 1, 2, \ldots, n$, let l_k denote the line segment joining w_k to w_{k+1}. Then the successive line segments $l_1, l_2, \ldots, l_n$ form a continuous chain known as a *polygonal path* that joins w_1 to w_{n+1}.

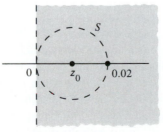

Figure 1.18 Interior point.

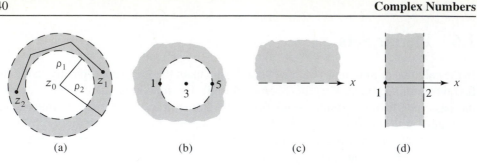

Figure 1.19 Open sets.

An open set S is said to be *connected* if every pair of points z_1, z_2 in S can be joined by a polygonal path that lies entirely in S [see Fig. 1.19(a)]. Roughly speaking, this means that S consists of a "single piece." Each of the sets in Fig. 1.19 is connected. The set consisting of all those points in the plane that do *not* lie on the circle $|z| = 1$ is an example of an open set that is not connected; indeed, if z_1 is a point inside the circle and z_2 is a point outside, then every polygonal path that joins z_1 and z_2 must intersect the circle.

We call an open connected set a *domain*. Therefore, all the sets in Fig. 1.19 are domains.

In the calculus of functions of a single real variable a useful and familiar fact is that, on an interval, the vanishing of the derivative implies that the function is identically constant. We now present an extension of this result to functions of two real variables, which underscores the importance of the notion of a domain.

Theorem 1. Suppose $u(x, y)$ is a real-valued function defined in a domain D. If the first partial derivatives of u satisfy

$$\frac{\partial u}{\partial x} = \frac{\partial u}{\partial y} = 0 \tag{1}$$

at all points of D, then $u \equiv$ constant in D.

Proof. Notice that the assumption $\partial u / \partial x = 0$ implies that u remains constant along any horizontal line segment contained in D; indeed, on such a segment, u is a function of a single variable (namely, x) whose derivative vanishes. Similarly, the assumption $\partial u / \partial y = 0$ means that u is constant along any vertical line segment that lies in D. Putting these facts together we see that u remains unchanged along any polygonal path in D that has all its segments parallel to the coordinate axes. Now from the definition of connectedness we know that any pair of points in D can be joined by some polygonal path lying entirely in D. The catch is that this path may have some segments that are neither vertical nor horizontal. However, it turns out from topological considerations (see Prob. 22) that any such segment can be replaced

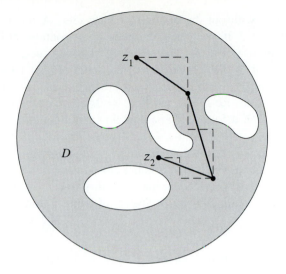

Figure 1.20 Polygonal path with vertical and horizontal segments.

by a chain of small horizontal and vertical segments lying in D (see Fig. 1.20). Thus, Theorem 1 follows.   ■

For the reader who prefers to avoid such topological arguments, an alternative proof of Theorem 1 can be given with the aid of the chain rule (see Prob. 24).

What is crucial for Theorem 1 is the connectedness property of domains; in fact, the theorem is no longer true if D is merely assumed to be an open set, because then "piecewise constant" functions would satisfy the hypothesis (see Prob. 19).

Example 1

A real-valued function $u(x, y)$ satisfies

$$\frac{\partial u}{\partial x} = 3 \quad \text{and} \quad \frac{\partial u}{\partial y} = 6 \tag{2}$$

at every point in the open unit disk $D = \{z : |z| < 1\}$. Show that $u(x, y) = 3x + 6y + c$ for some constant c.

Solution.   Let $v(x, y) = 3x + 6y$, and consider the function $w(x, y) := u(x, y) - v(x, y)$. From (2) and the definition of $v(x, y)$, we have

$$\frac{\partial w}{\partial x} = 3 - 3 = 0 \quad \text{and} \quad \frac{\partial w}{\partial y} = 6 - 6 = 0$$

at each point of D. Since D is a domain, Theorem 1 asserts that $w(x, y)$ is constant in D, say $w(x, y) = c$. Thus $u(x, y) = v(x, y) + w(x, y) = v(x, y) + c$, which is the requested formula for u.   ■

We now continue with our discussion of planar sets. A point z_0 is said to be a *boundary point* of a set S if every neighborhood of z_0 contains at least one point of S and at least one point not in S. The set of all boundary points of S is called the *boundary* or *frontier* of S. The boundaries of the sets in Fig. 1.19 are as follows: (a) the two circles $|z - z_0| = \rho_1$ and $|z - z_0| = \rho_2$, (b) the circle $|z - 3| = 2$, (c) the real axis, and (d) the two lines Re $z = 1$ and Re $z = 2$. Since each point of a domain D is an interior point of D, it follows that a domain cannot contain any of its boundary points.

A set S is said to be *closed* if it contains all of its boundary points. (As requested in Prob. 13, the reader should verify that S being closed is equivalent to its complement $\mathbf{C} \backslash S$ being open.) The set described by the inequality $0 < |z| \le 1$ is not closed since it does not contain the boundary point 0. The set of points z that satisfy the inequality

$$|z - z_0| \le \rho \qquad (\rho > 0)$$

is a closed set, for it contains its boundary $|z - z_0| = \rho$. Therefore, we call this set a *closed disk*.

A set of points S is said to be *bounded* if there exists a positive real number R such that $|z| < R$ for every z in S. In other words, S is bounded if it is contained in some neighborhood of the origin. An *unbounded* set is one that is not bounded. Of the sets in Fig. 1.19 only (a) is bounded. A set that is both closed and bounded is said to be *compact*.

A *region* is a domain together with some, none, or all of its boundary points. In particular, every domain is a region.

EXERCISES 1.6

1. Prove that the neighborhood $|z - z_0| < \rho$ is an open set. [HINT: Show that if z_1 belongs to the neighborhood, then so do all points z that satisfy $|z - z_1| < R$, where $R = \rho - |z_1 - z_0|$.]

Problems 2–8 refer to the sets described by the following inequalities:

(a) $|z - 1 + i| \le 3$ (b) $|\operatorname{Arg} z| < \pi/4$
(c) $0 < |z - 2| < 3$ (d) $-1 < \operatorname{Im} z \le 1$
(e) $|z| \ge 2$ (f) $(\operatorname{Re} z)^2 > 1$

2. Sketch each of the given sets.

3. Which of the given sets are open?

4. Which of the given sets are domains?

5. Which of the given sets are bounded?

6. Describe the boundary of each of the given sets.

7. Which of the given sets are regions?

8. Which of the given sets are closed regions?

9. Prove that any set consisting of finitely many points is bounded.

10. Prove that the closed disk $|z - z_0| \leq \rho$ is bounded.

11. Let S be the set consisting of the points $1, 1/2, 1/3, \ldots$. What is the boundary of S?

12. Let z_0 be a point of the set S. Prove that if z_0 is not an interior point of S, then z_0 must be a boundary point of S.

13. Let S be a subset of $\mathbf{C}$. Prove that S is closed if and only if its complement $\mathbf{C} \backslash S$ is an open set.

14. A point z_0 is said to be an *accumulation point* of a set S if every neighborhood of z_0 contains infinitely many points of the set S. Prove that a closed region contains all its accumulation points.

Problems 15–18 refer to the following definitions: Let S and T be sets. The set consisting of all points belonging to S or T or both S and T is called the *union* of S and T and is denoted by $S \cup T$. The set consisting of all points belonging to both S and T is called the *intersection* of S and T and is denoted by $S \cap T$.

15. Let S and T be the sets described by $|z + 1| < 2$ and $|z - i| < 1$, respectively. Sketch the sets $S \cup T$ and $S \cap T$.

16. If S and T are open sets, prove that $S \cup T$ is an open set.

17. If S and T are domains, is $S \cap T$ necessarily a domain?

18. Prove that if S and T are domains that have at least one point in common, then $S \cup T$ is a domain.

19. Let

$$u(x, y) := \begin{cases} 1 & \text{for } |z| < 1, \\ 0 & \text{for } |z| > 2. \end{cases}$$

Show that $\partial u / \partial x = \partial u / \partial y = 0$ in the open set

$$D := \{z : |z| < 1\} \cup \{z : |z| > 2\},$$

but u is not constant in D. Why doesn't this contradict Theorem 1?

20. Suppose $u(x, y)$ is a real-valued function defined in a domain D. If

$$\frac{\partial u}{\partial x} = y \quad \text{and} \quad \frac{\partial u}{\partial y} = x$$

at all points of D, prove that $u(x, y) = xy + c$ for some constant c.

21. A real-valued function $u(x, y)$ satisfies

$$\frac{\partial u}{\partial x} = \frac{2x}{x^2 + y^2} \quad \text{and} \quad \frac{\partial u}{\partial y} = \frac{2y}{x^2 + y^2}$$

at every point in the annulus $A = \{z : 1 < |z| < 3\}$. Determine $u(x, y)$ up to an additive constant.

22. Let D be a domain and l be a closed line segment lying in D. In elementary topology it is shown that l can be covered by a finite number of open disks that lie in D and have their centers on l. Use this fact to prove that any two points in a domain D can be joined by a polygonal path in D having all its segments parallel to the coordinate axes.

23. The notion of "connectedness" also applies to closed sets. We say that a closed set $S \subset \mathbf{C}$ is *connected* if it cannot be written as the union of two nonempty disjoint (nonintersecting) closed sets. A closed connected set is called a *continuum*. Determine which of the following sets is a continuum:

  (a) $\{z : |z - 3| = 4\}$

  (b) $\{z : |z| = 1\} \cup \{z : |z| = 3\}$

  (c) $\{1, -1, i\}$

  (d) $\{z : |z - 1| \geq 2\}$

24. Prove Theorem 1 by completing the following steps.

  (a) Show that any line segment can be parametrized by

$$x = at + b, \qquad y = ct + d,$$

  where a, b, c, and d are real constants and t ranges between 0 and 1. Hence the values of u along a line segment lying in D are given by

$$U(t) := u(at + b, ct + d), \qquad 0 \leq t \leq 1.$$

  (b) Use assumption (1) of Theorem 1 and the chain rule to show that $dU/dt = 0$ for $0 \leq t \leq 1$ and thus conclude that u is constant on any line segment in D.

  (c) By appealing to the definition of connectedness, argue that u must have the same value at any two points of D.

1.7   The Riemann Sphere and Stereographic Projection

For centuries cartographers have struggled with the problem of how to represent the spherical-like surface of Earth on a flat sheet of paper, and a variety of useful projections have resulted. In this section we describe one such method that identifies points on the surface of a sphere with points in the complex plane; namely, the so-called *stereographic projection*. While this projection is not one typically found in atlases, it is of considerable importance in the theory of complex variables.

To describe the stereographic projection, we consider the unit sphere in 3 dimensions (x_1, x_2, x_3) whose equation is given by

$$x_1^2 + x_2^2 + x_3^2 = 1.$$

A sketch of this sphere and its equatorial plane is given in Fig. 1.21.

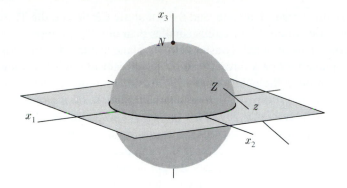

Figure 1.21 The Riemann sphere.

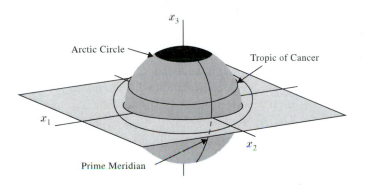

Figure 1.22 The Riemann geosphere.

Our goal is to associate with each point z in the equatorial plane a unique point Z on the sphere. For this purpose we construct the line that passes through the north pole $N = (0, 0, 1)$ of the sphere and the given point z in the x_1x_2-plane. This line pierces the spherical surface in exactly one point Z as shown in Fig. 1.21, and we say that the point Z is the stereographic projection of the point z. If we identify the equatorial plane as the complex plane (or z-plane), the unit sphere is called the *Riemann sphere*.[†]

Continuing with this geographical interpretation of the Riemann sphere, we note that under stereographic projection, points on the unit circle $|z| = 1$ (in the z-plane) remain fixed (that is, $z = Z$), forming the equator. Points outside the unit circle (for which $|z| > 1$) project to points in the northern hemisphere, while those inside the unit circle (for which $|z| < 1$) project to the southern hemisphere. In particular, the origin of the z-plane projects to the south pole of the Riemann sphere.

Figure 1.22 displays other features of the stereographic projection. Circles of latitude (parallel to the equator) on the sphere are the projections of circles centered at the origin in the z plane; the Arctic Circle and the Tropic of Cancer are projections

---

[†]Bernhard Riemann (1826–1866).

of circles of radius greater than one, and the Antarctic Circle and the Tropic of Capricorn are projections of circles of radius smaller than one. The temperate zones on the sphere are projections of annuli (washers) in the plane, centered at the origin. Straight lines in the z-plane that pass through the origin project to circles of longitude (great circles passing through the poles) on the sphere. (If we imagine the projection of the positive y-axis as the prime meridian through Greenwich, England, then the negative y-axis projects to the International Date Line, halfway around the globe.)[†]

Stereographic projection preserves the angles of intersection of these curves: circles centered at the origin intersect lines through the origin at right angles in the z-plane, and longitudes intersect latitudes at right angles on the sphere. This is no mere coincidence for it is a beautiful property of stereographic projection that *it preserves angles*—a useful feature for analyzing images.

To be more specific about the correspondence between the points in the z-plane and those on the Riemann sphere, it is imperative that we have a formula for the projection.

Example 1

Show that if $Z = (x_1, x_2, x_3)$ is the projection on the Riemann sphere of the point $z = x + iy$ in the complex plane, then

$$x_1 = \frac{2\,\mathrm{Re}\,z}{|z|^2 + 1}, \qquad x_2 = \frac{2\,\mathrm{Im}\,z}{|z|^2 + 1}, \qquad x_3 = \frac{|z|^2 - 1}{|z|^2 + 1}. \tag{1}$$

Solution.   The line through the north pole $N = (0, 0, 1)$ and $(x, y, 0)$ is given by the parametric equations

$$x_1 = tx, \quad x_2 = ty, \quad x_3 = 1 - t, \quad -\infty < t < \infty. \tag{2}$$

This line cuts the sphere precisely when t satisfies

$$1 = x_1^2 + x_2^2 + x_3^2 = t^2 x^2 + t^2 y^2 + (1 - t)^2$$

or

$$1 = t^2 (x^2 + y^2 + 1) + 1 - 2t,$$

whose roots are $t = 0$ (the north pole) and

$$t = \frac{2}{x^2 + y^2 + 1} = \frac{2}{|z|^2 + 1}. \tag{3}$$

Substitution of this value of t into (2) yields Eqs. (1).   ∎

[†]Most college students perform a crude stereographic projection when they wash their laundry. They toss their dirty clothes onto the bedsheet, then bring up the four corners of the sheet and knot them to form a laundry bag. The flat bedsheet corresponds to the z-plane, and the "bag" to the Riemann sphere.

As an example, the point $z = 1 + \sqrt{3}i$ corresponds to the point $Z = (\frac{2}{5}, \frac{2\sqrt{3}}{5}, \frac{3}{5})$ on the Riemann sphere.

Conversely, if we start with the point (x_1, x_2, x_3) on the Riemann sphere, we get from (2) that its corresponding point $x + iy$ in the z-plane is given by

$$x = x_1/t, \qquad y = x_2/t, \qquad t = 1 - x_3,$$

and by eliminating t we derive

$$\boxed{x = \frac{x_1}{1 - x_3}, \qquad y = \frac{x_2}{1 - x_3}.} \tag{4}$$

We have observed that circles centered at the origin and lines through the origin, in the z-plane, become circles of latitude or longitude on the Riemann sphere. With the help of the formulas we can generalize:

Example 2

Show that *all* lines and circles in the z-plane correspond under stereographic projection to circles on the Riemann sphere (see Fig. 1.23).

Solution. The general equation for a circle or line in the $z = x + iy$ plane is given by

$$A(x^2 + y^2) + Cx + Dy + E = 0. \tag{5}$$

When we substitute our formulas (4) into (5) we get

$$A\left[\left(\frac{x_1}{1-x_3}\right)^2 + \left(\frac{x_2}{1-x_3}\right)^2\right] + \frac{Cx_1}{1-x_3} + \frac{Dx_2}{1-x_3} + E = 0,$$

which simplifies to

$$A(x_1^2 + x_2^2) + Cx_1(1 - x_3) + Dx_2(1 - x_3) + E(1 - x_3)^2 = 0.$$

Recalling that $x_1^2 + x_2^2 + x_3^2 = 1$, we rewrite the last equation as

$$A(1 - x_3^2) + Cx_1(1 - x_3) + Dx_2(1 - x_3) + E(1 - x_3)^2 = 0,$$

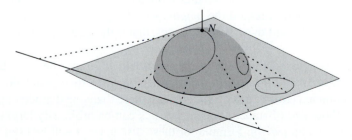

Figure 1.23 Circles on the Riemann sphere.

and dividing by $(1 - x_3)$ we obtain

$$A(1 + x_3) + Cx_1 + Dx_2 + E(1 - x_3) = 0$$

or

$$Cx_1 + Dx_2 + (A - E)x_3 + A + E = 0. \tag{6}$$

But (6) is just the equation of a plane in 3-dimensional space. We have thus shown that the projection of a line or circle in the z-plane must lie in the plane described by (6), as well as on the Riemann sphere. The intersection of a plane and a sphere is, of course, a circle. ■

The argument can be reversed; every circle on the Riemann sphere is the projection of either a line or circle in the z-plane (Prob. 10).

What is the role of the north pole $N = (0, 0, 1)$ in this picture? Clearly N does not arise as the projection of any point in the complex plane; equally clear is the exclusion of $x_3 = 1$ from the formulas (4). Nonetheless, we can give meaning to this exceptional point if we think of points in the complex plane that are very large in modulus (that is, are far from the origin). Such points project onto points near the north pole and, as $|z| \to +\infty$, their projections tend to this pole. In this context we associate with N the *extended* complex number "∞," and call

$$\widehat{\mathbf{C}} = \mathbf{C} \cup \{\infty\}$$

the extended complex plane.[†]

The nature of the "point at ∞" in the extended complex plane is quite different from the customary interpretation for the one-dimensional real line. In the latter case we consider $+\infty$ and $-\infty$ as *distinct* metaphors, and distinguish between situations where points on the line become unbounded by getting large positively or large negatively. In the present context, however, although points in the plane can become unbounded from many different directions (along the real axis, along the imaginary axis, along an expanding spiral, etc.), they all "approach infinity."

Carrying this notion a little further, we observe that the spherical cap formed by the interior of the Arctic Circle constitutes a *neighborhood* of the north pole N (see Fig. 1.22), and the interiors of the smaller circles of latitude form smaller neighborhoods. Thus we are led to regard the *neighborhoods of* ∞ in the extended complex plane $\widehat{\mathbf{C}}$ as the preimages of these spherical caps, that is the *exteriors* of circles centered at the origin.

This highlights the fact that there is a considerable difference between the distance between two points in the plane and the distance between their projections. Indeed, the sphere is bounded (two points on it cannot be more than one diameter apart), while the plane is unbounded (distances between points can be arbitrarily large). The next calculation gives a comparison formula for these distances; it will be used later in the book.

[†]In topological jargon, one refers to $\widehat{\mathbf{C}}$ as a one-point compactification of the plane.

Example 3

Show that the distance (in 3-space) between the projections Z, W of the points z, w in the complex plane is given by

$$\text{dist}(Z, W) = \frac{2|z - w|}{\sqrt{1 + |z|^2}\sqrt{1 + |w|^2}} = \frac{2\,\text{dist}(z, w)}{\sqrt{1 + |z|^2}\sqrt{1 + |w|^2}}. \tag{7}$$

Solution.   This formula can be proved geometrically; Probs. 6 and 7 guide the reader through the derivation. Here we'll simply grind out the algebra. Let (x_1, x_2, x_3) and $(\widehat{x}_1, \widehat{x}_2, \widehat{x}_3)$ denote the coordinates of Z and W, and set $d := \text{dist}(Z, W)$. Then

$$d^2 = (x_1 - \widehat{x}_1)^2 + (x_2 - \widehat{x}_2)^2 + (x_3 - \widehat{x}_3)^2.$$

Since $x_1^2 + x_2^2 + x_3^2 = 1$ and $\widehat{x}_1^2 + \widehat{x}_2^2 + \widehat{x}_3^2 = 1$, after expansion the formula for d^2 becomes

$$d^2 = 2[1 - (x_1\widehat{x}_1 + x_2\widehat{x}_2 + x_3\widehat{x}_3)]. \tag{8}$$

Next we use formulas (1) for $z = x + iy$ and $w = u + iv$ to write

$$1 - (x_1\widehat{x}_1 + x_2\widehat{x}_2 + x_3\widehat{x}_3) =$$

$$1 - \frac{4xu}{(|z|^2 + 1)(|w|^2 + 1)} - \frac{4yv}{(|z|^2 + 1)(|w|^2 + 1)} - \left(\frac{|z|^2 - 1}{|z|^2 + 1}\right)\left(\frac{|w|^2 - 1}{|w|^2 + 1}\right)$$

$$= \frac{2|z|^2 + 2|w|^2 - 4xu - 4yv}{(|z|^2 + 1)(|w|^2 + 1)}$$

$$= \frac{2(x^2 + y^2) + 2(u^2 + v^2) - 4xu - 4yv}{(|z|^2 + 1)(|w|^2 + 1)}$$

$$= 2\frac{(x - u)^2 + (y - v)^2}{(|z|^2 + 1)(|w|^2 + 1)}. \tag{9}$$

Substituting this expression into (8) and taking square roots yields (7).   ∎

The Euclidean distance between the projections Z and W that we have just computed is called the *chordal distance* between the (original) complex numbers z and w and is denoted by the Greek symbol χ ("chi"), that is,

$$\chi[z, w] := \frac{2|z - w|}{\sqrt{1 + |z|^2}\sqrt{1 + |w|^2}}. \tag{10}$$

Like the ordinary Euclidean distance between points in the plane, χ is a *metric* in the sense that it satisfies the triangle inequality

$$\chi[z_1, z_2] \leq \chi[z_1, w] + \chi[w, z_2] \tag{11}$$

and the familiar identities

$$\chi[z, z] = 0, \qquad \chi[z, w] = \chi[w, z],$$

for any points in the plane.

The χ-metric is also meaningful for the extended complex plane $\widehat{\mathbf{C}}$; we calculate the chordal distance from z to ∞ by manipulating Eq. (10),

$$\chi[z, \infty] = \lim_{|w| \to \infty} \frac{2|z - w|}{\sqrt{1 + |z|^2}\sqrt{1 + |w|^2}} = \lim_{|w| \to \infty} \frac{2|z/w - 1|}{\sqrt{1 + |z|^2}\sqrt{1/|w|^2 + 1}},$$

to conclude

$$\chi[z, \infty] = \frac{2}{\sqrt{1 + |z|^2}}. \tag{12}$$

This quantifies our earlier image of neighborhoods of ∞ in $\widehat{\mathbf{C}}$; the spherical cap described by $\mathrm{dist}(Z, N) \equiv \chi[z, \infty] < \rho$ for ρ less than 2 (the diameter of the sphere) is the projection of all points in the plane that lie *outside* the circle $|z| = \sqrt{(4/\rho^2) - 1}$.

EXERCISES 1.7

1. For each of the following points in $\mathbf{C}$, determine its stereographic projection on the Riemann sphere.

   (a) i **(b)** $6 - 8i$ **(c)** $-\frac{3}{10} + \frac{2}{5}i$.

2. **(a)** Show that the stereographic projections of the points z and $1/\bar{z}$ are reflections of each other in the equatorial plane of the Riemann sphere.

   (b) Show that the stereographic projections of the points z and $-1/\bar{z}$ are diametrically opposite points on the Riemann sphere.

3. If Z and W are two distinct points on the Riemann sphere, then the plane through these points and the origin cuts the sphere in a "great circle," that is, a circle with maximum diameter (2, for a unit sphere). Show that this great circle corresponds to the unique circle (or line) in the plane that passes through the points z, w, and $-1/\bar{z}$, where Z, W are the projections of z, w respectively. [HINT: See Prob. 2.]

4. By considering their stereographic projections, show that the (unique) circle (or line) through the three points z, w, and $-1/\bar{z}$ in the plane is the same as the circle (or line) through the three points z, w, and $-1/\bar{w}$, where $z \neq w$. [HINT: See Prob. 3.]

5. Describe the projections on the Riemann sphere of the following sets in the complex plane:

   (a) the right half-plane $\{z : \mathrm{Re}\, z > 0\}$

   (b) the disk $\{z : |z| < 1/2\}$

    (c) the annulus $\{z : 1 < |z| < 2\}$

    (d) the set $\{z : |z| > 3\}$

    (e) the line $y = x$ (including the point at ∞).

6. Give a geometric argument based on similar triangles to show that the chordal distance between a point z and ∞ (that is, the distance between the projection Z and the north pole) is given by

$$\chi[z, \infty] = 2/\sqrt{1 + |z|^2}.$$

[HINT: Draw a cross section of the Riemann sphere and the z-plane containing the north pole, the point z, and its projection Z.]

7. Establish formula (10) for the chordal distance by using a geometric argument. [HINT: Draw a figure displaying the north pole, z, Z, w, and W. Identify all the segments whose lengths are known; see Prob. 6. Use the law of cosines to write formulas for $|z - w|$ and $|Z - W|$, and note that same angle occurs in both formulas.]

8. For the χ-metric, verify from formula (10) that for any two points z, $w \in \widehat{\mathbf{C}}$, there holds

$$\chi[1/z, 1/w] = \chi[z, w].$$

(We adopt the convention that $1/0 = \infty$ and $1/\infty = 0$.) Now give a geometric argument (based on the Riemann sphere) as to why these two chordal distances should be the same.

9. Without performing any computations, explain why the triangle inequality (11) holds.

10. By reversing the steps in Example 2, show that *every* circle on the Riemann sphere is the projection of either a circle or a line in the z-plane, where we regard ∞ as belonging to every line.

SUMMARY

The complex number system is an extension of the real number system and consists of all expressions of the form $a + bi$, where a and b are real and $i^2 = -1$. The operations of addition, subtraction, multiplication, and division with complex numbers are performed in a manner analogous to "computing with radicals." Geometrically, complex numbers can be represented by points or vectors in the plane. Thus certain theorems from geometry, such as the triangle inequality, can be translated into the language of complex numbers. Associated with a complex number $z = a + bi$ are its absolute value, given by $|z| = \sqrt{a^2 + b^2}$, and its complex conjugate, given by $\bar{z} = a - bi$. The former is the distance from the point z to the origin, while the latter is the reflection of the point z in the x-axis. The numbers z, $\bar{z}$, and $|z|$ are related by $z\bar{z} = |z|^2$.

Every nonzero complex number z can be written in the polar form $z = r(\cos\theta + i\sin\theta)$, where $r = |z|$ and θ is the angle of inclination of the vector z. Any of the equivalent angles $\theta + 2k\pi$, $k = 0, \pm 1, \pm 2, \ldots$, is called the argument of z ($\arg z$). The polar form is useful in finding powers and roots of z.

For $z = x + iy$, the complex exponential e^z is defined by $e^z = e^x(\cos y + i\sin y)$. In particular, if θ is real, Euler's equation states that $e^{i\theta} = \cos\theta + i\sin\theta$. Moreover, the polar form of a complex number can be written simply as $z = re^{i\theta}$.

Special terminology is used in describing point sets in the plane. Important is the concept of a domain D. Such a set is characterized by two properties: (i) each point z of D is the center of an open disk completely contained in D; (ii) each pair of points z_1 and z_2 in D can be joined by a polygonal path that lies entirely in D. If some, all, or none of the boundary points are adjoined to a domain, the resulting set is called a region.

Complex numbers can be visualized as points on the unit sphere in 3-space (the Riemann sphere) via stereographic projection, which associates with a point z in the equatorial plane the point at which the line through z and the north pole cuts the sphere. The extended complex number ∞ is identified with the north pole, and $\mathbf{C} \cup \{\infty\}$ is called the extended complex plane.

Suggested Reading

Introductory Level

[1] Boas, R.P. *Invitation to Complex Analysis*. Random House/Birkhäuser Math Series, New York, 1987.

Advanced Level

[2] Ahlfors, L.V. *Complex Analysis*, 3rd ed. McGraw-Hill Book Company, New York, 1979.

[3] Hille, E. *Analytic Function Theory*, Vol. I, 2nd ed., Chelsea, New York, 1973.

[4] Nehari, Z. *Conformal Mapping*. Dover Publishing, New York, 1975.

Extending Number Fields

[5] Fraleigh, John B. *A First Course in Abstract Algebra*, 4th ed. Addison-Wesley Publishing Company, Reading, MA, 1989.

Mechanisms

[6] Martin, George H. *Kinematics and Dynamics of Machines*, 2nd ed. McGraw-Hill Book Company, New York, 1982.

Chapter 2

Analytic Functions

2.1 Functions of a Complex Variable

The concept of a complex number z was introduced in Chapter 1 in order to solve certain algebraic equations. We shall now study functions $f(z)$ defined on these complex variables. Our objective is to mimic the concepts, theorems, and mathematical structure of calculus; we want to differentiate and integrate $f(z)$. The notion of a derivative is far more subtle in the complex case because of the intrinsically two-dimensional nature of the complex variable, and the exposition of this point will consume all of Chapter 2. The payoff is enormous, however, and the remainder of the book will be devoted to developing the mathematical consequences and demonstrating their applications to physical problems.

Let us begin with a careful review of the basics. Recall that a *function* f is a rule that assigns to each element in a set A one and only one element in a set B. If f assigns the value b to the element a in A, we write

$$b = f(a)$$

and call b the *image* of a under f. The set A is the *domain of definition* of f (even if A is not a domain in the sense of Chapter 1), and the set of all images $f(a)$ is the *range* of f. We sometimes refer to f as a *mapping* of A into B.

Here we are concerned with complex-valued functions of a complex variable, so that the domains of definition and the ranges are subsets of the complex numbers. If $f(z)$ is expressed by a formula such as

$$f(z) = \frac{z^2 - 1}{z^2 + 1},$$

then, unless stated otherwise, we take the domain of f to be the set of all z for which the formula is well defined. (Thus the domain for this f comprises all z except for $\pm i$.)

If w denotes the value of the function f at the point z, we then write $w = f(z)$. Just as z decomposes into real and imaginary parts as $z = x + iy$, the real and imaginary parts of w are each (real-valued) functions of z or, equivalently, of x and y, and so we customarily write

$$w = u(x, y) + iv(x, y),$$

with u and v denoting the real and imaginary parts, respectively, of w. Thus a complex-valued function of a complex variable is, in essence, a pair of real functions of two real variables.

Example 1

Write the function $w = f(z) = z^2 + 2z$ in the form $w = u(x, y) + iv(x, y)$.

   Solution.   Setting $z = x + iy$ we obtain

$$w = f(z) = (x + iy)^2 + 2(x + iy) = x^2 - y^2 + i2xy + 2x + i2y.$$

Hence $w = (x^2 - y^2 + 2x) + i(2xy + 2y)$ is the desired form.   ■

Unfortunately, it is generally impossible to draw the graph of a complex function; to display two real functions of two real variables graphically would require four dimensions. We can, however, visualize some of the properties of a complex function $w = f(z)$ by sketching its domain of definition in the z-plane and its range in the w-plane, and depicting the relationship as in Fig. 2.1.

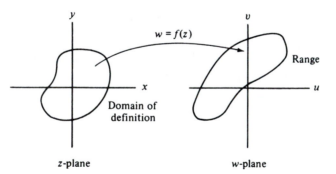

Figure 2.1 Representation of a complex function.

Example 2

Describe the range of the function $f(z) = x^2 + 2i$ defined on the closed unit disk $|z| \le 1$.

   Solution.   We have $u(x, y) = x^2$ and $v(x, y) = 2$. Thus as z varies over the closed unit disk, u varies between 0 and 1, and v is constant. The range is therefore the line segment from $w = 2i$ to $w = 1 + 2i$.   ■

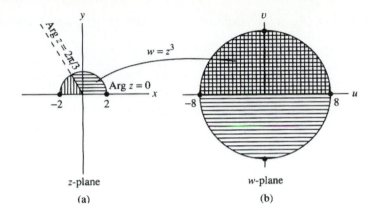

Figure 2.2 Mapping of a semidisk under $f(z) = z^3$.

Example 3

Describe the function $f(z) = z^3$ for z in the semidisk given by $|z| \leq 2$, Im $z \geq 0$ [see Fig. 2.2(a)].

Solution.   From Sec. 1.5 we know that the points z in the sector of the semidisk from Arg $z = 0$ to Arg $z = 2\pi/3$, when cubed, cover the entire disk $|w| \leq 8$. The cubes of the remaining z-points also fall in this disk, overlapping it in the upper half-plane, as depicted in Fig. 2.2(b).   ■

The function $f(z) = 1/z$ is called the *inversion mapping*. It is an example of a one-to-one function because it maps distinct points to distinct points, i.e., if $z_1 \neq z_2$, then $f(z_1) \neq f(z_2)$.

Example 4

Show that the inversion mapping $w = 1/z$ corresponds to a rotation of the Riemann sphere by 180° about the x_1-axis (see Fig. 1.21 on page 44).

Solution.   Let $Z = (x_1, x_2, x_3)$ denote the stereographic projection of the point z and $W = (\hat{x}_1, \hat{x}_2, \hat{x}_3)$ denote the projection of $1/z$. We need to show that W can be obtained by a 180° rotation of Z about the x_1-axis.

Referring to the formulas (1) in Sec. 1.7 we have

$$x_1 = \frac{2\,\mathrm{Re}(z)}{|z|^2 + 1}, \quad x_2 = \frac{2\,\mathrm{Im}(z)}{|z|^2 + 1}, \quad x_3 = \frac{|z|^2 - 1}{|z|^2 + 1};$$

$$\hat{x}_1 = \frac{2\,\mathrm{Re}(1/z)}{|1/z|^2 + 1}, \quad \hat{x}_2 = \frac{2\,\mathrm{Im}(1/z)}{|1/z|^2 + 1}, \quad \hat{x}_3 = \frac{|1/z|^2 - 1}{|1/z|^2 + 1}.$$

Strictly speaking, this function is not defined at $z = 0$, but in light of the discussion of the point at ∞ in Sec. 1.7, it is natural to define $f(0) = \infty$ and, moreover, to regard f as a function in the extended complex plane $\widehat{\mathbf{C}} = \mathbf{C} \cup \{\infty\}$, with $f(\infty) = 0$.

Since $\operatorname{Re}(1/z) = (\operatorname{Re} z)/|z|^2$ and $\operatorname{Im}(1/z) = -(\operatorname{Im} z)/|z|^2$, we get after simplification that

$$\hat{x}_1 = \frac{2\operatorname{Re} z}{1 + |z|^2} = x_1, \quad \hat{x}_2 = \frac{-2\operatorname{Im} z}{1 + |z|^2} = -x_2, \quad \text{and} \quad \hat{x}_3 = \frac{1 - |z|^2}{1 + |z|^2} = -x_3.$$

A rotation about the x_1-axis preserves x_1 while negating x_2 and x_3; so indeed W is the stated rotation of Z. ∎

One nice consequence of this example is the fact that an inversion mapping preserves the class of circles and lines (see Prob. 17)

EXERCISES 2.1

1. Write each of the following functions in the form $w = u(x, y) + iv(x, y)$.

   (a) $f(z) = 3z^2 + 5z + i + 1$ **(b)** $g(z) = 1/z$

   (c) $h(z) = \dfrac{z + i}{z^2 + 1}$ **(d)** $q(z) = \dfrac{2z^2 + 3}{|z - 1|}$

   (e) $F(z) = e^{3z}$ **(f)** $G(z) = e^z + e^{-z}$

2. Find the domain of definition of each of the functions in Prob. 1.

3. Describe the range of each of the following functions.

   (a) $f(z) = z + 5$ for $\operatorname{Re} z > 0$

   (b) $g(z) = z^2$ for z in the first quadrant, $\operatorname{Re} z \geq 0$, $\operatorname{Im} z \geq 0$

   (c) $h(z) = \dfrac{1}{z}$ for $0 < |z| \leq 1$

   (d) $p(z) = -2z^3$ for z in the quarter-disk $|z| < 1$, $0 < \operatorname{Arg} z < \dfrac{\pi}{2}$

4. Show that the inversion mapping $w = f(z) = 1/z$ maps

   (a) the circle $|z| = r$ onto the circle $|w| = 1/r$;

   (b) the ray $\operatorname{Arg} z = \theta_0$, $-\pi < \theta_0 < \pi$, onto the ray $\operatorname{Arg} w = -\theta_0$;

   (c) the circle $|z - 1| = 1$ onto the vertical line $x = 1/2$.

5. For the complex exponential function $f(z) = e^z$ defined in Sec. 1.4:

   (a) Describe the domain of definition and the range.

   (b) Show that $f(-z) = 1/f(z)$.

   (c) Describe the image of the vertical line $\operatorname{Re} z = 1$.

   (d) Describe the image of the horizontal line $\operatorname{Im} z = \pi/4$.

   (e) Describe the image of the infinite strip $0 \leq \operatorname{Im} z \leq \pi/4$.

6. The *Joukowski mapping* is defined by

$$w = J(z) = \frac{1}{2}\left(z + \frac{1}{z}\right).$$

Show that

    (a) $J(z) = J(1/z)$.

    (b) J maps the unit circle $|z| = 1$ onto the real interval $[-1, 1]$.

    (c) J maps the circle $|z| = r$ $(r > 0, r \neq 1)$ onto the ellipse

$$\frac{u^2}{\left[\frac{1}{2}\left(r + \frac{1}{r}\right)\right]^2} + \frac{v^2}{\left[\frac{1}{2}\left(r - \frac{1}{r}\right)\right]^2} = 1,$$

    which has foci at ± 1.

7. A function of the form $F(z) = z + c$, where c is a complex constant, generates a *translation mapping*. Sketch the image of the semidisk $|z| \leq 2$, $\mathrm{Im}\, z \geq 0$, [see Fig. 2.2(a)] under F when **(a)** $c = 3$; **(b)** $c = 2i$; **(c)** $c = -1 - i$.

8. A function of the form $G(z) = e^{i\phi}z$, where ϕ is a real constant, generates a *rotation mapping*. Sketch the image of the semidisk $|z| \leq 2$, $\mathrm{Im}\, z \geq 0$ [see Fig. 2.2(a)] under G when **(a)** $\phi = \pi/4$; **(b)** $\phi = -\pi/4$; **(c)** $\phi = 3\pi/4$.

9. A function of the form $H(z) = \rho z$, where ρ is a positive real constant, generates a *magnification mapping* when $\rho > 1$ and a *reduction mapping* when $\rho < 1$. Sketch the image of the semidisk $|z| \leq 2$, $\mathrm{Im}\, z \geq 0$ [see Fig. 2.2(a)] under H when **(a)** $\rho = 3$; **(b)** $\rho = 1/2$.

10. Let $F(z) = z + i$, $G(z) = e^{i\pi/4}z$, and $H(z) = z/2$. Sketch the image of the semidisk $|z| \leq 2$, $\mathrm{Im}\, z \geq 0$ [see Fig. 2.2(a)] under each of the following composite mappings:

    (a) $G(F(z))$ **(b)** $G(H(z))$

    (c) $H(F(z))$ **(d)** $F(G(H(z)))$

11. Let $F(z) = z - 3$, $G(z) = -iz$, and $H(z) = 2z$. Sketch the image of the circle $|z| = 1$ under each of the following composite mappings:

    (a) $G(F(z))$ **(b)** $G(H(z))$

    (c) $H(F(z))$ **(d)** $F(G(H(z)))$

12. A function of the form $f(z) = az + b$, where a and b are complex constants, is called a *linear transformation*. Show that every linear transformation can be expressed as the composition of a magnification (or reduction; Prob. 9), a rotation (Prob. 8), and a translation (Prob. 7). Deduce from this that a linear transformation maps lines to lines and circles to circles. [HINT: Write a in polar form.]

13. Show that the function $w = z^2$ maps

  (a) the line $x = 1$,

  (b) the hyperbola $xy = 1$, and

  (c) the circle $|z - 1| = 1$,

into a *parabola*, a *straight line*, and the *cardioid* $w = 2(1 + \cos\theta)e^{i\theta}$, respectively, in the w-plane.

14. (*Rotation of the Riemann Sphere*) Referring to Fig. 1.21 and Eqs. (1) in Sec. 1.7 for stereographic projection, show each of the following:

  (a) The mapping $w = e^{i\phi}z$ corresponds to a rotation of the Riemann sphere about the x_3-axis through an angle ϕ.

  (b) The mapping $w = -1/z$ corresponds to a 180° rotation of the Riemann sphere about the x_2-axis (the imaginary axis).

15. Show that the mapping $w = (1 + z)/(1 - z)$ corresponds to a 90° counterclockwise rotation of the Riemann sphere about the x_2-axis.

16. Describe the mapping $w = (1 - iz)/(z - i)$ in terms of a suitable rotation of the Riemann sphere.

17. Use the result of Example 4 and properties of stereographic projection to show that the inversion $w = 1/z$ maps any circle in the z-plane to either a circle or a line in the w-plane, and the same holds for the mapping of any line in the z-plane. (Regard ∞ as a point on every line.)

2.2   Limits and Continuity

As we observed in Chapter 1, the definition of absolute value can be used to designate the distance between two complex numbers. Having a concept of distance, we can proceed to introduce the notions of limit and continuity.

Informally, when we have an infinite sequence $z_1, z_2, z_3, \ldots$ of complex numbers, we say that the number z_0 is the limit of the sequence if the z_n eventually (i.e., for large enough n) stay arbitrarily close to z_0. More precisely, we state

---

Definition 1. A sequence of complex numbers $\{z_n\}_1^\infty$ is said to have the **limit** z_0 or to **converge** to z_0, and we write

$$\lim_{n \to \infty} z_n = z_0$$

or, equivalently,

$$z_n \to z_0 \quad \text{as} \quad n \to \infty,$$

if for any $\varepsilon > 0$ there exists an integer N such that $|z_n - z_0| < \varepsilon$ for all $n > N$.

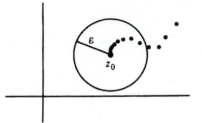

Figure 2.3 A convergent sequence.

Geometrically, this means that each term z_n, for $n > N$, lies in the open disk of radius ε about z_0 (see Fig. 2.3).

Example 1

Find the limit (if it exists) of the sequence

$$\textbf{(a)}\ z_n = \left(\frac{i}{3}\right)^n; \qquad \textbf{(b)}\ z_n = \frac{2+in}{1+3n}; \qquad \textbf{(c)}\ z_n = i^n.$$

Solution.   We use methods familiar from elementary calculus that can be rigorously justified from Definition 1.

(a) Since $|(i/3)^n| = 1/3^n \to 0$, it follows (see Prob. 5) that

$$\lim_{n\to\infty} \left(\frac{i}{3}\right)^n = 0.$$

(b) Dividing numerator and denominator by n we get

$$\frac{2+in}{1+3n} = \frac{(2/n)+i}{(1/n)+3} \to \frac{0+i}{0+3} = \frac{i}{3} \quad \text{as} \quad n \to \infty.$$

(c) The sequence i^n consists of infinitely many repetitions of i, -1, $-i$, and 1. Thus this sequence does not have a limit.   ∎

A related concept is the limit of a complex-valued function $f(z)$. Roughly speaking, we say that the number w_0 is the limit of the function $f(z)$ as z approaches z_0, if $f(z)$ stays arbitrarily close to w_0 whenever z is sufficiently near z_0. In precise terms we give

Definition 2. Let f be a function defined in some neighborhood of z_0, with the possible exception of the point z_0 itself. We say that the **limit of** $f(z)$ **as** z **approaches** z_0 **is the number** w_0 and write

$$\lim_{z\to z_0} f(z) = w_0$$

or, equivalently,

$$f(z) \to w_0 \quad \text{as} \quad z \to z_0,$$

if for any $\varepsilon > 0$ there exists a positive number δ such that

$$|f(z) - w_0| < \varepsilon \quad \text{whenever} \quad 0 < |z - z_0| < \delta.$$

Geometrically, this says that *any* neighborhood of w_0 contains all the values assumed by f in some full neighborhood of z_0, except possibly the value $f(z_0)$; see Fig. 2.4.

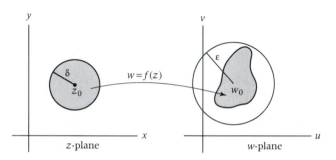

Figure 2.4 Mapping property of a function with limit w_0 as $z \to z_0$.

Example 2

Use Definition 2 to prove that $\lim_{z \to i} z^2 = -1$.

Solution. We must show that for given $\varepsilon > 0$ there is a positive number δ such that

$$|z^2 - (-1)| < \varepsilon \quad \text{whenever} \quad 0 < |z - i| < \delta.$$

So we express $|z^2 - (-1)|$ in terms of $|z - i|$:

$$z^2 - (-1) = z^2 + 1 = (z - i)(z + i) = (z - i)(z - i + 2i).$$

It follows from the properties of absolute value derived in Sec. 1.3 (in particular, the triangle inequality) that

$$\left| z^2 - (-1) \right| = |z - i||z - i + 2i| \le |z - i|(|z - i| + 2). \tag{1}$$

Now if $|z - i| < \delta$ the right-hand member of (1) is less than $\delta(\delta + 2)$; so to ensure that it is less than ε, we choose δ to be smaller than each either of the numbers $\varepsilon/3$ and 1:

$$|z - i|(|z - i| + 2) < \frac{\varepsilon}{3}(1 + 2) = \varepsilon. \quad \blacksquare$$

There is an obvious relation between the limit of a function and the limit of a sequence; namely, if $\lim_{z \to z_0} f(z) = w_0$, then for every sequence $\{z_n\}_1^\infty$ converging to z_0 ($z_n \neq z_0$) the sequence $\{f(z_n)\}_1^\infty$ converges to w_0. The converse of this statement is also valid and is left as an exercise.

The condition of continuity is expressed in

Definition 3. Let f be a function defined in a neighborhood of z_0. Then f is **continuous at z_0** if

$$\lim_{z \to z_0} f(z) = f(z_0).$$

In other words, for f to be continuous at z_0, it must have a limiting value at z_0, and this limiting value must be $f(z_0)$.

A function f is said to be *continuous on a set S* if it is continuous at each point of S.

Clearly the definitions of this section are direct analogues of concepts introduced in elementary calculus. In fact, one can show that $f(z)$ approaches a limit precisely when its real and imaginary parts approach limits (see Prob. 18); similarly, the continuity of the latter functions is equivalent to the continuity of f. Because of the analogy, many of the familiar theorems on real sequences, limits, and continuity remain valid in the complex case. Two such theorems are stated here.

Theorem 1. If $\lim_{z \to z_0} f(z) = A$ and $\lim_{z \to z_0} g(z) = B$, then

   (i) $\lim_{z \to z_0} (f(z) \pm g(z)) = A \pm B,$

   (ii) $\lim_{z \to z_0} f(z)g(z) = AB,$

   (iii) $\lim_{z \to z_0} \dfrac{f(z)}{g(z)} = \dfrac{A}{B}$ if $B \neq 0.$

Theorem 2. If $f(z)$ and $g(z)$ are continuous at z_0, then so are $f(z) \pm g(z)$ and $f(z)g(z)$. The quotient $f(z)/g(z)$ is also continuous at z_0 provided $g(z_0) \neq 0$.

(Theorem 2 is an immediate consequence of Theorem 1.)

One can easily verify that the constant functions as well as the function $f(z) = z$ are continuous on the whole plane **C**. Thus from Theorem 2 we deduce that the *polynomial functions* in z, i.e., functions of the form

$$a_0 + a_1 z + a_2 z^2 + \cdots + a_n z^n,$$

where the a_i are constants, are also continuous on the whole plane. *Rational functions in z*, which are defined as quotients of polynomials, i.e.,

$$\frac{a_0 + a_1 z + \cdots + a_n z^n}{b_0 + b_1 z + \cdots + b_m z^m},$$

are therefore continuous at each point where the denominator does not vanish. These considerations provide a much simpler solution for problems such as Example 2, as we illustrate next.

Example 3

Find the limits, as $z \to 2i$, of the functions $f_1(z) = z^2 - 2z + 1$, $f_2(z) = (z + 2i)/z$, and $f_3(z) = (z^2 + 4)/z(z - 2i)$.

Solution. Since $f_1(z)$ and $f_2(z)$ are continuous at $z = 2i$, we simply evaluate them there, i.e.,

$$\lim_{z \to 2i} f_1(z) = f_1(2i) = (2i)^2 - 2(2i) + 1 = -3 - 4i,$$

$$\lim_{z \to 2i} f_2(z) = f_2(2i) = \frac{2i + 2i}{2i} = 2.$$

The function $f_3(z)$ is not continuous at $z = 2i$ because it is not defined there (the denominator vanishes). However, for $z \neq 2i$ and $z \neq 0$ we have

$$f_3(z) = \frac{(z + 2i)(z - 2i)}{z(z - 2i)} = \frac{z + 2i}{z} = f_2(z),$$

and so

$$\lim_{z \to 2i} f_3(z) = \lim_{z \to 2i} f_2(z) = 2. \quad \blacksquare$$

Note that in the preceding example the discontinuity of $f_3(z)$ at $z = 2i$ can be removed by suitably defining the function at this point [set $f_3(2i) = 2$]. In general, if a function can be defined or redefined at a single point z_0 so as to be continuous there, we say that this function has a *removable discontinuity* at z_0.

We shall see that limits involving infinity are very useful in describing the behavior of certain sequences and functions. We say "$z_n \to \infty$" if, for each positive number M (no matter how large), there is an integer N such that $|z_n| > M$ whenever $n > N$; similarly "$\lim_{z \to z_0} f(z) = \infty$" means that for each positive number M (no matter how large), there is a $\delta > 0$ such that $|f(z)| > M$ whenever $0 < |z - z_0| < \delta$. Essentially we are saying that complex numbers approach infinity when their *magnitudes* approach infinity. Therefore

$$\lim_{z \to 3i} \frac{z}{z^2 + 9} = \infty, \qquad \lim_{z \to \infty} \frac{iz - 2}{4z + i} = \frac{i}{4}, \qquad \lim_{z \to \infty} \frac{z^3 + 3i}{z^2 + 5z} = \infty.$$

In fact, the concept of a *point at infinity*, as introduced in Sec. 1.7, quantifies this notion nicely: see Probs. 23–25.

In closing this section, we wish to emphasize an important distinction between the concepts of limit in the (one-dimensional) real and complex cases. For the latter situation, observe that a sequence $\{z_n\}_1^\infty$ may approach a limit z_0 from *any* direction in the plane, or even along a spiral, etc. Thus the manner in which a sequence of numbers approaches its limit can be much more complicated in the complex case.

EXERCISES 2.2

1. Sketch the first five terms of the sequence $(i/2)^n$, $n = 1, 2, 3, \ldots$, and then describe the convergence of this sequence.

2. Sketch the first five terms of the sequence $(2i)^n$, $n = 1, 2, 3, \ldots$, and then describe the divergence of this sequence.

3. Using Definition 1, prove that the sequence of complex numbers $z_n = x_n + iy_n$ converges to $z_0 = x_0 + iy_0$ if and only if x_n converges to x_0 and y_n converges to y_0. [HINT: $|x_n-x_0| \le |z_n-z_0|$, $|y_n-y_0| \le |z_n-z_0|$, and $|z_n-z_0| \le |x_n-x_0|+|y_n-y_0|$.]

4. Prove that $z_n \to z_0$ if and only if $\overline{z_n} \to \overline{z_0}$ as $n \to \infty$.

5. Prove that $\lim_{n\to\infty} z_n = 0$ if and only if $\lim_{n\to\infty} |z_n| = 0$ using Definition 1.

6. Prove that if $|z_0| < 1$, then $z_0^n \to 0$ as $n \to \infty$. Also prove that if $|z_0| > 1$, then the sequence z_0^n diverges.

7. Decide whether each of the following sequences converges, and if so, find its limit.

   (a) $z_n = \dfrac{i}{n}$ **(b)** $z_n = i(-1)^n$ **(c)** $z_n = \text{Arg}\left(-1 + \dfrac{i}{n}\right)$

   (d) $z_n = \dfrac{n(2+i)}{n+1}$ **(e)** $z_n = \left(\dfrac{1-i}{4}\right)^n$ **(f)** $z_n = \exp\left(\dfrac{2n\pi i}{5}\right)$

8. Use Definition 2 to prove that $\lim_{z\to 1+i}(6z - 4) = 2 + 6i$.

9. Use Definition 2 to prove that $\lim_{z\to -i} 1/z = i$.

10. Use Theorem 1 to prove Theorem 2.

11. Find each of the following limits.

   (a) $\lim_{z\to 2+3i} (z - 5i)^2$ **(b)** $\lim_{z\to 2} \dfrac{z^2 + 3}{iz}$

   (c) $\lim_{z\to 3i} \dfrac{z^2 + 9}{z - 3i}$ **(d)** $\lim_{z\to i} \dfrac{z^2 + i}{z^4 - 1}$

   (e) $\lim_{\Delta z\to 0} \dfrac{(z_0 + \Delta z)^2 - z_0^2}{\Delta z}$ **(f)** $\lim_{z\to 1+2i} |z^2 - 1|$

12. Show that the function Arg z is discontinuous at each point on the nonpositive real axis.

13. Let $f(z)$ be defined by

$$f(z) = \begin{cases} 2z/(z+1) & \text{if } z \neq 0, \\ 1 & \text{if } z = 0. \end{cases}$$

At which points does $f(z)$ have a finite limit, and at which points is it continuous? Which of the discontinuities of $f(z)$ are removable?

14. Prove that the function $g(z) = \bar{z}$ is continuous on the whole plane.

15. Prove that if $f(z)$ is continuous at z_0, then so are the functions $\overline{f(z)}$, Re $f(z)$, Im $f(z)$, and $|f(z)|$. [HINT: To show that $|f(z)|$ is continuous at z_0 use inequality (2) in Sec. 1.3.]

16. Let g be a function defined in a neighborhood of z_0 and let f be a function defined in a neighborhood of the point $g(z_0)$. Show that if g is continuous at z_0 and f is continuous at $g(z_0)$, then the composite function $f(g(z))$ is continuous at z_0.

17. Let $f(z) = \left[x^2/(x^2 + y^2)\right] + 2i$. Does f have a limit at $z = 0$? [HINT: Investigate $\{f(z_n)\}$ for sequences $\{z_n\}$ approaching 0 along the real and imaginary axes separately.]

18. Let $f(z) = u(x, y) + iv(x, y)$, $z_0 = x_0 + iy_0$, and $w_0 = u_0 + iv_0$. Prove that

$$\lim_{z \to z_0} f(z) = w_0$$

if, and only if,

$$\lim_{\substack{x \to x_0 \\ y \to y_0}} u(x, y) = u_0 \quad \text{and} \quad \lim_{\substack{x \to x_0 \\ y \to y_0}} v(x, y) = v_0.$$

[HINT: First show that $f(z) \to w_0$ if and only if $\overline{f(z)} \to \overline{w_0}$ as $z \to z_0$, and then use Theorem 1.]

19. Use Prob. 18 to find $\lim_{z \to 1-i} \left[x / (x^2 + 3y)\right] + ixy$.

20. Use Prob. 18 to prove that $f(z) = e^z$ is continuous everywhere.

21. Find each of the following limits:

(a) $\lim_{z \to 0} e^z$

(b) $\lim_{z \to 2\pi i} \left(e^z - e^{-z}\right)$

(c) $\lim_{z \to \pi i/2} (z + 1)e^z$

(d) $\lim_{z \to -\pi i} \exp\left(\dfrac{z^2 + \pi^2}{z + \pi i}\right)$

22. Show that if $\lim_{n \to \infty} f(z_n) = w_0$ for every sequence $\{z_n\}_1^\infty$ converging to z_0 ($z_n \neq z_0$), then $\lim_{z \to z_0} f(z) = w_0$. [HINT: Show that if the latter were not true, then one could construct a sequence $\{z_n\}_1^\infty$ violating the hypothesis.]

23. Show that the definition of "$z_n \to \infty$" in the text is equivalent to the following: A sequence of complex numbers $\{z_n\}_1^\infty$ is said to have the limit ∞ if $\lim_{n \to \infty} \chi[z_n, \infty] = 0$, where χ denotes the chordal distance (chi metric) (see page 49).

24. Show that the definition of "$\lim_{z\to z_0} f(z) = \infty$" in the text is equivalent to: The limit of $f(z)$ as $z \to z_0$ is ∞ if $\lim_{z\to z_0} \chi[f(z), \infty] = 0$, where χ denotes the chordal distance (as in the previous problem).

25. Find each of the following limits involving infinity.

(a) $\displaystyle \lim_{z\to 2i} \frac{z^2 + 9}{2z^2 + 8}$ **(b)** $\displaystyle \lim_{z\to\infty} \frac{3z^2 - 2z}{z^2 - iz + 8}$

(c) $\displaystyle \lim_{z\to 5} \frac{3z}{z^2 - (5 - i)z - 5i}$ **(d)** $\displaystyle \lim_{z\to\infty} (8z^3 + 5z + 2)$ **(e)** $\displaystyle \lim_{z\to\infty} e^z$

2.3 Analyticity

Now that we have a secure notion of functions of a complex variable, we are ready to turn to the main topic of this book—the theory of *analytic* functions. Before we proceed with the rigorous exposition, however, it will prove useful for the reader's perspective if we give an informal preview of what it is we want to achieve.

So far we have viewed a complex function of a complex variable, $f(z)$, as nothing more than an arbitrary mapping from the xy-plane to the uv-plane. We have individual names for the real and imaginary parts of z (x and y, respectively) and for the real and imaginary parts of f (u and v); and *any* pair $u(x, y)$ and $v(x, y)$ of two-variable functions gives us a complex function ($u + iv$) in this sense. But notice that there is something special about the pair

$$u_1(x, y) = x^2 - y^2, \qquad v_1(x, y) = 2xy,$$

as opposed to (say)

$$u_2(x, y) = x^2 - y^2, \qquad v_2(x, y) = 3xy;$$

namely, the complex function $u_1 + iv_1$ treats $z = x + iy$ as a single "unit," because it equals $x^2 - y^2 + i2xy = (x + iy)^2$ and thus it respects the complex structure of $z = x + iy$. However (apparently, at least), the formulation of $u_2 + iv_2$ requires us to break apart the real and imaginary parts of z.

In (real) calculus we don't deal with functions that look at a number like $3 + 4\sqrt{2}$ and square the 3 but cube the 4. The interesting calculus functions treat the number as an indivisible module. We seek to classify the complex functions that behave this same way with regard to their complex argument. Thus we want to admit functions such as

$$
\begin{array}{ll}
z = x + iy & \text{(admissible)}, \\
z^2 = x^2 - y^2 + i2xy & \text{(admissible)}, \\
z^3 = x^3 - 3xy^2 + i(3x^2y - y^3) & \text{(admissible)}, \\
1/z = \frac{x}{x^2+y^2} - i\frac{y}{x^2+y^2} & \text{(admissible)},
\end{array}
$$

and their basic arithmetic combinations (sums, products, quotients, powers, and roots) but ban such functions as

$$\text{Re } z = x \qquad \text{(inadmissible)},$$
$$\text{Im } z = y \qquad \text{(inadmissible)},$$
$$x^2 - y^2 + i3xy \quad \text{(inadmissible)}.$$

Notice that we will have to ban the conjugate function $\bar{z}$, because if we admit it we will open the gate to x [$= (z + \bar{z})/2$] and y [$= (z - \bar{z})/2i$]:

$$\bar{z} = x - iy \qquad \text{(inadmissible)}.$$

Similarly, admitting the modulus $|z|$ would be a mistake as well, since $\bar{z} = |z|^2/z$:

$$|z| \qquad \text{(inadmissible)}.$$

One could criticize our "inadmissible" classification of $u_2 + iv_2 = x^2 - y^2 + i3xy$, because we have not yet *proved* that it cannot be written in terms of z alone. The following computation is instructive: we set

$$x = (z + \bar{z})/2, \qquad y = (z - \bar{z})/2i \tag{1}$$

in $u_2 + iv_2$ and obtain, after some algebra,

$$u_2 + iv_2 = x^2 - y^2 + i3xy = \frac{(z + \bar{z})^2}{4} - \frac{(z - \bar{z})^2}{(-4)} + i3 \frac{(z + \bar{z})}{2} \frac{(z - \bar{z})}{2i}$$

$$= \frac{5}{4}z^2 - \frac{1}{4}\bar{z}^2.$$

Now we see that if we admit $u_2 + iv_2$, we would have to admit $\bar{z}^2$ [since it equals $5z^2 - 4(u_2 + iv_2)$] and its undesirable square root $\bar{z}$.

Example 1

Express the following functions in terms of z and $\bar{z}$:

$$f_1(z) = \frac{x - 1 - iy}{(x - 1)^2 + y^2}, \qquad f_2(z) = x^2 + y^2 + 3x + 1 + i3y.$$

Solution.   Using relations (1) we obtain

$$f_1(z) = \frac{\dfrac{z + \bar{z}}{2} - 1 - i\dfrac{z - \bar{z}}{2i}}{\left(\dfrac{z + \bar{z}}{2} - 1\right)^2 + \left(\dfrac{z - \bar{z}}{2i}\right)^2}$$

$$= \frac{\bar{z} - 1}{z\bar{z} - z - \bar{z} + 1} = \frac{1}{z - 1},$$

$$f_2(z) = \frac{(z + \bar{z})^2}{4} + \frac{(z - \bar{z})^2}{4i^2} + 3\left(\frac{z + \bar{z}}{2}\right) + 1 + i3\left(\frac{z - \bar{z}}{2i}\right)$$

$$= z\bar{z} + 3z + 1. \quad \blacksquare$$

Clearly we will want to accept f_1 as admissible, but the presence of $\bar{z}$ in f_2 disqualifies it. However, this procedure—the disqualification of functions with $\bar{z}$ in their formulas—does not lead to a workable criterion. For instance, who among us would recognize that the function

$$\frac{z^2 \bar{z}^2 + z^2 + \bar{z}^2 - 2\bar{z}z^2 - 2\bar{z} + 1}{10\bar{z} + z\bar{z}^2 - 2z\bar{z} - 5\bar{z}^2 + z - 5}$$

has a canceling common factor of $(\bar{z} - 1)^2$ in its numerator and denominator and thus is admissible?

The function e^z is even more vexing. The definition we have adopted separates the real and imaginary parts of z:

$$e^z = e^x (\cos y + i \sin y). \tag{2}$$

But recall (Example 1, Sec. 1.4) that this definition was shown to be consistent with the Taylor expansion for e^x,

$$e^z = 1 + z + \frac{z^2}{2!} + \frac{z^3}{3!} + \cdots . \tag{3}$$

As the right-hand side of (3) appears to respect the complex structure of z, we suspect e^z to be admissible. This is indeed the case, but we postpone the official verification until the next section.

Over the next four chapters we will see that the criterion we are seeking—the test that will distinguish the admissible functions from the others—can be expressed simply in terms of *differentiability*. The following definition is a straightforward extension of the definition in the real case and appears innocuous enough.

Definition 4. Let f be a complex-valued function defined in a neighborhood of z_0. Then the **derivative** of f at z_0 is given by

$$\frac{df}{dz}(z_0) \equiv f'(z_0) := \lim_{\Delta z \to 0} \frac{f(z_0 + \Delta z) - f(z_0)}{\Delta z},$$

provided this limit exists. (Such an f is said to be *differentiable at z_0*.)

The catch here is that Δz is a complex number, so it can approach zero in many different ways (from the right, from below, along a spiral, etc.); but the difference quotient must tend to a *unique* limit $f'(z_0)$ independent of the manner in which $\Delta z \to 0$. Let us see why this notion disqualifies $\bar{z}$.

Example 2
Show that $f(z) = \bar{z}$ is nowhere differentiable.

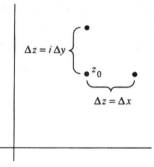

Figure 2.5 Horizontal and vertical approach to zero of Δz.

Solution.   The difference quotient for this function takes the form

$$\frac{f(z_0 + \Delta z) - f(z_0)}{\Delta z} = \frac{\overline{(z_0 + \Delta z)} - \overline{z_0}}{\Delta z} = \frac{\overline{\Delta z}}{\Delta z}.$$

Now if $\Delta z \to 0$ through real values, then $\Delta z = \Delta x$ (see Fig. 2.5) and $\overline{\Delta z} = \Delta z$, so the difference quotient is 1. On the other hand, if $\Delta z \to 0$ from above, then $\Delta z = i\Delta y$ and $\overline{\Delta z} = -\Delta z$, so the quotient is -1. Consequently there is no way of assigning a unique value to the derivative of $\bar{z}$ at any point. Hence $\bar{z}$ is not differentiable.   ■

A similar analysis demonstrates that neither x, y, nor $|z|$ is differentiable (see Prob. 4).

Let us reassure ourselves that the elementary functions such as sums, products, and quotients of powers of z are differentiable.

Example 3

Show that, for any positive integer n,

$$\frac{d}{dz} z^n = nz^{n-1}. \tag{4}$$

Solution.   Using the binomial formula (Prob. 27 in Exercises 1.1) we find

$$\frac{(z + \Delta z)^n - z^n}{\Delta z} = \frac{nz^{n-1}\Delta z + \frac{n(n-1)}{2} z^{n-2}(\Delta z)^2 + \cdots + (\Delta z)^n}{\Delta z}.$$

Thus

$$\frac{d}{dz} z^n = \lim_{\Delta z \to 0} [nz^{n-1} + \frac{n(n-1)}{2} z^{n-2} \Delta z + \cdots + (\Delta z)^{n-1}] = nz^{n-1}. \quad ■$$

Notice that the proof was just the same as for the real-variable case. In fact the validity of any of the following rules can be proven from Definition 4 by mimicking the corresponding proof from elementary calculus.

> **Theorem 3.** If f and g are differentiable at z, then
>
> $$(f \pm g)'(z) = f'(z) \pm g'(z), \tag{5}$$
>
> $$(cf)'(z) = cf'(z) \qquad \text{(for any constant } c\text{)}, \tag{6}$$
>
> $$(fg)'(z) = f(z)g'(z) + f'(z)g(z), \tag{7}$$
>
> $$\left(\frac{f}{g}\right)'(z) = \frac{g(z)f'(z) - f(z)g'(z)}{g(z)^2} \qquad \text{if } g(z) \neq 0. \tag{8}$$
>
> If g is differentiable at z and f is differentiable at $g(z)$, then the *chain rule* holds:
>
> $$\frac{d}{dz} f(g(z)) = f'(g(z))g'(z). \tag{9}$$

The reader will also not be surprised to learn that differentiability implies continuity, as in the real case (see Prob. 3).

It follows from Example 3 and rules (5) and (6) that any polynomial in z,

$$P(z) = a_n z^n + a_{n-1} z^{n-1} + \cdots + a_1 z + a_0,$$

is differentiable in the whole plane and that its derivative is given by

$$P'(z) = n a_n z^{n-1} + (n-1)a_{n-1} z^{n-2} + \cdots + a_1.$$

Consequently, from rule (8), any rational function of z is differentiable at every point in its domain of definition. We see then that for purposes of differentiation, polynomial and rational functions in z can be treated as if z were a real variable.

Example 4

Compute the derivative of

$$f(z) = \left(\frac{z^2 - 1}{z^2 + 1}\right)^{100}.$$

Solution. Unless $z = \pm i$ (where the denominator is zero), the usual calculus rules apply. Thus

$$f'(z) = 100 \left(\frac{z^2 - 1}{z^2 + 1}\right)^{99} \frac{(z^2 + 1)2z - (z^2 - 1)2z}{(z^2 + 1)^2} = 400z \frac{(z^2 - 1)^{99}}{(z^2 + 1)^{101}}. \qquad \blacksquare$$

As we demonstrate in Prob. 10, it is possible for a complex function to be differentiable solely at isolated points. Of course, this also occurs in real analysis. Such functions are treated there as exceptional cases, while the general theorems usually apply only to functions differentiable over open intervals of the real line. By analogy, then, we distinguish a special class of complex functions in

---

Definition 5. A complex-valued function $f(z)$ is said to be **analytic** on an open set G if it has a derivative at every point of G.

We emphasize that analyticity is a property defined over open sets, while differentiability could conceivably hold at one point only. Occasionally, however, we shall use the abbreviated phrase "$f(z)$ *is analytic at the point* z_0" to mean that $f(z)$ is analytic in some neighborhood of z_0. A point where f is not analytic but which is the limit of points where f is analytic is known as a *singular point* or *singularity*. Thus we can say that a rational function of z is analytic at every point for which its denominator is nonzero, and the zeros of the denominator are singularities. If $f(z)$ is analytic on the whole complex plane, then it is said to be *entire*. For example, all polynomial functions of z are entire.

As we shall see in the next few chapters, analyticity is the criterion that we have been seeking, for functions to respect the complex structure of the variable z. In fact, Sec. 5.2 will demonstrate that all analytic functions can be written in terms of z alone (no x, y, or $\bar{z}$).

When a function is given in terms of real and imaginary parts as $u(x, y) + iv(x, y)$, it may be very tedious to apply the definition to determine if f is analytic. In the next section we will establish a test that is easier to use. Also, we will verify the analyticity of e^z. We will then have no further occasion for using the substitution method based on Eqs. (1).

EXERCISES 2.3

1. Let $f(z)$ be defined in a neighborhood of z_0. Show that finding

$$\lim_{\Delta z \to 0} [f(z_0 + \Delta z) - f(z_0)] / \Delta z$$

   is equivalent to finding

$$\lim_{z \to z_0} [f(z) - f(z_0)]/(z - z_0).$$

2. Prove that if $f(z)$ is differentiable at z_0, then

$$f(z) = f(z_0) + f'(z_0)(z - z_0) + \lambda(z)(z - z_0),$$

   where $\lambda(z) \to 0$ as $z \to z_0$.

3. Prove that if $f(z)$ is differentiable at z_0, then it is continuous at z_0. [HINT: Use the result of Prob. 2.]

4. Using Definition 4, show that each of the following functions is nowhere differentiable.

Some authors use the words *holomorphic* or *regular* instead of analytic. The terminology "analytic function" was first used by Marquis de Condorcet (1743–1794.)

(a) Re z (b) Im z (c) $|z|$

5. Prove rules (5) and (7).

6. Prove that formula (4) is also valid for negative integers n.

7. Use rules (5)–(9) to find the derivatives of the following functions.

(a) $f(z) = 6z^3 + 8z^2 + iz + 10$

(b) $f(z) = \left(z^2 - 3i\right)^{-6}$

(c) $f(z) = \dfrac{z^2 - 9}{iz^3 + 2z + \pi}$

(d) $f(z) = \dfrac{(z + 2)^3}{\left(z^2 + iz + 1\right)^4}$

(e) $f(z) = 6i \left(z^3 - 1\right)^4 \left(z^2 + iz\right)^{100}$

8. (*Geometric Interpretation of f'*) Suppose that f is analytic at z_0 and $f'(z_0) \neq 0$. Show that

$$\lim_{z \to z_0} \frac{|f(z) - f(z_0)|}{|z - z_0|} = |f'(z_0)|$$

and

$$\lim_{z \to z_0} \{\arg [f(z) - f(z_0)] - \arg(z - z_0)\} = \arg f'(z_0).$$

Thus, on setting $w = f(z)$ and $w_0 = f(z_0)$ we see that for z near z_0, the mapping f dilates distances by the factor $|f'(z_0)|$:

$$|w - w_0| \approx |f'(z_0)| \times |z - z_0|.$$

Also, f rotates vectors emanating from z_0 by an angle of $\arg f'(z_0)$:

$$\arg(w - w_0) \approx \arg(z - z_0) + \arg f'(z_0).$$

In other words, for z near z_0 the mapping $w = f(z)$ behaves like the linear transformation

$$w = f(z_0) + f'(z_0)(z - z_0)$$
$$= c + e^{i\phi}\rho(z - z_0).$$

(See Prob. 12, Exercises 2.1.)

9. For each of the following determine the points at which the function is not analytic.

(a) $\dfrac{1}{z - 2 + 3i}$

(b) $\dfrac{iz^3 + 2z}{z^2 + 1}$

(c) $\dfrac{3z - 1}{z^2 + z + 4}$

(d) $z^2 \left(2z^2 - 3z + 1\right)^{-2}$

10. Let $f(z) = |z|^2$. Use Definition 4 to show that f is differentiable at $z = 0$ but is not differentiable at any other point. [HINT: Write

$$\frac{|z_0 + \Delta z|^2 - |z_0|^2}{\Delta z} = \frac{(z_0 + \Delta z)(\overline{z_0} + \overline{\Delta z}) - z_0 \overline{z_0}}{\Delta z}$$
$$= \overline{z_0} + \overline{\Delta z} + z_0 \frac{\overline{\Delta z}}{\Delta z}.]$$

11. Discuss the analyticity of each of the following functions.

(a) $8\bar{z} + i$ (b) $\dfrac{z}{\bar{z} + 2}$ (c) $\dfrac{z^3 + 2z + i}{z - 5}$ (d) $x^2 - y^2 + 2xyi$

(e) $x^2 + y^2 + y - 2 + ix$ (f) $\left(x + \dfrac{x}{x^2 + y^2}\right) + i\left(y - \dfrac{y}{x^2 + y^2}\right)$

(g) $|z|^2 + 2z$ (h) $\dfrac{|z| + z}{2}$

12. Let $P(z) = (z - z_1)(z - z_2) \cdots (z - z_n)$. Show by induction on n that

$$\frac{P'(z)}{P(z)} = \frac{1}{z - z_1} + \frac{1}{z - z_2} + \cdots + \frac{1}{z - z_n}.$$

[NOTE: $P'(z)/P(z)$ is called the *logarithmic derivative* of $P(z)$.]

13. Let $f(z)$ and $g(z)$ be entire functions. Decide which of the following statements are always true.

(a) $f(z)^3$ is entire.   (b) $f(z)g(z)$ is entire.
(c) $f(z)/g(z)$ is entire.   (d) $5f(z) + ig(z)$ is entire.
(e) $f(1/z)$ is entire.   (f) $g(z^2 + 2)$ is entire.
(g) $f(g(z))$ is entire.

14. Prove *L'Hôpital's rule*: If $f(z)$ and $g(z)$ are analytic at z_0 and $f(z_0) = g(z_0) = 0$, but $g'(z_0) \neq 0$, then

$$\lim_{z \to z_0} \frac{f(z)}{g(z)} = \frac{f'(z_0)}{g'(z_0)}.$$

[HINT: Write

$$\frac{f(z)}{g(z)} = \frac{f(z) - f(z_0)}{z - z_0} \left/ \frac{g(z) - g(z_0)}{z - z_0} \right. .]$$

15. Use L'Hôpital's rule to find $\lim_{z \to i}(1 + z^6)/(1 + z^{10})$.

16. Let $f(z) = z^3 + 1$, and let $z_1 = (-1 + \sqrt{3}i)/2$, $z_2 = (-1 - \sqrt{3}i)/2$. Show that there is no point w on the line segment from z_1 to z_2 such that

$$f(z_2) - f(z_1) = f'(w)(z_2 - z_1).$$

This shows that the mean-value theorem of calculus does not extend to complex functions.

17. Let $F(z) = f(z)g(z)h(z)$, where f, g, and h are each differentiable at z_0. Prove that

$$F'(z_0) = f'(z_0)g(z_0)h(z_0) + f(z_0)g'(z_0)h(z_0) + f(z_0)g(z_0)h'(z_0).$$

Guillaume De L'Hôpital (1661–1704) wrote the first textbook on differential calculus.

2.4 The Cauchy-Riemann Equations

The property of analyticity for a function indicates some type of connection between its real and imaginary parts. The precise expression of this kinship is easily derived, as we shall see shortly, by letting Δz approach zero from the right and from above in Definition 4. In this section we shall explore the nature of this relationship.

If the function $f(z) = u(x, y) + iv(x, y)$ is differentiable at $z_0 = x_0 + iy_0$, then the limit

$$f'(z_0) = \lim_{\Delta z \to 0} \frac{f(z_0 + \Delta z) - f(z_0)}{\Delta z}$$

can be computed by allowing $\Delta z \ (= \Delta x + i\Delta y)$ to approach zero from any convenient direction in the complex plane. If it approaches horizontally, then $\Delta z = \Delta x$, and we obtain

$$f'(z_0) = \lim_{\Delta x \to 0} \frac{u(x_0 + \Delta x, y_0) + iv(x_0 + \Delta x, y_0) - u(x_0, y_0) - iv(x_0, y_0)}{\Delta x}$$

$$= \lim_{\Delta x \to 0} \left[\frac{u(x_0 + \Delta x, y_0) - u(x_0, y_0)}{\Delta x} \right] + i \lim_{\Delta x \to 0} \left[\frac{v(x_0 + \Delta x, y_0) - v(x_0, y_0)}{\Delta x} \right].$$

(It may be helpful to consult Fig. 2.5 again.) Since the limits of the bracketed expressions are just the first partial derivatives of u and v with respect to x, we deduce that

$$f'(z_0) = \frac{\partial u}{\partial x}(x_0, y_0) + i \frac{\partial v}{\partial x}(x_0, y_0). \tag{1}$$

On the other hand, if Δz approaches zero vertically, then $\Delta z = i\Delta y$ and we have

$$f'(z_0) = \lim_{\Delta y \to 0} \left[\frac{u(x_0, y_0 + \Delta y) - u(x_0, y_0)}{i\Delta y} \right] + i \lim_{\Delta y \to 0} \left[\frac{v(x_0, y_0 + \Delta y) - v(x_0, y_0)}{i\Delta y} \right].$$

Hence

$$f'(z_0) = -i\frac{\partial u}{\partial y}(x_0, y_0) + \frac{\partial v}{\partial y}(x_0, y_0). \tag{2}$$

But the right-hand members of Eqs. (1) and (2) are equal to the same complex number $f'(z_0)$, so by equating real and imaginary parts we see that the equations

$$\boxed{\frac{\partial u}{\partial x} = \frac{\partial v}{\partial y}, \quad \frac{\partial u}{\partial y} = -\frac{\partial v}{\partial x}} \tag{3}$$

must hold at $z_0 = x_0 + iy_0$. Equations (3) are called the *Cauchy-Riemann* equations. We have thus established

---

Theorem 4. A necessary condition for a function $f(z) = u(x, y) + iv(x, y)$ to be differentiable at a point z_0 is that the Cauchy-Riemann equations hold at z_0.

Consequently, if f is analytic in an open set G, then the Cauchy-Riemann equations must hold at every point of G.

Augustin-Louis Cauchy (1789–1867), Bernhard Riemann (1826–1866).

There's an easy way to recall the Cauchy-Riemann equations. Simply remember that the horizontal derivative must equal the vertical derivative and write

$$\frac{\partial f}{\partial x} = \frac{\partial f}{\partial (iy)} \quad \text{or}$$

$$\frac{\partial (u + iv)}{\partial x} = \frac{1}{i} \frac{\partial (u + iv)}{\partial y},$$

and equate the real and imaginary parts:

$$\frac{\partial u}{\partial x} = \frac{\partial v}{\partial y}, \quad \frac{\partial v}{\partial x} = -\frac{\partial u}{\partial y}.$$

Example 1

Show that the function $f(z) = (x^2 + y) + i(y^2 - x)$ is not analytic at any point.

Solution. Since $u(x, y) = x^2 + y$ and $v(x, y) = y^2 - x$, we have

$$\frac{\partial u}{\partial x} = 2x, \quad \frac{\partial v}{\partial y} = 2y,$$

$$\frac{\partial u}{\partial y} = 1, \quad \frac{\partial v}{\partial x} = -1.$$

Hence the Cauchy-Riemann equations are simultaneously satisfied only on the line $x = y$ and therefore in no open disk. Thus by Theorem 4 the function $f(z)$ is nowhere analytic. ∎

To be mathematically precise, we point out that the Cauchy-Riemann equations alone are *not* sufficient to ensure differentiability; one needs the additional hypothesis of continuity of the first partial derivatives of u and v. The complete story is given in the following theorem.

Theorem 5. Let $f(z) = u(x, y) + iv(x, y)$ be defined in some open set G containing the point z_0. If the first partial derivatives of u and v exist in G, are continuous at z_0, and satisfy the Cauchy-Riemann equations at z_0, then f is differentiable at z_0.

Consequently, if the first partial derivatives are continuousand satisfy the Cauchy-Riemann equations at all points of G, then f is analytic in G.

Proof. The difference quotient for f at z_0 can be written in the form

$$\frac{f(z_0 + \Delta z) - f(z_0)}{\Delta z}$$

$$= \frac{[u(x_0 + \Delta x, y_0 + \Delta y) - u(x_0, y_0)] + i[v(x_0 + \Delta x, y_0 + \Delta y) - v(x_0, y_0)]}{\Delta x + i\Delta y} \quad (4)$$

Albeit far from obvious, it has been shown that the continuity assumption can be removed in this part of the theorem.

where $z_0 = x_0 + iy_0$ and $\Delta z = \Delta x + i\Delta y$. The above expressions are well defined if $|\Delta z|$ is so small that the closed disk with center z_0 and radius $|\Delta z|$ lies entirely in G. Let us rewrite the difference

$$u(x_0 + \Delta x, y_0 + \Delta y) - u(x_0, y_0)$$

as

$$[u(x_0 + \Delta x, y_0 + \Delta y) - u(x_0, y_0 + \Delta y)] + [u(x_0, y_0 + \Delta y) - u(x_0, y_0)]. \quad (5)$$

Because the partial derivatives exist in G, the mean-value theorem says that there is a number x^* between x_0 and $x_0 + \Delta x$ such that

$$u(x_0 + \Delta x, y_0 + \Delta y) - u(x_0, y_0 + \Delta y) = \Delta x \frac{\partial u}{\partial x}(x^*, y_0 + \Delta y).$$

Furthermore, since the partial derivatives are continuous at (x_0, y_0), we can write

$$\frac{\partial u}{\partial x}(x^*, y_0 + \Delta y) = \frac{\partial u}{\partial x}(x_0, y_0) + \varepsilon_1,$$

where the function $\varepsilon_1 \to 0$ as $x^* \to x_0$ and $\Delta y \to 0$ (in particular, as $\Delta z \to 0$). Thus the first bracketed expression in (5) can be written as

$$u(x_0 + \Delta x, y_0 + \Delta y) - u(x_0, y_0 + \Delta y) = \Delta x \left[\frac{\partial u}{\partial x}(x_0, y_0) + \varepsilon_1\right].$$

The second bracketed expression in (5) is treated similarly, introducing the function ε_2. Then working the same strategy for the v-difference in Eq. (4), we ultimately have

$$\frac{f(z_0 + \Delta z) - f(z_0)}{\Delta z} = \frac{\Delta x \left[\frac{\partial u}{\partial x} + \varepsilon_1 + i\frac{\partial v}{\partial x} + i\varepsilon_3\right] + \Delta y \left[\frac{\partial u}{\partial y} + \varepsilon_2 + i\frac{\partial v}{\partial y} + i\varepsilon_4\right]}{\Delta x + i\Delta y},$$

where each partial derivative is evaluated at (x_0, y_0) and where each $\varepsilon_i \to 0$ as $\Delta z \to 0$. Now we use the Cauchy-Riemann equations to express the difference quotient as

$$\frac{\Delta x \left[\frac{\partial u}{\partial x} + i\frac{\partial v}{\partial x}\right] + i\Delta y \left[\frac{\partial u}{\partial x} + i\frac{\partial v}{\partial x}\right]}{\Delta x + i\Delta y} + \frac{\lambda}{\Delta x + i\Delta y}, \quad (6)$$

where $\lambda := \Delta x(\varepsilon_1 + i\varepsilon_3) + \Delta y(\varepsilon_2 + i\varepsilon_4)$. Since

$$\left|\frac{\lambda}{\Delta x + i\Delta y}\right| \leq \left|\frac{\Delta x}{\Delta x + i\Delta y}\right| |\varepsilon_1 + i\varepsilon_3| + \left|\frac{\Delta y}{\Delta x + i\Delta y}\right| |\varepsilon_2 + i\varepsilon_4|$$
$$\leq |\varepsilon_1 + i\varepsilon_3| + |\varepsilon_2 + i\varepsilon_4|,$$

we see that the last term in (6) approaches zero as $\Delta z \to 0$, and so

$$\lim_{\Delta z \to 0} \frac{f(z_0 + \Delta z) - f(z_0)}{\Delta z} = \frac{\partial u}{\partial x}(x_0, y_0) + i\frac{\partial v}{\partial x}(x_0, y_0);$$

i.e., $f'(z_0)$ exists.   ∎

It follows from Theorem 4 that the nowhere analytic function $f(z)$ of Example 1 is, nonetheless, differentiable at each point on the line $x = y$.

As promised in Sec. 1.4, we now offer one last vindication of our definition of the complex exponential, by demonstrating its analyticity.

Example 2

Prove that the function $f(z) = e^z = e^x \cos y + i e^x \sin y$ is entire, and find its derivative.

Solution.   Since we have $\partial u/\partial x = e^x \cos y$, $\partial v/\partial y = e^x \cos y$, $\partial u/\partial y = -e^x \sin y$, and $\partial v/\partial x = e^x \sin y$, the first partial derivatives are continuous and satisfy the Cauchy-Riemann equations at every point in the plane. Hence $f(z)$ is entire. From Eq. (1) we see that

$$f'(z) = \frac{\partial u}{\partial x} + i \frac{\partial v}{\partial x} = e^x \cos y + i e^x \sin y.$$

Not surprisingly, $f'(z) = f(z)$.   ∎

As a further application of these techniques, let us prove the following theorem whose analogue in the real case is well known.

Theorem 6. If $f(z)$ is analytic in a domain D and if $f'(z) = 0$ everywhere in D, then $f(z)$ is constant in D.

Before we proceed with the proof, we observe that the *connectedness* property of the domain is essential. Indeed, if $f(z)$ is defined by

$$f(z) = \begin{cases} 0 & \text{if } |z| < 1, \\ 1 & \text{if } |z| > 2, \end{cases}$$

then f is analytic and $f'(z) = 0$ on its *domain of definition* (which is not a *domain!*), yet f is not constant.

Proof of Theorem 6.   Since $f'(z) = 0$ in D, we see from Eqs. (1) and (2) that all the first partial derivatives of u and v vanish in D; that is,

$$\frac{\partial u}{\partial x} = \frac{\partial u}{\partial y} = \frac{\partial v}{\partial x} = \frac{\partial v}{\partial y} = 0.$$

Thus, by Theorem 1 in Sec. 1.6 (see page 40), we have $u = $ constant and $v = $ constant in D. Consequently, $f = u + iv$ is also constant in D.   ∎

One easy consequence of Theorem 6 is the fact that if f and g are two functions analytic in a domain D whose derivatives are identical in D, then $f = g + $ constant in D (see Prob. 7).

Using Theorem 6 and the Cauchy-Riemann equations, one can further show that an analytic function $f(z)$ must be constant when any one of the following conditions hold in a domain D:

$$\text{Re } f(z) \text{ is constant;}$$
$$\text{Im } f(z) \text{ is constant;}$$
$$|f(z)| \text{ is constant.} \qquad (7)$$

The proofs are left as problems.

EXERCISES 2.4

1. Use the Cauchy-Riemann equations to show that the following functions are nowhere differentiable.

   (a) $w = \bar{z}$ **(b)** $w = \text{Re } z$ **(c)** $w = 2y - ix$

2. Show that $h(z) = x^3 + 3xy^2 - 3x + i(y^3 + 3x^2y - 3y)$ is differentiable on the coordinate axes but is nowhere analytic.

3. Use Theorem 5 to show that $g(z) = 3x^2 + 2x - 3y^2 - 1 + i(6xy + 2y)$ is entire. Write this function in terms of z.

4. Let
   $$f(z) = \begin{cases} (x^{4/3}y^{5/3} + ix^{5/3}y^{4/3})/(x^2 + y^2) & \text{if } z \neq 0, \\ 0 & \text{if } z = 0. \end{cases}$$
   Show that the Cauchy-Riemann equations hold at $z = 0$ but that f is not differentiable at this point. [HINT: Consider the difference quotient $f(\Delta z)/\Delta z$ for $\Delta z \to 0$ along the real axis and along the line $y = x$.]

5. Show that the function $f(z) = e^{x^2 - y^2}[\cos(2xy) + i \sin(2xy)]$ is entire, and find its derivative.

6. If u and v are expressed in terms of polar coordinates (r, θ), show that the Cauchy-Riemann equations can be written in the form
   $$\frac{\partial u}{\partial r} = \frac{1}{r}\frac{\partial v}{\partial \theta}, \qquad \frac{\partial v}{\partial r} = -\frac{1}{r}\frac{\partial u}{\partial \theta}.$$
   [HINT: Consider the difference quotient $(f(z) - f(z_0))/(z - z_0)$, as $z \to z_0 = r_0 e^{i\theta_0}$ along the ray $\arg z = \theta_0$ and along the circle $|z| = r_0$.]

7. Show that if two analytic functions f and g have the same derivative throughout a domain D, then they differ only by an additive constant. [HINT: Consider $f - g$.]

8. Show that if f is analytic in a domain D and either $\text{Re } f(z)$ or $\text{Im } f(z)$ is constant in D, then $f(z)$ must be constant in D.

9. Show, by contradiction, that the function $F(z) = |z^2 - z|$ is nowhere analytic because of condition (7).

10. Show that if $f(z)$ is analytic and real-valued in a domain D, then $f(z)$ is constant in D.

11. Suppose that $f(z)$ and $\overline{f(z)}$ are analytic in a domain D. Show that $f(z)$ is constant in D.

12. Show that if f is analytic in a domain D and $|f(z)|$ is constant in D, then the function $f(z)$ is constant in D. [HINT: $|f|^2$ is constant, so $\partial|f|^2/\partial x = \partial|f|^2/\partial y = 0$ throughout D. Using these two relations and the Cauchy-Riemann equations, deduce that $f'(z) = 0$.]

13. Given that $f(z)$ and $|f(z)|$ are each analytic in a domain D, prove that $f(z)$ is constant in D.

14. Show that if the analytic function $w = f(z)$ maps a domain D onto a portion of a line, then f must be constant throughout D.

15. The *Jacobian* of a mapping

$$u = u(x, y), \qquad v = v(x, y)$$

from the xy-plane to the uv-plane is defined to be the determinant

$$J(x_0, y_0) := \begin{vmatrix} \dfrac{\partial u}{\partial x} & \dfrac{\partial u}{\partial y} \\[2mm] \dfrac{\partial v}{\partial x} & \dfrac{\partial v}{\partial y} \end{vmatrix},$$

where the partial derivatives are all evaluated at (x_0, y_0). Show that if $f = u + iv$ is analytic at $z_0 = x_0 + iy_0$, then $J(x_0, y_0) = |f'(z_0)|^2$.

16. The notion of analyticity as discussed in the preceding section requires that the function $f(x, y) = u(x, y) + iv(x, y)$ can be written in terms of $(x + iy)$ alone, without using $\bar{z} = (x - iy)$. To make this concept more explicit, we introduce the change of variables

$$\begin{cases} \xi = x + iy \\ \eta = x - iy \end{cases} \text{ or, equivalently, } \begin{cases} x = (\xi + \eta)/2 \\ y = (\xi - \eta)/2i \end{cases}$$

producing the function

$$\tilde{f}(\xi, \eta) := f(x(\xi, \eta), y(\xi, \eta)).$$

(a) Using the chain rule show formally that

$$\frac{\partial \tilde{f}}{\partial \xi} = \frac{1}{2}\left(\frac{\partial u}{\partial x} + \frac{\partial v}{\partial y}\right) + \frac{i}{2}\left(\frac{\partial v}{\partial x} - \frac{\partial u}{\partial y}\right),$$

$$\frac{\partial \tilde{f}}{\partial \eta} = \frac{1}{2}\left(\frac{\partial u}{\partial x} - \frac{\partial v}{\partial y}\right) + \frac{i}{2}\left(\frac{\partial u}{\partial y} + \frac{\partial v}{\partial x}\right).$$

That is, "z-bar" is barred!

(b) Since η is the same as $\bar{z}$, the statement "f is independent of $\bar{z}$" is equivalent
to

$$\frac{\partial \tilde{f}}{\partial \eta} = \frac{\partial \tilde{f}}{\partial \bar{z}} = 0.$$

Show that this condition is the same as the Cauchy-Riemann equations for f.

2.5   Harmonic Functions

Solutions of the two-dimensional *Laplace equation*

$$\nabla^2 \phi := \frac{\partial^2 \phi}{\partial x^2} + \frac{\partial^2 \phi}{\partial y^2} = 0 \tag{1}$$

are among the most important functions in mathematical physics. The electrostatic
potential solves Eq. (1) in two-dimensional free space, as does the scalar magnetostatic
potential; the corresponding field in any direction is given by the directional derivative
of $\phi(x, y)$. Two-dimensional fluid flow problems are described by such functions
under certain idealized conditions, and ϕ can also be interpreted as the displacement
of a membrane stretched across a loop of wire, if the loop is nearly flat. In the next
section we shall discuss equilibrium temperature distributions as models for solutions
to Eq. (1).

One of the most important applications of analytic function theory to applied mathematics is the abundance of solutions of Eq. (1) that it supplies. We shall adopt the
following standard terminology for these solutions.

Definition 6. A real-valued function $\phi(x, y)$ is said to be **harmonic** in a domain
D if all its second-order partial derivatives are continuous in D and if, at each
point of D, ϕ satisfies Laplace's equation (1).

The sources of these harmonic functions are the real and imaginary parts of analytic functions, as we prove in the next theorem.

Theorem 7. If $f(z) = u(x, y) + iv(x, y)$ is analytic in a domain D, then each
of the functions $u(x, y)$ and $v(x, y)$ is harmonic in D.

Proof.   In a later chapter we shall show that the real and imaginary parts of any
analytic function have continuous partial derivatives of all orders. Assuming this fact,

---

Marquis Pierre-Simon de Laplace, 1749–1827.

we recall from elementary calculus that under such conditions mixed partial derivatives can be taken in any order; i.e.,

$$\frac{\partial}{\partial y}\frac{\partial u}{\partial x} = \frac{\partial}{\partial x}\frac{\partial u}{\partial y}. \tag{2}$$

Using the Cauchy-Riemann equations for the first derivatives, we transform Eq. (2) into

$$\frac{\partial^2 v}{\partial y^2} = -\frac{\partial^2 v}{\partial x^2}.$$

which is equivalent to Eq. (1). Thus v is harmonic in D, and a similar computation proves that u is also. ∎

Conversely, if we are given a function $u(x, y)$ harmonic in, say, an open disk, then we can find another harmonic function $v(x, y)$ so that $u + iv$ is an analytic function of z in the disk. Such a function v is called a *harmonic conjugate* of u. The construction of v is effected by exploiting the Cauchy-Riemann equations, as we illustrate in the following example.

Example 1

Construct an analytic function whose real part is $u(x, y) = x^3 - 3xy^2 + y$.

Solution.   First we verify that

$$\frac{\partial^2 u}{\partial x^2} + \frac{\partial^2 u}{\partial y^2} = 6x - 6x = 0,$$

and so u is harmonic in the whole plane. Now we have to find a mate, $v(x, y)$, for u such that the Cauchy-Riemann equations are satisfied. Thus we must have

$$\frac{\partial v}{\partial y} = \frac{\partial u}{\partial x} = 3x^2 - 3y^2 \tag{3}$$

and

$$\frac{\partial v}{\partial x} = -\frac{\partial u}{\partial y} = 6xy - 1. \tag{4}$$

If we hold x constant and integrate Eq. (3) with respect to y, we get

$$v(x, y) = 3x^2 y - y^3 + \text{constant},$$

but the "constant" could conceivably be any differentiable function of x; it need only be independent of y. Therefore, we write

$$v(x, y) = 3x^2 y - y^3 + \psi(x).$$

We can find $\psi(x)$ by plugging this last expression into Eq. (4);

$$\frac{\partial v}{\partial x} = 6xy + \psi'(x) = 6xy - 1. \tag{5}$$

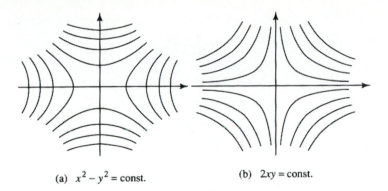

(a) $x^2 - y^2 = $ const.   (b) $2xy = $ const.

Figure 2.6 Level curves of real and imaginary parts of z^2.

This yields $\psi'(x) \equiv -1$, and so $\psi(x) = -x + a$, where a is some (genuine) constant. It follows that a harmonic conjugate of $u(x, y)$ is given by

$$v(x, y) = 3x^2 y - y^3 - x + a,$$

and the analytic function

$$f(z) = x^3 - 3xy^2 + y + i\left(3x^2 y - y^3 - x + a\right),$$

which we recognize as $z^3 - i(z - a)$, solves the problem. ∎

This procedure will always work for an $u(x, y)$ harmonic in a *disk*, as is shown in Prob. 20. Thus we can learn a great deal about analytic functions by studying harmonic functions and vice versa.

The harmonic functions forming the real and imaginary parts of an analytic function $f(z)$ each generate a family of curves in the xy-plane, namely, the *level curves* or *isotimic curves*

$$u(x, y) = \text{constant} \qquad (6)$$

and

$$v(x, y) = \text{constant}. \qquad (7)$$

If u is interpreted as an electrostatic potential, then the curves (6) are the *equipotentials*. If u is temperature, (6) describes the *isotherms*.

For the function $f(z) = z^2 = x^2 - y^2 + i2xy$, the level curves $u = x^2 - y^2 = $ constant are hyperbolas asymptotic to the lines $y = \pm x$, as shown in Fig. 2.6(a). The curves $v = 2xy = $ constant are also hyperbolas, asymptotic to the coordinate axes; see Fig. 2.6(b).

We caution the reader that finding a harmonic conjugate in an *arbitrary* domain may not always be possible. See Prob. 21 for an example of this unfortunate circumstance when the domain is a *punctured* disk.

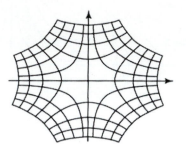

Figure 2.7 Level curves of Fig. 2.6 superimposed.

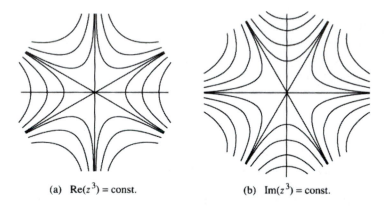

(a) $\text{Re}(z^3) = \text{const.}$ (b) $\text{Im}(z^3) = \text{const.}$

Figure 2.8 Level curves of real and imaginary parts of z^3.

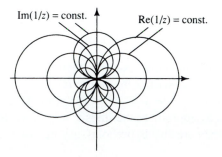

Figure 2.9 Level curves of real and imaginary parts of $1/z$.

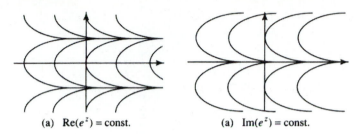

(a) Re(e^z) = const.   (a) Im(e^z) = const.

Figure 2.10 Level curves of real and imaginary parts of e^z.

Notice that if the two families of curves are superimposed as in Fig. 2.7 they appear to intersect at right angles. The same effect occurs with the level curves for the analytic functions z^3 (Fig. 2.8), $1/z$ (Fig. 2.9), and e^z (Fig. 2.10). This is no accident; the level curves of the real and imaginary parts of an analytic function $f(z)$ will always intersect at right angles—unless $f'(z) = 0$ at the point of intersection. This can be seen from the Cauchy-Riemann equations as follows.

Recall that the vector with components $[\partial u/\partial x, \partial u/\partial y]$ is the *gradient* of u and is normal to the level curves of u. Similarly, $[\partial v/\partial x, \partial v/\partial y]$ is normal to the level curves of v. The scalar (dot) product of these gradient vectors is

$$\frac{\partial u}{\partial x}\frac{\partial v}{\partial x} + \frac{\partial u}{\partial y}\frac{\partial v}{\partial y} = \frac{\partial v}{\partial y}\frac{\partial v}{\partial x} - \frac{\partial v}{\partial x}\frac{\partial v}{\partial y} = 0$$

by the Cauchy-Riemann equations. Thus *if these gradients are nonzero*, they are perpendicular, and hence so are the level curves. *Level curves of harmonic functions and their harmonic conjugates intersect at right angles.*

The following examples illustrate how analytic function theory can be used to solve Laplace's equation in regions whose boundaries are identifiable as level curves.

Example 2

Find a function $\phi(x, y)$ that is harmonic in the region of the right half-plane between the curves $x^2 - y^2 = 2$ and $x^2 - y^2 = 4$ and takes the value 3 on the left edge and the value 7 on the right edge (Fig. 2.11).

Solution. We recognize $x^2 - y^2$ as the real part of z^2, so the boundary curves are level curves of a known harmonic function. To meet the specified boundary conditions, we add some flexibility by considering

$$\phi(x, y) = A\left(x^2 - y^2\right) + B = \text{Re}\left(Az^2 + B\right), \quad A, B \text{ real,}$$

and adjust A and B accordingly. When $x^2 - y^2 = 2$, we require $\phi = 3$;

$$A(2) + B = 3.$$

When $x^2 - y^2 = 4$, we want $\phi = 7$;

$$A(4) + B = 7.$$

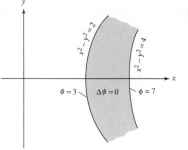

Figure 2.11 Laplace's equation for the region of Example 2.

Solving for A and B we find the solution to be

$$\phi(x, y) = 2\left(x^2 - y^2\right) - 1. \quad \blacksquare$$

This example was clearly contrived. In Chapters 3 and 7 we shall consider more profound applications of this idea.

EXERCISES 2.5

1. Verify directly that the real and imaginary parts of the following analytic functions satisfy Laplace's equation.

    (a) $f(z) = z^2 + 2z + 1$ **(b)** $g(z) = \dfrac{1}{z}$

    (c) $h(z) = e^z$

2. Find the most general harmonic polynomial of the form $ax^2 + bxy + cy^2$.

3. Verify that each given function u is harmonic (in the region where it is defined) and then find a harmonic conjugate of u.

    (a) $u = y$ **(b)** $u = e^x \sin y$
    (c) $u = xy - x + y$ **(d)** $u = \sin x \cosh y$
    (e) $u = \ln |z|$ for $\operatorname{Re} z > 0$ **(f)** $u = \operatorname{Im} e^{z^2}$

4. Show that if $v(x, y)$ is a harmonic conjugate of $u(x, y)$ in a domain D, then *every* harmonic conjugate of $u(x, y)$ in D must be of the form $v(x, y) + a$, where a is a real constant.

5. Show that if v is a harmonic conjugate for u, then $-u$ is a harmonic conjugate for v.

6. Show that if v is a harmonic conjugate of u in a domain D, then uv is harmonic in D.

7. Find a function $\phi(x, y)$ that is harmonic in the infinite vertical strip

$$\{z: -1 \leq \text{Re } z \leq 3\}$$

and takes the value 0 on the left edge and the value 4 on the right edge.

8. Suppose that the functions u and v are harmonic in a domain D.

   (a) Is the sum $u + v$ necessarily harmonic in D?

   (b) Is the product uv necessarily harmonic in D?

   (c) Is $\partial u/\partial x$ harmonic in D? (You may use the fact—which we will prove in Chapter 4—that harmonic functions have continuous partial derivatives of all orders.)

9. Find a function $\phi(x, y)$ that is harmonic in the region of the first quadrant between the curves $xy = 2$ and $xy = 4$ and takes the value 1 on the lower edge and the value 3 on the upper edge. [HINT: Begin by considering z^2.]

10. Show that in polar coordinates (r, θ) Laplace's equation becomes

$$\frac{\partial^2 \phi}{\partial r^2} + \frac{1}{r}\frac{\partial \phi}{\partial r} + \frac{1}{r^2}\frac{\partial^2 \phi}{\partial \theta^2} = 0.$$

11. Let $f(z) = z + 1/z$. Show that the level curve $\text{Im } f(z) = 0$ consists of the real axis (excluding $z = 0$) and the circle $|z| = 1$. [The level curves $\text{Im } f(z) = \text{constant}$ can be interpreted as streamlines for fluid flow around a cylindrical obstacle.]

12. Prove that if r and θ are polar coordinates, then the functions $r^n \cos n\theta$ and $r^n \sin n\theta$, where n is an integer, are harmonic as functions of x and y. [HINT: Recall De Moivre's formula.]

13. Find a function harmonic inside the wedge bounded by the nonegative x-axis and the half-line $y = x$ $(x \geq 0)$ that goes to zero on these sides but is not identically zero. [HINT: See Prob. 12.] The level curves for this function can be interpreted as streamlines for a fluid flowing inside this wedge, under certain idealized conditions.

14. Suppose that $f(z)$ is analytic and nonzero in a domain D. Prove that $\ln |f(z)|$ is harmonic in D.

15. Find a function $\phi(z)$, harmonic within the annulus (ring domain) bounded by the concentric circles $|z| = 1$ and $|z| = 2$, such that $\phi = 0$ on the inner circle and $\phi(2e^{i\theta}) = 5\cos 3\theta$ on the outer circle. [HINT: Think of z^n and z^{-n}.]

16. Find a function harmonic outside the circle $|z| = 3$ that goes to zero on $|z| = 3$ but is not identically zero. [HINT: See Prob. 14]

17. Find a function $\phi(x, y)$ harmonic in the upper half-plane $\operatorname{Im} z > 0$ and continuous on $\operatorname{Im} z \geq 0$ such that

    (a) $\phi(x, 0) = x^2 + 5x + 1$ for all x.

    (b) $\phi(x, 0) = 2x^3/(x^2 + 4)$ for all x.

[HINT: $\phi = \operatorname{Re}[2z^3/(z^2 + 4)]$ won't work because $2z^3/(z^2 + 4)$ is not analytic at $z = 2i$ in the upper half-plane. Instead, write

$$\frac{2x^3}{x^2 + 4} = \frac{x^2}{x - 2i} + \frac{x^2}{x + 2i} = 2\operatorname{Re}\frac{x^2}{x + 2i},$$

which suggests the proper choice for ϕ.]

18. Show that if $\phi(x, y)$ is harmonic, then $\phi_x - i\phi_y$ is analytic. (You may assume that ϕ has continuous partial derivatives of all orders.)

19. Find a function $\phi(z)$ harmonic *outside* the unit circle $|z| = 1$, satisfying

$$\phi(e^{i\theta}) = \cos^2\theta, \quad 0 \leq \theta \leq 2\pi,$$

such that $\phi(re^{i\theta})$ approaches the constant value $1/2$ along all large radii r. [HINT: Recall that z^{-n} is analytic outside the unit circle and goes to zero along large radii r.]

20. By tracing the steps in Example 1, show that every function $u(x, y)$ harmonic in a disk has a harmonic conjugate $v(x, y)$. [HINT: The only difficulty which could occur is in the step corresponding to Eq. (5), where in order to find $\psi'(x)$ we must be certain that all appearances of the variable y cancel. Show that this is guaranteed because u is harmonic.]

21. Show that although $u = \ln|z|$ is harmonic in the complex plane except at $z = 0$ (i.e., in the domain $\mathbf{C}\backslash\{0\}$), u does *not* have a harmonic conjugate v throughout $\mathbf{C}\backslash\{0\}$. In other words, show that there is no function v such that $\ln|z| + iv(z)$ is analytic in $\mathbf{C}\backslash\{0\}$. [HINT: Show that if $\ln|z| + iv(z)$ is analytic in $\mathbf{C}\backslash\{0\}$, then $v(z) = \operatorname{Arg} z + a$ except along the nonpositive real axis.]

22. Show that if $\phi(x, y)$ and $\psi(x, y)$ are harmonic, then u and v defined by

$$u(x, y) = \phi_x\phi_y + \psi_x\psi_y$$

and

$$v(x, y) = \tfrac{1}{2}\left(\phi_x^2 + \psi_x^2 - \phi_y^2 - \psi_y^2\right)$$

satisfy the Cauchy-Riemann equations.

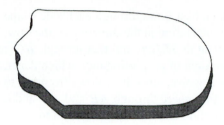

Figure 2.12 Slab of thermally conducting material.

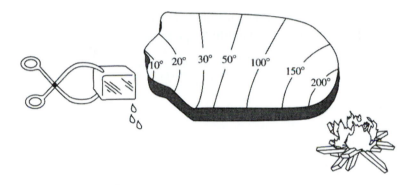

Figure 2.13 Sinks and sources.

2.6   *Steady-State Temperature as a Harmonic Function

It is useful to have a familiar physical model for harmonic functions as an aid in visualizing and remembering their properties. The equilibrium temperatures in a slab, as we shall see, fill this role nicely.

Figure 2.12 depicts a uniform slab of a thermally-conducting material, such as a copper plate or a ceramic substrate for microelectronic circuitry. It has constant thickness, so its top and bottom surfaces lie parallel to the xy-plane. We assume that these surfaces are also insulated, and no heat flows in the vertical direction. As a result the equilibrium temperature T is a function of x and y;

$$T = T(x, y).$$

This temperature distribution is maintained by heat sources (or sinks) and insulation placed around the edges, so that the isotherms appear as illustrated in Fig. 2.13.

Now once the temperature has reached equilibrium, $T(x, y)$ will be a harmonic function:

$$\frac{\partial^2 T}{\partial x^2} + \frac{\partial^2 T}{\partial y^2} = 0. \tag{1}$$

The physical reason for this is as follows. Focusing attention on a small square in the slab as depicted in Fig. 2.14, we call upon Fourier's law of heat conduction, which states that the rate at which heat flows through each side of the square is proportional to the rate of change of temperature in the direction of the flow. Thus the flow through AB and CD is proportional to $\partial T/\partial x$, and that through BC and AD is proportional to $\partial T/\partial y$. (In fact the constant of proportionality, which depends on the cross-section area and the material, is negative, since heat flows from hot to cold!) The heat flows are depicted as *entering* the square through AB and AD and *exiting* through BC and CD. Therefore, the net *out*flow of heat is proportional to

$$\left.\frac{\partial T}{\partial x}\right|_{CD} - \left.\frac{\partial T}{\partial x}\right|_{AB} + \left.\frac{\partial T}{\partial y}\right|_{BC} - \left.\frac{\partial T}{\partial y}\right|_{AD}.$$

For small dimensions s the difference in the first derivatives can be approximated by the second derivative, and the net outflux is proportional to

$$\frac{\partial^2 T}{\partial x^2} s + \frac{\partial^2 T}{\partial y^2} s. \tag{2}$$

At equilibrium the temperature has settled; the square has finished cooling down (or heating up), and the net outflux will be zero. Dividing expression (2) by s, then, we conclude that $T(x, y)$ satisfies Eq. (1).

The fact that harmonic functions arise as temperature distributions permits us to anticipate some of their mathematical properties. For example, look at the isothermal curves in Fig. 2.15. They indicate a "hot spot" in the interior of the slab. This cannot occur at equilibrium, because heat would flow away from the hot spot and it would cool down. Of course, this pattern *could* be maintained by an external source underneath the slab, but we have precluded this by assuming that such sources are located only on the edge. We conclude that the temperature distribution can never exhibit such an *interior* maximum. The rigorous formulation and generalization of this observation to harmonic functions is identified in Chapter 4 as the *maximum principle*, which says that a harmonic function cannot take its maximum in the interior of a region, except in the trivial case when it is constant throughout.

As another example consider the following experiment. The edges of the slab in Fig. 2.16 are maintained at fixed temperatures by external heat sources. except for a small section along which *we* can control the temperature to our liking (using some type of adjustable furnace). Then, on physical grounds we would expect to be able, by turning up the furnace sufficiently, to raise the temperature of an arbitrary interior point to any specified value—although, of course, we couldn't guarantee to replicate a whole *pattern* of temperatures across the slab.

Our expectation is premised on the intuitive feeling that the interior temperatures are completely determined by the edge temperature distribution. This is actually an instance of the *boundary value property* of harmonic functions, and in fact in Chapter 4 we shall study *Poisson's formulas*, which express the explicit relationship between the interior and boundary values of such functions for certain geometries.

Note also that the thermodynamic reality of a zero absolute temperature inhibits our ability to *cool* interior points arbitrarily.

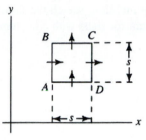

Figure 2.14 Heat flow.

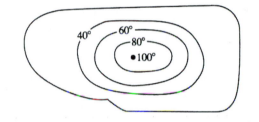

Figure 2.15 Isotherms.

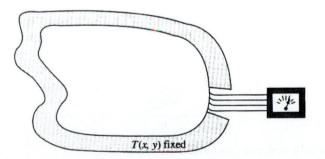

Figure 2.16 Adjustable boundary temperatures.

EXERCISES 2.6

1. Using only your physical intuition, sketch the family of isotherms that you would expect to see at equilibrium for slabs with edge temperatures maintained as shown in Fig. 2.17.

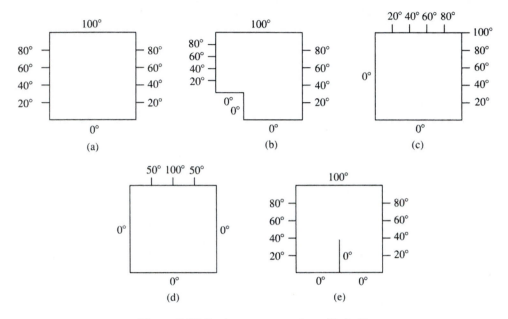

Figure 2.17 Isotherm constructions (Prob. 1).

2. Sketch the isotherms for the edge-temperature distribution in Fig. 2.18. Does this configuration violate the maximum principle?

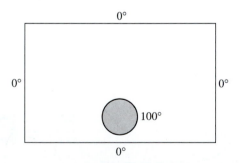

Figure 2.18 Isotherm construction (Prob. 2).

3. Sketch the isotherms for the edge-temperature distribution in Fig. 2.19. Does this configuration violate the maximum principle?

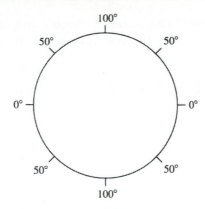

Figure 2.19 Isotherm construction (Prob. 3).

2.7 *Iterated Maps: Julia and Mandelbrot Sets

What happens if one enters a number into a calculator, pushes a function key like x^2, and then pushes it again and again? The calculator squares the number, then squares the result, then squares *that* result, and so on; one has *iterated* the function $f(x) = x^2$. Now this iteration process can be very interesting when it is performed with complex numbers; the sequence of points $z_0,\ f(z_0),\ f(f(z_0)),\ f(f(f(z_0))), \ldots$ becomes an *orbit* in the complex plane.

If we iterate the complex function $f(z) = z^2$, it is easy to predict many of the orbits. When the starting point or "seed" z_0 lies within the unit circle, that is $|z_0| < 1$, the orbit stays bounded (because the squares get smaller in modulus) and converges to $z = 0$. If $|z_0| > 1$, the iterates get larger in modulus and the orbit is unbounded.

Example 1 describes a test for bounded orbits that is easy to apply.

Example 1
Show that if

    (i) $f(z)$ is analytic in a neighborhood of $z = \zeta$,

    (ii) $f(\zeta) = \zeta$, and

    (iii) $|f'(\zeta)| < 1$,

then there is a disk around ζ with the property that all orbits launched from inside the disk remain confined to the disk and converge to ζ.

    Solution.   Since

$$\lim_{z \to \zeta} \left| \frac{f(z) - f(\zeta)}{z - \zeta} \right| = |f'(\zeta)| < 1$$

and $f(\zeta) = \zeta$, we can pick a real number ρ lying between $|f'(\zeta)|$ and 1 such that

$$|f(z) - f(\zeta)| \equiv |f(z) - \zeta| \le \rho |z - \zeta| \tag{1}$$

for all z in a sufficiently small disk around ζ. Such a disk meets the specifications; indeed, if any point z_0 in this disk is the seed for an orbit $z_1 = f(z_0)$, $z_2 = f(z_1)$, ..., then by (1) we have

$$|z_n - \zeta| \leq \rho|z_{n-1} - \zeta| \leq \cdots \leq \rho^n|z_0 - \zeta|.$$

Since $\rho < 1$, the point z_n lies closer to ζ than z_{n-1} and, in fact, $\lim_{n \to \infty} z_n = \zeta$. ∎

If $f(\zeta) = \zeta$, then ζ is called a *fixed point* of the function f. A fixed point meeting the conditions of Example 1 is called an *attractor*, and the set of seed points whose orbits converge to ζ is called its *basin of attraction*. Thus $\zeta = 0$ is an attractor for $f(z) = z^2$ (since $0 = 0^2 = f(0)$ and $|f'(0)| = 0 < 1$) whose basin is the open disk $|z| < 1$. Example 1 shows that every attractor has a basin containing, at least, a small disk.

The other fixed point for $f(z) = z^2$, $\zeta = 1$, is a *repellor*. Its properties are explored in Prob. 2.

For the function $f(z) = z^2$, if z_0 lies *on* the unit circle $|z_0| = 1$, so does the entire orbit launched from z_0. In fact if $z_0 = 1$ or -1 the orbit quickly settles down to the fixed point $z = 1$. If $z_0 = e^{i2\pi/3}$ (a primitive cube root of unity in the parlance of Sec. 1.5) the orbit oscillates between two points $e^{\pm i2\pi/3}$ and is called a *2-cycle*, with *period* 2; note that this is equivalent to saying that $e^{i2\pi/3}$ is a fixed point of $f(f(z))$. The seed $e^{i2\pi/7}$ gives birth to a 3-cycle, and $e^{i2\pi/15}$ to a 4-cycle. Do you see the pattern (cf. Prob. 6)?

It can be shown that the seed choice $z_0 = e^{i\alpha 2\pi}$, for irrational α, generates an orbit whose points never repeat (cf. Prob. 4) and, in fact, permeate the unit circle densely. So the unit circle, which separates the seeds of orbits converging to zero from those of unbounded orbits, contains a variety of orbits itself.

Definition 7. The **filled Julia set** for a polynomial function $f(z)$ is defined to be the set of points that launch bounded orbits through iteration of f; the **Julia set** is the boundary of the filled Julia set.

So the Julia set for z^2 is the unit circle, and the filled Julia set is the closed unit disk.

The Julia set for the function $f(z) = z^2 - 2$ consists of the real interval $[-2, 2]$ (which is already "filled"). Indeed, one immediately sees that if $-2 \leq x \leq 2$, then $0 \leq x^2 \leq 4$ and $-2 \leq x^2 - 2 \equiv f(x) \leq 2$, so orbits launched from $[-2, 2]$ remain bounded. The proof that *other* values of z are seeds of unbounded orbits will have to wait until we have studied the *Joukowski transformation* (Prob. 8a, Exercises 7.7).

G. Julia investigated these sets in 1918.

From these considerations, one might conclude that the Julia sets for all functions of the form $f(z) = z^2 + c$ are disks, segments, or other mundane configurations. Nothing could be further from the truth! An astonishing assortment of exotic patterns result when we locate the seeds of bounded orbits of $z^2 + c$, for various complex values of the constant c. Fig. 2.20 displays some of the dragons, rabbits, fern leaves, and shoreline patterns that have been discovered to be Julia sets, together with the corresponding values of c. Many of these patterns are *fractals*; objects that typically have dimension neither one nor two, but some fraction in between. Some also enjoy the property of *self-similarity*, in that if one zooms in to see the details of a small subset, a replica of the original pattern reappears.

The most enjoyable way to explore Julia sets is with software. Although functional iteration is trivial to code, the graphics displays are best left to experts, and the references at the end of this chapter contain pointers to some websites and software packages that provide this facility. We invite the reader to try to replicate the designs depicted in Fig. 2.20, using software.

Note that some of the filled Julia sets consist of single connected components, while the others appear to be totally disconnected. In 1982, Benoit Mandelbrot was inspired to investigate which values of c give rise to (filled) Julia sets that are connected. The answer was the astonishing *Mandelbrot set*, depicted in Fig. 2.21. The point c lies in the Mandelbrot set if the filled Julia set for $f(z) = z^2 + c$ is connected; it lies outside if the filled Julia set is disconnected. It is amusing to use software to see what happens to the Julia sets for a family of values of c tracing a path from the interior to the exterior of the Mandelbrot set; the Julia sets "vaporize" as they metamorphose from connected to disconnected. (See references.) It should not surprise the reader that some *phase transitions* in physics have been modeled using the Mandelbrot set.

The interplay between rigorous, theoretical analyses of these sets and computer experimentation has generated some interesting insights into the latter. For example, we can't trust a computer to tell us if an orbit is unbounded, because the computer can only distinguish a finite set of numbers; eventually it will recycle (or overflow). So we have to make a judgment as to how many iterations we must simulate before deciding that an orbit is unbounded. Secondly, virtually every time the computer iterates the function $f(z)$ it commits a roundoff error, and these errors accumulate as we simulate long orbits; so we have to carefully evaluate the credibility of our calculations. Despite these forebodings, in 1987 Hammel, Yorke, and Grebogi proved that every computed orbit is arbitrarily close to *some* true orbit!

The Julia sets (and their analogs) for more complicated functions are subjects of continuing mathematical research. The software listed at the end of the chapter will guide the reader to explore the beautiful convergence patterns resulting from iterating the complex trigonometric functions (defined in Chapter 3) and others. In fact a international exhibition of patterns generated using the Mandelbrot set, titled "Frontiers of Chaos," toured many museums in the late 1980s; see the reference by Peitgen and Richter.

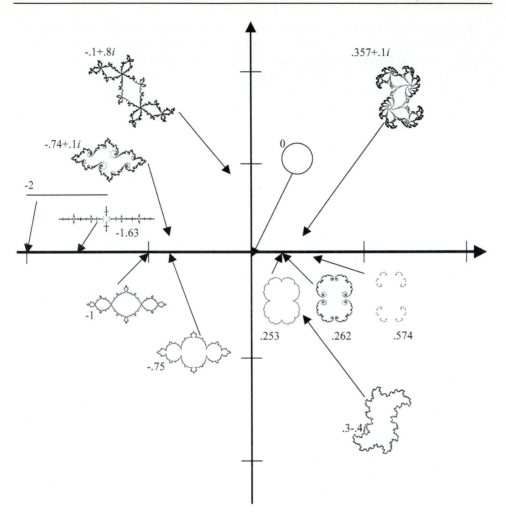

Figure 2.20 Julia sets.

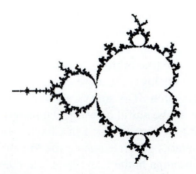

Figure 2.21 Mandelbrot set.

EXERCISES 2.7

1. Find the fixed points and attractors for the function $f(z) = z^2 + c$, in terms of the real constant c.

2. If $f(z)$ is entire and ζ is a fixed point of f such that $|f'(\zeta)| > 1$, then ζ is said to be a *repellor* for f. Show that there is a disk around ζ such that all orbits launched from within the disk, other than the 1-cycle launched from ζ itself, eventually leave the disk.

3. For each of the following functions, determine the fixed points and decide which are attractors, repellors (see Prob. 2), or neither.

   (a) $f(z) = z^2 + z + 1$ (b) $f(z) = z^3 + z^2 + (3/4)z - 1/4$.

4. Prove that if α is a real irrational number, the seed $z_0 = e^{i2\pi\alpha}$ generates an orbit under $f(z) = z^2$ whose points never repeat.

5. For the function $f(z) = 1/(z+1)$, determine the fixed points and decide which are attractors, repellors (see Prob. 2), or neither.

6. Derive the formula $z_0 = e^{i2\pi/(2^n-1)}$ for a seed launching an n-cycle for orbits formed by iterating $f(z) = z^2$.

7. The seed point $z_0 = e^{i2\pi/5}$ launches a 4-cycle for iterates of $f(z) = z^2$, but it does not fit into the pattern of Prob. 6. Explain.

8. Use software to generate the Julia sets in Fig. 2.20.

9. Determine the filled Julia set for the mapping $f(z) = \alpha z$, where α is a complex constant. Consider separately the cases where $|\alpha| \le 1$ and $|\alpha| > 1$.

10. In general, *Newton's method* for approximating the zeros of the entire function $F(z)$ can be described as forming the orbits of the function

$$f(z) = z - \frac{F(z)}{F'(z)} .$$

Show that the fixed points of $f(z)$ are the same as the zeros of $F(z)$, with the possible exception of the points where $F'(z) = 0$. Then show that *every* zero of $F(z)$ (other than ones where $F'(z) = 0$) is an attractor for $f(z)$.

SUMMARY

A complex-valued function f of a complex variable $z = x + iy$ can be considered as a pair of real functions of two real variables in accordance with $f(z) = u(x, y) + iv(x, y)$. The definitions of limit, continuity, and derivative for such functions are direct analogues of the corresponding concepts introduced in calculus, but the greater freedom of z to vary in two dimensions lends added strength to these conditions. In particular, the existence of a derivative, defined as the limit of

$$\frac{f(z + \Delta z) - f(z)}{\Delta z} \text{ as } \Delta z \to 0,$$

implies a strong relationship between the functions u and v, namely, the Cauchy-Riemann equations

$$\frac{\partial u}{\partial x} = \frac{\partial v}{\partial y}, \qquad \frac{\partial u}{\partial y} = -\frac{\partial v}{\partial x}.$$

If the function f is differentiable in an open set, it is said to be analytic. This property can be established by showing that the first-order partial derivatives of u and v are continuous and satisfy the Cauchy-Riemann equations on the open set. Analyticity of a function f is the mathematical expression of the intuitive condition that f respects the complex structure of z; i.e., f can be computed using x and y only in the combination $(x + iy)$. If f is given in terms of z alone, the basic formulas of calculus can be used to find its derivative.

The real and imaginary parts of an analytic function are harmonic; i.e., they satisfy Laplace's equation

$$\frac{\partial^2 \phi}{\partial x^2} + \frac{\partial^2 \phi}{\partial y^2} = 0,$$

and their second-order partial derivatives are continuous. Furthermore, the level curves of the real part intersect those of the imaginary part orthogonally. Given a harmonic function $u(x, y)$ in a disk it is possible to construct another harmonic function $v(x, y)$ so that $u(x, y) + iv(x, y)$ is analytic in that disk; such a function v is called a harmonic conjugate of u. Harmonic functions can be physically interpreted as equilibrium temperature distributions.

Suggested Reading

In addition to the references following Chapter 1, the following texts, articles, websites, and software packages may be helpful for special topics:

Harmonic Functions

[1] Davis, H., and Snider, A.D. *Introduction to Vector Analysis*, 7th ed. Quant Systems, Charleston, SC, 1994.

[2] Hille, E. *Analytic Function Theory*, Vol. II. Chelsea, New York, 1973.

[3] Snider, A.D. *Partial Differential Equations: Sources and Solutions*, Prentice-Hall, Upper Saddle River, NJ, 1999.

Julia and Mandelbrot Sets

[1] Devaney, R. L. *A First Course in Chaotic Dynamical Systems*, Addison-Wesley Publishing Co., Reading, MA, 1992.

[2] Hammel, S. M., Yorke, J. A., and Grebogi, C. "Do numerical orbits of chaotic dynamical processes represent true orbits?," *J. of Complexity* 3 (1987), 136–145.

[3] Peitgen, H.-O. and Richter, P. H. *The Beauty of Fractals*, Springer-Verlag, Berlin, 1986. (Describes the exhibit "Frontiers of Chaos.")

[4] *http://math.bu.edu/DYSYS/explorer/tour4.html*, a web site maintained at Boston University by R. L. Devaney, contains much material on the Julia and Mandelbrot sets. The "evaporating" Julia sets mentioned in the text are visible here.

[5] *http://www.unca.edu/ mcmcclur/java/Julia/* allows the user to click on an arbitrary value for c and view the Julia set for $z^2 + c$.

Chapter 3

Elementary Functions

3.1 Polynomials and Rational Functions

As we indicated in Sec. 2.2, *polynomial* functions of z are functions of the form

$$p_n(z) = a_0 + a_1 z + a_2 z^2 + \cdots + a_n z^n, \tag{1}$$

and *rational functions* are ratios of polynomials:

$$R_{m,n}(z) = \frac{a_0 + a_1 z + a_2 z^2 + \cdots + a_m z^m}{b_0 + b_1 z + b_2 z^2 + \cdots + b_n z^n}. \tag{2}$$

The *degree* of the polynomial (1) is n if the complex constant a_n is nonzero. The rational function (2) has *numerator degree m* and *denominator degree n*, if $a_m \neq 0$ and $b_n \neq 0$. The analyticity of these functions is quite transparent: polynomials are entire, and rational functions are analytic everywhere except for the zeros of their denominators. These families are ideal for launching our survey of the elementary analytic functions.

To begin, we take a specific polynomial of degree three as a prototype for our analysis:

$$p_3(z) = 12 + 10z - 4z^2 - 2z^3 \equiv -2z^3 - 4z^2 + 10z + 12. \tag{3}$$

We are going to study how $p_3(z)$ is characterized by its zeros and by the values of its derivatives at a single point.

It is easy to verify (by substitution) that the zeros of $p_3(z)$ are 2, -1, and -3. The reader is no doubt acquainted with the fact that this implies $p_3(z)$ can be expressed in *factored form* as

$$p_3(z) = -2(z - 2)(z + 1)(z + 3). \tag{4}$$

A familiar line of reasoning that leads to the factored form goes roughly as follows. One can always divide a "dividend" polynomial by a "divisor" polynomial to obtain a

---

The identically zero polynomial is assigned a degree of $-\infty$. This has little significance other than to avoid "exceptions" (see Prob. 22).

"quotient" polynomial and a "remainder" polynomial whose degree is less than that of the divisor:

$$\text{dividend} = \text{divisor} \times \text{quotient} + \text{remainder} . \tag{5}$$

If z_1 is any arbitrary complex number, then division of $p_n(z)$ in (1) by the degree-one polynomial $z - z_1$ must result in a remainder of lower degree: in other words, a constant,

$$p_n(z) = (z - z_1) \, p_{n-1}(z) + constant , \tag{6}$$

where the quotient polynomial $p_{n-1}(z)$ has degree $n - 1$. But suppose z_1 happens to be a zero of $p_n(z)$. Then by setting $z = z_1$ in (6) we deduce that the (constant) remainder is zero. Thus (6) displays how $(z - z_1)$ has been factored out from $p_n(z)$; we say $p_n(z)$ has been "deflated." For our prototype $p_3(z)$, factoring out the first zero $z_1 = 2$ results in

$$-2z^3 - 4z^2 + 10z + 12 = (z - 2)(-2z^2 - 8z - 6).$$

Now if z_2 is a zero of the quotient $p_{n-1}(z)$ (and consequently of $p_n(z)$) we can deflate further by factoring out $(z - z_2)$, and so on, until we run out of zeros, leaving us with the factorization

$$p_n(z) = (z - z_1)(z - z_2) \cdots (z - z_k) \, p_{n-k}(z). \tag{7}$$

Since we knew 3 zeros for $p_3(z)$, Eq. (4) displays its "complete" deflation down to factors of degree one (and the degree-zero factor -2).

Example 1

Carry out the deflation of the polynomial $z^3 + (2 - i)z^2 - 2iz$.

Solution.   Obviously $z_1 = 0$ is a zero, and the first deflation is trivial:

$$z^3 + (2 - i)z^2 - 2iz = z \, (z^2 + (2 - i)z - 2i).$$

The quadratic formula (Example 3, Sec. 1.5) provides 2 zeros of the degree-two polynomial:

$$z_2, \, z_3 = \frac{-(2 - i) \pm \sqrt{(2 - i)^2 - 4(1)(-2i)}}{2} = -2, \, i$$

(see Prob. 21, Exercises 1.5, for the square root) and the factored form is

$$z^3 + (2 - i)z^2 - 2z = z(z + 2)(z - i). \quad \blacksquare$$

If we contemplate the deflation of the degree-six polynomial $p_6(z) = z^6 + z^4 - 4iz^2 - 4z + 4 - 3i$, we are confronted with the question as to how we find a zero of

---

The proper name for this division procedure is the *Division Algorithm*.

The reader may be aware of the fact that, in theory, deflation can *always* be continued until $p_{n-k}(z)$ has been reduced to a constant. This will be discussed shortly.

$p_6(z)$. A logically prior question is: *how do we know $p_6(z)$* **has** *any zeros?* Recall that in Sec. 1.1 we observed that $x^2 + 1$ had no zeros among the real numbers, and that we had to extend our number system to the complex numbers **C** to create the zeros $\pm i$. This raises the chilling possibility that there is a horde of polynomials with coefficients in **C**, which don't have zeros in **C**, and that we will have to keep extending the number system.

The hero of the saga is Gauss. In his doctoral dissertation of 1799 he proved the *Fundamental Theorem of Algebra*, demonstrating that no further extensions are necessary:

Theorem 1. Every nonconstant polynomial with complex coefficients has at least one zero in **C**.

We immediately conclude that a polynomial of degree n has, in fact, n zeros, since we can continue to factor out zeros in the deflation process until we reach the final, constant, quotient. Repeated zeros are counted according to their multiplicities; for example, $z^4 + 2z^2 + 1 = (z - i)^2(z + i)^2$ has zeros in two points, each of multiplicity 2, and we count them as 4 zeros.

Gauss constructed four proofs of the Fundamental Theorem in his lifetime. The easiest one is based on complex integration, which we study in Chapter 4; the details of the proof are postponed to Sec. 4.6.

Returning to the deflation process, with the issue of *existence* of zeros for the quotients settled we have a complete factorization of any polynomial

$$p_n(z) = a_n(z - z_1)(z - z_2) \cdots (z - z_n). \tag{8}$$

The display (8) conveys a lot of information. It demonstrates that a polynomial $p_n(z)$ of degree n has n zeros: no less (if we count multiplicities) and no more (why?); and that $p_n(z)$ is completely determined by its zeros, up to a constant multiple (a_n). If two polynomials of degree n have the same (n) zeros, then they are simply constant multiples of each other. Furthermore, the factorization shows that z_0 is a zero of $p(z)$ of multiplicity precisely k if and only if

$$p_n(z) = (z - z_0)^k q(z),$$

where $q(z)$ is a polynomial with $q(z_0) \neq 0$.

Example 2

Show that if the polynomial $p(z)$ has real coefficients, it can be expressed as a product of linear and quadratic factors, each having real coefficients.

Solution.   The nonreal zeros of a polynomial with real coefficients occur in complex conjugate pairs (Prob. 17, Exercises 1.2). So if, say, z_2 is the conjugate of z_1, we can combine $(z - z_1)(z - z_2)$ in (8) to obtain

$$p_n(z) = a_n(z - z_1)(z - z_2) \cdots (z - z_n)$$
$$= a_n \left(z^2 - (z_1 + \overline{z_1})z + z_1\overline{z_1} \right) \cdots (z - z_n)$$
$$= a_n \left(z^2 - 2(\operatorname{Re} z_1)z + |z_1|^2 \right) \cdots (z - z_n).$$

The pair of complex factors of degree one has been replaced by a real factor of degree two. Combining the remaining complex factors similarly (and noting that the real zeros, if any, generate real degree-one factors) we achieve the required display. For example, recalling the fifth roots of unity, the reader can verify that

$$z^5 - 1 = (z - 1)(z^2 - 2\cos(2\pi/5)z + 1)(z^2 - 2\cos(4\pi/5)z + 1). \quad \blacksquare$$

Of course the Fundamental Theorem only tells us that there *are* zeros; it doesn't tell us how to find them. We do know how to find all zeros of polynomials of degree one (trivial!) and two (the quadratic formula). The references at the end of this chapter describe algorithms for finding zeros of polynomials of degrees three and four. Remarkably, Abel and Galois ended a centuries-long search by proving that no similar such algorithm exists for polynomials of degrees five or higher!

Therefore to implement the factored form of a polynomial in general we have to rely on numerical approximations for the zeros. Newton's method (Prob. 10, Exercises 2.7) is an excellent zero-finder for *any* analytic function if a good initial estimate is available. The references describe other algorithms that are specifically designed for polynomials.

As motivation for the next topic, consider the following example.

Example 3

Express the polynomial $p_3(z)$ of Eq. (3) in terms of powers of $(z - 1)$ (instead of powers of z).

Solution. The task is to find coefficients d_0, d_1, d_2, and d_3 so that

$$p_3(z) = 12 + 10z - 4z^2 - 2z^3 \tag{9}$$
$$= d_0 + d_1(z - 1) + d_2(z - 1)^2 + d_3(z - 1)^3. \tag{10}$$

A brute-force solution would be to expand (10) in powers of z, match the coefficients with (9), and solve the resulting 4 equations in 4 unknowns. Somewhat craftier would be to note that $z = (z - 1) + 1$; call this $\zeta + 1$, and expand $p_3(z) = p_3(\zeta + 1)$ in powers of ζ, replacing ζ by $(z - 1)$ at the end:

$$p_3(\zeta + 1) = 12 + 10(\zeta + 1) - 4(\zeta + 1)^2 - 2(\zeta + 1)^3$$
$$= 12 + 10\zeta + 10 - 4\zeta^2 - 8\zeta - 4 - 2\zeta^3 - 6\zeta^2 - 6\zeta - 2$$
$$= 16 - 4\zeta - 10\zeta^2 - 2\zeta^3;$$
$$p_3(z) = 16 - 4(z - 1) - 10(z - 1)^2 - 2(z - 1)^3. \quad \blacksquare \tag{11}$$

Abel, Niels Henrik (1802–1829); Galois, Evariste (1811–1832).

From this example we readily generalize to conclude that any polynomial can be rewritten in powers of $(z - z_0)$, for arbitrary z_0. But there's an easier way of accomplishing this "re-expansion." First note that the coefficients of $p_3(z)$, as displayed in (10), are directly expressible in terms of the values of p_3 and its derivatives at $z = 1$. For example, if we differentiate (10) twice, we eliminate d_0 and d_1; and if we set $z = 1$ in what's left, we eliminate d_3; only d_2 (times 2) survives. So from (10) we get via successive differentiation

$$p_3(1) = d_0,$$
$$p_3'(1) = 1d_1,$$
$$p_3''(1) = 2 \cdot 1d_2,$$
$$p_3'''(1) = 3 \cdot 2 \cdot 1d_3,$$
$$p_3^{(4)}(1) = p_3^{(5)}(1) = \cdots = 0.$$

The pattern is clear. The coefficient of $(z - z_0)^k$, in the expansion of a polynomial $p_n(z)$ in powers of $(z - z_0)$, is given by its kth derivative, evaluated at z_0, and divided by k factorial:

$$p_n(z) = \frac{p_n(z_0)}{0!} + \frac{p_n'(z_0)}{1!}(z - z_0)^1 + \frac{p_n''(z_0)}{2!}(z - z_0)^2 + \cdots + \frac{p_n^{(n)}(z_0)}{n!}(z - z_0)^n$$

$$= \sum_{k=0}^{n} \frac{p_n^{(k)}(z_0)}{k!}(z - z_0)^k. \tag{12}$$

This is known as the **Taylor form** of the polynomial $p_n(z)$, *centered at z_0*. Of course, we can compute the Taylor form of $p_3(z)$ centered at 1 directly from (9):

$$p_3(1) = 12 + 10 \cdot 1^1 - 4 \cdot 1^2 - 2 \cdot 1^3 = \mathbf{16} = d_0,$$
$$p_3'(1) = 10 - 4 \cdot 2 \cdot 1^1 - 2 \cdot 3 \cdot 1^2 = \mathbf{-4} = d_1,$$

etc. (compare (11)). The "standard" form (3) of the polynomial $p_3(z)$ is, then, its Taylor form centered at $z_0 = 0$. Usually one uses the nomenclature **Maclaurin form** for the Taylor form centered at 0; the standard form (9) is thus its Maclaurin form.

If the Taylor form of a polynomial starts with a term of degree one

$$p_n(z) = \frac{p_n'(z_0)}{1!}(z - z_0) + \frac{p_n''(z_0)}{2!}(z - z_0)^2 + \cdots + \frac{p_n^{(n)}(z_0)}{n!}(z - z_0)^n,$$

it clearly signals that z_0 is a zero of the polynomial. For instance, the Taylor form of $p_3(z)$ centered at its zero $z_0 = 2$ is

$$p_3(z) = (0) - \frac{30}{1!}(z - 2) - \frac{32}{2!}(z - 2)^2 - \frac{12}{3!}(z - 2)^3. \tag{13}$$

Remember that $0! = 1$.

Brook Taylor (1685–1731) published his discovery in 1715, but it had already been anticipated by James Gregory (1638–1675) 40 years earlier.

Colin Maclaurin, 1698–1746.

In fact, not only does the absence of the constant term $p_3(2)$ highlight the fact that $p_3(z)$ is zero when $z = 2$, but (13) also provides an alternative demonstration of how $(z - 2)$ can be factored out. When more of the leading terms in the Taylor form for the general polynomial $p_n(z)$ are missing, it looks like

$$p_n(z) = \frac{p_n^{(k)}(z_0)}{k!}(z - z_0)^k + \frac{p_n^{(k+1)}(z_0)}{(k + 1)!}(z - z_0)^{k+1} + \cdots + \frac{p_n^{(n)}(z_0)}{n!}(z - z_0)^n$$

$$= (z - z_0)^k \left[\frac{p_n^{(k)}(z_0)}{k!} + \frac{p_n^{(k+1)}(z_0)}{(k + 1)!}(z - z_0) + \cdots + \frac{p_n^{(n)}(z_0)}{n!}(z - z_0)^{n-k} \right].$$

$$(14)$$

Thus we see that if $p_n(z)$ has a zero of multiplicity precisely k at z_0, then $p_n^{(k)}(z_0) \neq 0$, while $p_n^{(j)}(z_0) = 0$ for $0 \leq j < k$.

Now we direct our attention to rational functions. Since they are ratios of polynomials, all of the previous considerations can be applied to their numerators and denominators separately. Probably the most enlightening display comes from the factored form:

$$R_{m,n}(z) = \frac{a_m(z - z_1)(z - z_2) \cdots (z - z_m)}{b_n(z - \zeta_1)(z - \zeta_2) \cdots \cdots (z - \zeta_n)}, \tag{15}$$

where $\{z_k\}$ designates the zeros of the numerator and $\{\zeta_k\}$ designates those of the denominator. We assume that *common zeros have been cancelled.* The zeros of the numerator are, of course, zeros of $R_{m,n}(z)$; zeros of the denominator are called *poles* of $R_{m,n}(z)$. (Zeros and poles can, of course, be multiple.) Clearly the magnitude of $R_{m,n}(z)$ grows without bound as z approaches a pole.

Example 4

Find all the poles and their multiplicities for

$$R(z) = \frac{(3z + 3i)(z^2 - 4)}{(z - 2)(z^2 + 1)^2}.$$

Solution.   The zeros of the denominator are 2, i, and $-i$, which are *candidates* for the poles. To determine whether they truly are, we factor both numerator and denominator and cancel common terms:

$$\frac{(3z + 3i)(z^2 - 4)}{(z - 2)(z^2 + 1)^2} = \frac{3(z + i)(z - 2)(z + 2)}{(z - 2)(z - i)^2(z + i)^2}$$

$$= \frac{3(z + 2)}{(z - i)^2(z + i)}.$$

Thus we see that the only poles of $R(z)$ are at $z = i$ of multiplicity 2 and $z = -i$ of multiplicity 1.   ∎

A rational function maps neighborhoods of its poles into neighborhoods of infinity, in the sense of Sec. 1.7.

Knowledge of the poles of $R_{m,n}(z)$ enables its expression in terms of *partial fractions*. Some examples of the partial fraction expansion are

$$\frac{3z^2 + 4z - 5}{(z-2)(z+1)(z+3)} = \frac{1}{z-2} + \frac{1}{z+1} + \frac{1}{z+3} \tag{16}$$

and, for multiple poles,

$$\frac{4z+4}{z(z-1)(z-2)^2} = \frac{-1}{z} + \frac{8}{z-1} + \frac{-7}{z-2} + \frac{6}{(z-2)^2} \ . \tag{17}$$

Note that the partial fraction decomposition focuses on the poles of the rational function, conveying little information about its zeros.

Theorem 2. If

$$R_{m,n}(z) = \frac{a_0 + a_1 z + a_2 z^2 + \cdots + a_m z^m}{b_n(z-\zeta_1)^{d_1}(z-\zeta_2)^{d_2}\cdots(z-\zeta_r)^{d_r}} \tag{18}$$

is a rational function whose denominator degree $n = d_1 + d_2 + \cdots + d_r$ exceeds its numerator degree m, then $R_{m,n}(z)$ has a **partial fraction decomposition** of the form

$$R_{m,n}(z) = \frac{A_0^{(1)}}{(z-\zeta_1)^{d_1}} + \frac{A_1^{(1)}}{(z-\zeta_1)^{d_1-1}} + \cdots + \frac{A_{d_1-1}^{(1)}}{(z-\zeta_1)}$$

$$+ \frac{A_0^{(2)}}{(z-\zeta_2)^{d_2}} + \cdots + \frac{A_{d_2-1}^{(2)}}{(z-\zeta_2)}$$

$$+ \cdots + \frac{A_0^{(r)}}{(z-\zeta_r)^{d_r}} + \cdots + \frac{A_{d_r-1}^{(r)}}{(z-\zeta_r)}, \tag{19}$$

where the $\{A_s^{(j)}\}$ are constants. (The ζ_k's are assumed distinct.)

The reader may have had some experience with partial fractions in the evaluation of integrals. Advanced applications will be discussed in later chapters.

Before proceeding with the proof of Theorem 2, assume for the moment that every suitable rational function *has* a partial fraction decomposition and let us consider how to find it.

The brute-force procedure consists in rearranging the proposed form (19) over a common denominator and comparing the resulting numerator, term by term, with the original numerator of $R_{m,n}(z)$. This results in a system of linear equations for the unknown coefficients $\{A_s^{(j)}\}$. A quicker, more sophisticated method for evaluating the $\{A_s^{(j)}\}$ is illustrated in the following example.

In most applications, when the numerator degree of a rational function equals or exceeds its denominator degree, one performs long division to extract the "polynomial" part of the function and applies the partial fraction decomposition to the residual fraction.

Example 5

Reproduce the partial fraction decomposition of the rational function in (17).

Solution. The desired form is

$$R(z) = \frac{4z + 4}{z(z-1)(z-2)^2} = \frac{A_0^{(1)}}{z} + \frac{A_0^{(2)}}{z-1} + \frac{A_0^{(3)}}{(z-2)^2} + \frac{A_1^{(3)}}{z-2}. \tag{20}$$

Note that if the proposed display (20) is multiplied by z, canceling the z in the first denominator, and then the result is evaluated at $z = 0$, only the number $A_0^{(1)}$ would survive; in other words, $A_0^{(1)} = \lim_{z \to 0} z R(z)$. But this limit is readily evaluated from the original formula for $R(z)$:

$$A_0^{(1)} = \lim_{z \to 0} z R(z) = \frac{4 \cdot 0 + 4}{(0-1)(0-2)^2} = -1.$$

Similarly, we read off

$$A_0^{(2)} = \lim_{z \to 1} (z-1) R(z) = \frac{4 \cdot 1 + 4}{1(1-2)^2} = 8$$

$$A_0^{(3)} = \lim_{z \to 2} (z-2)^2 R(z) = \frac{4 \cdot 2 + 4}{2(2-1)} = 6.$$

To get $A_1^{(3)}$, we multiply (20) by $(z-2)^2$, *then we differentiate to kill off the term $A_0^{(3)}$*:

$$\frac{d}{dz}\left[(z-2)^2 R(z)\right] = \frac{d}{dz}\left[\frac{A_0^{(1)}(z-2)^2}{z} + \frac{A_0^{(2)}(z-2)^2}{z-1} + A_1^{(3)}(z-2) + A_0^{(3)}\right]$$

$$= \frac{d}{dz}\left[\frac{A_0^{(1)}(z-2)^2}{z} + \frac{A_0^{(2)}(z-2)^2}{z-1}\right] + A_1^{(3)}.$$

After the differentiation, everything but $A_1^{(3)}$ will still have at least one factor of $(z-2)$, so

$$A_1^{(3)} = \lim_{z \to 2} \frac{d}{dz}\left[(z-2)^2 R(z)\right] = \lim_{z \to 2} \frac{d}{dz}\left(\frac{4z+4}{z^2-z}\right)$$

$$= \frac{(2^2-2)4 - (4 \cdot 2 + 4)(2 \cdot 2 - 1)}{(2^2-2)^2} = -7. \quad \blacksquare$$

From the method employed in the above example, it can be seen that *if $R_{m,n}(z)$ can be written in the form* (19), then a general expression for the coefficients is

$$\boxed{A_s^{(j)} = \lim_{z \to \zeta_j} \frac{1}{s!} \frac{d^s}{dz^s}[(z-\zeta_j)^{d_j} R_{m,n}(z)].} \tag{21}$$

This formula will enable us to give the

Proof of Theorem 2. Let the constants $A_s^{(j)}$ be defined by formula (21) and consider, at first, the sum of those partial fractions in (19) that are associated with the pole ζ_1 (the so-called *singular part* of $R_{m,n}(z)$ around ζ_1). We claim that the difference

$$R_{m,n}(z) - \sum_{s=0}^{d_1-1} \frac{A_s^{(1)}}{(z-\zeta_1)^{d_1-s}} = R_{m,n}(z) - \frac{1}{(z-\zeta_1)^{d_1}} \sum_{s=0}^{d_1-1} A_s^{(1)}(z-\zeta_1)^s \qquad (22)$$

has *no* pole at ζ_1. Writing

$$T(z) := \sum_{s=0}^{d_1-1} A_s^{(1)}(z-\zeta_1)^s, \qquad (23)$$

$$P(z) := a_0 + a_1 z + a_2 z^2 + \cdots + a_m z^m,$$

$$Q(z) := b_n(z-\zeta_2)^{d_2}(z-\zeta_3)^{d_3} \cdots (z-\zeta_r)^{d_r},$$

the expression (22) becomes

$$\frac{P(z)}{(z-\zeta_1)^{d_1} Q(z)} - \frac{T(z)}{(z-\zeta_1)^{d_1}} = \frac{P(z) - Q(z)T(z)}{(z-\zeta_1)^{d_1} Q(z)}.$$

Thus our claim will be proved if we show that the polynomial $P - QT$ has a zero at ζ_1 of order at least d_1, that is

$$(P - QT)^{(s)}(\zeta_1) = 0, \qquad \text{for } s = 0, 1, \ldots, d_1 - 1. \qquad (24)$$

For this purpose observe that (23) is the Taylor form of the polynomial T centered at ζ_1. Hence

$$T^{(s)}(\zeta_1)/s! = A_s^{(1)}, \qquad \text{for } s = 0, 1, \ldots, d_1 - 1.$$

From the defining formula (21) we also know that

$$A_s^{(1)} = \frac{1}{s!} g^{(s)}(\zeta_1), \quad \text{where} \quad g(z) := (z-\zeta_1)^{d_1} R_{m,n}(z) = \frac{P(z)}{Q(z)}.$$

Thus

$$T^{(s)}(\zeta_1) = g^{(s)}(\zeta_1), \qquad \text{for } s = 0, 1, \ldots, d_1 - 1.$$

But these last equations, when written in the equivalent form $(g-T)^{(s)}(\zeta_1) = 0$, imply (24) since

$$(P - QT) = Q(g - T)$$
$$(P - QT)' = Q(g - T)' + Q'(g - T)$$
$$(P - QT)'' = Q(g - T)'' + 2Q'(g - T)' + Q''(g - T),$$

etc. This verifies the claim that the difference (22) has no pole at ζ_1.

Similarly, we deduce that the difference between $R_{m,n}(z)$ and its singular part around ζ_2 has no pole at ζ_2, and so on. From this it follows that

$$R_{m,n} - \sum_{s=0}^{d_1-1} \frac{A_s^{(1)}}{(z-\zeta_1)^{d_1-s}} - \cdots - \sum_{s=0}^{d_r-1} \frac{A_s^{(r)}}{(z-\zeta_r)^{d_r-s}} \tag{25}$$

is a *rational function with no poles*, i.e., a polynomial. But every term in (25) approaches 0 as $|z|$ becomes unbounded, so this must be the zero polynomial, which yields equality (19). ■

EXERCISES 3.1

1. A polynomial $p(z)$ of degree 4 has zeros at the points -1, $3i$, and $-3i$ of respective multiplicities 2, 1, and 1. If $p(1) = 80$, find $p(z)$.

2. Show that if the polynomial $p(z) = a_n z^n + a_{n-1} z^{n-1} + \cdots + a_0$ is written in factored form as $p(z) = a_n(z-z_1)^{d_1}(z-z_2)^{d_2} \cdots (z-z_r)^{d_r}$, then

    (a) $n = d_1 + d_2 + \cdots + d_r$,

    (b) $a_{n-1} = -a_n(d_1 z_1 + d_2 z_2 + \cdots + d_r z_r)$,

    (c) $a_0 = a_n(-1)^n z_1^{d_1} z_2^{d_2} \cdots z_r^{d_r}$.

3. Write the following polynomials in factored form:

    (a) $z^5 + (2+2i)z^4 + 2iz^3$ (b) $z^4 - 16$

    (c) $1 + z + z^2 + z^3 + z^4 + z^5 + z^6$

4. Show that if $p(z) = z^n + a_{n-1}z^{n-1} + \cdots + a_0$ is a polynomial of degree $n \geq 1$ and $|a_0| > 1$, then $p(z)$ has at least one zero outside the unit circle. [HINT: Notice that the leading coefficient $a_n = 1$ and consider the factored form of p.]

5. Write the following polynomials in the Taylor form, centered at $z = 2$:

    (a) $z^5 + 3z + 4$ (b) z^{10} (c) $(z-1)(z-2)^3$

6. If $p(z) = a_n z^n + a_{n-1} z^{n-1} + \cdots + a_0$ ($a_n \neq 0$), then its *reverse polynomial* $p^*(z)$ is given by
    $$p^*(z) = \overline{a_n} + \overline{a_{n-1}}z + \cdots + \overline{a_0}z^n.$$

    (a) Show that $p^*(z) = z^n \overline{p(1/\overline{z})}$.

    (b) Show that if $p(z)$ has a zero at $z_0(\neq 0)$, then $p^*(z)$ has a zero at $1/\overline{z_0}$.

    (c) Show that for $|z| = 1$, we have $|p(z)| = |p^*(z)|$.

7. Prove that if the polynomial $p(z)$ has a zero of order m at z_0, then $p'(z)$ has a zero of order $m-1$ at z_0.

8. Show that if the polynomials $p(z)$ and $q(z)$ have, respectively, zeros of order m and k at z_0, then the product polynomial $p(z)q(z)$ has a zero of order $m + k$ at z_0.

9. Show that if z_0 is a zero of $p_n(z)$ of order d, then for all z sufficiently near z_0, there are positive constants c_1 and c_2 such that $c_1|z - z_0|^d \leq |p_n(z)| \leq c_2|z - z_0|^d$.

10. Show that if $p_n(z)$ has degree n, then for all z with $|z|$ sufficiently large, there are positive constants c_1 and c_2 such that $c_1|z|^n < |p_n(z)| < c_2|z|^n$.

11. For each of the following rational functions find all its poles and their multiplicities.

(a) $\dfrac{3z^2 + 1}{z^3(z^2 + 2iz + 1)}$

(b) $\dfrac{z^2 + 4}{(z - 2)(z - 3)^2}$

(c) $\left(\dfrac{2z + 3}{z^2 + 4z + 4}\right)^3$

(d) $\dfrac{2z}{z^2 + 3z + 2} + \dfrac{2}{z + 1}$

12. Let $R_{m,n}(z) = P(z)/Q(z)$ and $r_{m,n} = p(z)/q(z)$ be two rational functions each with numerator degree m and denominator degree n. Show that if $R_{m,n}$ and $r_{m,n}$ agree at $m + n + 1$ distinct points, then $R_{m,n} = r_{m,n}$ for all z.

13. Use formula (21) to find the partial fraction decompositions of each of the following rational functions:

(a) $\dfrac{3 + i}{z(z + 1)(z + 2)}$

(b) $\dfrac{2z + i}{z^3 + z}$

(c) $\dfrac{z}{(z^2 + z + 1)^2}$

(d) $\dfrac{5z^4 + 3z^2 + 1}{2z^2 + 3z + 1}$ [HINT: First apply the division algorithm to obtain a rational function of suitable form.]

14. Show that if the rational function $R(z)$ has a pole of order m at z_0, then its derivative $R'(z)$ has a pole of order $m + 1$ at z_0.

15. **(Residue)** Let $R = P/Q$ be a rational function with deg $P <$ deg Q. If ζ is a pole of R, then the coefficient of $1/(z - \zeta)$ in the partial fraction expansion of R is called the *residue* of $R(z)$ at ζ and is denoted by Res(ζ). Using formula (21), compute each of the following residues:

(a) Res(i) for $R(z) = \dfrac{2z + 3}{(z - i)(z^2 + 1)}$

(b) Res(-1) for $R(z) = \dfrac{z^3 + 4z + 9}{(2z + 2)(z - 3)^5}$

(c) Res(0) for $R(z) = \dfrac{2z^2 + 3}{z^2(z^2 + 2z + i)}$

(d) Res($3i$) for $R(z) = \dfrac{z^2 - 9}{(z^2 + 9)^2}$

(e) Res(0) for $R(z) = \dfrac{2z^3 + 3}{z^3(z + 1)}$

16. Show that if $R_{m,n}(z)$ is a rational function with numerator degree m and denominator degree n, then for all $|z|$ sufficiently large, there are positive constants c_1 and c_2 such that $c_1 |z|^{m-n} < |R_{m,n}(z)| < c_2 |z|^{m-n}$.

17. Show that if $p(z) = a_n (z - z_1)^{d_1} (z - z_2)^{d_2} \cdots (z - z_r)^{d_r}$, then the partial fraction expansion of the *logarithmic derivative* p'/p is given by

$$\frac{p'(z)}{p(z)} = \frac{d_1}{z - z_1} + \frac{d_2}{z - z_2} + \cdots + \frac{d_r}{z - z_r}.$$

[HINT: Generalize from the formula $(fgh)' = f'gh + fg'h + fgh'$.]

18. Show that if

$$R(z) = \frac{d_1}{z - z_1} + \frac{d_2}{z - z_2} + \cdots + \frac{d_r}{z - z_r},$$

where each d_i is real and positive and each z_k lies in the upper half-plane $\operatorname{Im} z > 0$, then $R(z)$ has no zeros in the lower half-plane $\operatorname{Im} z < 0$. [HINT: Write $\overline{R(z)} = \frac{d_1 (z - z_1)}{|z - z_1|^2} + \cdots + \frac{d_r (z - z_r)}{|z - z_r|^2}$. Then sketch the vectors $(z - z_k)$ for $\operatorname{Im} z_k > 0$ and $\operatorname{Im} z < 0$. Argue from the sketch that any linear combination of these vectors with real, positive coefficients $(d_k/|z - z_k|^2)$ must have a negative (and hence nonzero) imaginary part. Alternatively, show directly that $\operatorname{Im} R(z) > 0$ for $\operatorname{Im} z < 0$.]

19. Show that if all the zeros of a polynomial $p(z)$ lie in the upper half-plane, then the same is true for the zeros of $p'(z)$. [HINT: See Probs. 17 and 18.]

20. Generalize the geometric argument in Prob. 19 to show that if all the zeros of a polynomial $p(z)$ lie on one side of *any* line, then the same is true for the zeros of $p'(z)$.

21. When a set of points lies entirely on one side of a straight line, let us say that the line "shelters" the point set. If $z_1, z_2, \ldots, z_r$ is any finite set of points in the plane, their *convex hull* is the common "territory" that is sheltered by every straight line that shelters the $\{z_j\}$. Argue that Prob. 20 proves the *Gauss-Lucas Theorem*: if $p(z)$ is any polynomial, all the zeros of $p'(z)$ lie in the convex hull of the zeros of $p(z)$.

22. Show how exceptions to the statement "the degree of the product of two polynomials equals the sum of the degrees of each factor" are finessed by the choice $(-\infty)$ as the degree of the identically zero polynomial.

3.2   The Exponential, Trigonometric, and Hyperbolic Functions

The complex exponential function e^z plays a prominent role in analytic function theory, not only because of its own important properties but because it is used to define the

---

Francois Edouard Anatole Lucas (1842–1891) invented the Tower of Hanoi puzzle.

complex trigonometric and hyperbolic functions. Recall the definition from Chapter 1: If $z = x + iy$,

$$\boxed{e^z = e^x(\cos y + i \sin y).} \qquad (1)$$

As a consequence of Example 2 in Sec. 2.4, we know that e^z is an *entire* function and that

$$\frac{d}{dz} e^z = e^z.$$

The polar components of e^z are readily derived from (1):

$$|e^z| = e^x, \qquad (2)$$

$$\arg e^z = y + 2k\pi \qquad (k = 0, \pm 1, \pm 2, \ldots). \qquad (3)$$

It follows that e^z is never zero. However, e^z does assume every other complex value (see Prob. 4).

Recall that a function f is *one-to-one* on a set S if the equation $f(z_1) = f(z_2)$, where z_1 and z_2 are in S, implies that $z_1 = z_2$. As is shown in calculus, the exponential function is one-to-one on the real axis. However it is *not* one-to-one on the complex plane. In fact, we have the following.

Theorem 3.

   (i) The equation $e^z = 1$ holds if, and only if, $z = 2k\pi i$, where k is an integer.

   (ii) The equation $e^{z_1} = e^{z_2}$ holds if, and only if, $z_1 = z_2 + 2k\pi i$, where k is an integer.

Proof of (i). First suppose that $e^z = 1$, with $z = x + iy$. Then we must have

$$|e^z| = |e^{x+iy}| = e^x = 1,$$

and so $x = 0$. This implies that

$$e^z = e^{iy} = \cos y + i \sin y = 1,$$

or, equivalently,

$$\cos y = 1, \qquad \sin y = 0.$$

These two simultaneous equations are clearly satisfied only when $y = 2k\pi$ for some integer k; i.e., $z = 2k\pi i$.

Conversely, if $z = 2k\pi i$, where k is an integer, then

$$e^z = e^{2k\pi i} = e^0(\cos 2k\pi + i \sin 2k\pi) = 1.$$

Proof of (ii). It follows from the division rule that

$$e^{z_1} = e^{z_2} \quad \text{if, and only if,} \quad e^{z_1 - z_2} = 1.$$

But, by part (i), the last equation holds precisely when $z_1 - z_2 = 2k\pi i$, where k is an integer. ■

One important consequence of Theorem 3 is the fact that e^z is periodic. In general, a function f is said to be *periodic* in a domain D if there exists a nonzero constant λ such that the equation $f(z + \lambda) = f(z)$ holds for every z in D. Any constant λ with this property is called a *period* of f. Since, for all z,

$$e^{z+2\pi i} = e^z,$$

we see that e^z is periodic with complex period $2\pi i$. Consequently, if we divide up the z-plane into the infinite horizontal strips

$$S_n := \{x + iy | -\infty < x < \infty, (2n - 1)\pi < y \le (2n + 1)\pi\} \quad (n = 0, \pm 1, \pm 2, \ldots),$$

as shown in Fig. 3.1, then e^z will behave in the same manner on each strip. Furthermore, from part (ii) of Theorem 3, it follows that e^z is one-to-one on each strip S_n. For these reasons any one of these strips is called a *fundamental region* for e^z.

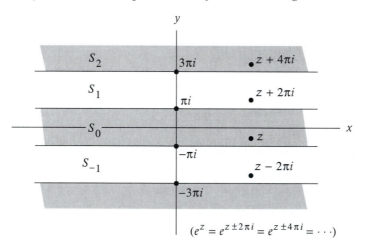

Figure 3.1 Fundamental regions for e^z.

From the identity

$$e^{iy} = \cos y + i \sin y,$$

and its obvious consequence

$$e^{-iy} = \cos y - i \sin y,$$

we deduce, by subtracting and adding these equations, that

$$\sin y = \frac{e^{iy} - e^{-iy}}{2i}, \quad \cos y = \frac{e^{iy} + e^{-iy}}{2}.$$

These real variable formulas suggest the following extensions of the trigonometric functions to complex "angles."

Definition 1. Given any complex number z, we define

$$\sin z := \frac{e^{iz} - e^{-iz}}{2i}, \quad \cos z := \frac{e^{iz} + e^{-iz}}{2}.$$

Since e^{iz} and e^{-iz} are entire functions, so are $\sin z$ and $\cos z$. In fact,

$$\frac{d}{dz} \sin z = \frac{d}{dz} \left(\frac{e^{iz} - e^{-iz}}{2i} \right) = \frac{1}{2i}(ie^{iz} - (-i)e^{-iz}) = \cos z, \tag{4}$$

and, similarly,

$$\frac{d}{dz} \cos z = -\sin z. \tag{5}$$

We recognize that Eqs. (4) and (5) agree with the familiar formulas derived in calculus. Some further identities that remain valid in the complex case are listed below:

$$\sin(z + 2\pi) = \sin z, \quad \cos(z + 2\pi) = \cos z. \tag{6}$$

$$\sin(-z) = -\sin z, \quad \cos(-z) = \cos z. \tag{7}$$

$$\sin^2 z + \cos^2 z = 1. \tag{8}$$

$$\sin(z_1 \pm z_2) = \sin z_1 \cos z_2 \pm \sin z_2 \cos z_1. \tag{9}$$

$$\cos(z_1 \pm z_2) = \cos z_1 \cos z_2 \mp \sin z_1 \sin z_2. \tag{10}$$

$$\sin 2z = 2 \sin z \cos z, \quad \cos 2z = \cos^2 z - \sin^2 z. \tag{11}$$

The proofs of these identities follow directly from the properties of the exponential function and are left to the exercises. Notice that Eqs. (6) imply that $\sin z$ and $\cos z$ are both periodic with period 2π.

Example 1

Prove that $\sin z = 0$ if, and only if, $z = k\pi$, where k is an integer.

Solution. If $z = k\pi$, then clearly $\sin z = 0$. Now suppose, conversely, that $\sin z = 0$. Then we have

$$\frac{e^{iz} - e^{-iz}}{2i} = 0,$$

or, equivalently,

$$e^{iz} = e^{-iz}.$$

By Theorem 3(ii) it follows that

$$iz = -iz + 2k\pi i,$$

which implies that $z = k\pi$ for some integer k. ∎

Thus the only zeros of $\sin z$ are its real zeros. The same is true of the function $\cos z$; i.e.,

$$\cos z = 0 \quad \text{if and only if} \quad z = \frac{\pi}{2} + k\pi,$$

where k is an integer.

The other four complex trigonometric functions are defined by

$$\tan z := \frac{\sin z}{\cos z}, \quad \cot z := \frac{\cos z}{\sin z}, \quad \sec z := \frac{1}{\cos z}, \quad \csc z := \frac{1}{\sin z}.$$

Notice that the functions $\cot z$ and $\csc z$ are analytic except at the zeros of $\sin z$, i.e., the points $z = k\pi$, whereas the functions $\tan z$ and $\sec z$ are analytic except at the points $z = \pi/2 + k\pi$, where k is any integer. Furthermore, the usual rules for differentiation remain valid for these functions:

$$\frac{d}{dz} \tan z = \sec^2 z, \quad \frac{d}{dz} \sec z = \sec z \tan z,$$

$$\frac{d}{dz} \cot z = -\csc^2 z, \quad \frac{d}{dz} \csc z = -\csc z \cot z.$$

The preceding discussion has emphasized the similarity between the real trigonometric functions and their complex extensions. However, this analogy should not be carried too far. For example, the real cosine function is bounded by 1, i.e.,

$$|\cos x| \leq 1, \qquad \text{for all real} \quad x,$$

but

$$|\cos(iy)| = \left| \frac{e^{-y} + e^y}{2} \right| = \cosh y,$$

which is unbounded and, in fact, is never less than 1!

The complex *hyperbolic* functions are defined by a natural extension of their definitions in the real case:

Definition 2. For any complex number z we define

$$\sinh z := \frac{e^z - e^{-z}}{2}, \qquad \cosh z := \frac{e^z + e^{-z}}{2}. \qquad (12)$$

Notice that the functions (12) are entire and satisfy

$$\frac{d}{dz} \sinh z = \cosh z, \qquad \frac{d}{dz} \cosh z = \sinh z. \qquad (13)$$

One nice feature of the complex variable perspective is that it reveals the intimate connection between hyperbolic functions and their trigonometric analogues. The reader can readily verify that

$$\sin iz = i \sinh z, \ \sinh iz = i \sin z, \ \cos iz = \cosh z, \ \cosh iz = \cos z, \qquad (14)$$

and using these formulas one can transform the trigonometric identities (6)–(11) into their hyperbolic versions. For example, replacing z by iz in the identity (8) yields the familiar hyperbolic identity $\cosh^2 z - \sinh^2 z = 1$.

The four remaining complex hyperbolic functions are given by

$$\tanh z := \frac{\sinh z}{\cosh z}, \quad \coth z := \frac{\cosh z}{\sinh z}, \quad \operatorname{sech} z := \frac{1}{\cosh z}, \quad \operatorname{csch} z := \frac{1}{\sinh z}.$$

EXERCISES 3.2

1. Show that $e^z = (1+i)/\sqrt{2}$ if and only if $z = (\pi/4 + 2k\pi)i$, $k = 0, \pm 1, \pm 2, \ldots$.

2. Let $f(z) = e^z - (1 + z + z^2/2 + z^3/6)$. Show that $f^{(3)}(0) = 0$.

3. Find the sum $\sum_{k=0}^{100} e^{kz}$.

4. Let $\omega \, (\neq 0)$ have the polar representation $\omega = re^{i\theta}$. Show that $\exp(\log r + i\theta) = \omega$ (base e logarithm).

5. Write each of the following numbers in the form $a + bi$.

   (a) $\exp(2 + i\pi/4)$

   (b) $\dfrac{\exp(1 + i3\pi)}{\exp(-1 + i\pi/2)}$

   (c) $\sin(2i)$

   (d) $\cos(1 - i)$

   (e) $\sinh(1 + \pi i)$

   (f) $\cosh(i\pi/2)$

6. Establish the trigonometric identities (8) and (9).

7. Show that the formula $e^{iz} = \cos z + i \sin z$ holds for all complex numbers z.

8. Verify the differentiation formulas (13).

9. Find dw/dz for each of the following.

   (a) $w = \exp(\pi z^2)$

   (b) $w = \cos(2z) + i \sin\left(\dfrac{1}{z}\right)$

   (c) $w = \exp[\sin(2z)]$

   (d) $w = \tan^3 z$

   (e) $w = [\sinh z + 1]^2$

   (f) $w = \tanh z$

10. Explain why the function $f(z) = \sin(z^2) + e^{-z} + iz$ is entire.

11. Explain why the function $\operatorname{Re}\left(\dfrac{\cos z}{e^z}\right)$ is harmonic in the whole plane.

12. Establish the following hyperbolic identities by using the relations (14) and the corresponding trigonometric identities.

    (a) $\cosh^2 z - \sinh^2 z = 1$

    (b) $\sinh(z_1 + z_2) = \sinh z_1 \cosh z_2 + \cosh z_1 \sinh z_2$

(c) $\cosh(z_1 + z_2) = \cosh z_1 \cosh z_2 + \sinh z_1 \sinh z_2$

13. Show the following.

(a) $\sin(x + iy) = \sin x \cosh y + i \cos x \sinh y$

(b) $\cos(x + iy) = \cos x \cosh y - i \sin x \sinh y$

14. Prove the following.

(a) e^{iz} is periodic with period 2π.

(b) $\tan z$ is periodic with period π.

(c) $\sinh z$ and $\cosh z$ are both periodic with period $2\pi i$.

(d) $\tanh z$ is periodic with period πi.

15. Prove that $\cos z = 0$ if and only if $z = \pi/2 + k\pi$, where k is an integer.

16. Verify the identity

$$\sin z_2 - \sin z_1 = 2 \cos \left(\frac{z_2 + z_1}{2} \right) \sin \left(\frac{z_2 - z_1}{2} \right),$$

and use it to show that $\sin z_1 = \sin z_2$ if and only if $z_2 = z_1 + 2k\pi$ or $z_2 = -z_1 + (2k + 1)\pi$, where k is an integer.

17. Find all numbers z (if any) such that

(a) $e^{4z} = 1$

(b) $e^{iz} = 3$

(c) $\cos z = i \sin z$

18. Prove the following.

(a) $\lim\limits_{z \to 0} \dfrac{\sin z}{z} = 1$ (b) $\lim\limits_{z \to 0} \dfrac{\cos z - 1}{z} = 0$

[HINT: Use the fact that $f'(0) = \lim\limits_{z \to 0} [f(z) - f(0)]/z$.]

19. Prove that the function e^z is one-to-one on any open disk of radius π.

20. Show that the function $w = e^z$ maps the shaded rectangle in Fig. 3.2(a) one-to-one onto the semi-annulus in Fig. 3.2(b).

21.   (a) Show that the mapping $w = \sin z$ is one-to-one in the semi-infinite strip

$$S_1 = \{x + iy \mid -\pi < x < \pi, \ y > 0\}$$

and find the image of this strip. [HINT: See Prob. 16]

(b) For $w = \sin z$, what is the image of the smaller semi-infinite strip

$$S_2 = \{x + iy \mid -\pi/2 < x < \pi/2, \ y > 0\}?$$

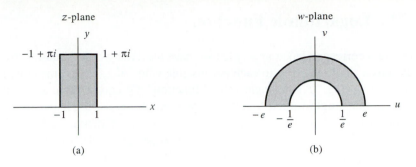

Figure 3.2 Mapping of a rectangle under $w = e^z$.

22. Prove that for any m distinct complex numbers $\lambda_1, \lambda_2, \ldots, \lambda_m$ ($\lambda_i \neq \lambda_j$ for $i \neq j$), the functions $e^{\lambda_1 z}, e^{\lambda_2 z}, \ldots, e^{\lambda_m z}$ are *linearly independent* on **C**. In other words, show that if $c_1 e^{\lambda_1 z} + c_2 e^{\lambda_2 z} + \cdots + c_m e^{\lambda_m z} = 0$ for all z, then $c_1 = c_2 = \cdots = c_m = 0$. [HINT: Proceed by induction on m. In the inductive step, divide by one of the exponentials and then take the derivative.]

23. Below is an outline of an alternative proof that $\sin^2 z + \cos^2 z = 1$ for all z. Justify each step in the proof.

    (a) The function $f(z) = \sin^2 z + \cos^2 z$ is entire.

    (b) $f'(z) = 0$ for all z.

    (c) $f(z)$ is a constant function.

    (d) $f(0) = 1$.

    (e) $f(z) = 1$ for all z.

24. Using only real arithmetic operations, write a computer program that, when given as input the real and imaginary parts (x, y) of $z = x + iy$, produces as output the real and imaginary parts of **(a)** e^z, **(b)** $\sin z$, and **(c)** $\cosh z$.

25. The behavior of the function $e^{1/z}$ around $z = 0$ is extremely erratic; in Sec. 5.6 this point is classified as an "essential singularity." Find values of z, all located in the tiny disk $|z| \leq 0.001$, where $e^{1/z}$ takes the value **(a)** i, **(b)** -1, **(c)** 6.02×10^{23} *(Avogadro's number)*, **(d)** 1.6×10^{-19} *(the electronic charge in Coulombs)*.

3.3   The Logarithmic Function

In discussing a correspondence $w = f(z)$ we have used the word "function" to mean that f assigns a single value w to each permissible value of z. Sometimes this fact is emphasized by saying "f is a single-valued function." Of course there are equations that do not define single-valued functions; for example, $w = \arg z$ and $w = z^{1/2}$. Indeed, for each nonzero z there are an infinite number of distinct values of $\arg z$ and two distinct values of $z^{1/2}$. In general, if for some values of z there corresponds more than one value of $w = f(z)$, then we say that $w = f(z)$ is a *multiple-valued function*. We commonly obtain multiple-valued functions by taking the inverses of single-valued functions that are not one-to-one. This is, in fact, how we obtain the complex logarithmic function $\log z$.

So we want to define $\log z$ as the inverse of the exponential function; i.e.,

$$w = \log z \quad \text{if} \quad z = e^w. \tag{1}$$

Since e^w is never zero, we presume that $z \neq 0$. To find $\log z$ explicitly, let us write z in polar form as $z = re^{i\theta}$ and write w in standard form as $w = u + iv$. Then the equation $z = e^w$ becomes

$$re^{i\theta} = e^{u+iv} = e^u e^{iv}. \tag{2}$$

Taking magnitudes of both sides of (2) we deduce that $r = e^u$, or that u is the ordinary (real) logarithm of r:

$$u = \mathrm{Log}\, r.$$

(We capitalize Log here to distinguish the natural logarithmic function of real variables.) The equality of the remaining factors in Eq. (2), namely, $e^{i\theta} = e^{iv}$, identifies v as the (multiple-valued) polar angle $\theta = \arg z$:

$$v = \arg z = \theta.$$

Thus $w = \log z$ is also a multiple-valued function. The explicit definition is as follows.

Definition 3. If $z \neq 0$, then we define $\log z$ to be the set of infinitely many values

$$\begin{aligned} \log z : &= \mathrm{Log}\, |z| + i \arg z \\ &= \mathrm{Log}\, |z| + i \, \mathrm{Arg}\, z + i2k\pi \quad (k = 0, \pm 1, \pm 2, \ldots). \end{aligned} \tag{3}$$

All logarithms in this text are taken to the base e and are hereafter abbreviated log or Log. The notations ln and Ln are not used in this book. Leonhard Euler (1707–1783) conceived the idea that the logarithm could be extended to negative and complex numbers.

The multiple-valuedness of $\log z$ simply reflects the fact that the imaginary part of the logarithm is the polar angle θ; the real part is single-valued. As examples, consider

$$\log 3 = \operatorname{Log} 3 + i \arg 3 = (1.908\ldots) + i2k\pi,$$
$$\log(-1) = \operatorname{Log} 1 + i \arg(-1) = i(2k+1)\pi,$$
$$\log(1+i) = \operatorname{Log} |1+i| + i \arg(1+i)$$
$$= \frac{1}{2} \operatorname{Log} 2 + i \left(\frac{\pi}{4} + 2k\pi\right),$$

where $k = 0, \pm 1, \pm 2, \ldots$.

The familiar properties of the real logarithmic function extend to the complex case, but the precise statements of these extensions are complicated by the fact that $\log z$ is multiple-valued. For example, if $z \neq 0$, then we have

$$z = e^{\log z},$$

but

$$\log e^z = z + 2k\pi i \qquad (k = 0, \pm 1, \pm 2, \ldots).$$

Furthermore, using the representation (3) and the equations

$$\arg z_1 z_2 = \arg z_1 + \arg z_2, \tag{4}$$

$$\arg \left(\frac{z_1}{z_2}\right) = \arg z_1 - \arg z_2, \tag{5}$$

one can readily verify that

$$\log z_1 z_2 = \log z_1 + \log z_2, \tag{6}$$

and that

$$\log \left(\frac{z_1}{z_2}\right) = \log z_1 - \log z_2. \tag{7}$$

As with Eqs. (5), we must interpret Eqs. (6) and (7) to mean that if particular values are assigned to any two of their terms, then one can find a value of the third term so that the equation is satisfied. For example, if $z_1 = z_2 = -1$ and we select πi to be the value of $\log z_1$ and $\log z_2$, then Eq. (6) will be satisfied if we use the particular value $2\pi i$ for $\log(z_1 z_2) = \log 1$.

Recall that in Sec. 1.3 we used the notion of a *branch cut* to resolve the ambiguity in the designation of the polar angle $\theta = \arg z$. We took Arg z to be the principal value of $\arg z$, the value in the interval $(-\pi, \pi]$, which jumps by 2π as z crosses the branch cut along the negative real axis. Other branches $\arg_\tau z$ resulted from restricting the values of $\arg z$ to $(\tau, \tau + 2\pi]$ and shifting the 2π-discontinuities to the ray $\theta = \tau$.

Clearly the same artifice will generate single-valued *branches* of $\log z$. The *principal value of the logarithm* Log z is the value inherited from the principal value of the argument:

$$\boxed{\operatorname{Log} z := \operatorname{Log} |z| + i \operatorname{Arg} z.} \tag{8}$$

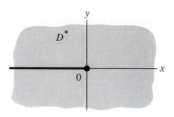

Figure 3.3 Analyticity domain for Log z.

[Notice that we can use the same convention (Log with a capital L) for the principal value as for the usual (real) value, since Arg $z = 0$ if z is positive real.] Log z also inherits, from Arg z, the discontinuities along the branch cut; it jumps by $2\pi i$ as z crosses the negative real axis. However, at all points off the nonpositive real axis, Log z is continuous, and this fact enables us to prove the next theorem.

Theorem 4. The function Log z is analytic in the domain D^* consisting of all points of the complex plane except those lying on the nonpositive real axis (see Fig. 3.3). Furthermore,

$$\frac{d}{dz} \operatorname{Log} z = \frac{1}{z}, \qquad \text{for } z \text{ in } D^*. \tag{9}$$

Proof. Let us set $w = \operatorname{Log} z$. Our goal is to prove that, for z_0 in D^* and $w_0 = \operatorname{Log} z_0$, the limit of the difference quotient

$$\lim_{z \to z_0} \frac{w - w_0}{z - z_0} \tag{10}$$

exists and equals $1/z_0$. We are guided in this endeavor by the knowledge that $z = e^w$ and that the exponential function *is* analytic so that

$$\lim_{w \to w_0} \frac{z - z_0}{w - w_0} = \left. \frac{dz}{dw} \right|_{w = w_0} = e^{w_0} = z_0.$$

Observe that we will have accomplished our goal if we can show that

$$\lim_{z \to z_0} \frac{w - w_0}{z - z_0} = \lim_{w \to w_0} \frac{1}{\frac{z - z_0}{w - w_0}}, \tag{11}$$

It is convenient here to use form (10) instead of the usual form

$$\lim_{\Delta z \to 0} \left[\operatorname{Log}(z_0 + \Delta z) - \operatorname{Log} z_0 \right] / \Delta z.$$

The equivalence of these limits can be seen by putting $\Delta z = z - z_0$.

because the limit on the right exists and equals $1/z_0$. But (11) will follow from the trivial identity

$$\frac{w - w_0}{z - z_0} = \frac{1}{\frac{z - z_0}{w - w_0}}, \tag{12}$$

provided we show that

(a) As z approaches z_0, w must approach w_0, and

(b) For $z \neq z_0$, w will not coincide with w_0 [so that the terms in Eq. (11) are meaningful].

Condition (a) follows from the continuity, in D^*, of $w = \text{Log}\, z$. Condition (b) is even more immediate; if w coincided with w_0, z would have to equal z_0, since $z = e^w$. Thus $w = \text{Log}\, z$ is differentiable at every point in D^* and hence is analytic there. ∎

Corollary 1. The function $\text{Arg}\, z$ is harmonic in the domain D^* of Theorem 4.

Corollary 2. The real function $\text{Log}\, |z|$ is harmonic in the entire plane with the exception of the origin. (See Prob. 8.)

Example 1

Determine the domain of analyticity for the function $f(z) := \text{Log}(3z - i)$. Compute $f'(z)$.

Solution.   Since f is the composition of Log with the function $g(z) = 3z - i$, the chain rule asserts that f will be differentiable at each point z for which $3z - i$ lies in the domain D^* of Theorem 4. Thus points where $3z - i$ is negative or zero are disallowed; a little thought shows that these points lie on the horizontal ray $x \leq 0$, $y = \frac{1}{3}$ (see Fig. 3.4). In this slit plane, then, from Eq. (9):

$$f'(z) = \frac{d}{dz} \text{Log}(3z - i) = \frac{1}{3z - i} \frac{d}{dz}(3z - i) = \frac{3}{3z - i}. \quad ∎$$

Other branches of $\log z$ can be employed if the location of the discontinuities on the negative axis is inconvenient. Clearly the specification

$$\mathcal{L}_\tau(z) := \text{Log}\, |z| + i \arg_\tau z \tag{13}$$

Advanced readers will observe that the same proof could be applied to *any* function $f(z)$ that is analytic and one-to-one around z_0 and for which $f'(z_0) \neq 0$, to conclude that the *inverse function* is analytic and has derivative $1/f'(z_0)$.

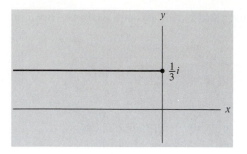

Figure 3.4 Analyticity domain for Log$(3z - i)$.

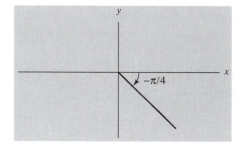

Figure 3.5 Domain of $\mathcal{L}_{-\pi/4}(z)$.

results in a single-valued function whose imaginary part lies in the interval $(\tau, \tau + 2\pi]$. Moreover, the same reasoning used in the proof of Theorem 4 shows that this function is analytic in the complex plane excluding the ray $\theta = \tau$ and the origin, and in this domain,

$$\frac{d}{dz}\mathcal{L}_\tau(z) = \frac{1}{z}.$$

Figure 3.5 depicts the domain of analyticity for $\mathcal{L}_{-\pi/4}(z)$. Of course, no branch of $\log z$ is analytic at the origin, which is called a *branch point* for $\log z$.

Thus far we have used the phrase "branch of $\log z$" in a somewhat informal manner to denote specific values for this multiple-valued function. To make matters more precise, we give the following definition.

Definition 4. $F(z)$ is said to be a **branch** of a multiple-valued function $f(z)$ in a domain D if $F(z)$ is single-valued and continuous in D and has the property that, for each z in D, the value $F(z)$ is one of the values of $f(z)$.

For example, Arg z is a branch of arg z and Log z is a branch of $\log z$ in the plane slit along the negative real axis. In this same domain, the function defined by $e^{(1/2)\operatorname{Log} z}$ gives a branch of $z^{1/2}$ whose values all lie in the right half-plane.

Example 2

Determine a branch of $f(z) = \log(z^3 - 2)$ that is analytic at $z = 0$, and find $f(0)$ and $f'(0)$.

Solution.   The multiple-valued function $f(z)$ is the composition of the logarithm with the analytic function $g(z) = z^3 - 2$. Thus, by the chain rule, it suffices to choose any branch of the logarithm that is analytic at $g(0) = -2$. In particular, $F(z) = \mathcal{L}_{-\pi/4}(g(z))$ solves the problem. For this choice,

$$F(0) = \mathcal{L}_{-\pi/4}(0^3 - 2) = \text{Log } 2 + i\pi,$$

$$F'(0) = \mathcal{L}'_{-\pi/4}(g(0))g'(0) = \frac{g'(0)}{g(0)} = 0. \quad \blacksquare$$

We conclude this section with a word of warning. When complex arithmetic is incorporated into computer packages, all functions must of necessity be programmed as single-valued. The complex logarithm, for instance, is usually programmed as our "principal value," Log z. This invalidates some identitites, such as Eq. (6), since it is not true in general that $\text{Log } z_1 z_2 = \text{Log } z_1 + \text{Log } z_2$. (See Prob. 3.)

EXERCISES 3.3

1. Evaluate each of the following.

   (a) $\log i$ (b) $\log(1 - i)$
   (c) $\text{Log}(-i)$ (d) $\text{Log}(\sqrt{3} + i)$

2. Verify formulas (6) and (7).

3. Show that if $z_1 = i$ and $z_2 = i - 1$, then y

$$\text{Log } z_1 z_2 \neq \text{Log } z_1 + \text{Log } z_2.$$

4. Prove that $\text{Log } e^z = z$ if and only if $-\pi < \text{Im } z \leq \pi$.

5. Solve the following equations.

   (a) $e^z = 2i$ (b) $\text{Log}(z^2 - 1) = \frac{i\pi}{2}$ (c) $e^{2z} + e^z + 1 = 0$

6. Find the error in the following "proof" that $z = -z$: Since $z^2 = (-z)^2$, it follows that $2 \text{Log } z = 2 \text{Log}(-z)$, and hence $\text{Log } z = \text{Log}(-z)$, which implies that $z = e^{\text{Log } z} = e^{\text{Log}(-z)} = -z$.

7. Use the polar form of the Cauchy-Riemann equations (Prob. 6 in Exercises 2.4) to give another proof of Theorem 4.

8. Without directly verifying Laplace's equation, explain why the function $\text{Log } |z|$ is harmonic in every domain that does not contain the origin.

9. Determine the domain of analyticity for $f(z) = \text{Log}(4 + i - z)$. Compute $f'(z)$.

10. Show that the function $\text{Log}(-z) + i\pi$ is a branch of $\log z$ analytic in the domain D_0 consisting of all points in the plane except those on the nonnegative real axis.

11. Determine a branch of $\log(z^2 + 2z + 3)$ that is analytic at $z = -1$, and find its derivative there.

12. Find a branch of $\log(z^2 + 1)$ that is analytic at $z = 0$ and takes the value $2\pi i$ there.

13. Find a branch of $\log(2z - 1)$ that is analytic at all points in the plane except those on the following rays.

  (a) $\{x + iy \mid x \leq \frac{1}{2}, \ y = 0\}$

  (b) $\{x + iy \mid x \geq \frac{1}{2}, \ y = 0\}$

  (c) $\{x + iy \mid x = \frac{1}{2}, \ y \geq 0\}$

14. Prove that there exists no function $F(z)$ analytic in the annulus $D : 1 < |z| < 2$ such that $F'(z) = 1/z$ for all z in D. [HINT: Assume that F exists and show that for z in D, z not a negative real number, $F(z) = \text{Log } z + c$, where c is a constant.]

15. Find a one-to-one analytic mapping of the upper half-plane $\text{Im } z > 0$ onto the infinite horizontal strip

$$\mathcal{H} := \{u + iv \mid -\infty < u < \infty, 0 < v < 1\}.$$

[HINT: Start by considering $w = \text{Log } z$.]

16. Sketch the level curves for the real and imaginary parts of $\text{Log } z$ and verify the orthogonality property discussed in Sec. 2.5.

17. Prove that any branch of $\log z$ (cf. Definition 4) is analytic in its domain and has derivative $1/z$.

18. Prove that if F is a branch of $\log z$ analytic in a domain D, then the totality of branches of $\log z$ analytic in D are the functions $F + 2k\pi i, \ k = 0, \pm 1, \pm 2, \ldots$. [HINT: Use the result of Prob. 17.]

19. How would you construct a branch of $\log z$ that is analytic in the domain D consisting of all points in the plane except those lying on the half-parabola $\{x + iy : \ x \geq 0, y = \sqrt{x}\}$?

20. Using only real arithmetic operations, write a computer program whose input (x, y) is the real and imaginary parts of $z = x + iy$ and whose output is the real and imaginary parts of

  (a) $\text{Log } z$ (b) $\mathcal{L}_{-\pi/4}(z)$ (c) $\mathcal{L}_0(z)$ (d) $\mathcal{L}_{4\pi}(z)$

21. Find a counterexample to the rule $\log(z_1 z_2) = \log z_1 + \log z_2$ for the software system your computer uses.

3.4 Washers, Wedges, and Walls

Now that we have established that the function $\log z$, when suitably restricted, is analytic, the discussion in Sec. 2.5 implies that we have a new pair of *harmonic* functions (the real and imaginary parts) that we can use to solve steady-state temperature problems.

The real part, $\mathrm{Log}\,|z|$ (or $\mathrm{Log}\,r$ in polar-coordinate jargon), is constant on circles centered at the origin (obviously!). So steady-state temperatures, or electrostatic voltages, or any quantity $\phi(x, y)$ that is governed by Laplace's equation $\phi_{xx}+\phi_{yy} = 0$ and is constant on concentric circles must vary logarithmically between the circles. Example 1 can be interpreted as describing a thermally conducting "heat pipe" separating two fluids at different temperatures.

Example 1

Find a function $\phi(x, y)$ that is harmonic in the washer-shaped region between the circles $|z| = 1$ and $|z| = 2$ and takes the values $\phi = 20$ on the inner circle and $\phi = 30$ on the outer circle. See Fig. 3.6.

Solution. As in Example 2, Sec. 2.5, we gain some flexibility by noting that for any value of the constants A and B the function $\phi(x, y) = A\,\mathrm{Log}\,|z| + B$ is harmonic, and we only have to adjust A and B so as to achieve the prescribed boundary conditions. We require that

$$A\,\mathrm{Log}\,1 + B = 20 \ \text{ and } A\,\mathrm{Log}\,2 + B = 30\,,$$

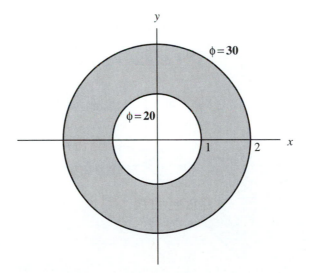

Figure 3.6 Boundary value problem for a heat pipe.

so $B = 20$, $A = 10/\operatorname{Log} 2$, and the solution is

$$\phi(z) = 10\frac{\operatorname{Log}|z|}{\operatorname{Log} 2} + 20 . \quad \blacksquare$$

If the circles were centered at z_0, we would simply use the solution form $A \operatorname{Log}|z - z_0| + B$.

Note that we did not worry about the branch cuts that encumber the $\operatorname{Log} z$ function for this application, since we only used the *real* part, $\operatorname{Log}|z|$, whose values do not depend on the choice of branch. We need to be a little more fastidious when we use the imaginary part $\arg z$, which is constant along rays emanating from the origin. The following example can be interpreted as describing the temperature distribution in a *wedge* whose sides are maintained at fixed temperatures.

Example 2

Find a function $\phi(x, y)$ that is harmonic in the wedge-shaped region depicted in Fig. 3.7 and takes the values $\phi = 20$ on the upper side and $\phi = 30$ on the lower side.

Solution. Strictly speaking, we cannot employ the form $A \operatorname{Arg} z + B$ because the nonpositive axis, which is the branch cut for $\operatorname{Arg} z$, lies in the region of interest. But the remedy is obvious: we simply use another branch of $\arg z$, such as $\arg_0 z$, whose discontinuities lie on the *positive* axis. In other words, we measure the polar angle $\theta = \arg z$ from 0 to 2π, instead of from $-\pi$ to $+\pi$. The boundary conditions are

$$A\frac{3\pi}{4} + B = 20 \text{ and } A\frac{5\pi}{4} + B = 30,$$

so $A = 20/\pi$, $B = 5$, and the solution is

$$\phi(z) = \frac{20}{\pi} \arg_0 z + 5 . \quad \blacksquare$$

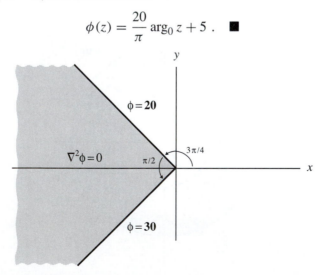

Figure 3.7 Boundary value problem for a wedge.

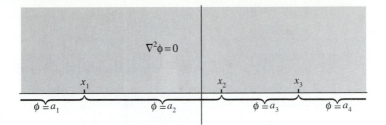

Figure 3.8 Boundary value problem for upper half-plane.

The more sophisticated boundary value problem depicted in Fig. 3.8 can be regarded as an alignment of several 180^0 wedges. If we create the function template

$$\phi(x, y) = A_1 \operatorname{Arg}(z - x_1) + A_2 \operatorname{Arg}(z - x_2) + \cdots + A_n \operatorname{Arg}(z - x_n) + B,$$

which is harmonic in the upper half-plane, we can adjust the constants so as to satisfy the boundary conditions. An example will make the process clear.

Example 3

Find a function $\phi(x, y)$ that is harmonic in the upper half-plane and takes the following values on the x-axis: $\phi(x, 0) = 0$ for $x > 1$, $\phi(x, 0) = 1$ for $-1 < x < 1$, and $\phi(x, 0) = 0$ for $x < -1$.

Solution. We enforce the given boundary conditions on the form

$$\phi(x, 0) = A_1 \operatorname{Arg}(z + 1) + A_2 \operatorname{Arg}(z - 1) + B.$$

For $z = x > 1$, $\operatorname{Arg}(z - 1) = \operatorname{Arg}(z + 1) = 0$, so

$$A_1 \cdot 0 + A_2 \cdot 0 + B = 0.$$

For $z = x$, $-1 < x < 1$, $\operatorname{Arg}(z - 1) = \pi$, $\operatorname{Arg}(z + 1) = 0$, so

$$A_1 \cdot 0 + A_2 \cdot \pi + B = 1.$$

For $z = x < -1$, $\operatorname{Arg}(z - 1) = \operatorname{Arg}(z + 1) = \pi$, so

$$A_1 \cdot \pi + A_2 \cdot \pi + B = 0.$$

Therefore $B = 0$, $A_1 = -1/\pi$, $A_2 = 1/\pi$, and the solution is

$$\phi(z) = -\frac{1}{\pi} \operatorname{Arg}(z + 1) + \frac{1}{\pi} \operatorname{Arg}(z - 1). \quad \blacksquare$$

In closing we note that the simple function $\phi(x, y) = x$ (the real part of $f(z) = z$) is harmonic and is constant along vertical lines. Thus the format $Ax + B$ can be used

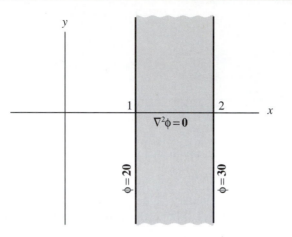

Figure 3.9 Boundary value problem for a slab.

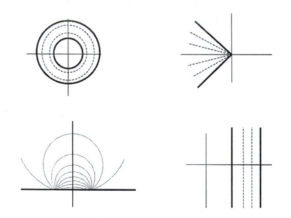

Figure 3.10 Isotherms.

to find steady state temperatures in a conducting wall, or slab, separating two temperature baths; the solution to the problem depicted in Fig. 3.9 is $\phi(x, y) = 10x + 10$. The level curves or isotherms for ϕ for these four geometries are depicted in Fig. 3.10.

Note that the boundary values are discontinuous for the wedge examples; tiny insulators would have to be embedded in the wedge vertices to maintain the temperature drops. Nonetheless, the solutions in the *interiors* are smooth functions of position. This a characteristic of Laplace's equation; its solutions are, in fact, infinitely differentiable except at the boundaries. It is interesting to note how the isotherms near the corners form "fan" patterns to mollify the discontinuities.

In Chapter 7 we will see that by using the techniques of complex analysis, one can often reduce the problem of solving Laplace's equation in domains with complicated geometries to solving it in a washer, wedge, or wall. So the tools developed in this section will be extremely significant.

EXERCISES 3.4

1. Find a solution to the boundary value problem depicted in Fig. 3.11 and evaluate it at (0,0).

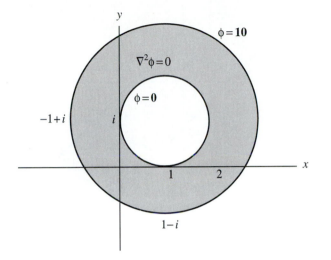

Figure 3.11 Boundary value problem for Prob. 1.

2. Find a solution to the boundary value problem depicted in Fig. 3.12 and evaluate it at (0,0).

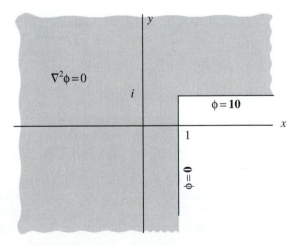

Figure 3.12 Boundary value problem for Prob. 2.

3. Find a solution to the boundary value problem depicted in Fig. 3.13 and evaluate it at (1,1).

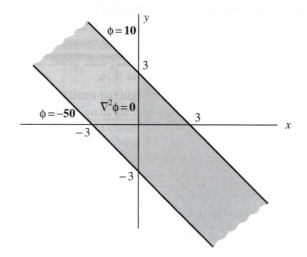

Figure 3.13 Boundary value problem for Prob. 3.

4. Find a solution to the boundary value problem depicted in Fig. 3.14 and evaluate it at (2,3).

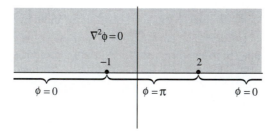

Figure 3.14 Boundary value problem for Prob. 4.

5. Find a general formula for a solution $\phi(x, y)$ to the boundary value problem inside the washer depicted in Fig. 3.15.

6. Find a general formula for a solution $\phi(x, y)$ to the boundary value problem inside the *disk* depicted in Fig. 3.16. (If you're not sure, apply the formula derived in the previous exercise and let the inner radius r_1 go to zero. Now don't you feel foolish?)

7. Prove that the function $(\arg z)(\text{Log } |z|)$ is harmonic. [HINT: It occurs as the imaginary part of what analytic function?]

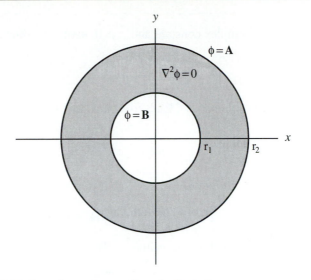

Figure 3.15 Boundary value problem for Prob. 5.

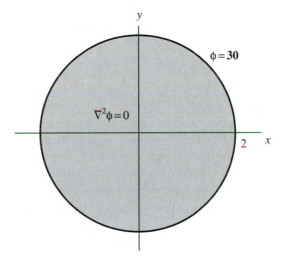

Figure 3.16 Boundary value problem for Prob. 6.

3.5 Complex Powers and Inverse Trigonometric Functions

One important theoretical use of the logarithmic function is to define complex powers of z. The definition is motivated by the identity

$$z^n = (e^{\log z})^n = e^{n \log z},$$

which holds for any integer n.

Definition 5. If α is a complex constant and $z \neq 0$, then we define z^α by

$$z^\alpha := e^{\alpha \log z}.$$

This means that each value of $\log z$ leads to a particular value of z^α.

Example 1

Find all the values of $(-2)^i$.

Solution. Since $\log(-2) = \text{Log } 2 + (\pi + 2k\pi)i$, we have

$$(-2)^i = e^{i \log(-2)} = e^{i \text{ Log } 2} e^{-\pi - 2k\pi} \qquad (k = 0, \pm 1, \pm 2, \dots).$$

Thus $(-2)^i$ has infinitely many different values. ■

Using the representations of Sec. 3.3, we can write

$$z^\alpha = e^{\alpha(\text{Log } |z| + i \text{ Arg } z + 2k\pi i)} = e^{\alpha(\text{Log } |z| + i \text{ Arg } z)} e^{\alpha 2k\pi i}, \tag{1}$$

where $k = 0, \pm 1, \pm 2, \dots$. The values of z^α obtained by taking $k = k_1$ and $k = k_2$ ($\neq k_1$) in Eq. (1) will therefore be the same when

$$e^{\alpha 2k_1 \pi i} = e^{\alpha 2k_2 \pi i}.$$

But by Theorem 3 of Section 3.2 this occurs only if

$$\alpha 2k_1 \pi i = \alpha 2k_2 \pi i + 2m\pi i,$$

where m is an integer. Solving the equation we get $\alpha = m/(k_1 - k_2)$; i.e., *formula (1) yields some identical values of z^α only when α is a real rational number.* Consequently, if α is not a real rational number, we obtain infinitely many different values for z^α, one for each choice of the integer k in Eq. (1). On the other hand, if $\alpha = m/n$, where m and $n > 0$ are integers having no common factor, then one can verify that there are exactly n distinct values of $z^{m/n}$, namely,

$$z^{m/n} = \exp\left(\frac{m}{n} \text{Log } |z|\right) \exp\left(i\frac{m}{n}(\text{Arg } z + 2k\pi)\right) \quad (k = 0, 1, \dots, n - 1). \tag{2}$$

This is entirely consistent with the theory of roots discussed in Sec. 1.5. In summary,

> z^α is single-valued when α is a real integer;
> z^α takes finitely many values when α is a real rational number;
> z^α takes infinitely many values in all other cases.

It is clear from Definitions 4 and 5 that each branch of $\log z$ yields a branch of z^α. For example, using the principal branch of $\log z$ we obtain the *principal branch of z^α*, namely, $e^{\alpha \text{ Log } z}$. Since e^z is entire and $\text{Log } z$ is analytic in the slit domain

$D^* = \mathbf{C}\backslash(-\infty, 0]$ of Theorem 4, the chain rule implies that *the principal branch of* z^α *is also analytic in* D^*. Furthermore, for z in D^*, we have

$$\frac{d}{dz}(e^{\alpha\, \mathrm{Log}\, z}) = e^{\alpha\, \mathrm{Log}\, z}\frac{d}{dz}(\alpha\, \mathrm{Log}\, z) = e^{\alpha\, \mathrm{Log}\, z}\frac{\alpha}{z}.$$

Other branches of z^α can be constructed by using other branches of $\log z$, and since each branch of the latter has derivative $1/z$, the formula

$$\frac{d}{dz}(z^\alpha) = \alpha z^\alpha \frac{1}{z} \tag{3}$$

is valid for each corresponding branch of z^α (provided the same branch is used on both sides of the equation). Observe that if z_0 is any given nonzero point, then by selecting a branch cut for $\log z$ that avoids z_0, we get a branch of z^α that is analytic in a neighborhood of z_0.

   "Branch chasing" with complicated functions is often a tedious task; fortunately, it is seldom necessary for elementary applications. Some of the subtleties are demonstrated in the following example.

Example 2

Define a branch of $(z^2 - 1)^{1/2}$ that is analytic in the exterior of the unit circle, $|z| > 1$.

   Solution.   Our task, restated, is to find a function $w = f(z)$ that is analytic outside the unit circle and satisfies

$$w^2 = z^2 - 1. \tag{4}$$

Note the principal branch of $(z^2 - 1)^{1/2}$, namely,

$$e^{(1/2)\,\mathrm{Log}(z^2-1)},$$

will not work; it has branch cuts wherever $z^2 - 1$ is negative real, and this constitutes the whole y-axis as well as a portion of the x-axis (see Fig. 3.17). But if we experiment with some alternative expressions for w, we are led to consider $z(1 - 1/z^2)^{1/2}$ as a solution to (4). The principal branch of $(1 - 1/z^2)^{1/2}$, i.e.,

$$e^{(1/2)\,\mathrm{Log}(1-1/z^2)},$$

has cuts where $1 - 1/z^2$ is negative real, and this occurs only when $1/z^2$ is real and greater than one—i.e., the cut is the segment $[-1, 1]$, as shown in Fig. 3.18. Thus

$$w = f(z) = ze^{(1/2)\,\mathrm{Log}(1-1/z^2)}$$

satisfies the required condition of analyticity outside the unit circle.   ∎

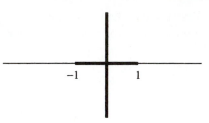

Figure 3.17 Branch cut for $e^{(1/2)\,\mathrm{Log}(z^2-1)}$.

Figure 3.18 Branch cut for $e^{(1/2)\,\mathrm{Log}(1-1/z^2)}$.

Now that we have exponentials expressed in terms of trig functions (Sec. 1.4), trig functions expressed as exponentials (Sec. 3.2), and logs interpreted as inverses of exponentials, the following example should come as no surprise. It demonstrates that the arcsine is, in fact, a logarithm.

Example 3

The inverse sine function $w = \sin^{-1} z$ is defined by the equation $z = \sin w$. Show that $\sin^{-1} z$ is a multiple-valued function given by

$$\sin^{-1} z = -i \log[iz + (1 - z^2)^{1/2}]. \tag{5}$$

Solution.   From the equation

$$z = \sin w = \frac{e^{iw} - e^{-iw}}{2i},$$

we deduce that

$$e^{2iw} - 2ize^{iw} - 1 = 0. \tag{6}$$

We remark that Eq. (3) can be written in the more familiar form

$$\frac{d}{dz}(z^\alpha) = \alpha z^{\alpha-1}$$

with the proviso that the branch of the logarithm used in defining z^α is the same as the branch of the logarithm used in defining $z^{\alpha-1}$.

Using the quadratic formula we can solve Eq. (6) for e^{iw}:

$$e^{iw} = iz + (1 - z^2)^{1/2},$$

where, of course, the square root is two-valued. Formula (5) now follows by taking logarithms.   ■

We can obtain a branch of the multiple-valued function $\sin^{-1} z$ by first choosing a branch of the square root and then selecting a suitable branch of the logarithm. Using the chain rule and formula (5) one can show that any such branch of $\sin^{-1} z$ satisfies

$$\frac{d}{dz}(\sin^{-1} z) = \frac{1}{(1 - z^2)^{1/2}} \qquad (z \neq \pm 1), \tag{7}$$

where the choice of the square root on the right must be the same as that used in the branch of $\sin^{-1} z$.

Example 4

Suppose z is real and lies in the interval $(-1, 1)$. If principal values are used in formula (5), what is the range of $\sin^{-1} z$?

Solution.   With principal values, Eq. (5) is realized as

$$\mathrm{Sin}^{-1} z = -i\, \mathrm{Log}[iz + e^{(1/2)\,\mathrm{Log}(1-z^2)}]. \tag{8}$$

For $|z| = |x| < 1$, clearly $1 - z^2$ lies in the interval $(0, 1]$, and its Log is real; hence the exponential in (8), which represents $(1 - z^2)^{1/2}$, is positive real. The term iz, on the other hand, is pure imaginary. Consequently, the bracketed expression in (8) lies in the right half-plane. As a matter of fact, it also lies on the unit circle, since

$$\left| iz + \left(1 - z^2\right)^{1/2} \right| = \sqrt{x^2 + \left(1 - x^2\right)} = 1$$

(see Fig. 3.19). Taking the Log, then, results in values $i\theta$, where $-\pi/2 < \theta < \pi/2$, and the leading factor $(-i)$ in (8) produces

$$-\frac{\pi}{2} < \mathrm{Sin}^{-1} x < \frac{\pi}{2}$$

(in keeping with the usual interpretation).   ■

For the inverse cosine and inverse tangent functions, calculations similar to those in Example 3 lead to the expressions

$$\cos^{-1} z = -i\, \log[z + (z^2 - 1)^{1/2}], \tag{9}$$

$$\tan^{-1} z = \frac{i}{2} \log \frac{i + z}{i - z} = \frac{i}{2} \log \frac{1 - iz}{1 + iz} \qquad (z \neq \pm i) \tag{10}$$

To ensure that our method has not introduced extraneous solutions, one should verify that every value w given by Eq. (5) satisfies $z = \sin w$.

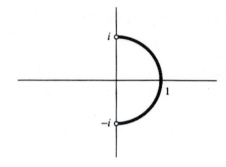

Figure 3.19 Points $ix + (1 - x^2)^{1/2}$ for $|x| < 1$.

and the formulas

$$\frac{d}{dz}(\cos^{-1} z) = \frac{-1}{(1 - z^2)^{1/2}} \qquad (z \neq \pm 1), \tag{11}$$

$$\frac{d}{dz}(\tan^{-1} z) = \frac{1}{1 + z^2} \qquad (z \neq \pm i). \tag{12}$$

Notice that the derivative in Eq. (12) is independent of the branch chosen for $\tan^{-1} z$, whereas the derivative in Eq. (11) depends on the choice of the square root used in the branch of $\cos^{-1} z$.

The same methods can be applied to the inverse hyperbolic functions. The results are

$$\sinh^{-1} z = \log\left[z + (z^2 + 1)^{1/2}\right], \tag{13}$$

$$\cosh^{-1} z = \log\left[z + (z^2 - 1)^{1/2}\right], \tag{14}$$

$$\tanh^{-1} z = \frac{1}{2}\log\frac{1 + z}{1 - z} \qquad (z \neq \pm 1). \tag{15}$$

EXERCISES 3.5

1. Find all the values of the following.

    (a) i^i (b) $(-1)^{2/3}$ (c) $2^{\pi i}$
    (d) $(1 + i)^{1-i}$ (e) $(1 + i)^3$

2. Show from Definition 5 that if $z \neq 0$, then $z^0 = 1$.

3. Find the principal value (i.e., the value given by the principal branch) of each of the following.

    (a) $4^{1/2}$ (b) i^{2i} (c) $(1 + i)^{1+i}$

4. Is 1 raised to any power always equal to 1?

5. Give an example to show that the principal value of $(z_1 z_2)^\alpha$ need not be equal to the product of principal values $z_1^\alpha z_2^\alpha$.

6. Let α and β be complex constants and let $z \neq 0$. Show that the following identities hold when each power function is given by its principal branch.

   (a) $z^{-\alpha} = 1/z^\alpha$ **(b)** $z^\alpha z^\beta = z^{\alpha+\beta}$ **(c)** $\dfrac{z^\alpha}{z^\beta} = z^{\alpha-\beta}$

7. Find the derivative of the principal branch of z^{1+i} at $z = i$.

8. Show that all solutions of the equation $\sin z = 2$ are given by $\pi/2 + 2k\pi \pm i\,\mathrm{Log}(2 + \sqrt{3})$, where $k = 0, \pm 1, \pm 2, \ldots$. [REMARK: This solution set can also be represented as $\pi/2 + 2k\pi - i\,\mathrm{Log}(2 \pm \sqrt{3}, k = 0, \pm 1, \pm 2, \ldots]$

9. Derive formulas (9) and (11) concerning $\cos^{-1} z$.

10. Show that all the solutions of the equation $\cos z = 2i$ are given by $\pi/2 + 2k\pi - i\,\mathrm{Log}(\sqrt{5} + 2)$, $-\pi/2 + 2k\pi + i\,\mathrm{Log}(\sqrt{5} + 2)$, for $k = 0, \pm 1, \pm 2, \ldots$. [REMARK: This solution set can also be represented as $\pi/2 + 2k\pi - i\,\mathrm{Log}(\sqrt{5} + 2)$, $-\pi/2 + 2k\pi - i\,\mathrm{Log}(\sqrt{5} - 2)$, for $k = 0, \pm 1, \pm 2, \ldots$.]

11. Find all solutions of the equation $\sin z = \cos z$.

12. Derive formulas (10) and (12) concerning $\tan^{-1} z$.

13. Derive formulas (13) and (14) for $\sinh^{-1} z$ and $\cosh^{-1} z$.

14. Derive the formula $d(\sinh^{-1} z)/dz = 1/(1+z^2)^{1/2}$ and explain the conditions under which it is valid.

15. Find a branch of each of the following multiple-valued functions that is analytic in the given domain:

   (a) $(z^2 - 1)^{1/2}$ in the unit disk, $|z| < 1$. [HINT: $z^2 - 1 = (z - 1)(z + 1)$.]

   (b) $(4 + z^2)^{1/2}$ in the complex plane slit along the imaginary axis from $-2i$ to $2i$.

   (c) $(z^4 - 1)^{1/2}$ in the exterior of the unit circle, $|z| > 1$.

   (d) $(z^3 - 1)^{1/3}$ in the exterior of the unit circle, $|z| > 1$.

16. According to Definition 5 the multiple-valued function c^z, where c is a nonzero constant, is given by $c^z = e^{z \log c}$. Show that by selecting a particular value of $\log c$ we obtain a branch of c^z that is entire. Find the derivative of such a branch.

17. Derive the identity

$$\sec^{-1} z = -i \log\left[\frac{1}{z} + \left(\frac{1}{z^2} - 1\right)^{1/2}\right].$$

Using principal values determine the range of $\mathrm{Sec}^{-1} x$ when $x > 1$, and when $x < -1$. Compare this with the ranges listed in standard mathematical handbooks:

$$0 \leq \mathrm{Sec}^{-1} x < \pi/2, \text{ for } x \geq 1; \quad -\pi \leq \mathrm{Sec}^{-1} x < -\pi/2, \text{ for } x \leq -1.$$

18. Using only real arithmetic operations, write a computer program whose input (x, y) is the real and imaginary parts of $z = x + iy$ and whose output is the real and imaginary parts of

    (a) the principal branch of z^α

    (b) $\mathrm{Sin}^{-1} z$

    (c) $\mathrm{Sec}^{-1} z$

    (d) the branch of $(z^2 - 1)^{1/2}$ discussed in Example 3.5

19. Determine the inverse of the function

$$w = q(z) := 2e^z + e^{2z}$$

explicitly in terms of the complex logarithm. Use your formula to find *all* values of z for which $q(z) = 3$.

3.6 *Application to Oscillating Systems

Many engineering problems are ultimately based upon the response of a system to a sinusoidal input. Naturally, all the parameters in such a situation are real, and the models can be analyzed using the techniques of real variables. However, the utilization of complex variables can greatly simplify the computations and lend some insight into the roles played by the various parameters. In this section we shall illustrate the technique for the analysis of a simple RLC (resistor-inductor-capacitor) circuit.

    The electric circuit is shown in Fig. 3.20. We suppose that the power supply is driving the system with a sinusoidal voltage V_s oscillating at a frequency of $\nu = \omega/(2\pi)$ cycles per second. To be precise, let us say that at time t

$$V_s = \cos \omega t. \tag{1}$$

Our goal is to find the current I_s drawn out of the power supply.

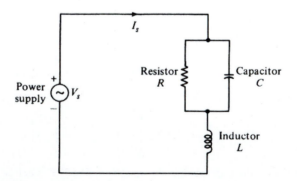

Figure 3.20 *RLC* circuit.

Across each element of the circuit there is a voltage drop V which is related to the current I flowing through the element. The relationships between these two quantities are as follows: for the resistor we have *Ohm's law*

$$V_r = I_r R, \tag{2}$$

where R is a constant known as the *resistance*; for the capacitor

$$C \frac{dV_c}{dt} = I_c, \tag{3}$$

where C is the *capacitance*; and for the inductor

$$V_l = L \frac{dI_l}{dt}, \tag{4}$$

where L is the *inductance*.

One can incorporate formulas (1)–(4), together with Kirchhoff's laws which express conservation of charge and energy, to produce a system of differential equations determining all the currents and voltages. The solution of this system is, however, a laborious process, and a simpler technique is desirable. The complex exponential function will provide this simplification. Before we demonstrate the utilization of complex variables in this problem, we shall make some observations based upon physical considerations.

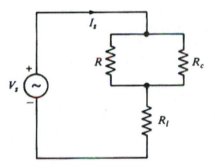

Figure 3.21 Resistor circuit.

First, notice that if the capacitor and inductor were replaced by resistors, as we illustrate in Fig. 3.21, the solution would be an elementary exercise in high school physics. The pair of resistances R and R_c are wired in parallel, so they can be replaced by an equivalent resistance $R_{||}$ given by the familiar law

$$\frac{1}{R_{||}} = \frac{1}{R} + \frac{1}{R_c},$$

George Simon Ohm, a professor of mathematics, published this law in 1827.
Gustav Robert Kirchhoff (1824–1887.)

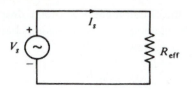

Figure 3.22 Equivalent resistor circuit.

or

$$R_{||} = \frac{R R_c}{R + R_c}.$$

(5)

This resistance then appears in series with R_l, yielding an effective total resistance R_{eff} given by

$$R_{eff} = R_{||} + R_l,$$

(6)

and (from the point of view of the power supply) the circuit is equivalently represented by Fig. 3.22. Equations (1) and (2) yield the current output I_s:

$$I_s = \frac{\cos \omega t}{R_{eff}} = \frac{\cos \omega t}{\frac{R R_c}{R+R_c} + R_l}.$$

Since this model is so easy to solve, it would clearly be advantageous to replace the capacitor and inductor by equivalent resistors.

The second point we wish to make involves the nature of the solution I_s for the original circuit. If the power supply is turned on at time $t = 0$, there will be a fairly complicated initial current response; this so-called "transient," however, eventually dies out, and the system enters a steady state in which all the currents and voltages oscillate sinuosoidally at the same frequency as the driving voltage. This "in-synch" behavior is common to all damped linear systems, and it will be familiar to anyone who has ridden a bicycle over railroad ties or mastered the art of dribbling a basketball. For many applications it is this steady-state response, which is independent of the initial state of the system, that is of interest.

The complex variable technique that we shall describe allows us to replace capacitors and inductors by resistors, and is ideally suited for finding the steady state response.

The technique is based upon exploiting Euler's identity $e^{i\omega t} = \cos \omega t + i \sin \omega t$, which enables us to express the power supply voltage $\cos \omega t$ in terms of the exponential,

$$\cos \omega t = \operatorname{Re} e^{i\omega t}.$$

(7)

In fact, the more general sinusoidal forms $a \cos \omega t + b \sin \omega t$ and $A \cos(\omega t + \gamma)$ can also be expressed in an exponential form $\operatorname{Re} \alpha e^{i\omega t}$:

$$a \cos \omega t + b \sin \omega t = \operatorname{Re}(ae^{i\omega t} - ibe^{i\omega t}) = \operatorname{Re} \alpha e^{i\omega t},$$

(8)

The widespread adoption of this technique in electrical engineering is traditionally attributed to Charles Proteus Steinmetz (1865–1923).

where $\alpha = a - ib$; or

$$A \cos(\omega t + \gamma) = \text{Re}[\alpha e^{i\omega t}], \tag{9}$$

where $\alpha = A e^{i\gamma}$. The complex factor $\alpha = a - ib = A e^{i\gamma}$ is known as a *phasor* in this context. (See Problem 2 for a generalization.)

The advantage of the exponential representation for these functions lies in its compactness, and in the simplicity of the expression for the derivative; differentiating $e^{i\omega t}$ (with respect to t) merely amounts to multiplying by $i\omega$. Therefore, the derivative of the general sinusoid in Eq. (8) can be computed by

$$\frac{d}{dt}(a \cos \omega t + b \sin \omega t) = \frac{d}{dt} \text{Re}\, \alpha e^{i\omega t} = \text{Re}\, \frac{d}{dt}(\alpha e^{i\omega t})$$

$$= \text{Re}\, i\omega \alpha e^{i\omega t}$$

$$(= -a\omega \sin \omega t + b\omega \cos \omega t).$$

To exploit these advantages, we must make one more observation about the circuit in Fig. 3.20. Each of the elements is a *linear* device in the sense that the superposition principle holds; that is, if an element responds to the excitation voltages $V_1(t)$ and $V_2(t)$ by producing the currents $I_1(t)$ and $I_2(t)$, respectively, then the response to the voltage $V_1(t) + \beta V_2(t)$ will be the current $I_1(t) + \beta I_2(t)$, for any constant β. This is a mathematical consequence of the linearity of Eqs. (2)–(4); it will still hold true, in the mathematical sense, even if the functions take complex values. Furthermore, since we know that the circuit responds to real voltages $V_1(t)$ and $V_2(t)$ with real currents, it follows that both $I_1(t)$ and $I_2(t)$ are real, so $I_1(t) = \text{Re}[I_1(t) + iI_2(t)]$. In short, *if the mathematical response to the complex voltage $V(t)$ is $I(t)$, then the response to the (real) voltage $\text{Re}\, V(t)$ will be $\text{Re}\, I(t)$.*

With these tools at hand, let us return to the solution of the problem depicted in Fig. 3.20. The supply voltage, given in Eq. (1), can be represented by

$$V_s(t) = \text{Re}\, e^{i\omega t}. \tag{10}$$

Our strategy will be to find the steady-state response of the circuit to the complex voltage $e^{i\omega t}$, and to take the real part of the answer as our solution.

From the earlier observations about the nature of the steady-state response and Eqs. (8, 9), we are led to propose that the sinusoidal current or voltage in any part of the circuit can be written as a (possibly complex) constant times $e^{i\omega t}$; furthermore, differentiation of such a function is equivalent to multiplication by the factor $i\omega$. Thus, in the situation at hand, Eq. (3) for the behavior of the capacitor becomes

$$i\omega C V_c = I_c, \tag{11}$$

and Eq. (4) for the inductor becomes

$$V_l = i\omega L I_l. \tag{12}$$

But the voltage-current relationships (11) and (12) now have the same form as Eq. (2) for a resistor; in other words, operated at the frequency $\nu = \omega/2\pi$, a capacitor behaves mathematically like a resistor with resistance

$$R_c = \frac{1}{i\omega C},$$ (13)

and an inductor behaves like a resistor with resistance

$$R_l = i\omega L.$$ (14)

These pure imaginary parameters are artifacts of our postulating a complex power-supply voltage. Engineers have adopted the term "impedance" to describe complex resistances.

Having replaced, formally, the capacitor and inductor by resistors, we are back in the situation of the simple circuit of Fig. 3.21. The effective impedance of the series-parallel arrangement is displayed in Eqs. (5) and (6), and substitution of the relations (13) and (14) yields the expression

$$R_{eff} = \frac{\frac{R}{i\omega C}}{R + \frac{1}{i\omega C}} + i\omega L.$$

The current output of the power supply is therefore

$$I_s = \operatorname{Re} \frac{e^{i\omega t}}{R_{eff}},$$

which after some manipulation can be written as

$$I_s = \frac{R \cos \omega t - [R^2 \omega C (1 - \omega^2 LC) - \omega L] \sin \omega t}{R^2 (1 - \omega^2 LC)^2 + \omega^2 L^2}.$$ (15)

Again complex variables can be used to provide a more meaningful interpretation of the answer. If we define ϕ_0 and R_0 by

$$\phi_0 := \operatorname{Arg} R_{eff}, \qquad R_0 := |R_{eff}|,$$

then I_s can be expressed as

$$I_s = \operatorname{Re} \frac{e^{i\omega t}}{R_0 e^{i\phi_0}} = \operatorname{Re} \frac{1}{R_0} e^{i(\omega t - \phi_0)} = \frac{1}{R_0} \cos(\omega t - \phi_0).$$ (16)

This displays some easily visualized properties of the output current in relation to the input voltage (1). The current is, as we indicated earlier, a sinusoid with the same frequency, but its amplitude differs from the voltage amplitude by the factor $1/R_0$. Furthermore, the two sinusoids are out of phase, with ϕ_0 measuring the phase difference in radians. The numbers R_0 and ϕ_0 can be computed in terms of the circuit parameters and the frequency, and the circuit problem is solved.

If the power supply voltage had been a more complicated sinusoid, like $a \cos \omega t + b \sin \omega t$ or $A \cos(\omega t - \gamma)$, we would express it as an exponential $\operatorname{Re} \alpha e^{i\omega t}$ using the prescription (8) or (9). The complex current response would simply then be multiplied by the constant ("phasor") α, and

$$I_s = \text{Re } \frac{\alpha e^{i\omega t}}{R_{eff}}.$$

In conclusion we reiterate the two advantages of using complex notation in analyzing linear sinusoidal systems. First, the representation of general sinusoids (8) is compact and leads naturally to a reinterpretation of the sinusoid in terms of amplitudes and phases, as is evidenced by comparing Eqs. (15) and (16). And second, the process of differentiation is replaced by simple multiplication. It is important to keep in mind, however, that only the real part of the solution corresponds to physical reality and that the condition of linearity is of the utmost importance.

EXERCISES 3.6

1. What is the steady-state current response of the circuit in Fig. 3.20 if the power supply voltage is $V_s = \sin \omega t$?

2. Find a formula for the complex constant ("phasor") α so that the general sinusoid at frequency $\nu = \omega/(2\pi)$,

$$a_1 \cos(\omega t + \gamma_1) + a_2 \cos(\omega t + \gamma_2) + \cdots + a_m \cos(\omega t + \gamma_m)$$
$$+ b_1 \sin(\omega t + \delta_1) + b_2 \sin(\omega t + \delta_2) + \cdots b_n \sin(\omega t + \delta_n)$$

is expressed as $\text{Re } \alpha e^{i\omega t}$.

3. Verify the expression in Eq. (15).

4. Using the techniques of this section, find the steady-state current output I_s of the circuits in Fig. 3.23.

5. In the limit of very low frequencies, a capacitor behaves like an open circuit (infinite resistance), while an inductor behaves like a short circuit (zero resistance). Draw and analyze the low-frequency limit of the circuit in Fig. 3.20, and verify expression (15) in this limit.

6. Repeat Prob. 5 for the high-frequency limit. (What are the behaviors of the capacitor and inductor in this case?)

7. The operation of synchronous and induction motors requires that a rotating magnetic flux be established inside the motor; that is, if we regard a cross section plane of the motor to be the complex z-plane, the motor needs a magnetic flux vector which rotates like the complex number $e^{i\omega t}$. Now a solenoid whose coils are wound around an axis directed along the complex number α produces a magnetic flux in the interior of the motor given by $I(t)\alpha$, where $I(t)$ is the *real* current in its windings. See Fig. 3.24.

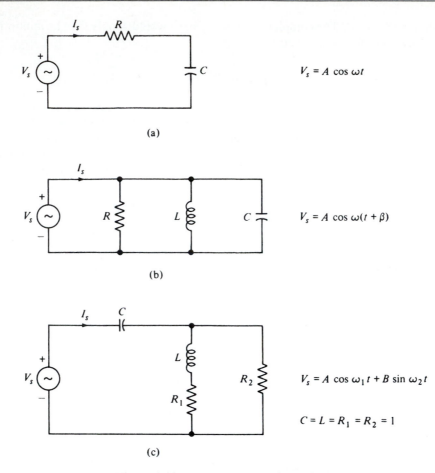

Figure 3.23 Electrical circuits for Prob. 4.

(a) Show that three inductors, wound around axes directed along the complex numbers 1, $e^{i2\pi/3}$, $e^{i4\pi/3}$, and carrying the "3-phase" currents $I_0 \cos \omega t$, $I_0 \cos(\omega t - 2\pi/3)$, $I_0 \cos(\omega t - 4\pi/3)$ respectively, will produce the desired rotating flux (by superposition of their individual fields). [HINT:You will find the algebra easier if you remember

$$\cos(\omega t - 2\pi/3) = \frac{e^{i(\omega t - 2\pi/3)} + e^{-i(\omega t - 2\pi/3)}}{2}. \,]$$

(b) Would two symmetrically placed 2-phase inductors work? Four? Five? Can you figure out the general rule?

8. In radio transmission, a carrier radio wave takes the form of a pure sinusoid, $A \cos \omega t$. The carrier wave can transmit an information signal $f(t)$ by using *modulation*. Two forms of modulation are

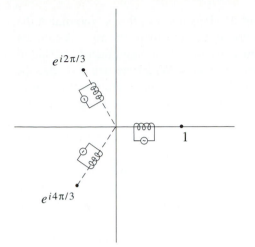

Figure 3.24 Rotating magnetic flux.

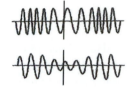

Figure 3.25 Modulation schemes.

Amplitude Modulation: the amplitude A is perturbed by a small multiple (β) of the information signal, so that the transmitted signal equals $[A + \beta f(t)] \cos \omega t$;

(Narrow Band) Frequency Modulation: the angular frequency ω is perturbed by a small multiple (β) of the information signal, so that the transmitted signal equals $A \cos \int [\omega + \beta f(t)] dt$. (Here the constant of integration is not important, and we are identifying the "instantaneous angular velocity" associated with the sinusoid $\cos g(t)$ as $g'(t)$.)

(a) Show that if $f(t)$ is also a sinusoid $\cos \omega_m t$, and if β is sufficiently small, the FM signal approximately equals the real part of $e^{i\omega t}[A + i(A\beta/\omega_m) \sin \omega_m t]$, while the AM signal (exactly) equals the real part of $e^{i\omega t}[A + \beta \cos \omega_m t]$. [HINT: $\sin \alpha \approx \alpha$ and $\cos \alpha \approx 1$ for small $|\alpha|$.]

(b) Which of the diagrams in Fig. 3.25 depicts AM, and which depicts FM, for a sinusoidal information signal?

(c) To get a feeling for AM and FM, use software to plot the transmitted AM and FM signals for $0 \le t \le 1$ with increments $\Delta t = 0.0005$, for $A = 1$, $\omega = 300$, $f(t) = \cos 18t$, and $\beta = 18$. (This value of β is chosen excessively large to highlight the modulations.)

SUMMARY

Polynomials and rational functions of z are analytic except at the poles of the latter. Every polynomial can be completely factored into degree-one polynomial factors with complex coefficients; the factorization displays the multiplicity of each zero. The

Taylor form of the polynomial displays all of the derivatives of the polynomial at the expansion "center", as well as the multiplicity of the zero there (if any). A rational function P/Q, where P and Q are polynomials having no common factors, is said to have poles at the points where the denominator vanishes. Whenever deg $P <$ deg Q, the rational can be written as the sum of constants times simple fractions of the form $1/(z - z_k)^j$, where $\{z_k\}$ are the poles of the rational function and the integer j ranges from 1 to the multiplicity of the pole.

The complex sine and cosine functions are defined in terms of the exponential functions: $\sin z = \left(e^{iz} - e^{-iz}\right)/2i$, $\cos z = \left(e^{iz} + e^{-iz}\right)/2$. When z is real these definitions agree with those given in calculus. Further, the usual differentiation formulas and trigonometric identities remain valid for these functions. Complex hyperbolic functions are also defined in terms of the exponential function, and, again, they retain many of their familiar properties.

The complex logarithmic function $\log z$ is the inverse of the exponential function and is given by $\log z = \text{Log}\, |z| + i \arg z$, where the capitalization signifies the natural logarithmic function of real variables. Since for each $z \neq 0$, $\arg z$ has infinitely many values, the same is true of $\log z$. Thus $\log z$ is an example of a multiple-valued function.

The real and imaginary parts of $\log z$, namely $\text{Log}\, r$ and θ in terms of the polar coordinates, are harmonic functions taking constant values on circles and wedges respectively. They are useful in solving boundary value problems for Laplace's equation.

The concept of analyticity for a multiple-valued function f is discussed in terms of branches of f. A single-valued function F is said to be a branch of f if it is continuous in some domain, at each point of which $F(z)$ coincides with one of the values of $f(z)$.

Branches of $\log z$ can be obtained by restricting $\arg z$ so that it is single-valued and continuous. For example, the function $\text{Log}\, z = \text{Log}\, |z| + i \,\text{Arg}\, z$, which is the principal branch of $\log z$, is analytic in the domain D^* consisting of all points in the plane except those on the nonpositive real axis. The formula $d(\log z)/dz = 1/z$ is valid in the sense that it holds for every branch of $\log z$.

Complex powers of z are defined by means of logarithms. Specifically, $z^\alpha = e^{\alpha \log z}$ for $z \neq 0$. Unless α is an integer, $w = z^\alpha$ is a multiple-valued function whose branches can be obtained by selecting branches of $\log z$. The inverse trigonometric and hyperbolic functions can also be expressed in terms of logarithms, and they too are multiple-valued.

The analysis of sinusoidally oscillating systems can be greatly simplified with the use of complex variables. In particular, the differentiation operation, applied to terms containing the factor $e^{i\omega t}$, reduces to simple multiplication (by $i\omega$).

Suggested Reading

The algorithms for extracting the zeros of degree-three and degree-four polynomials are presented in the following tables:

[1] *Standard Mathematical Tables*. The Chemical Rubber Company, Cleveland (continuing editions).

[2] *Handbook of Mathematical Tables and Formulas*. R.S. Burington. Handbook Publishers, Sandusky, Ohio (continuing editions).

The following text may be helpful for further study of boundary value problems for Laplace's equation (Sec. 3.4):

[3] Snider, A.D. *Partial Differential Equations: Sources and Solutions*, Prentice-Hall, Upper Saddle River, NJ, 1999.

The following texts may be helpful for further study of the concepts used in Sec. 3.6:

Electrical Circuits

[4] Scott, D.E. *An Introduction to Circuit Analysis: A Systems Approach*, McGraw-Hill Book Company, New York, 1987.

[5] Hayt, W.H., Kemmerly, J.E., and Durbin, S.M. *Engineering Circuit Analysis*, 6th ed. McGraw-Hill, New York, 2001.

Sinusoidal Analysis

[6] Lathi, B.P. *Signal Processing and Linear Systems*, Oxford University Press, New York, 2000.

[7] Guillemin, E.A. *Theory of Linear Physical Systems*. John Wiley & Sons, Inc., New York, 1963.

Chapter 4

Complex Integration

In Chapter 2 we saw that the *two*-dimensional nature of the complex plane required us to generalize our notion of a derivative because of the freedom of the variable to approach its limit along any of an infinite number of directions. This two-dimensional aspect will have an effect on the theory of integration as well, necessitating the consideration of integrals along general curves in the plane and not merely segments of the x-axis. Fortunately, such well-known techniques as using antiderivatives to evaluate integrals carry over to the complex case.

When the function under consideration is analytic the theory of integration becomes an instrument of profound significance in studying its behavior. The main result is the theorem of *Cauchy*, which roughly says that the integral of a function around a closed loop is zero if the function is analytic "inside and on" the loop. Using this result we shall derive the *Cauchy integral formula*, which explicitly displays many of the important properties of analytic function.

4.1 Contours

Let us turn to the problem of finding a mathematical explication of our intuitive concept of a *curve* in the xy-plane. Although most of the applications described in this book involve only two simple types of curves—line segments and arcs of circles—it will be necessary for proving theorems to nail down the definitions of more general curves.

It is helpful in this regard to visualize an artist actually tracing the curve γ on graph paper. At any particular instant of time t, a dot is drawn at, say, the point $z = x + iy$; the locus of dots generated over an interval of time $a \leq t \leq b$ constitutes the curve. Clearly we can interpret the artist's actions as generating z as a function of t, and then the curve γ is the range of $z(t)$ as t varies between a and b. In such a case $z(t)$ is called the *parametrization* of γ. Fig. 4.1 shows some examples of the types of curves we need to consider.

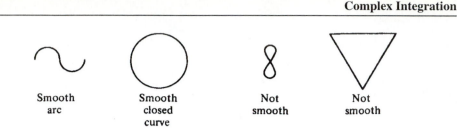

Smooth arc	Smooth closed curve	Not smooth	Not smooth

Figure 4.1 Examples of smooth and nonsmooth curves.

The mathematical description of curves is considerably simplified if the curves have no sharp corners or self intersections, so we begin by confining attention to the first two, "smooth," curves in the figure. The other curves will be dubbed "contours," and eventually we will deal with them by breaking them into smooth pieces. So here are the ground rules for the artist to follow in drawing a smooth curve γ.

First, we do not permit the pen to be lifted from the paper during the sketch; mathematically, we are requiring that $z(t)$ be continuous. Second, we insist that the curves be drawn with an even, steady stroke; specifically, the pen point must move with a well-defined (finite) velocity that, also, must vary continuously. Now the velocity of the point tracing out the trajectory $(x(t), y(t))$ is the vector

$$(dx/dt, dy/dt) = x'(t) + iy'(t).$$

It makes sense to call this vector $z'(t)$. Thus we insist that $z'(t)$ exist[†] and be continuous on $[a, b]$. Furthermore, to avoid the appearance of sharp corners (which would necessitate an abrupt interruption in the sketching process), we require that $z'(t) \neq 0$.

Finally, we stipulate that no point be drawn twice; in other words, $z(t)$ must be one-to-one. We do, however, allow the possibility that the initial and terminal points coincide, as for the second curve in Fig. 4.1.

Putting this all together, we now specify the class of *smooth curves*. These fall into two separate categories; *smooth arcs*, which have distinct endpoints, and *smooth closed curves*, whose endpoints coincide.

Definition 1. A point set γ in the complex plane is said to be a **smooth**[‡] **arc** if it is the range of some continuous complex-valued function $z = z(t), a \leq t \leq b$, that satisfies the following conditions:

   (i) $z(t)$ has a continuous derivative[††] on $[a, b]$,

   (ii) $z'(t)$ never vanishes on $[a, b]$,

[†] Observe, however, that since t is a real variable in this context, the existence of the derivative $z'(t)$ is not nearly so profound as it was in Chapter 2, where the independent variable was complex.

[‡] The term *regular* is sometimes used instead of smooth.

[††] At the endpoint $t = a$, $z'(t)$ denotes the right-hand derivative, while at $t = b$, $z'(t)$ is the left-hand derivative.

> (iii) $z(t)$ is one-to-one on $[a, b]$.
>
> A point set γ is called a **smooth closed curve** if it is the range of some continuous function $z = z(t)$, $a \le t \le b$, satisfying conditions (i) and (ii) and the following:
>
> (iii′) $z(t)$ is one-to-one on the half-open interval $[a, b)$, but
> $z(b) = z(a)$ and $z'(b) = z'(a)$.

The phrase "γ is a smooth curve" means that γ is either a smooth arc or a smooth closed curve.

In elementary calculus it is shown that the vector $(x'(t), y'(t))$, if it exists and is nonzero, can be interpreted geometrically as being tangent to the curve at the point $(x(t), y(t))$. Hence the conditions of Definition 1 imply that a smooth curve possesses a unique tangent at every point and that the tangent direction varies continuously along the curve. Consequently a smooth curve has no corners or cusps; see Fig. 4.1.

To show that a set of points γ in the complex plane is a smooth curve, we have to exhibit a parametrization function $z(t)$ whose range is γ, and is "admissible" in the sense that it meets the criteria of Definition 1. Actually, if a curve *is* smooth, it will have an infinite number of admissible parametrizations. The artist, for instance, can draw it forwards or backwards; if the curve is closed he can begin his sketch anywhere on the curve; and he can draw some parts fast and other parts slow. A given smooth curve γ will have many different admissible parametrizations, but we need produce only *one* admissible parametrization in order to show that a given curve is smooth.

In this book we shall deal only with explicit parametrizations for line segments or circular arcs, for the most part. The following example shows that these are quite elementary.

Example 1

Find an admissible parametrization for each of the following smooth curves:

(a) the horizontal line segment from $z = 1$ to $z = 8$,

(b) the vertical line segment from $z = 2 - 2i$ to $z = 2 + 2i$,

(c) the straight-line segment joining $-2 - 3i$ and $5 + 6i$,

(d) the circle of radius 2 centered at $1 - i$, and

(e) the graph of the function $y = x^3$ for $0 \le x \le 1$.

Solution. (a) The point set is described as $z = x$, $1 \le x \le 8$. So just let the parameter t be x, itself:

$$z(t) = t \quad (1 \le t \le 8).$$

In fact, there is no reason to insist that the parameter be called "t"; the formula

$$z(x) = x \quad (1 \le x \le 8)$$

is quite satisfactory.

(b) The point set is $z = 2 + iy$, $-2 \le y \le 2$, so simply take

$$z(y) = 2 + iy \quad (-2 \le y \le 2).$$

(c) Given any two distinct points z_1 and z_2, every point on the line segment joining z_1 and z_2 is of the form $z_1 + t(z_2 - z_1)$, where $0 \le t \le 1$ (see Prob. 18 in Exercises 1.3). Therefore, the given curve constitutes the range of

$$z(t) = -2 - 3i + t(7 + 9i) \quad (0 \le t \le 1).$$

(d) In Section 1.4 it was shown that any point on the unit circle centered at the origin can be written in the form $e^{i\theta} = \cos\theta + i\sin\theta$ for $0 \le \theta < 2\pi$; therefore, an admissible parametrization for this smooth closed curve is constructed by interpreting θ as the parameter: $z_0(\theta) = e^{i\theta}$, $0 \le \theta \le 2\pi$ (notice that the endpoints are joined properly). To parametrize the given circle (d) we simply shift the center and double the radius:

$$z(\theta) = 1 - i + 2e^{i\theta} \quad (0 \le \theta \le 2\pi).$$

Note that by suitably restricting the limits on θ we could generate a semicircle or any other circular arc.

(e) The parametrization of the graph of any function $y = f(x)$ is also easy; simply let x be the parameter and write $z(x) = x + if(x)$, and set the limits. This is an admissible parametrization as long as $f(x)$ is continuously differentiable. For the graph (e) we have

$$z(x) = x + ix^3 \quad (0 \le x \le 1).$$

The verification of conditions (i), (ii), and (iii) (or (iii′)) is immediate for each of these curves; thus these are admissible parametrizations. ■

Let us carry our analysis of curve sketching a little further. Suppose that the artist is to draw a smooth *arc* like that in Fig. 4.2 and is to abide by our ground rules (in particular it is illegal to retrace points). Then it is intuitively clear that the artist must either start at z_I and work toward z_{II}, or start at z_{II} and terminate at z_I. Either mode produces an ordering of the points along the curve (Fig. 4.3).

Thus we see that there are exactly two such "natural" orderings of the points of a smooth arc γ, and either one can be specified by declaring which endpoint of γ is the initial point. A smooth arc, together with a specific ordering of its points, is called a *directed smooth arc*. The ordering can be indicated by an arrow, as in Fig. 4.3.

The ordering that the artist generates while drawing γ is reflected in the parametrization function $z(t)$ that describes the pen's trajectory; specifically, the point $z(t_1)$

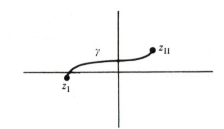

Figure 4.2 Smooth arc.

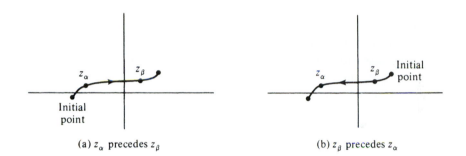

(a) z_α precedes z_β (b) z_β precedes z_α

Figure 4.3 Directed smooth arcs.

will precede the point $z(t_2)$ whenever $t_1 < t_2$. Since there are only two possible (natural) orderings, *any* admissible parametrization must fall into one of two categories, according to the particular ordering it respects. In general, if $z = z(t)$, $a \le t \le b$, is an admissible parametrization consistent with one of the orderings, then $z = z(-t)$, $-b \le t \le -a$, always corresponds to the opposite ordering.

The situation is slightly more complicated if the artist is to draw a smooth *closed* curve. First an initial point must be selected; then the artist must choose one of the two directions in which to trace the curve (see Fig. 4.4). Having made these decisions, the artist has established the ordering of the points of γ. Now, however, there is one anomaly; the initial point both precedes and is preceded by every other point, since it also serves as the terminal point. Ignoring this schizophrenic pest, we shall say that the points of a smooth closed curve have been ordered when (i) a designation of the initial point is made and (ii) one of the two "directions of transit" from this point is selected. A smooth closed curve whose points have been ordered is called a *directed smooth closed curve*.

As in the case of smooth arcs, the parametrization of the trajectory of the artist's pen reflects the ordering generated in sketching a smooth closed curve. If this parametrization is given by $z = z(t)$, $a \le t \le b$, then (i) the initial point must be $z(a)$ and (ii) the point $z(t_1)$ precedes the point $z(t_2)$ whenever $a < t_1 < t_2 < b$. Any other admissible parametrization having the same initial point must reflect either the same or the opposite ordering.

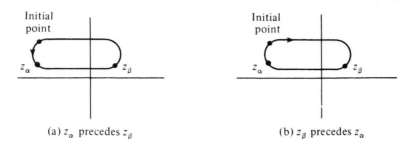

(a) z_α precedes z_β (b) z_β precedes z_α

Figure 4.4 Directed smooth closed curves.

The phrase *directed smooth curve* will be used to mean either a directed smooth arc or a directed smooth closed curve.

Now we are ready to specify the more general kinds of curves that will be used in the theory of integration. They are formed by joining directed smooth curves together end-to-end; this allows self-intersections, cusps, and corners. In addition, it will be convenient to include single isolated points as members of this class. Let us explore the possibilities uncovered by the following definition.

Definition 2. A **contour** Γ is either a single point z_0 or a finite sequence of directed smooth curves $(\gamma_1, \gamma_2, \ldots, \gamma_n)$ such that the terminal point of γ_k coincides with the initial point of γ_{k+1} for each $k = 1, 2, \ldots, n-1$. In this case one can write $\Gamma = \gamma_1 + \gamma_2 + \cdots + \gamma_n$.

Notice that a single directed smooth curve is a contour with $n = 1$.

Speaking loosely, we can say that the contour Γ inherits a direction from its components γ_k: If z_1 and z_2 lie on the same directed smooth curve γ_k, they are ordered by the direction on γ_k, and if z_1 lies on γ_i while z_2 lies on γ_j, we say that z_1 precedes z_2 if $i < j$. This is ambiguous because of the possibility that a point of self-intersection, say z_1, would lie on two different smooth curves, and therefore we must indicate which "occurrence" of z_1 is meant when we say z_1 precedes z_2.

Figure 4.5 illustrates four elementary examples of contours formed by joining directed smooth curves. In Fig. 4.5(d), if z_α is regarded as a point of γ_1 it precedes z_β, but regarded as a point of γ_3, it is preceded by z_β.

Figs. 4.6, 4.7, and 4.8 depict three interesting contours that will be employed when we study examples of contour integration in Chapter 6. Note that in Fig. 4.7 we retrace entire segments in the course of tracing the contour.

A parametrization of a contour is simply a "piecing together" of admissible parametrizations of its smooth-curve components. We will never have need to carry this out explicitly, because in practice we always break up a contour into its smooth-curve components. However the theory is much easier to express in terms of contour parametrizations, so let us spell it out once and for all. One says that $z = z(t)$, $a \leq t \leq b$, is a

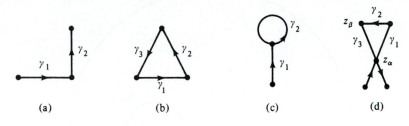

(a)   (b)   (c)   (d)

Figure 4.5 Examples of contours.

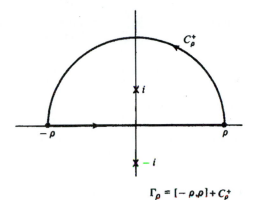

$$\Gamma_\rho = [-\rho,\rho] + C_\rho^+$$

Figure 4.6 Semicircular contour.

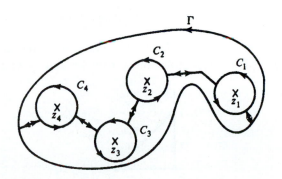

Figure 4.7 Contour with intrusions.

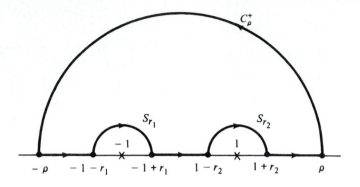

Figure 4.8 Contour with indentations.

parametrization of the contour $\Gamma = (\gamma_1, \gamma_2, \ldots, \gamma_n)$ if there is a subdivision of $[a, b]$ into n subintervals $[\tau_0, \tau_1], [\tau_1, \tau_2], \ldots, [\tau_{n-1}, \tau_n]$, where

$$a = \tau_0 < \tau_1 < \cdots < \tau_{n-1} < \tau_n = b,$$

such that on each subinterval $[\tau_{k-1}, \tau_k]$ the function $z(t)$ is an admissible parametrization of the smooth curve γ_k, consistent with the direction on γ_k. Since the endpoints of consecutive γ_k's are properly connected, $z(t)$ must be continuous on $[a, b]$. However $z'(t)$ may have jump discontinuities at the points τ_k.

The contour parametrization of a point is simply a constant function.

When we have admissible parametrizations of the components γ_k of a contour Γ, we can piece these together to get a contour parametrization for Γ by simply rescaling and shifting the parameter intervals for t. The technique is amply illustrated by the following example. (The general case is discussed in Prob. 6.)

Example 2

Parametrize the contour in Fig. 4.9, for t in the interval $0 \le t \le 1$.

Solution. We have already seen how to parametrize straight lines. The following functions are admissible parametrizations for γ_1, γ_2, and γ_3, consistent with their

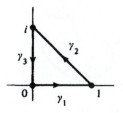

Figure 4.9 Contour for Example 2.

directions:

$$\gamma_1 : \quad z_1(t) = t \qquad\qquad (0 \leq t \leq 1),$$

$$\gamma_2 : \quad z_2(t) = 1 + t(i - 1) \quad (0 \leq t \leq 1),$$

$$\gamma_3 : \quad z_3(t) = i - ti \qquad (0 \leq t \leq 1).$$

Now we rescale so that γ_1 is traced as t varies between 0 and $\frac{1}{3}$, γ_2 is traced for $\frac{1}{3} \leq t \leq \frac{2}{3}$, and γ_3 is traced for $\frac{2}{3} \leq t \leq 1$. This is simply a matter of shifting and stretching the variable t.

For γ_1, observe that the range of the function $z_1(t) = t$, $0 \leq t \leq 1$, is the same as the range of $z_I(t) = 3t$, $0 \leq t \leq \frac{1}{3}$, and that $z_I(t)$ is an admissible parametrization corresponding to the same ordering. The curve γ_2 is the range of $z_2(t) = 1 + t(i - 1)$, $0 \leq t \leq 1$, and this is the same as the range of $z_{II}(t) = 1 + 3(t - \frac{1}{3})(i - 1)$, $\frac{1}{3} \leq t \leq \frac{2}{3}$, again preserving admissibility and ordering. Handling $z_3(t)$ similarly, we find

$$z(t) = \begin{cases} 3t & \left(0 \leq t \leq \tfrac{1}{3}\right), \\[2mm] 1 + 3\left(t - \tfrac{1}{3}\right)(i - 1) & \left(\tfrac{1}{3} \leq t \leq \tfrac{2}{3}\right), \\[2mm] i - 3\left(t - \tfrac{2}{3}\right)i & \left(\tfrac{2}{3} \leq t \leq 1\right). \quad \blacksquare \end{cases}$$

The (undirected) point set underlying a contour is known as a *piecewise smooth curve*. We shall use the symbol Γ ambiguously to refer to both the contour and its underlying curve, allowing the context to provide the proper interpretation.

Much of the terminology of directed smooth curves is readily applied to contours. The initial point of Γ is the initial point of γ_1, and its terminal point is the terminal point of γ_n; therefore Γ can be regarded as a path connecting these points. If the directions on all the components of Γ are reversed and the components are taken in the opposite order, the resulting contour is called the *opposite contour* and is denoted by $-\Gamma$ (see Fig. 4.10). Notice that if $z = z(t)$, $a \leq t \leq b$, is a parametrization of Γ, then $z = z(-t)$, $-b \leq t \leq -a$, parametrizes $-\Gamma$.

Γ is said to be a *closed contour* or a *loop* if its initial and terminal points coincide. A *simple closed contour* is a closed contour with no multiple points other than its

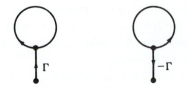

Figure 4.10 Oppositely oriented contours.

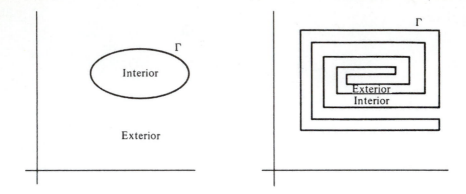

Figure 4.11 Jordan curve theorem.

initial-terminal point; in other words, if $z = z(t)$, $a \leq t \leq b$, is a parametrization of the closed contour, then $z(t)$ is one-to-one on the half-open interval $[a, b)$.

There is an alternative way of specifying the direction along a curve if the curve happens to be a simple closed contour. We employ the venerable *Jordan curve theorem*[†] from topology, which guarantees the intuitively transparent observation that such a curve has an inside and an outside (see Fig. 4.11).

Theorem 1. Any simple closed contour separates the plane into two domains, each having the curve as its boundary. One of these domains, called the *interior*, is bounded; the other, called the *exterior*, is unbounded.

Jordan's theorem actually holds for more general curves, but the proof is quite involved—even for contours.

Now given a simple closed contour Γ we can imagine a child bicycling around the curve and tracing out its points in the order specified by its direction. If the bicycle has training wheels and if it is small enough, then one of the training wheels will always remain in the interior domain of the contour, while the other remains in the exterior (otherwise we would have a path connecting these domains without crossing Γ, in contradiction to the Jordan curve theorem). Consequently the direction along Γ can be completely specified by declaring its initial-terminal point and stating which domain (interior or exterior) lies to the *left* of an observer tracing out the points in order. When the interior domain lies to the left, we say that Γ is *positively* oriented. Otherwise Γ is said to be oriented *negatively*. A positive orientation generalizes the concept of counterclockwise motion; see Fig. 4.12.

The final topic we want to discuss in this section is the length of a contour. We begin by considering a smooth curve γ, with any admissible parametrization $z =$

[†]Camille Jordan (1838–1922).

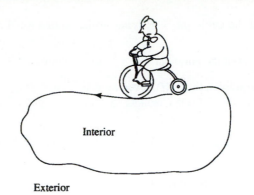

Figure 4.12 Jordan, age four.

$z(t) = x(t) + iy(t)$, $a \le t \le b$. Let $s(t)$ denote the length of the arc of γ traversed in going from the point $z(a)$ to the point $z(t)$. As shown in elementary calculus, we have

$$\frac{ds}{dt} = \sqrt{\left(\frac{dx}{dt}\right)^2 + \left(\frac{dy}{dt}\right)^2}\,,$$

i.e. $ds/dt = |dz/dt|$. Consequently, the length of the smooth curve is given by the important integral formula

$$\ell(\gamma) = \text{length of } \gamma = \int_a^b \frac{ds}{dt}\, dt = \int_a^b \left|\frac{dz}{dt}\right|\, dt. \tag{1}$$

This formula is established rigorously in the references. Sometimes the shorthand $\int_\gamma |dz|$ is used to indicate $\int_a^b |dz/dt|\, dt$; it emphasizes the intuitively evident fact that $\ell(\gamma)$ is a geometric quantity that depends only on the point set γ and is independent of the particular admissible parametrization used in the computation.

The *length of a contour* is simply defined to be the sum of the lengths of its component curves. For example, the length of the contour in Fig. 4.9 is just the sum $\ell(\gamma_1)+\ell(\gamma_2)+\ell(\gamma_3) = 1+\sqrt{2}+1 = 2+\sqrt{2}$. For a contour Γ that consists of two counterclockwise laps around the circle $C : |z - i| = 3$, we have $\ell(\Gamma) = 6\pi + 6\pi = 12\pi$.

EXERCISES 4.1

1. For each of the following smooth curves give an admissible parametrization that is consistent with the indicated direction.

   (a) the line segment from $z = 1 + i$ to $z = -2 - 3i$

   (b) the circle $|z - 2i| = 4$ traversed once in the clockwise direction starting from the point $z = 4 + 2i$

(c) the arc of the circle $|z| = R$ lying in the second quadrant, from $z = Ri$ to $z = -R$

(d) the segment of the parabola $y = x^2$ from the point $(1, 1)$ to the point $(3, 9)$

2. Show why the condition that $z'(t)$ never vanishes is necessary to ensure that smooth curves have no cusps. [HINT: Consider the curve traced by $z(t) = t^2 + it^3$, $-1 \le t \le 1$.]

3. Show that the ellipse $x^2/a^2 + y^2/b^2 = 1$ is a smooth curve by producing an admissible parametrization.

4. Show that the range of the function $z(t) = t^3 + it^6$, $-1 \le t \le 1$, is a smooth curve even though the given parametrization is not admissible.

5. Identify the interior of the simple closed contour Γ in Fig. 4.13. Is Γ positively oriented?

Figure 4.13 Contour for Prob. 5.

6. Let γ be a directed smooth curve. Show that if $z = z(t)$, $a \le t \le b$, is an admissible parametrization of γ consistent with the ordering on γ, then the same is true of

$$z_1(t) = z\left(\frac{b-a}{d-c}t + \frac{ad-bc}{d-c}\right) \quad (c \le t \le d).$$

7. Parametrize the contour consisting of the perimeter of the square with vertices $-1 - i$, $1 - i$, $1 + i$, and $-1 + i$ traversed once in that order. What is the length of this contour?

8. Parametrize the contour Γ indicated in Fig. 4.14. Also give a parametrization for the opposite contour $-\Gamma$.

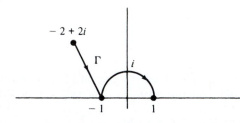

Figure 4.14 Contour for Prob. 8.

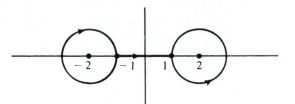

Figure 4.15 Contour for Prob. 9.

9. Parametrize the barbell-shaped contour in Fig. 4.15; it has initial point -1 and terminal point 1.

10. Using an admissible parametrization verify from formula (1) that

  (a) the length of the line segment from z_1 to z_2 is $|z_2 - z_1|$;

  (b) the length of the circle $|z - z_0| = r$ is $2\pi r$.

11. Find the length of the contour Γ parametrized by $z = z(t) = 5e^{3it}, 0 \le t \le \pi$.

12. Does formula (1) remain valid when γ is a *contour* parametrized by $z = z(t)$, $a \le t \le b$?

13. Interpreting t as time and the admissible parametrization $z = z(t), a \le t \le b$, as the position function of a moving particle, give the physical meaning of the following quantities.

  (a) $z'(t)$ **(b)** $|z'(t)|$
  (c) $|z'(t)\,dt|$ **(d)** $\int_a^b |z'(t)|\,dt$

14. Let $z = z_1(t)$ be an admissible parametrization of the smooth curve γ. If $\phi(s), c \le s \le d$, is a strictly increasing function such that (i) $\phi'(s)$ is positive and continuous on $[c, d]$ and (ii) $\phi(c) = a, \phi(d) = b$, then the function $z_2(s) = z_1(\phi(s)), c \le s \le d$, is also an admissible parametrization of γ. Verify that

$$\int_a^b |z_1'(t)|\,dt = \int_c^d |z_2'(s)|\,ds,$$

which demonstrates the invariance of formula (1).

4.2   Contour Integrals

In calculus the definite integral of a real-valued function f over an interval $[a, b]$ is defined as the limit of certain sums $\sum_{k=1}^{n} f(c_k)\Delta x_k$ (called *Riemann sums*). However, the fundamental theorem of calculus lets us evaluate integrals more directly when an antiderivative is known. The aim of the present section is to use this notion of Riemann sums to define the definite integral of a *complex*-valued function f along a *contour* Γ in

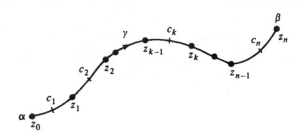

Figure 4.16 Partitioned curve.

the plane. We will accomplish this by first defining the integral along a single directed smooth curve and then defining integrals along a contour in terms of the integrals along its smooth components. When we are finished, however, we will once again obtain simple rules for evaluating integrals in terms of antiderivatives.

Consider then a function f defined over a directed smooth curve γ with initial point α and terminal point β (possibly coinciding with α). As in the previous section, the points on γ are ordered in accordance with the direction.

For any positive integer n, we define a *partition* $\mathcal{P}_n$ of γ to be a finite number of points $z_0, z_1, \ldots, z_n$ on γ such that $z_0 = \alpha$, $z_n = \beta$, and z_{k-1} precedes z_k on γ for $k = 1, 2, \ldots, n$ (see Fig. 4.16). If we compute the arc length along γ between every consecutive pair of points (z_{k-1}, z_k), the largest of these lengths provides a measure of the "fineness" of the subdivision; this maximum length is called the *mesh* of the partition and is denoted by $\mu(\mathcal{P}_n)$. It follows that if a given partition $\mathcal{P}_n$ has a "small" mesh, then n must be large and the successive points of the partition must be close to one another.

Now let $c_1, c_2, \ldots, c_n$ be any points on γ such that c_1 lies on the arc from z_0 to z_1, c_2 lies on the arc from z_1 to z_2, etc. Under these circumstances the sum $S(\mathcal{P}_n)$ defined by

$$S(\mathcal{P}_n) := f(c_1)(z_1 - z_0) + f(c_2)(z_2 - z_1) + \cdots + f(c_n)(z_n - z_{n-1})$$

is called a *Riemann sum* for the function f corresponding to the partition $\mathcal{P}_n$. On writing $z_k - z_{k-1} = \Delta z_k$, this becomes

$$S(\mathcal{P}_n) = \sum_{k=1}^{n} f(c_k)(z_k - z_{k-1}) = \sum_{k=1}^{n} f(c_k)\Delta z_k.$$

Now we can generalize the definition of definite integral given in calculus.

Definition 3. Let f be a complex-valued function defined on the directed smooth curve γ. We say that f **is integrable along** γ if there exists a complex number L that is the limit of *every* sequence of Riemann sums

$S(\mathcal{P}_1), S(\mathcal{P}_2), \ldots, S(\mathcal{P}_n), \ldots$ corresponding to *any* sequence of partitions of γ satisfying $\lim\limits_{n\to\infty} \mu(\mathcal{P}_n) = 0$; i.e.

$$\lim_{n\to\infty} S(\mathcal{P}_n) = L \quad \text{whenever} \quad \lim_{n\to\infty} \mu(\mathcal{P}_n) = 0.$$

The constant L is called the *integral of f along γ*, and we write

$$L = \lim_{n\to\infty} \sum_{k=1}^{n} f(c_k)\Delta z_k = \int_\gamma f(z)\,dz \quad \text{or} \quad L = \int_\gamma f.$$

Because Definition 3 is analogous to the definition of the integral given in calculus, certain familiar properties of the latter integrals carry over to the complex case. For example, if f and g are integrable along γ, then

$$\int_\gamma [f(z) \pm g(z)]\,dz = \int_\gamma f(z)\,dz \pm \int_\gamma g(z)dz, \tag{1}$$

$$\int_\gamma cf(z)\,dz = c \int_\gamma f(z)\,dz \quad (c \text{ any complex constant}), \tag{2}$$

and

$$\int_{-\gamma} f(z)\,dz = -\int_\gamma f(z)\,dz, \tag{3}$$

where $-\gamma$ denotes the curve directed opposite to γ.

As we know from calculus, not all functions f are integrable. However, if we require that f be continuous, then its integral must exist.

Theorem 2. If f is continuous[†] on the directed smooth curve γ, then f is integrable along γ.

For a proof of this theorem see Ref. 2 at the end of the chapter.

While Theorem 2 is of great theoretical importance, it gives us no information on how to compute the integral $\int_\gamma f(z)\,dz$. However, since we are already skilled in evaluating the definite integrals of calculus, it would certainly be advantageous if we could express the complex integral in terms of real integrals.

For this purpose we first consider the special case when γ is the real line segment $[a, b]$ directed from left to right. Notice that if f happened to be a real-valued function

---

[†]The meaning of continuity for a function f having an arbitrary set S as its domain of definition is as follows: f is continuous on the set S if for any point z_0 in S and for every $\epsilon > 0$, there exists a $\delta > 0$ such that $|f(z) - f(z_0)| < \epsilon$ *whenever z belongs to S and* $|z - z_0| < \delta$.

defined on $[a, b]$, the Definition 3 would agree precisely with the definition of the integral $\int_a^b f(t)\,dt$ given in calculus. Hence, even when f is complex-valued, we shall use the symbol

$$\int_a^b f(t)\,dt$$

to denote the integral of f along the directed real line segment. In this case, when $f(t)$ is a complex-valued function continuous on $[a, b]$, we can write $f(t) = u(t) + iv(t)$, where u and v are each real-valued and continuous on $[a, b]$. Then from properties (1) and (2) we have

$$\int_a^b f(t)\,dt = \int_a^b [u(t) + iv(t)]\,dt$$

$$= \int_a^b u(t)\,dt + i \int_a^b v(t)\,dt; \tag{4}$$

this expresses the complex integral in terms of two real integrals.

If $f(t)$ has the antiderivative $F(t) = U(t) + iV(t)$, then $U' = u$, $V' = v$, and Eq. (4) leads immediately to the following generalization of the fundamental theorem of calculus.

Theorem 3. If the complex-valued function f is continuous on $[a, b]$ and $F'(t) = f(t)$ for all t in $[a, b]$, then

$$\int_a^b f(t)\,dt = F(b) - F(a).$$

This result is illustrated in the following example.

Example 1

Compute $\int_0^\pi e^{it}\,dt$.

Solution. Since $F(t) = e^{it}/i$ is an antiderivative of $f(t) = e^{it}$, we have by Theorem 3

$$\int_0^\pi e^{it}\,dt = \frac{e^{it}}{i}\bigg|_0^\pi = \frac{e^{i\pi}}{i} - \frac{e^{i0}}{i} = \frac{-2}{i} = 2i. \quad \blacksquare$$

Now we move on to the general case where γ is any directed smooth curve along which f is continuous. We can obtain a formula for the integral $\int_\gamma f(z)\,dz$ by considering an admissible parametrization $z = z(t)$, $a \le t \le b$, for γ (consistent with its direction). Indeed, if $\mathcal{P}_n = \{z_0, z_1, \ldots, z_n\}$ is a partition of γ, then we can write

$$z_0 = z(t_0), \quad z_1 = z(t_1), \quad \ldots, \quad z_n = z(t_n),$$

where

$$a = t_0 < t_1 < \cdots < t_n = b.$$

Furthermore, since the function $z(t)$ has a continuous derivative on $[a, b]$, the difference $\Delta z_k = z(t_k) - z(t_{k-1})$ is approximately equal to $z'(t_k)(t_k - t_{k-1}) = z'(t_k)\Delta t_k$, the error going to zero faster than Δt_k. Hence we see that the sum

$$\sum_{k=1}^{n} f(z_k)\Delta z_k = \sum_{k=1}^{n} f(z(t_k))\Delta z_k,$$

which is a Riemann sum for f along γ, can be approximated by the sum

$$\sum_{k=1}^{n} f(z(t_k))z'(t_k)\Delta t_k,$$

which is a Riemann sum for the continuous function $f(z(t))z'(t)$ over the interval $[a, b]$. These considerations suggest the following theorem (and provide the essential ingredients for its justification):

Theorem 4. Let f be a function continuous on the directed smooth curve γ. Then if $z = z(t)$, $a \leq t \leq b$, is any admissible parametrization of γ consistent with its direction, we have

$$\int_{\gamma} f(z)\, dz = \int_{a}^{b} f(z(t))z'(t)\, dt. \tag{5}$$

The precise details of the proof of Theorem 4, though not difficult, are quite laborious and not particularly illuminating for our subject matter. Hence we shall omit them. A rigorous treatment of this theorem can be found in Ref. [2].

Since Eq. (5) is valid for all suitable parametrizations of γ and since the integral of f along γ was defined independently of any parametrization, we immediately deduce the following.

Corollary 1. If f is continuous on the directed smooth curve γ and if $z = z_1(t)$, $a \leq t \leq b$, and $z = z_2(t)$, $c \leq t \leq d$, are any two admissible parametrizations of γ consistent with its direction, then

$$\int_{a}^{b} f(z_1(t))\, z_1'(t)\, dt = \int_{c}^{d} f(z_2(t))\, z_2'(t)\, dt.$$

Example 2

Compute the integral $\int_{C_r} (z - z_0)^n \, dz$, with n an integer and C_r the circle $|z - z_0| = r$ traversed once in the counterclockwise direction,[†] as indicated in Fig. 4.17.

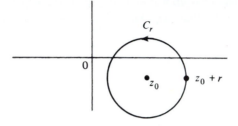

Figure 4.17 Directed smooth curve for Example 2.

Solution.   A suitable parametrization for C_r is given by $z(t) = z_0 + re^{it}$, $0 \le t \le 2\pi$. Setting $f(z) = (z - z_0)^n$, we have

$$f(z(t)) = (z_0 + re^{it} - z_0)^n = r^n e^{int}$$

and

$$z'(t) = ire^{it}.$$

Hence, by formula (5),

$$\int_{C_r} (z - z_0)^n \, dz = \int_0^{2\pi} \left(r^n e^{int} \right) \left(ire^{it} \right) dt = ir^{n+1} \int_0^{2\pi} e^{i(n+1)t} \, dt.$$

The evaluation of the last integral requires two separate computations. If $n \ne -1$, we obtain

$$ir^{n+1} \int_0^{2\pi} e^{i(n+1)t} \, dt = ir^{n+1} \left. \frac{e^{i(n+1)t}}{i(n+1)} \right|_0^{2\pi} = ir^{n+1} \left[\frac{1}{i(n+1)} - \frac{1}{i(n+1)} \right] = 0,$$

while if $n = -1$, then

$$ir^{n+1} \int_0^{2\pi} e^{i(n+1)t} \, dt = i \int_0^{2\pi} dt = 2\pi i.$$

Thus (regardless of the value of r)

$$\int_{C_r} (z - z_0)^n \, dz = \begin{cases} 0 & \text{for } n \ne -1, \\ 2\pi i & \text{for } n = -1. \end{cases} \quad \blacksquare \tag{6}$$

Integrals along a contour are computed according to the following definition.

[†] Occasionally we write

$$\oint_{C_r} f(z) \, dz$$

to emphasize the fact that the integration is taken in the positive direction.

Definition 4. Suppose that Γ is a contour consisting of the directed smooth curves $(\gamma_1, \gamma_2, \ldots, \gamma_n)$, and let f be a function continuous on Γ. Then the **contour integral of f along** Γ is denoted by the symbol $\int_\Gamma f(z)\,dz$ and is defined by the equation

$$\int_\Gamma f(z)\,dz := \int_{\gamma_1} f(z)\,dz + \int_{\gamma_2} f(z)\,dz + \cdots + \int_{\gamma_n} f(z)\,dz. \qquad (7)$$

If Γ consists of a single point, then for obvious reasons we set

$$\int_\Gamma f(z)\,dz := 0.$$

Example 3

Compute

$$\int_\Gamma \frac{1}{z - z_0}\,dz,$$

where Γ is the circle $|z - z_0| = r$ traversed twice in the counterclockwise direction starting from the point $z_0 + r$.

Solution.   Letting C_r denote the circle traversed once in the counterclockwise direction, we have $\Gamma = (C_r, C_r)$. Hence, from formula (6) obtained in the solution of Example 2, there follows

$$\int_\Gamma \frac{dz}{z - z_0} = \int_{C_r} \frac{dz}{z - z_0} + \int_{C_r} \frac{dz}{z - z_0} = 2\pi i + 2\pi i = 4\pi i. \quad \blacksquare$$

Example 4

Compute $\int_\Gamma \bar{z}^2 dz$ along the simple closed contour Γ of Fig. 4.18.

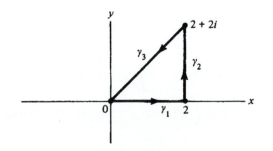

Figure 4.18 Contour for Example 4.

Solution.   According to Definition 4 we have

$$\int_\Gamma \bar{z}^2\, dz = \int_{\gamma_1} \bar{z}^2\, dz + \int_{\gamma_2} \bar{z}^2\, dz + \int_{\gamma_3} \bar{z}^2\, dz.$$

Suitable parametrizations for the line segments γ_k are

$$
\begin{array}{llll}
\gamma_1: & z_1(t) = t & (0 \le t \le 2), \\
\gamma_2: & z_2(t) = 2 + ti & (0 \le t \le 2), \\
\gamma_3: & z_3(t) = -t(1+i) & (-2 \le t \le 0),
\end{array}
$$

and so by Theorem 4 we have

$$\int_{\gamma_1} \bar{z}^2\, dz = \int_0^2 \overline{z_1(t)}^2 z_1'(t)\, dt = \int_0^2 t^2\, dt = \left.\frac{t^3}{3}\right|_0^2 = \frac{8}{3},$$

$$\int_{\gamma_2} \bar{z}^2\, dz = \int_0^2 \overline{z_2(t)}^2 z_2'(t)\, dt = \int_0^2 (2 - ti)^2 i\, dt$$

$$= \left.\frac{i(2 - ti)^3}{-3i}\right|_0^2 = \frac{-(2 - 2i)^3}{3} + \frac{8}{3},$$

and

$$\int_{\gamma_3} \bar{z}^2\, dz = \int_{-2}^0 \overline{z_3(t)}^2 z_3'(t)\, dt = \int_{-2}^0 [-t(1 - i)]^2 [-(1+i)]\, dt$$

$$= -(1+i)(1-i)^2 \int_{-2}^0 t^2\, dt = -(1+i)(1-i)^2 \frac{8}{3}.$$

Therefore

$$\int_\Gamma \bar{z}^2\, dz = \frac{8}{3} + \left[\frac{-(2-2i)^3}{3} + \frac{8}{3}\right] + \left[-(1+i)(1-i)^2\frac{8}{3}\right],$$

which after some computation turns out to equal $16/3 + 32i/3$.   ∎

Using Definition 4 it is easy to see that the results discussed previously for integrals along a directed smooth curve carry over to integrals along a contour. In particular, we have

$$\int_\Gamma [f(z) \pm g(z)]\, dz = \int_\Gamma f(z)\, dz \pm \int_\Gamma g(z)\, dz, \tag{8}$$

$$\int_\Gamma cf(z)\, dz = c \int_\Gamma f(z)\, dz \quad (c \text{ any complex constant}), \tag{9}$$

and

$$\int_{-\Gamma} f(z)\, dz = -\int_\Gamma f(z)\, dz, \tag{10}$$

where f and g are both continuous on the contour Γ.

Furthermore, if we have a parametrization $z = z(t)$, $a \leq t \leq b$, for the whole contour $\Gamma = (\gamma_1, \gamma_2, \ldots, \gamma_n)$, then we know that there is a subdivision

$$a = \tau_0 < \tau_1 < \cdots < \tau_{n-1} < \tau_n = b$$

such that the function $z(t)$ restricted to the kth subinterval $\left[\tau_{k-1}, \tau_k\right]$ constitutes a suitable parametrization of γ_k. Hence by formula (5)

$$\int_{\gamma_k} f(z)\, dz = \int_{\tau_{k-1}}^{\tau_k} f(z(t)) z'(t)\, dt \quad (k = 1, 2, \ldots, n),$$

and so

$$\int_{\Gamma} f(z)\, dz = \sum_{k=1}^{n} \int_{\tau_{k-1}}^{\tau_k} f(z(t)) z'(t)\, dt,$$

which we can write as

$$\int_{\Gamma} f(z)\, dz = \int_a^b f(z(t)) z'(t)\, dt.$$

Using this formula it is not difficult to prove that integration around simple closed contours is independent of the choice of the initial-terminal point (see Prob. 18). Consequently, in problems dealing with integrals along such contours, we need only specify the direction of transit, not the starting point.

Many times in theory and in practice, it is not actually necessary to evaluate a contour integral. What may be required is simply a good upper bound on its magnitude. We therefore turn to the problem of estimating contour integrals.

Suppose that the function f is continuous on the directed smooth curve γ and that $f(z)$ is bounded by the constant M on γ; i.e., $|f(z)| \leq M$ for all z on γ. If we consider a Riemann sum $\sum_{k=1}^{n} f(c_k) \Delta z_k$ corresponding to a partition $\mathcal{P}_n$ of γ, then we have, by the generalized triangle inequality,

$$\left| \sum_{k=1}^{n} f(c_k) \Delta z_k \right| \leq \sum_{k=1}^{n} |f(c_k)| |\Delta z_k| \leq M \sum_{k=1}^{n} |\Delta z_k|.$$

Furthermore, notice that the sum of the chordal lengths $\sum_{k=1}^{n} |\Delta z_k|$ cannot be greater than the length of γ. Hence

$$\left| \sum_{k=1}^{n} f(c_k) \Delta z_k \right| \leq M \ell(\gamma). \tag{11}$$

Since inequality (11) is valid for all Riemann sums of $f(z)$, it follows by taking the limit [as $\mu(\mathcal{P}_n) \to 0$] that

$$\left| \int_{\gamma} f(z)\, dz \right| \leq M \ell(\gamma). \tag{12}$$

Applying this fact and the triangle inequality to the Eq. (7) defining a contour integral, we deduce

Theorem 5. If f is continuous on the contour Γ and if $|f(z)| \leq M$ for all z on Γ, then

$$\left| \int_\Gamma f(z)\, dz \right| \leq M\ell(\Gamma), \tag{13}$$

where $\ell(\Gamma)$ denotes the length of Γ. In particular, we have

$$\left| \int_\Gamma f(z)\, dz \right| \leq \max_{z \text{ on } \Gamma} |f(z)| \cdot \ell(\Gamma). \tag{14}$$

Example 5

Find an upper bound for $\left| \int_\Gamma e^z / (z^2 + 1)\, dz \right|$, where Γ is the circle $|z| = 2$ traversed once in the counterclockwise direction.

Solution.   First observe that the path of integration has length $\ell = 4\pi$. Next we seek an upper bound M for the function $e^z / (z^2 + 1)$ when $|z| = 2$. Writing $z = x+iy$ we have

$$\left| e^z \right| = \left| e^{x+iy} \right| = e^x \leq e^2, \qquad \text{for } |z| = \sqrt{x^2 + y^2} = 2,$$

and by the triangle inequality

$$\left| z^2 + 1 \right| \geq |z|^2 - 1 = 4 - 1 = 3, \qquad \text{for } |z| = 2.$$

Hence $\left| e^z / (z^2 + 1) \right| \leq e^2 / 3$ for $|z| = 2$, and so by the theorem

$$\left| \int_\Gamma \frac{e^z}{z^2 + 1}\, dz \right| \leq \frac{e^2}{3} \cdot 4\pi. \quad \blacksquare$$

In concluding this section we remark that although the real definite integral can be interpreted, among other things, as an area, no corresponding geometric visualization is available for contour integrals. Nevertheless, the latter integrals are extremely useful in applied problems, as we shall see in subsequent chapters.

EXERCISES 4.2

1. Let γ be a directed smooth curve with initial point α and terminal point β. Show directly from Definition 3 that $\int_\gamma c\, dz = c(\beta - \alpha)$, where c is any complex constant. Does the same formula hold for integration along an arbitrary contour joining α to β?

2. Using Definition 3, prove properties (1), (2) and (3).

3. Evaluate each of the following integrals.

(a) $\displaystyle\int_0^1 \left(2t + it^2\right) dt$ (b) $\displaystyle\int_{-2}^0 (1 + i)\cos(it)\, dt$

(c) $\displaystyle\int_0^1 (1 + 2it)^5\, dt$ (d) $\displaystyle\int_0^2 \frac{t}{\left(t^2 + i\right)^2}\, dt$

4. Furnish the details of the proof of Theorem 3.

5. Utilize Example 2 to evaluate

$$\int_C \left[\frac{6}{(z - i)^2} + \frac{2}{z - i} + 1 - 3(z - i)^2\right] dz,$$

where C is the circle $|z - i| = 4$ traversed once counterclockwise.

6. Compute $\int_\Gamma \bar{z}\, dz$, where

   (a) Γ is the circle $|z| = 2$ traversed once counterclockwise.

   (b) Γ is the circle $|z| = 2$ traversed once clockwise.

   (c) Γ is the circle $|z| = 2$ traversed three times clockwise.

7. Compute $\int_\Gamma \operatorname{Re} z\, dz$ along the directed line segment from $z = 0$ to $z = 1 + 2i$.

8. Let C be the perimeter of the square with vertices at the points $z = 0$, $z = 1$, $z = 1 + i$, and $z = i$ traversed once in that order. Show that

$$\int_C e^z\, dz = 0.$$

9. Evaluate $\int_\Gamma (x - 2xyi)\, dz$ over the contour $\Gamma: z = t + it^2$, $0 \le t \le 1$, where $x = \operatorname{Re} z$, $y = \operatorname{Im} z$.

10. Compute $\int_C \bar{z}^2\, dz$ along the perimeter of the square in Prob. 8.

11. Evaluate $\int_\Gamma (2z + 1)\, dz$, where Γ is the following contour from $z = -i$ to $z = 1$:

    (a) the simple line segment.

    (b) two simple line segments, the first from $z = -i$ to $z = 0$ and the second from $z = 0$ to $z = 1$.

    (c) the circular arc $z = e^{it}$, $-\pi/2 \le t \le 0$.

12. True or false: $\displaystyle\oint_{|z|=1} \bar{z}\, dz = \oint_{|z|=1} \frac{1}{z}\, dz.$

13. Compute $\int_\Gamma \left(|z - 1 + i|^2 - z\right) dz$ along the semicircle $z = 1 - i + e^{it}$, $0 \le t \le \pi$.

14. For each of the following, use Theorem 5 to establish the indicated estimate.

    (a) If C is the circle $|z| = 3$ traversed once, then

$$\left|\int_C \frac{dz}{z^2 - i}\right| \le \frac{3\pi}{4}.$$

(b) If γ is the vertical line segment from $z = R$ (> 0) to $z = R + 2\pi i$, then

$$\left| \int_\gamma \frac{e^{3z}}{1 + e^z} \, dz \right| \leq \frac{2\pi e^{3R}}{e^R - 1}.$$

(c) If Γ is the arc of the circle $|z| = 1$ that lies in the first quadrant, then

$$\left| \int_\Gamma \operatorname{Log} z \, dz \right| \leq \frac{\pi^2}{4}.$$

(d) If γ is the line segment from $z = 0$ to $z = i$, then

$$\left| \int_\gamma e^{\sin z} \, dz \right| \leq 1.$$

15. Let f be a continuous complex-valued function on the real interval $[a, b]$. Prove that

$$\left| \int_a^b f(t) \, dt \right| \leq \int_a^b |f(t)| \, dt.$$

[HINT: Consider the Riemann sums of f over $[a, b]$.]

16. Let γ be a directed smooth curve with initial point α and terminal point β. Use formula (5) and Theorem 3 to show that

$$\int_\gamma z \, dz = \frac{\beta^2 - \alpha^2}{2}.$$

17. Using the result of Prob. 16, prove that for any closed contour Γ

$$\int_\Gamma z \, dz = 0.$$

18. Let Γ_1 be a closed contour parametrized by $z = z_1(t)$, $a \leq t \leq b$. We can shift the initial-terminal point of Γ_1 by choosing a number c in the interval (a, b) and letting Γ_2 be the contour parametrized by

$$z_2(t) = \begin{cases} z_1(t) & \text{if } c \leq t \leq b, \\ z_1(t - b + a) & \text{if } b \leq t \leq b - a + c. \end{cases}$$

Prove that

$$\int_{\Gamma_1} f(z) \, dz = \int_{\Gamma_2} f(z) \, dz$$

for any function f continuous on the points of Γ_1.

4.3   Independence of Path

One of the important results in the theory of complex analysis is the extension of the Fundamental Theorem of Calculus to *contour* integrals. It implies that in certain situations the integral of a function is independent of the particular path joining the initial and terminal points; in fact, it completely characterizes the conditions under which this property holds. In this section, we shall explore this phenomenon in detail. We begin with the Fundamental Theorem, which enables us to evaluate integrals without introducing parametrizations, provided that an antiderivative of the integrand is known.

> **Theorem 6.** Suppose that the function $f(z)$ is continuous in a domain D and has an antiderivative $F(z)$ throughout D; i.e., $dF(z)/dz = f(z)$ for each z in D. Then for any contour Γ lying in D, with initial point z_I and terminal point z_T, we have
>
> $$\int_\Gamma f(z)\,dz = F(z_T) - F(z_I). \qquad (1)$$

[Notice that the conditions of the theorem imply that $F(z)$ is analytic and hence continuous in D. The function $\operatorname{Log} z$, for example, is *not* an antiderivative for $1/z$ in any domain containing points of the negative real axis.]

Before proceeding with the proof we shall show how Theorem 6 can greatly facilitate the computation of certain contour integrals.

Example 1

Compute the integral $\int_\Gamma \cos z\,dz$, where Γ is the contour shown in Fig. 4.19.

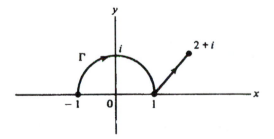

Figure 4.19 Contour for Example 1.

Solution. There is no need to parametrize Γ since the integrand has the antiderivative $F(z) = \sin z$ for all z. Hence by Theorem 6, the value of the integral can be computed using only the endpoints of Γ:

$$\int_\Gamma \cos z\,dz = \sin z \Big|_{-1}^{2+i} = \sin(2+i) - \sin(-1). \quad \blacksquare$$

Proof of Theorem 6.   The demonstration is quite straightforward; once we write the integral in terms of a parametrization, the conclusion will follow as a result of the chain rule and Theorem 3 of the last section.

So suppose that Γ is a contour in D joining z_I to z_T. We select a suitable parametrization $z = z(t)$, $a \leq t \leq b$, for Γ and, as in the previous section, let $\{\tau_k\}_0^n$ denote the values of t corresponding to the endpoints of the smooth components $\{\gamma_j\}_1^n$ of Γ [in particular, $z(\tau_0) = z_I$ and $z(\tau_n) = z_T$]. Then we have

$$\int_\Gamma f(z)\, dz = \sum_{j=1}^n \int_{\gamma_j} f(z)\, dz$$

$$= \sum_{j=1}^n \int_{\tau_{j-1}}^{\tau_j} f(z(t))z'(t)\, dt. \tag{2}$$

Using the fact that F is an antiderivative of f, it is possible to rewrite the integrands appearing in Eq. (2). For this purpose we recall that on each separate interval $[\tau_{j-1}, \tau_j]$ the derivative dz/dt exists and is continuous. Hence the chain rule implies that

$$\frac{d}{dt}[F(z(t))] = \frac{dF}{dz}\frac{dz}{dt} = f(z(t))z'(t) \quad (\tau_{j-1} \leq t \leq \tau_j),$$

and so, by Theorem 3,

$$\int_{\tau_{j-1}}^{\tau_j} f(z(t))z'(t)\, dt = \int_{\tau_{j-1}}^{\tau_j} \frac{d}{dt}[F(z(t))]\, dt$$

$$= F(z(\tau_j)) - F(z(\tau_{j-1})).$$

Therefore, we have

$$\int_\Gamma f(z)\, dz = \sum_{j=1}^n \left[F(z(\tau_j)) - F(z(\tau_{j-1})) \right]$$

$$= [F(z(\tau_1)) - F(z(\tau_0))] + [F(z(\tau_2)) - F(z(\tau_1))] \tag{3}$$

$$+ \cdots + \left[F(z(\tau_n)) - F(z(\tau_{n-1})) \right].$$

But this sum telescopes, leaving

$$\int_\Gamma f(z)\, dz = F(z(\tau_n)) - F(z(\tau_0))$$

$$= F(z_T) - F(z_I). \quad \blacksquare$$

Example 2

Compute $\int_\Gamma 1/z\, dz$, where **(a)** Γ is the contour shown in Fig. 4.20 and **(b)** Γ is the contour indicated in Fig. 4.21.

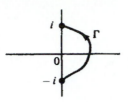

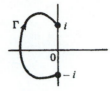

Figure 4.20 Contour for Example 2(a).   **Figure 4.21** Contour for Example 2(b).

Solution.   **(a)** At each point of the contour Γ of Fig. 4.20 the function $1/z$ is the derivative of the principal branch of $\log z$ (cf. Sec. 3.3). Hence

$$\int_\Gamma \frac{dz}{z} = \text{Log}\, z \Big|_{-i}^{i} = \frac{\pi}{2}i - \left(-\frac{\pi}{2}i\right) = \pi i.$$

(b) For the contour Γ of Fig. 4.21 we cannot employ the function $\text{Log}\, z$, since its branch cut intersects Γ. We use instead $\mathcal{L}_0(z) = \text{Log}\,|z| + i \arg z$, $0 < \arg z < 2\pi$, which is a branch of the logarithm with cut along the nonnegative x-axis. Thus

$$\int_\Gamma \frac{dz}{z} = \mathcal{L}_0(z) \Big|_{-i}^{i} = \frac{\pi}{2}i - \frac{3\pi}{2}i = -\pi i. \quad \blacksquare$$

Since the endpoints of a *loop*, i.e., a closed contour, are equal, we have the following immediate consequence of Theorem 6.

Corollary 2. If f is continuous in a domain D and has an antiderivative throughout D, then $\int_\Gamma f(z)\, dz = 0$ for all loops Γ lying in D.

Corollary 2 provides an alternative solution to the problem of evaluating the integral $\int_{C_r} (z - z_0)^n\, dz$ of Example 2 in Sec. 4.2 when $n \neq -1$. For if we set $f(z) = (z - z_0)^n$, then $f(z)$ is the derivative of the function $F(z) = (z - z_0)^{n+1}/(n + 1)$, which is analytic in the domain D consisting of all points in the plane except $z = z_0$. (Actually the point z_0 need be excluded only in the case when n is negative. Why?) Since C_r is a closed contour which lies in D, we deduce from the corollary that $\int_{C_r} (z - z_0)^n\, dz = 0, n \neq -1$.

Another important conclusion that can be drawn from Eq. (1) is that when a function f has an antiderivative throughout a domain D, its integral along a contour in D depends only on the endpoints z_I and z_T; i.e., the integral is independent of the path Γ joining these two points! For instance, in Fig. 4.22 all the integrals $\int_{\Gamma_1} f(z)\, dz$, $\int_{\Gamma_2} f(z)\, dz$, and $\int_{\Gamma_3} f(z)\, dz$ are equal under this condition. As a matter of fact, we shall establish that the three properties we have discussed in this section amount to logically equivalent statements when applied to a continuous function $f(z)$.

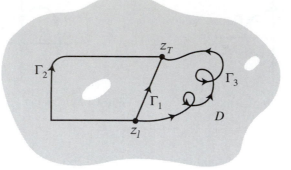

Figure 4.22 Independence of path.

Theorem 7. Let f be continuous in a domain D. Then the following are equivalent:

   (i) f has an antiderivative in D.

  (ii) Every loop integral of f in D vanishes [i.e., if Γ is any loop in D, then $\int_\Gamma f(z)dz = 0$].

 (iii) The contour integrals of f are independent of path in D [i.e., if Γ_1 and Γ_2 are any two contours in D sharing the same initial and terminal points, then $\int_{\Gamma_1} f(z)\,dz = \int_{\Gamma_2} f(z)\,dz$].

Proof. We have already seen from Theorem 6 that statement (i) implies (ii) [as well as (iii)]. Thus Theorem 7 will be proved if we can show that (ii) implies (iii) and that (iii) implies (i).

So assume that statement (ii) is true, and let Γ_1 and Γ_2 be any two contours in D sharing the same initial point z_I and terminal point z_T. Now define Γ to be the contour generated by proceeding first along Γ_1 from z_I to z_T and then backwards from z_T to z_I along $-\Gamma_2$. Then by Eq. (10) of Sec. 4.2, we have

$$\int_\Gamma f(z)\,dz = \int_{\Gamma_1} f(z)\,dz + \int_{-\Gamma_2} f(z)\,dz = \int_{\Gamma_1} f(z)\,dz - \int_{\Gamma_2} f(z)\,dz.$$

On the other hand, since Γ is closed, (ii) implies that

$$\int_\Gamma f(z)\,dz = 0.$$

Thus we deduce statement (iii).

We now show that whenever property (iii) holds, so does (i). To prove that f has an antiderivative, we must define some function $F(z)$ and show that its derivative is

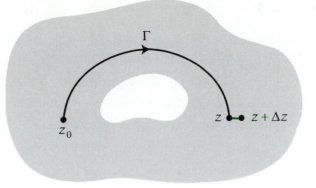

Figure 4.23 Path of integration for Theorem 7.

$f(z)$. The clue as to where to look for $F(z)$ is provided by the earlier considerations; if f *had* an antiderivative, Eq. (1) would hold. So we use Eq. (1) to *define* the function $F(z)$ and show that it is, indeed, an antiderivative.

Accordingly, we fix some point z_0 in D. Then for any point z in D, let $F(z)$ be the integral of f along some contour Γ in D joining z_0 to z. Since D is connected, we know that there will be at least one such contour (a directed polygonal path), and by condition (iii) it does not matter which contour we choose; all the possible paths will yield the same value for $F(z)$. Hence $F(z)$ is a well-defined single-valued function in D. To prove (i) we compute $F(z + \Delta z) - F(z)$.

We prudently elect to evaluate $F(z+\Delta z)$ by first integrating f along the contour Γ from z_0 to z and then along the straight-line segment from z to $z + \Delta z$. This segment will lie in D if Δz is small enough, because D is an open set; see Fig. 4.23. But now the difference $F(z + \Delta z) - F(z)$ is simply the integral of f along this segment. Parametrizing the latter by $z(t) = z + t\Delta z, 0 \leq t \leq 1$, we have

$$F(z + \Delta z) - F(z) = \int_0^1 f(z + t\Delta z)\, \Delta z\, dt,$$

and thus

$$\frac{F(z + \Delta z) - F(z)}{\Delta z} = \int_0^1 f(z + t\Delta z)\, dt.$$

Since f is continuous, it is easy to see (Prob. 10) that as $\Delta z \to 0$ the last integral approaches $\int_0^1 f(z)\, dt = f(z)$. Thus $F'(z)$ exists and equals $f(z)$. This concludes the proof of the equivalence. ■

Theorem 7 probably appears useless at present; you may wonder how in the world one tests whether the integral of a function around *every* closed curve is zero. In the next section our efforts will be vindicated thanks to a surprising result known as Cauchy's theorem, which gives a simple condition for this property to hold. For now, we shall simply summarize by saying that a given continuous function has an antiderivative in D *if and only if* its integral around every loop in D is zero.

EXERCISES 4.3

1. Calculate each of the following integrals along the indicated contours. (Observe that a standard table of integrals can be used. Explain why.)

    (a) $\int_\Gamma (3z^2 - 5z + i)\, dz$ along the line segment from $z = i$ to $z = 1$.

    (b) $\int_\Gamma e^z\, dz$ along the upper half of the circle $|z| = 1$ from $z = 1$ to $z = -1$.

    (c) $\int_\Gamma 1/z\, dz$ for any contour in the right half-plane from $z = -3i$ to $z = 3i$.

    (d) $\int_\Gamma \csc^2 z\, dz$ for any closed contour that avoids the points $0, \pm\pi, \pm 2\pi, \ldots$.

    (e) $\int_\Gamma \sin^2 z \cos z\, dz$ along the contour in Fig. 4.24.

    (f) $\int_\Gamma e^z \cos z\, dz$ along the contour in Fig. 4.24.

    (g) $\int_\Gamma z^{1/2}\, dz$ for the principal branch of $z^{1/2}$ along the contour in Fig. 4.24.

    (h) $\int_\Gamma (\text{Log } z)^2\, dz$ along the line segment from $z = 1$ to $z = i$.

    (i) $\int_\Gamma 1/(1 + z^2)\, dz$ along the line segment from $z = 1$ to $z = 1 + i$.

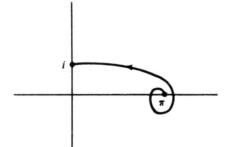

Figure 4.24 Contour for Prob. 1(e), (f), and (g).

2. If $P(z)$ is a polynomial and Γ is any closed contour, explain why $\int_\Gamma P(z)\, dz = 0$.

3. In Chapter 5 we shall show that if f is entire and Γ is any contour, then for each $\epsilon > 0$ there exists a polynomial $P(z)$ such that

$$|f(z) - P(z)| < \epsilon \quad \text{for all } z \text{ on } \Gamma.$$

    Assuming this fact, prove that if f is entire, then

    (a) $\int_\Gamma f(z)\, dz = 0$ for all closed contours Γ. [HINT: Use the result of Prob. 2.]

    (b) f is the derivative of an entire function.

4. True or false: If f is analytic at each point of a closed contour Γ, then $\int_\Gamma f(z)\, dz = 0$.

5. Explain why Example 2 shows that the function $f(z) = 1/z$ has no antiderivative in the punctured plane $\mathbf{C} \setminus \{0\}$.

6. Although Corollary 2 does not apply to the function $1/(z-z_0)$ in the plane punctured at z_0, Theorem 6 can be used as follows to show that

$$\int_C \frac{dz}{z-z_0} = 2\pi i$$

for any circle C traversed once in the positive direction surrounding the point z_0. Introduce a horizontal branch cut from z_0 to ∞ as in Fig. 4.25. In the resulting "slit plane" the function $1/(z-z_0)$ has the antiderivative Log $(z-z_0)$. Apply Theorem 6 to compute the integral along the portion of C from α to β as indicated in Fig. 4.25. Now let α and β approach the point τ on the cut to evaluate the given integral over all of C.

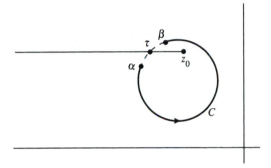

Figure 4.25 Contour for Prob. 6.

7. Show that if C is a positively oriented circle and z_0 lies outside C, then

$$\int_C \frac{dz}{z-z_0} = 0.$$

8. Show directly that property (iii) implies (ii) in Theorem 7.

9. As we know, an antiderivative is only specified up to a constant. How was this flexibility reflected in the proof that (iii) implies (i) in Theorem 7?

10. Verify the statement made in the text that if f is continuous at the point z, then

$$\int_0^1 f(z+t\Delta z)\,dt \to f(z) \quad \text{as } \Delta z \to 0.$$

[HINT: Estimate the difference

$$\int_0^1 f(z+t\Delta z)\,dt - f(z) = \int_0^1 [f(z+t\Delta z) - f(z)]\,dt. \ ]$$

11. Prove the *integration-by-parts formula*: If f and g have continuous first derivatives in a domain containing the contour Γ, then

$$\int_\Gamma f'(z)g(z)\,dz = f(z)g(z)\Big|_{z_I}^{z_T} - \int_\Gamma f(z)g'(z)\,dz,$$

where z_I and z_T are the initial and terminal points of Γ. [HINT: Use Theorem 6 on the function $d(fg)/dz$.]

12. Let f be an analytic function with a continuous derivative satisfying $\left| f'(z) \right| \leq M$ for all z in the disk $D : |z| < 1$. Show that

$$|f(z_2) - f(z_1)| \leq M |z_2 - z_1| \quad (z_1, z_2 \text{ in } D).$$

[HINT: Observe that $f(z_2) - f(z_1) = \int f'(z)\, dz$, where the integration can be taken along the line segment from z_1 to z_2.]

4.4   Cauchy's Integral Theorem

The essential content of this section is the Cauchy integral theorem.[†] We feel that a clear and intuitive approach to this subject is provided by the concept of continuous deformations of one contour into another. On the other hand, some instructors may feel that the theorem is better handled by appealing to vector analysis, in particular Green's theorem. Accordingly, we have provided the reader with two alternative sections, 4.4a and 4.4b, and either one may be studied without affecting the subsequent development.

In order that each section may be self-contained, some duplication of the text appears; for instance Theorem 12 in Sec. 4.4b restates Theorem 9 in Sec. 4.4a, and many of the same examples occur in both sections (though the methods of solution are different).

Exercises 4.4 are divided into three parts: problems appropriate to Sec. 4.4a, problems appropriate to Sec. 4.4b, and problems for all readers.

4.4a   Deformation of Contours Approach

In the last section we saw that if a continuous function f possesses an (analytic) antiderivative in a domain D, its integral around any loop in D is zero and vice versa. Now we are going to show how this property ties in with the analyticity of f itself. Our first task will be to develop the necessary geometry.

The critical notion in this regard is the *continuous deformation of one loop into another*, in a given domain of the plane. Deformations are quite easily visualized but somewhat harder to express in precise mathematical language. Most of the time, however, the visualization alone will suffice for our purposes. With this in mind, we first give an intuitive definition of deformations.

We say that the loop Γ_0 can be continuously deformed into the loop Γ_1 in the domain D if Γ_0 (considered as an elastic string with indicated orientation) can be continuously moved about the plane, *without leaving D*, in such a manner that it ultimately coincides with Γ_1 (in position as well as direction).

The following examples serve to illustrate this notion:

[†]Gauss actually discovered this theorem in 1811, a few years before Cauchy published it.

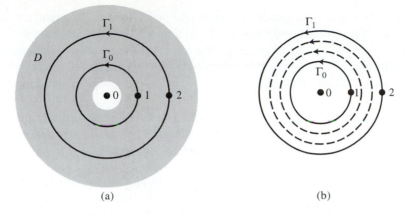

(a)   (b)

Figure 4.26 Expanding circular contours.

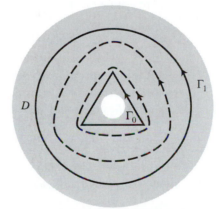

Figure 4.27 Deformation of a triangle.

(a) Let D be the annulus and let Γ_0 and Γ_1 be the circular contours indicated in Fig. 4.26(a). Since both circles are positively oriented, the "elastic" circle Γ_0 can be continuously deformed to Γ_1 in D by expanding the radius of Γ_0 from 1 to 2; i.e., we visualize a continuum of concentric circles varying in radii from 1 to 2. The dashed circles in Fig. 4.26(b) depict some of the intermediate loops; notice that all of them lie in D.

(b) Let D be the annulus, Γ_0 the triangular contour, and Γ_1 the circular contour of Fig. 4.27. Then Γ_0 can be deformed to Γ_1 in D by expanding Γ_0 and simultaneously making its sides more circular. Some intermediate loops are sketched in Fig. 4.27. Again they all remain in D.

(c) Let D be the whole plane, Γ_0 the loop indicated in Fig. 4.28, and Γ_1 the point contour $z = 0$. Then Γ_0 can be continuously deformed to Γ_1 in D by simply shrinking and shifting, as indicated in Fig. 4.28.

(d) Let D be the first quadrant and let Γ_0 and Γ_1 be the circular contours in Fig. 4.29(a). Notice that merely moving Γ_0 to the right will not yield the desired deformation, for while Γ_0 will eventually coincide in position with Γ_1, the orientations will be different. To circumvent this difficulty we first shrink Γ_0 to a point (which has no direction) and then expand the point to Γ_1, always remaining in D, as indicated in Fig. 4.29(b).

(e) Let D be the plane minus the points $\pm i$, and let Γ_0 be the circular contour and Γ_1 be the "barbell" contour of Fig. 4.30. Then Γ_0 can be continuously deformed to Γ_1 in D, as illustrated by the dashed-line intermediate loops in Fig. 4.30.

Now let us be more precise. The preceding examples show that Γ_0 can be continuously deformed to Γ_1 in D if Γ_0 and Γ_1 belong to a continuum of loops $\{\Gamma_s\}$, $0 \le s \le 1$, each lying in D, such that any pair $\Gamma_{s'}$ and $\Gamma_{s''}$ can be made "arbitrarily close" by taking s' sufficiently near s''. Thus there must be parametrizations $\{z_s(t)\}$ for the contours $\{\Gamma_s\}$ which are continuous in the variable s. Using the standard $[0, 1]$ parametric interval for the loops and rewriting $z_s(t)$ as $z(s, t)$, we formalize these ideas in the following definition.

Definition 5. The loop Γ_0 is said to be **continuously deformable**[†] to the loop Γ_1 **in the domain** D if there exists a function $z(s, t)$ continuous on the unit square $0 \le s \le 1, 0 \le t \le 1$, that satisfies the following conditions:

(i) For each fixed s in $[0, 1]$, the function $z(s, t)$ parametrizes a loop lying in D.

(ii) The function $z(0, t)$ parametrizes the loop Γ_0.

(iii) The function $z(1, t)$ parametrizes the loop Γ_1. (See Fig. 4.31)

Example 1

By exhibiting a deformation function $z(s, t)$, *prove* that the loop $\Gamma_0 : z = e^{2\pi i t}$, $0 \le t \le 1$, can be continuously deformed to the loop $\Gamma_1 : z = 2e^{2\pi i t}$, $0 \le t \le 1$, in the domain D consisting of the annulus $1/2 < |z| < 3$.

Solution.   This is precisely the problem illustrated in Fig. 4.26; the intermediate loops Γ_s, $0 \le s \le 1$, are concentric circles with radii varying from 1 to 2. The function

$$z(s, t) = (1 + s)e^{2\pi i t} \qquad (0 \le s \le 1, \ 0 \le t \le 1)$$

therefore effects the deformation.   ■

[†]The word *homotopic* is sometimes used.

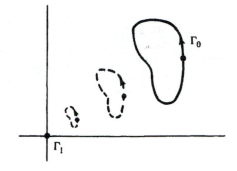

Figure 4.28 Shrinking a contour.

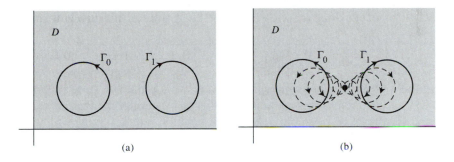

(a)

(b)

Figure 4.29 Reversal of orientation by shrinking to a point.

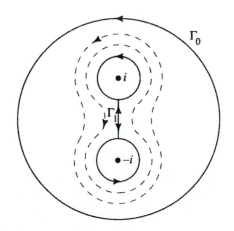

Figure 4.30 Circle deforming to a barbell.

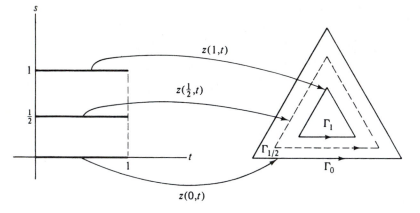

Figure 4.31 Parametrization of deformation.

Example 2

Exhibit a deformation function that shows that in the domain consisting of the whole plane any loop can be shrunk to the point contour $z = 0$.

Solution. This is the situation of Fig. 4.28. If Γ_0 is parametrized by $z = z_0(t)$, $0 \le t \le 1$, then the shrinking can be accomplished by multiplying $z_0(t)$ by a scaling factor that varies from 1 to 0. The deformation function is therefore given by

$$z(s, t) = (1 - s)z_0(t) \qquad (0 \le s \le 1, \, 0 \le t \le 1). \quad \blacksquare$$

A few elementary observations about continuous deformations are in order. First, notice that if $z(s, t)$ generates a deformation of loop Γ_0 into Γ_1, then $z(1-s, t)$ deforms Γ_1 into Γ_0. Furthermore, if in a given domain Γ_0 can be deformed into a single point and Γ_1 can also be deformed into a point, then Γ_0 can be deformed into Γ_1 in the domain (see Prob. 2).

As we have observed in Example 2, in the domain D consisting of the entire complex plane *any* loop can be deformed into the single point $z = 0$. (Consequently, any two loops can be deformed one into the other in this domain.) There are many other domains with this property, e.g., interiors of circles, interiors of regular polygons, half-planes, etc. We categorize such domains as follows.

Definition 6. Any domain D possessing the property that every loop in D can be continuously deformed in D to a point is called a **simply connected domain**.

Roughly speaking, we say that a simply connected domain cannot have any "holes," for if there were a hole in D, then a loop surrounding it could not be shrunk to a point without leaving D. It is shown in topology that if γ is any simple closed contour, then

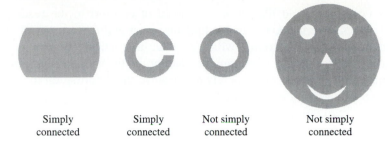

Simply connected   Simply connected   Not simply connected   Not simply connected

Figure 4.32 Examples of simply connected and multiply connected domains.

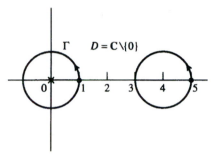

Figure 4.33 Nondeformable loops in punctured plain.

its interior is a simply connected domain. Indeed, this fact is often regarded as part of the Jordan Curve Theorem. These considerations provide us with a quick method of identifying some simply connected domains (see Fig. 4.32).

Interesting situations arise when the domain is not simply connected (such a domain is called *multiply connected*). For example, let D be the complex plane with the origin deleted, and let Γ be the unit circle $|z| = 1$, traversed once in the counterclockwise direction starting from the point $z = 1$. We list some loops which are not deformable to Γ in D:

(a) The circle parametrized by $z(t) = 4 + e^{2\pi it}$, $0 \le t \le 1$, cannot be deformed into Γ, because some intermediate loop would have to pass through $z = 0$ (see Fig. 4.33).

(b) Γ cannot be shrunk to a point in D. [However, the circle in (a) can be so deformed.]

(c) Γ cannot be continuously deformed into the opposite contour, $-\Gamma$. (The reader should mentally try to devise a family of loops linking these two, to see why the quarantining of the origin inhibits the deformation.)

(d) Γ cannot be deformed into the same unit circle circumscribed *twice* in the positive direction.

At this point we would like to insert a word of comfort to the reader, who is probably developing some anxiety concerning his or her ability to construct the deformation function $z(s, t)$. Theorem 8 will show that, in practice, only the *existence* of the continuous deformation is important (at least for the theory of analytic functions). Consequently, we shall be content merely to visualize the deformation in most cases, without officially verifying its existence.

We are now ready to state the main theorem of this section.

Theorem 8. (*Deformation Invariance Theorem*) Let f be a function analytic in a domain D containing the loops Γ_0 and Γ_1. If these loops can be continuously deformed into one another in D, then

$$\int_{\Gamma_0} f(z)\, dz = \int_{\Gamma_1} f(z)\, dz.$$

A rigorous proof of Theorem 8 involves procedures that take us far afield from the basic techniques of complex analysis. Here we shall only prove a weaker version of this theorem for the special case when Γ_0 and Γ_1 are linked by a deformation function $z(s, t)$ whose second-order partial derivatives are continuous. We shall also assume that $f'(z)$ is continuous (recall that analyticity merely requires that f' *exist*).

The fact that one need *not* assume continuity for f' was first demonstrated by the mathematician Edouard Goursat; see Ref. 2. The subsequent extension of the restricted theorem to the general statement of Theorem 8 can be effected by techniques of approximation theory (Ref. 5).

Proof of Weak Version of Theorem 8.   As mentioned before, we shall assume that the deformation function $z(s, t)$ has continuous partial derivatives up to order 2 for $0 \le s \le 1, 0 \le t \le 1$, and that $f'(z)$ is continuous. Now, for each fixed s the equation $z = z(s, t), 0 \le t \le 1$, defines a loop Γ_s in D. Let $I(s)$ be the integral of f along this loop, so that

$$I(s) := \int_{\Gamma_s} f(z)\, dz = \int_0^1 f(z(s, t)) \frac{\partial z(s, t)}{\partial t}\, dt. \tag{1}$$

We wish to take the derivative of $I(s)$ with respect to s. The assumptions guarantee that the integrand in Eq. (1) is continuously differentiable in s, so *Leibniz's rule for integrals* (Ref. 6) sanctions differentiation under the integral sign. Using the chain rule we obtain

$$\frac{dI(s)}{ds} = \int_0^1 \left[f'(z(s, t)) \frac{\partial z}{\partial s} \cdot \frac{\partial z}{\partial t} + f(z(s, t)) \frac{\partial^2 z}{\partial s\, \partial t} \right] dt. \tag{2}$$

On the other hand observe that

$$\frac{\partial}{\partial t} \left[f(z(s, t)) \frac{\partial z}{\partial s} \right] = f'(z(s, t)) \frac{\partial z}{\partial t} \cdot \frac{\partial z}{\partial s} + f(z(s, t)) \frac{\partial^2 z}{\partial t\, \partial s}.$$

Because of the continuity conditions the mixed partials of $z(s, t)$ are equal, so the last expression is the same as the integrand in Eq. (2). Thus

$$\frac{dI(s)}{ds} = \int_0^1 \frac{\partial}{\partial t}\left[f(z(s, t))\frac{\partial z}{\partial s}\right]dt$$

$$= f(z(s, 1))\frac{\partial z}{\partial s}(s, 1) - f(z(s, 0))\frac{\partial z}{\partial s}(s, 0).$$

But since each Γ_s is closed, we have $z(s, 1) = z(s, 0)$ for all s, from which we also see that the derivatives of these functions are identical. Consequently $I(s)$ is constant. In particular $I(0) = I(1)$, which is merely a disguised form of the conclusion

$$\int_{\Gamma_0} f(z)\,dz = \int_{\Gamma_1} f(z)\,dz. \quad\blacksquare$$

An easy consequence of Theorem 8 is the following, familiarly known as *Cauchy's integral theorem.*

Theorem 9. If f is analytic in a simply connected domain D and Γ is any loop (closed contour) in D, then

$$\int_{\Gamma} f(z)\,dz = 0. \tag{3}$$

Proof. The proof is immediate; in a simply connected domain any loop can be shrunk to a point. The integral of a continuous function over a shrinking loop converges, of course, to zero. $\quad\blacksquare$

It can be shown by topological methods that if Γ is a simple closed contour and f is analytic at each point on and inside Γ, then f must be analytic in some simply connected domain containing Γ. Thus, by Theorem 9, **the integral along Γ must vanish whenever the integrand is analytic "inside and on Γ."**

Cauchy's theorem links the considerations of the last section with the property of analyticity. We can conclude the following.

Theorem 10. In a simply connected domain, an analytic function has an antiderivative, its contour integrals are independent of path, and its loop integrals vanish.

In an earlier section we showed that if C is any circle centered at z_0 and n is an integer, then

$$\oint_C (z - z_0)^n\,dz = \begin{cases} 0 & \text{for } n \neq -1, \\ 2\pi i & \text{for } n = -1 \end{cases} \tag{4}$$

(see Example 2, Sec. 4.2). Equation (4) with $z_0 = 0$ neatly exemplifies the theory we have been discussing. If n is a positive integer or zero, z^n is analytic in the whole plane, which is simply connected; thus Theorem 10 applies, z^n has an antiderivative [the function $z^{n+1} / (n + 1)$], and its loop integrals are zero.

If n is negative, z^n is analytic only in the *punctured* plane, with the origin deleted. This domain is not simply connected, so Theorem 10 does not apply. In fact, for $n = -1$, the function z^n does not even have an antiderivative in the punctured plane (since any branch of $\log z$ will be discontinuous on the branch cut), and sure enough the loop integral (4) fails to vanish. On the other hand, if $n \leq -2$, then z^n regains its antiderivative $z^{n+1} / (n + 1)$ in the punctured plane, and the loop integrals (4) are zero. Thus either case can occur in multiply connected domains.

The main value of the Deformation Invariance Theorem is that it allows us to replace complicated contours with more familiar ones, for the purpose of integration. We shall illustrate this point with several examples.

Example 3

Evaluate $\int_\Gamma 1/z \, dz$, where Γ is the ellipse defined by $x^2 + 4y^2 = 1$, traversed once in the positive sense, as indicated in Fig. 4.34.

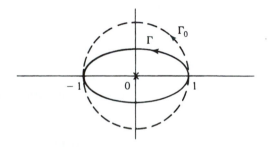

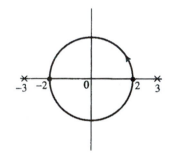

Figure 4.34 Contour for Example 3.   **Figure 4.35** Contour for Example 4.

Solution. The integrand $1/z$ is analytic in the plane with the origin deleted. Furthermore, it is obvious that without passing through the origin Γ can be continuously deformed into the unit circle Γ_0, oriented positively. Thus, by the Deformation Invariance Theorem and Eq. (4),

$$\int_\Gamma \frac{1}{z} \, dz = \int_{\Gamma_0} \frac{1}{z} \, dz = 2\pi i. \quad \blacksquare$$

Example 4

Evaluate

$$\oint_{|z|=2} \frac{e^z}{z^2 - 9} \, dz.$$

Solution. The notation employed signifies that the contour of integration is the circle $|z| = 2$ traversed once counterclockwise. The integrand $e^z/(z^2 - 9)$ is analytic everywhere except at $z = \pm 3$, where the denominator vanishes. From Fig. 4.35 we immediately see that the contour can be shrunk to a point in the domain of analyticity, and thus the integral is zero. (Alternatively, Cauchy's theorem can be applied to this example.) ∎

Example 5

Determine the possible values for $\int_\Gamma 1/(z - a)\, dz$, where Γ is any circle not passing through $z = a$, traversed once in the counterclockwise direction.

Solution. The integrand is analytic in the domain D consisting of the plane with the point $z = a$ deleted. If this point lies exterior to Γ, then Γ can be continuously deformed to a point in D, and so the integral vanishes. If a lies in the interior of Γ, the contour can be continuously deformed in D to a positively oriented circle centered at $z = a$, and thus the integral is $2\pi i$ by Eq. (4). Summarizing, we have

$$\int_\Gamma \frac{dz}{z - a} = \begin{cases} 0 & \text{if } a \text{ lies outside } \Gamma, \\ 2\pi i & \text{if } a \text{ lies inside } \Gamma. \end{cases} \tag{5}$$

∎

Example 6

Compute

$$\int_\Gamma (3z - 2) \Big/ (z^2 - z)\, dz,$$

where Γ is the simple closed contour indicated in Fig. 4.36.

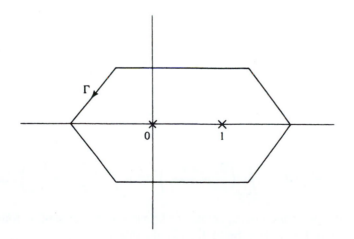

Figure 4.36 Contour for Example 6.

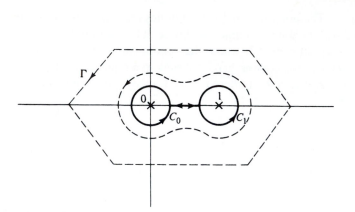

Figure 4.37 Deformation of contour in Fig. 4.36.

Solution. We don't need an exact description of Γ; since the integrand $f(z) = (3z - 2)/(z^2 - z)$ is analytic except at $z = 0$ and $z = 1$, the contour can be deformed to the "barbell"-shaped contour of Fig. 4.37 without affecting the value of the integral. This can be further simplified by observing that the integration along the line segment proceeds forward and backward, the results canceling each other. Thus

$$\int_\Gamma f(z)\,dz = \int_{C_0} f(z)\,dz + \int_{C_1} f(z)\,dz,$$

where the circles C_0 and C_1 are as indicated in Fig. 4.37.

We shall derive a powerful method for evaluating these integrals in Sec. 6.1. For now we merely use the partial fraction expansion to rewrite the integrand as

$$\frac{3z - 2}{z^2 - z} = \frac{A}{z} + \frac{B}{z - 1}. \tag{6}$$

In Eq. (6) the constants A and B are determined by formula (21) of Sec. 3.1:

$$A = \lim_{z \to 0} z \frac{3z - 2}{z^2 - z} = 2, \quad B = \lim_{z \to 1} (z - 1) \frac{3z - 2}{z^2 - z} = 1.$$

Thus

$$\int_\Gamma \frac{3z - 2}{z(z - 1)}\,dz = \int_{C_0} \left(\frac{2}{z} + \frac{1}{z - 1} \right) dz + \int_{C_1} \left(\frac{2}{z} + \frac{1}{z - 1} \right) dz.$$

The right-hand side of the last equation can be viewed as the sum of four integrals, each of the form of Example 5. So by Eq. (5), the integral is

$$2(2\pi i) + 0 + 2 \cdot 0 + 2\pi i = 6\pi i. \quad \blacksquare$$

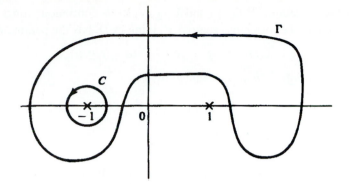

Figure 4.38 Contour deformation for Example 7.

Example 7

Evaluate $\int_{\Gamma} 1 / (z^2 - 1)\, dz$, where Γ is depicted as in Fig. 4.38.

Solution. Observing that $1 / (z^2 - 1)$ fails to be analytic only at $z = \pm 1$, we see that without passing through these points Γ can be continuously deformed into a small positively oriented circle C around $z = -1$. Again we use partial fractions to find

$$\frac{1}{z^2 - 1} = \frac{1}{2(z - 1)} - \frac{1}{2(z + 1)}.$$

Hence

$$\int_{\Gamma} \frac{1}{z^2 - 1}\, dz = \int_{C} \left[\frac{1}{2(z - 1)} - \frac{1}{2(z + 1)} \right] dz,$$

which, by Example 5, equals

$$\frac{1}{2} \cdot 0 - \frac{1}{2}(2\pi i) = -\pi i. \quad \blacksquare$$

Note: Exercises appear at the end of Sec. 4.4b.

4.4b Vector Analysis Approach

In Sec. 4.3 we deduced that a continuous function $f(z)$ possesses an (analytic) antiderivative in a domain D if, and only if, its integral around every loop in D is zero. Now we are going to show how this property ties in with the analyticity of $f(z)$ itself. To do this we shall employ some concepts and theorems from vector analysis (Ref. 8). First we demonstrate that our definition of the integral of $f(z)$ over a contour Γ can be related to the vector concept of a *line integral*.

Suppose that we have a two-dimensional vector $\mathbf{V} = (V_1, V_2)$ defined at every point (x, y) in some domain D in the plane; i.e., $\mathbf{V}$ is a *vector field*

$$\mathbf{V} = \mathbf{V}(x, y) = (V_1(x, y), V_2(x, y)).$$

For our purposes we require $V_1(x, y)$ and $V_2(x, y)$ to be continuous functions. Suppose furthermore that the (oriented) contour Γ, lying in D, has the parametrization

$$x = x(t), \qquad y = y(t) \qquad (a \le t \le b). \tag{7}$$

Then the *line integral of* $\mathbf{V}$ *along* Γ, denoted by

$$\int_\Gamma (V_1\, dx + V_2\, dy),$$

is given by

$$\int_\Gamma (V_1\, dx + V_2\, dy) := \int_a^b \left[V_1(x(t), y(t)) \frac{dx}{dt} + V_2(x(t), y(t)) \frac{dy}{dt} \right] dt.$$

Students of physics can interpret this as the work done by a force $\mathbf{V}(x, y)$ exerted on a particle as it traverses the contour Γ.

To see how line integrals relate to complex integration we shall write out $\int_\Gamma f(z)\, dz$ in terms of its real and imaginary parts, utilizing the parametrization (7). With the usual notation $f(z) = u(x, y) + iv(x, y)$, we have

$$\int_\Gamma f(z)\, dz = \int_a^b f(z(t)) \frac{dz(t)}{dt}\, dt$$

$$= \int_a^b [u(x(t), y(t)) + iv(x(t), y(t))] \left(\frac{dx}{dt} + i\frac{dy}{dt} \right) dt$$

$$= \int_a^b \left[u(x(t), y(t)) \frac{dx}{dt} - v(x(t), y(t)) \frac{dy}{dt} \right] dt$$

$$+ i \int_a^b \left[v(x(t), y(t)) \frac{dx}{dt} + u(x(t), y(t)) \frac{dy}{dt} \right] dt;$$

that is,

$$\int_\Gamma f(z)\, dz = \int_\Gamma (u\, dx - v\, dy) + i \int_\Gamma (v\, dx + u\, dy). \tag{8}$$

From this equation we can see that *the real part of* $\int_\Gamma f(z)\, dz$ *equals the line integral over* Γ *of the vector field* $(u, -v)$, *and that its imaginary part equals the line integral of the vector field* (v, u).[†]

Now that we can express complex integrals in terms of line integrals, we can translate many of the theorems of vector analysis into theorems about complex analysis. In Probs. 6, 7 and 8, the reader is guided in rediscovering some of the earlier theorems by this route. Our immediate goal is to uncover the consequences of *Green's theorem* in the context of the complex integral.

To apply Green's theorem it is convenient to introduce one new geometric concept; the simply connected domain. Roughly speaking, a domain is said to be simply connected if it has no holes, e.g., the inside of a simple closed contour (recall the Jordan curve theorem). One way of characterizing such domains is given in Definition 7.

[†] Observe that the vector $(u, -v)$ corresponds to the complex number $u - iv = \overline{f}$ and that (v, u) corresponds to $i\overline{f}$.

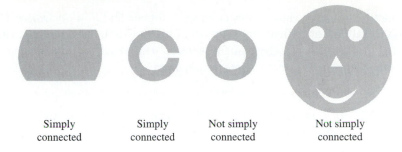

Simply   Simply   Not simply   Not simply
connected   connected   connected   connected

Figure 4.39 Examples of simply connected and multiply connected domains.

Definition 7.[†] A **simply connected domain** D is a domain having the following property: If Γ is any simple closed contour lying in D, then the domain interior to Γ lies wholly in D. (See Fig. 4.39.)

In this context, one statement of Green's theorem is given as follows (cf. Ref. 8).

Theorem 11. Let $\mathbf{V} = (V_1, V_2)$ be a continuously differentiable[‡] vector field defined on a simply connected domain D, and let Γ be a positively oriented simple closed contour in D. Then the line integral of $\mathbf{V}$ around Γ equals the integral of $(\partial V_2/\partial x - \partial V_1/\partial y)$, integrated with respect to area over the domain D' interior to Γ; i.e.,

$$\int_\Gamma (V_1\, dx + V_2\, dy) = \iint_{D'} \left(\frac{\partial V_2}{\partial x} - \frac{\partial V_1}{\partial y} \right) dx\, dy.$$

Let's apply this to the line integrals that occur in $\int_\Gamma f(z)\, dz$. Using Eq. (8), we have

$$\int_\Gamma f(z)\, dz = \int_\Gamma (u\, dx - v\, dy) + i \int_\Gamma (v\, dx + u\, dy)$$

$$= \iint_{D'} \left(-\frac{\partial v}{\partial x} - \frac{\partial u}{\partial y} \right) dx\, dy + i \iint_{D'} \left(\frac{\partial u}{\partial x} - \frac{\partial v}{\partial y} \right) dx\, dy. \tag{9}$$

Observe that we have assumed that u and v are continuously differentiable.

Now we take the big step. If $f(z)$ is *analytic* in D, the double integrals in Eq. (9) are zero because of the Cauchy-Riemann equations! In other words, we have shown

---

[†]Definition 7 is equivalent to Definition 6.

[‡]Recall that this means the partials $\partial V_1/\partial x$, $\partial V_1/\partial y$, $\partial V_2/\partial x$, $\partial V_2/\partial y$ exist and are continuous.

that if a function is analytic in a simply connected domain and *if its derivative $f'(z)$ is continuous* (recall that analyticity only stipulates that f' exist), then its integral around any simple closed contour in the domain is zero. This result, in a somewhat more general form, is known as *Cauchy's integral theorem.*

Theorem 12.[†] If f is analytic in a simply connected domain D and Γ is any loop (closed contour) in D, then

$$\int_\Gamma f(z)\, dz = 0.$$

Observe that we have generalized in two directions. First, we require only that Γ be a loop, not necessarily a simple closed curve. This is justified by the geometrically obvious fact that integration over a loop can always be decomposed into integrations over simple closed curves—see Fig. 4.40 for an illustration.

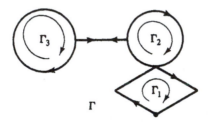

Figure 4.40 Decomposition of a loop integral.

The second generalization is that we have dropped the assumption that $f'(z)$ is continuous. The fact that this is possible was first demonstrated by the mathematician Edouard Goursat; see Ref. 2.

We remark that it can be shown by topological methods that if Γ is a simple closed contour and f is analytic at each point on and inside Γ, then f must be analytic in some simply connected domain containing Γ. Thus, by Theorem 12, **the integral along Γ must vanish whenever the integrand is analytic "inside and on Γ."**

Cauchy's theorem links the considerations of Sec. 4.3 with the property of analyticity. Combining Theorems 7 and 12 yields the following.

Theorem 13.[‡] In a simply connected domain, an analytic function has an antiderivative, its contour integrals are independent of path, and its loop integrals vanish.

[†]Theorem 12 is the same as Theorem 9.

In an earlier section we showed that if C is any circle centered at z_0 and n is an integer, then

$$\oint_C (z - z_0)^n \, dz = \begin{cases} 0 & \text{for } n \neq -1, \\ 2\pi i & \text{for } n = -1 \end{cases} \tag{10}$$

(see Example 2, Sec. 4.2). Equation (10) with $z_0 = 0$ neatly exemplifies the theory we have been discussing. If n is a positive integer or zero, z^n is analytic in the whole plane, which is simply connected; thus Theorem 13 applies, z^n has an antiderivative [the function $z^{n+1}/(n+1)$], and its loop integrals are zero.

If n is negative, z^n is analytic only in the *punctured* plane, with the origin deleted. This domain is not simply connected, so Theorem 13 does not apply. In fact, for $n = -1$, the function z^n does not even have an antiderivative in the punctured plane (since any branch of $\log z$ will be discontinuous on the branch cut), and sure enough the loop integral (10) fails to vanish. On the other hand, if $n \leq -2$, then z^n regains its antiderivative $z^{n+1}/(n+1)$ in the punctured plane, and the loop integrals (10) are zero. Thus either case can occur in a domain that is not simply connected (such a domain is called *multiply connected*).

Example 1

Evaluate

$$\oint_{|z|=2} \frac{e^z}{z^2 - 9} \, dz.$$

Solution. The notation employed signifies that the contour of integration is the circle $|z| = 2$ traversed once counterclockwise. The integrand is analytic everywhere except at $z = \pm 3$, where the denominator vanishes. Since these points lie exterior to the contour, the integral is zero, by Cauchy's integral theorem. ∎

Theorems 12 and 13 can often be used to change the contour of integration, as the following examples demonstrate:

Example 2

Evaluate $\int_\Gamma 1/z \, dz$, where Γ is the ellipse defined by $x^2 + 4y^2 = 1$, traversed once in the positive sense, as indicated in Fig. 4.41(a).

Solution. We shall show that one can change the contour from Γ to the positively oriented unit circle without changing the integral. With reference to Fig. 4.41(b), observe that the complex plane, slit down the negative y-axis from the origin, constitutes a simply connected domain in which the function $1/z$ is analytic. Hence, by Theorem 13,

$$\int_{\gamma_2} \frac{1}{z} \, dz = \int_{\gamma_1} \frac{1}{z} \, dz.$$

Similarly, by considering the plane slit along the positive y-axis we have

$$\int_{\gamma_3} \frac{1}{z} \, dz = \int_{\gamma_4} \frac{1}{z} \, dz.$$

---

‡ Theorem 13 is the same as Theorem 10.

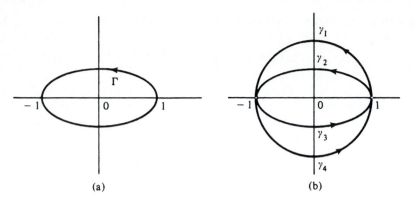

Figure 4.41 Contours for Example 2.

Hence

$$\int_\Gamma \frac{1}{z}\,dz = \int_{\gamma_2+\gamma_3} \frac{1}{z}\,dz = \int_{\gamma_1+\gamma_4} \frac{1}{z}\,dz = \oint_{|z|=1} \frac{1}{z}\,dz,$$

and by Eq. (10) the answer is $2\pi i$. ∎

This technique is easily generalized in the next example.

Example 3

Determine the possible values for $\int_\Gamma 1/(z-a)\,dz$, where Γ is any positively oriented simple closed contour not passing through $z = a$.

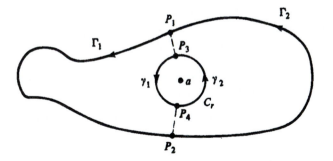

Figure 4.42 Contours for Example 3.

Solution.   Observe that the integrand is analytic everywhere except at the point $z = a$. Thus if a lies exterior to Γ, Cauchy's theorem yields the answer zero for the integral. If a lies inside Γ, we choose a small circle C_r centered at a and lying within Γ, and we draw two segments from Γ to C_r; see Fig. 4.42 for the construction. Then the endpoints P_1 and P_2 of the segments divide Γ into two contours Γ_1 and Γ_2, and the endpoints P_3 and P_4 divide C_r into γ_1 and γ_2. Now observe that both the contour

Γ_1 and the composite contour consisting of the directed segment P_1P_3, γ_1, and the segment P_4P_2 can be enclosed in a simply connected domain that excludes the point a. Thus we deduce from Theorem 13 that the integral is the same along these contours:

$$\int_{\Gamma_1} \frac{dz}{z-a} = \left(\int_{P_1P_3} + \int_{\gamma_1} + \int_{P_4P_2} \right) \frac{dz}{z-a}.$$

Similarly,

$$\int_{\Gamma_2} \frac{dz}{z-a} = \left(\int_{P_2P_4} + \int_{\gamma_2} + \int_{P_3P_1} \right) \frac{dz}{z-a}.$$

Adding these and taking account of the cancellations along the line segments we find that

$$\int_{\Gamma} \frac{dz}{z-a} = \left(\int_{\Gamma_1} + \int_{\Gamma_2} \right) \frac{dz}{z-a} = \left(\int_{\gamma_1} + \int_{\gamma_2} \right) \frac{dz}{z-a} = \oint_{C_r} \frac{dz}{z-a} = 2\pi i$$

(recall Eq. (10)).

Summarizing, we have

$$\int_{\Gamma} \frac{dz}{z-a} = \begin{cases} 0 & \text{if } a \text{ lies outside } \Gamma, \\ 2\pi i & \text{if } a \text{ lies inside } \Gamma. \end{cases} \tag{11}$$

■

Example 4

Find $\int_{\Gamma} (3z-2)\big/(z^2-z)\,dz$, where Γ is the simple closed contour indicated in Fig. 4.43(a).

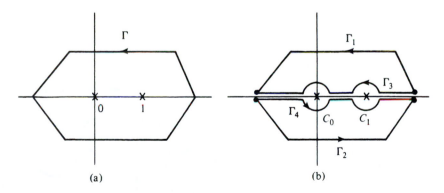

Figure 4.43 Contours for Example 4.

Solution.   The integrand $f(z) = (3z-2)\big/(z^2-z)$ is, of course, analytic everywhere except for the zeros of the denominator, $z = 0$ and $z = 1$. Referring to Fig. 4.43(b), we begin by enclosing these points in small circles C_0 and C_1, respectively, and observe by Theorem 13 that the integral over Γ_1, the upper portion of Γ, equals the integral over the contour indicated as Γ_3. Similarly, integration over Γ_2 can be replaced by integration over Γ_4. Combining these and taking into account the cancellations along the segments of the real axis we find

$$\int_\Gamma f(z)\,dz = \left(\int_{\Gamma_1} + \int_{\Gamma_2} \right) f(z)\,dz = \oint_{C_0} f(z)\,dz + \oint_{C_1} f(z)\,dz.$$

We shall derive a powerful method for evaluating these integrals in Sec. 6.1. For now we merely use the partial fraction expansion to write the integrand as

$$\frac{3z-2}{z^2-z} = \frac{A}{z} + \frac{B}{z-1}. \tag{12}$$

In Eq. (12) the constants A and B are determined by formula (21) of Sec. 3.1:

$$A = \lim_{z\to 0} z\frac{3z-2}{z^2-z} = 2, \ \ B = \lim_{z\to 1}(z-1)\frac{3z-2}{z^2-z} = 1.$$

Thus

$$\int_\Gamma \frac{3z-2}{z(z-1)}\,dz = \oint_{C_0}\left(\frac{2}{z} + \frac{1}{z-1}\right)dz + \oint_{C_1}\left(\frac{2}{z} + \frac{1}{z-1}\right)dz.$$

The right-hand side of the last equation can be viewed as the sum of four integrals, each of the form of Example 3. So by Eq. (11), the integral is

$$2(2\pi i) + 0 + 2\cdot 0 + 2\pi i = 6\pi i. \quad \blacksquare$$

Example 5

Evaluate $\int_\Gamma 1\left/\left(z^2 - 1\right)\right. dz$, where Γ is depicted in Fig. 4.44(a).

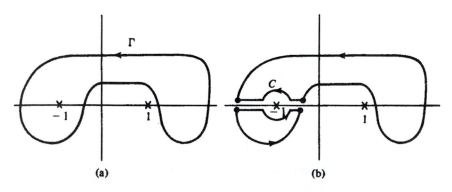

Figure 4.44 Contours for Example 5.

Solution. Observing that $1\left/\left(z^2 - 1\right)\right.$ fails to be analytic only at $z = \pm 1$, the reader should be able by now to use the construction indicated in Fig. 4.44(b) to argue that the integral over Γ is the same as the integral over the small circle C enclosing -1. Using partial fractions again we find

$$\frac{1}{z^2 - 1} = \frac{1}{2(z-1)} - \frac{1}{2(z+1)}.$$

Hence

$$\int_\Gamma \frac{dz}{z^2 - 1} = \int_C \frac{dz}{z^2 - 1} = \int_C \left[\frac{1}{2(z-1)} - \frac{1}{2(z+1)} \right] dz,$$

which, by Example 3, equals

$$\frac{1}{2} \cdot 0 - \frac{1}{2}(2\pi i) = -\pi i. \quad \blacksquare$$

EXERCISES 4.4

Problems 1–5 refer to Sec. 4.4a.

1. Let D be the domain consisting of the complex plane with the three points 0, $2i$, and 4 deleted and let Γ be the (solid-line) contour shown in Fig. 4.45. Decide which of the following contours are continuously deformable to Γ in D.

   (a) the dashed line contour Γ_0 in Fig. 4.45

   (b) the circle $|z| = 3$ traversed once in the positive direction starting from the point $z = 3$

   (c) the circle $|z| = 10^4$ traversed once in the positive direction starting from the point $z = 10^4$

   (d) the circle $|z - 2| = 1$ traversed once in the positive direction starting from the point $z = 3$

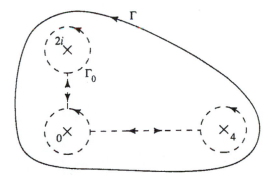

Figure 4.45 Contour for Prob. 1.

2. Prove the statement made in the text: If the contours Γ_0 and Γ_1 can each be shrunk to points in the domain D, then Γ_0 can be continuously deformed into Γ_1 in D. (Do not assume that Γ_0 and Γ_1 are deformable to the *same* point.)

3. Let D be the annulus $1 < |z| < 5$, and let Γ be the circle $|z - 3| = 1$ traversed once in the positive direction starting from the point $z = 4$. Decide which of the following contours are continuously deformable to Γ in D.

   (a) the circle $|z - 3| = 1$ traversed once in the positive direction starting from the point $z = 2$

   (b) the point $z = 3i$

   (c) the circle $|z| = 2$ traversed once in the positive direction starting from the point $z = 2$

   (d) the circle $|z + 3| = 1$ traversed once in the positive direction starting from the point $z = -2$

   (e) the circle $|z - 3| = 1$ traversed twice in the negative direction starting from the point $z = 4$

4. Let Γ_0 be the unit circle $|z| = 1$ traversed once counterclockwise and then once clockwise, starting from $z = 1$. Construct a function $z(s, t)$ which deforms Γ_0 to the single point $z = 1$ in *any* domain D containing the unit circle. Verify directly that the conclusion of Theorem 8 is true for these two contours.

5. Write down a function $z(s, t)$ deforming Γ_0 into Γ_1 in the domain D, where Γ_0 is the ellipse $x^2/4 + y^2/9 = 1$ traversed once counterclockwise starting from $(2, 0)$, Γ_1 is the circle $|z| = 1$ traversed once counterclockwise starting from $(1, 0)$ and D is the annulus $1/2 < |z| < 4$. [HINT: Start with the parametrization $x(t) = 2 \cos 2\pi t$, $y(t) = 3 \sin 2\pi t$, $0 \le t \le 1$, for Γ_0.]

Problems 6–8 refer to Sec. 4.4b.

6. It is well known from potential theory that if the line integrals of a vector field $\mathbf{V}(x, y)$ are independent of path (i.e., if $\mathbf{V}$ is a "conservative" field), then there is a scalar function of position $\phi(x, y)$ such that $V_1 = \partial\phi/\partial x$ and $V_2 = \partial\phi/\partial y$ (under such conditions we say that ϕ is a *potential* for $\mathbf{V}$). Apply this result to the vector fields $\overline{f(z)}$ and $i\,\overline{f(z)}$ to prove that property (iii) implies (i) in Theorem 7. What is the relationship between the (analytic) antiderivative $F(z)$ and the potentials?

7. In vector analysis a vector field $\mathbf{V} = (V_1, V_2)$ is said to be *irrotational* if its components satisfy

$$\frac{\partial V_1}{\partial y} = \frac{\partial V_2}{\partial x};$$

it is called *solenoidal* if

$$\frac{\partial V_1}{\partial x} = -\frac{\partial V_2}{\partial y}.$$

(a) Show that, if $f(z)$ is an analytic function, the vector field corresponding to $\overline{f(z)}$ is both irrotational and solenoidal.

(b) Prove the converse to part (a), if the vector field is continuously differentiable.

8. An important result from potential theory says that if a vector field $\mathbf{V}$ is irrotational (see Prob. 7) in a simply connected domain D, then there is a potential function for $\mathbf{V}(x, y)$ in D. Applying this fact to $\overline{f(z)}$ and $i\,\overline{f(z)}$, prove the first assertion in Theorem 13.

Problems 9–19 are for both Secs. 4.4a and 4.4b.

9. Which of the following domains are simply connected?

(a) the horizontal strip $|\operatorname{Im} z| < 1$

(b) the annulus $1 < |z| < 2$

(c) the set of all points in the plane except those on the nonpositive x-axis

(d) the interior of the ellipse $4x^2 + y^2 = 1$

(e) the exterior of the ellipse $4x^2 + y^2 = 1$

(f) the domain D in Fig. 4.46.

Figure 4.46 Is D simply connected?.

10. Determine the domain of analyticity for each of the given functions f and explain why

$$\oint_{|z|=2} f(z)\, dz = 0.$$

(a) $f(z) = \dfrac{z}{z^2 + 25}$

(b) $f(z) = e^{-z}(2z + 1)$

(c) $f(z) = \dfrac{\cos z}{z^2 - 6z + 10}$

(d) $f(z) = \operatorname{Log}(z + 3)$

(e) $f(z) = \sec\left(\dfrac{z}{2}\right)$

11. Explain why the function e^{z^2} has an antiderivative in the whole plane.

12. Given that D is a domain containing the closed contour Γ, that z_0 is a point not in D, and that $\int_\Gamma (z - z_0)^{-1}\, dz \neq 0$, explain why D is not simply connected.

13. Evaluate $\int 1/(z^2 + 1)\, dz$ along the three closed contours Γ_1, Γ_2, Γ_3 in Fig. 4.47.

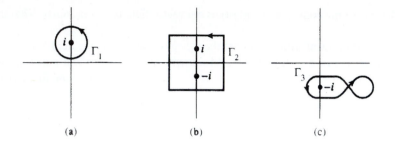

(a)   (b)   (c)

Figure 4.47 Contours for Prob. 13.

14. Consider the shaded domain D in Fig. 4.48 bounded by the simple closed positively oriented contours C, C_1, C_2 and C_3. If $f(z)$ is analytic on D and on its boundary, explain why

$$\int_C f(z)\, dz = \int_{C_1} f(z)\, dz + \int_{C_2} f(z)\, dz + \int_{C_3} f(z)\, dz.$$

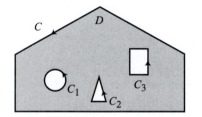

Figure 4.48 Domain for Prob. 14.

15. Evaluate

$$\int_\Gamma \frac{z}{(z+2)(z-1)}\, dz,$$

where Γ is the circle $|z| = 4$ traversed twice in the clockwise direction.

16. Show that if f is of the form

$$f(z) = \frac{A_k}{z^k} + \frac{A_{k-1}}{z^{k-1}} + \cdots + \frac{A_1}{z} + g(z) \quad (k \geq 1),$$

where g is analytic inside and on the circle $|z| = 1$, then

$$\oint_{|z|=1} f(z)\, dz = 2\pi i A_1.$$

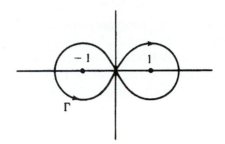

Figure 4.49 Contour for Prob. 17.   **Figure 4.50** Contour for Prob. 20.

17. Evaluate
$$\int_\Gamma \frac{2z^2 - z + 1}{(z-1)^2(z+1)}\, dz,$$
where Γ is the figure-eight contour traversed once as shown in Fig. 4.49. [HINT: Use the partial fraction expansion $\dfrac{A}{(z-1)^2} + \dfrac{B}{(z-1)} + \dfrac{C}{(z+1)}$; see Sec. 3.1.]

18. Let
$$I := \oint_{|z|=2} \frac{dz}{z^2(z-1)^3}.$$
Below is an outline of a proof that $I = 0$. Justify each step.

  (a) For every $R > 2$, $I = I(R)$, where
$$I(R) := \oint_{|z|=R} \frac{1}{z^2(z-1)^3}\, dz.$$
  (b) $|I(R)| \le \dfrac{2\pi}{R(R-1)^3}$ for $R > 2$.
  (c) $\lim\limits_{R \to +\infty} I(R) = 0.$
  (d) $I = 0.$

19. Using the method of proof in Prob. 18, establish the following theorem. If P is a polynomial of degree at least 2 and P has all its zeros inside the circle $|z| = r$, then
$$\oint_{|z|=r} \frac{1}{P(z)}\, dz = 0.$$

20. Let Γ denote the four-leaf clover path traversed once as shown in Fig. 4.50. Show that
$$\int_\Gamma \frac{1}{z^4 - 1}\, dz = 0$$
in two ways; first, by using partial fractions and second, by using the result of Prob. 19.

4.5   Cauchy's Integral Formula and Its Consequences

Given f analytic inside and on the simple closed contour Γ, we know from Cauchy's theorem that $\int_\Gamma f(z)\,dz = 0$. However, if we consider the integral $\int_\Gamma f(z)/(z - z_0)\,dz$, where z_0 is a point in the interior of Γ, then there is no reason to expect that this integral is zero, because the integrand has a singularity inside the contour Γ. In fact, as the primary result of this section, we shall show that for all z_0 inside Γ the value of this integral is proportional to $f(z_0)$.

Theorem 14. *(Cauchy's Integral Formula)* Let Γ be a simple closed positively oriented contour. If f is analytic in some simply connected domain D containing Γ and z_0 is any point inside Γ, then

$$f(z_0) = \frac{1}{2\pi i} \int_\Gamma \frac{f(z)}{z - z_0}\,dz. \tag{1}$$

Proof.   The function $f(z)/(z - z_0)$ is analytic everywhere in D except for the point z_0. Hence by the methods of Sec. 4.4 the integral over Γ can be equated to the integral over some small positively oriented circle $C_r : |z - z_0| = r$; Fig. 4.51(a) illustrates the continuous deformation method of Sec. 4.4a, while Fig. 4.51(b) shows the construction appropriate to Sec. 4.4b. So we can write

$$\int_\Gamma \frac{f(z)}{z - z_0}\,dz = \int_{C_r} \frac{f(z)}{z - z_0}\,dz.$$

It is now convenient to express the right-hand side as the sum of two integrals:

$$\int_{C_r} \frac{f(z)}{z - z_0}\,dz = \int_{C_r} \frac{f(z_0)}{z - z_0}\,dz + \int_{C_r} \frac{f(z) - f(z_0)}{z - z_0}\,dz.$$

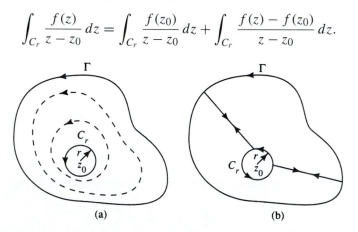

(a)   (b)

Figure 4.51 Contours for Cauchy's integral formula.

From our earlier deliberations we know that

$$\int_{C_r} \frac{f(z_0)}{z - z_0} \, dz = f(z_0) \int_{C_r} \frac{dz}{z - z_0} = f(z_0) \, 2\pi i;$$

consequently,

$$\int_{\Gamma} \frac{f(z)}{z - z_0} \, dz = f(z_0) \, 2\pi i + \int_{C_r} \frac{f(z) - f(z_0)}{z - z_0} \, dz. \tag{2}$$

Now observe that the first two terms in Eq. (2) are constants independent of r, and so the value of the last term does not change if we allow r to decrease to zero; i.e.,

$$\int_{\Gamma} \frac{f(z)}{z - z_0} \, dz = f(z_0) \, 2\pi i + \lim_{r \to 0^+} \int_{C_r} \frac{f(z) - f(z_0)}{z - z_0} \, dz. \tag{3}$$

Therefore, Cauchy's formula will follow if we can prove that the last limit is zero.

For this purpose set $M_r := \max\{|f(z) - f(z_0)| : z \text{ on } C_r\}$. Then for z on C_r, we have

$$\left| \frac{f(z) - f(z_0)}{z - z_0} \right| = \frac{|f(z) - f(z_0)|}{r} \leq \frac{M_r}{r},$$

and hence by Theorem 5 in Sec. 4.2,

$$\left| \int_{C_r} \frac{f(z) - f(z_0)}{z - z_0} \, dz \right| \leq \frac{M_r}{r} \, \ell(C_r) = \frac{M_r}{r} \, 2\pi r = 2\pi M_r.$$

But since f is continuous at the point z_0, we know that $\lim_{r \to 0^+} M_r = 0$. Thus

$$\lim_{r \to 0^+} \int_{C_r} \frac{f(z) - f(z_0)}{z - z_0} \, dz = 0,$$

and so Eq. (3) reduces to (1).   ■

One remarkable consequence of Cauchy's formula is that by merely knowing the values of the analytic function f *on* Γ we can compute the integral in Eq. (1) and hence all the values of f *inside* Γ. In other words, the behavior of a function analytic in a region is completely determined by its behavior on the boundary.

We shall now present some examples that employ Cauchy's formula to evaluate certain integrals. The reader should keep in mind, however, that Chapter 6 will be devoted to more efficient and powerful techniques for computing integrals and that the present examples are intended primarily to illustrate the integral formula (1).

Example 1

Compute the integral

$$\int_{\Gamma} \frac{e^z + \sin z}{z} \, dz,$$

where Γ is the circle $|z - 2| = 3$ traversed once in the counterclockwise direction.

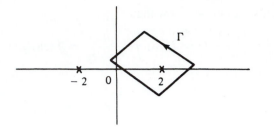

Figure 4.52 Contour for Example 2.

Solution.   Observe that the function $f(z) = e^z + \sin z$ is analytic inside and on Γ and that the point $z_0 = 0$ lies inside this circle. Hence the integral has the format of formula (1) and the desired value is

$$2\pi i f(0) = 2\pi i \left[e^0 + \sin 0 \right] = 2\pi i. \quad \blacksquare$$

Example 2

Evaluate the integral

$$\int_\Gamma \frac{\cos z}{z^2 - 4}\, dz$$

along the contour sketched in Fig. 4.52.

Solution.   We first notice that the integrand fails to be analytic at the points $z = \pm 2$. However, only one of these, $z = 2$, occurs inside Γ. Thus if we write

$$\frac{\cos z}{z^2 - 4} = \frac{(\cos z)/(z + 2)}{z - 2},$$

we again have the format of Eq. (1). Hence the integral is

$$\int_\Gamma \frac{\cos z}{z^2 - 4}\, dz = 2\pi i \ \cos z/(z + 2)\Big|_{z=2} = \frac{2\pi i \cos 2}{4}. \quad \blacksquare$$

Example 3

Compute

$$\int_C \frac{z^2 e^z}{2z + i}\, dz,$$

where C is the unit circle $|z| = 1$ traversed once in the clockwise direction.

Solution.   Two minor difficulties inhibit an immediate application of Cauchy's formula. First, the denominator is not of the form $z - z_0$, and second, the contour is *negatively* oriented. The former difficulty is easily resolved by writing

$$\frac{z^2 e^z}{2z + i} = \frac{\frac{1}{2} z^2 e^z}{z + i/2}$$

(notice that the singular point $z = -i/2$ lies *inside* C). And to compensate for the negative orientation of C we have to introduce a minus sign in formula (1):

$$\int_C \frac{z^2 e^z}{2z + i} \, dz = \int_C \frac{\frac{1}{2} z^2 e^z}{z + i/2} \, dz = -2\pi i \cdot \frac{1}{2} z^2 e^z \Big|_{z=-i/2} = \frac{\pi i}{4} e^{-i/2}. \quad \blacksquare$$

If in Cauchy's formula (1) we replace z by ζ and z_0 by z, then we obtain

$$f(z) = \frac{1}{2\pi i} \int_\Gamma \frac{f(\zeta)}{\zeta - z} \, d\zeta \quad (z \text{ inside } \Gamma). \tag{4}$$

The advantage of this representation is that it suggests a formula for the derivative $f'(z)$, obtained by formally differentiating with respect to z under the integral sign. Thus we are led to suspect that

$$f'(z) = \frac{1}{2\pi i} \int_\Gamma \frac{f(\zeta)}{(\zeta - z)^2} \, d\zeta \quad (z \text{ inside } \Gamma). \tag{5}$$

In verifying this equation we shall actually establish a more general theorem.

Theorem 15. Let g be continuous on the contour Γ, and for each z not on Γ set

$$G(z) := \int_\Gamma \frac{g(\zeta)}{\zeta - z} \, d\zeta. \tag{6}$$

Then the function G is analytic at each point not on Γ, and its derivative is given by

$$G'(z) = \int_\Gamma \frac{g(\zeta)}{(\zeta - z)^2} \, d\zeta. \tag{7}$$

(Observe that we have generalized in two directions; we have not assumed that Γ is closed or that g is analytic. Furthermore, we make no claim about the limiting values of $G(z)$ as z approaches a point on Γ; see Prob. 13.)

Proof of Theorem 15. Let z be any fixed point not on Γ. To prove the existence of $G'(z)$ and the formula (7) we need to show that

$$\lim_{\Delta z \to 0} \frac{G(z + \Delta z) - G(z)}{\Delta z} = \int_\Gamma \frac{g(\zeta)}{(\zeta - z)^2} \, d\zeta,$$

or, equivalently, that the difference

$$J := \frac{G(z + \Delta z) - G(z)}{\Delta z} - \int_\Gamma \frac{g(\zeta)}{(\zeta - z)^2} \, d\zeta \tag{8}$$

approaches zero as $\Delta z \to 0$. This is accomplished by first writing J in a convenient form obtained as follows.

Using Eq. (6) we have

$$\frac{G(z + \Delta z) - G(z)}{\Delta z} = \frac{1}{\Delta z} \int_{\Gamma} \left[\frac{1}{\zeta - (z + \Delta z)} - \frac{1}{\zeta - z} \right] g(\zeta) \, d\zeta$$

$$= \int_{\Gamma} \frac{g(\zeta) \, d\zeta}{(\zeta - z - \Delta z)(\zeta - z)},$$

where Δz is chosen sufficiently small so that $z + \Delta z$ also lies off of Γ. Then from Eq. (8) and some elementary algebra we find

$$J = \int_{\Gamma} \frac{g(\zeta) \, d\zeta}{(\zeta - z - \Delta z)(\zeta - z)} - \int_{\Gamma} \frac{g(\zeta)}{(\zeta - z)^2} \, d\zeta$$

$$= \Delta z \int_{\Gamma} \frac{g(\zeta) \, d\zeta}{(\zeta - z - \Delta z)(\zeta - z)^2}.$$

$$(9)$$

To verify that $J \to 0$ as $\Delta z \to 0$, we estimate the last expression. In this regard, let M equal the maximum value of $|g(\zeta)|$ on Γ, and set d equal to the shortest distance from z to Γ, so that $|\zeta - z| \geq d > 0$ for all ζ on Γ. Since we are letting Δz approach zero, we may assume that $|\Delta z| < d/2$. Then, by the triangle inequality,

$$|\zeta - z - \Delta z| \geq |\zeta - z| - |\Delta z| \geq d - \frac{d}{2} = \frac{d}{2} \quad (\zeta \text{ on } \Gamma)$$

(see Fig. 4.53) and so

$$\left| \frac{g(\zeta)}{(\zeta - z - \Delta z)(\zeta - z)^2} \right| \leq \frac{M}{\frac{d}{2} \cdot d^2} = \frac{2M}{d^3}$$

for all ζ on Γ. Hence from Theorem 5, Sec. 4.2, we see that

$$|J| = \left| \Delta z \int_{\Gamma} \frac{g(\zeta) d\zeta}{(\zeta - z - \Delta z)(\zeta - z)^2} \right| \leq \frac{|\Delta z| 2M \ell(\Gamma)}{d^3},$$

where $\ell(\Gamma)$ denotes the length of Γ. Thus J must approach zero as $\Delta z \to 0$. This implies that formula (7) is valid and completes the proof. ∎

The preceding argument can be carried further; namely, starting with the function

$$H(z) := \int_{\Gamma} \frac{g(\zeta)}{(\zeta - z)^2} \, d\zeta \quad (z \text{ not on } \Gamma), \quad (10)$$

it can be shown that H is analytic off of Γ and that H' is given by the formula

$$H'(z) = 2 \int_{\Gamma} \frac{g(\zeta)}{(\zeta - z)^3} \, d\zeta \quad (z \text{ not on } \Gamma), \quad (11)$$

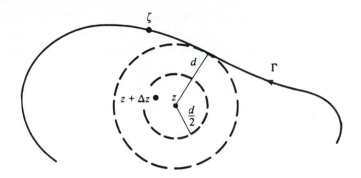

Figure 4.53 Contour for Theorem 15.

obtained formally from (10) by differentiation under the integral sign. The proof of this fact parallels the proof of Theorem 15 and is left as an exercise.

One important consequence of these results is that the derivative of an analytic function is again analytic. For suppose that f is analytic at the point z_0. We wish to argue that f' is also analytic at z_0; i.e., f' itself has a derivative in some neighborhood of z_0. To this end, we choose a positively oriented circle $C : |\zeta - z_0| = r$ so small that f is analytic inside and on C. Since f has the Cauchy integral representation

$$f(z) = \frac{1}{2\pi i} \int_C \frac{f(\zeta)}{\zeta - z} \, d\zeta \qquad (z \text{ inside } C),$$

it follows from Theorem 15 that

$$f'(z) = \frac{1}{2\pi i} \int_C \frac{f(\zeta)}{(\zeta - z)^2} \, d\zeta \qquad (z \text{ inside } C),$$

But the right-hand side is a function of the form (10) and hence has a derivative at each point inside C. Thus f' must be analytic at z_0.

The same reasoning can be applied to the function f' to deduce that *its* derivative, f'', is also analytic at z_0. More generally, the analyticity of $f^{(n)}$ implies that of $f^{(n+1)}$, and so by induction we obtain the following result.

Theorem 16. If f is analytic in a domain D, then all its derivatives $f', f'', \ldots,$ $f^{(n)}, \ldots$ exist and are analytic in D.

Theorem 16 is particularly surprising in light of the fact that its analogue in calculus fails to be true; for example, the function $f(x) = x^{5/3}$ is differentiable for all real x, but $f'(x) = 5x^{2/3}/3$ has no derivative at $x = 0$.

Recall that if the analytic function f is written in the form $f(z) = u(x, y) + iv(x, y)$, then as explained in Sec. 2.4 we have the alternative expression

$$f'(z) = \frac{\partial u}{\partial x} + i\frac{\partial v}{\partial x} \text{ and } f'(z) = \frac{\partial v}{\partial y} - i\frac{\partial u}{\partial y}. \tag{12}$$

We now know that f' is analytic and hence continuous. Therefore, from (12), all the first-order partial derivatives of u and v must be continuous. Similarly, since f'' exists, the formulas (12) together with the Cauchy-Riemann equations for f' lead to the expressions

$$f''(z) = \frac{\partial^2 u}{\partial x^2} + i \frac{\partial^2 v}{\partial x^2} = \frac{\partial^2 v}{\partial y \partial x} - i \frac{\partial^2 u}{\partial y \partial x}$$

$$f''(z) = \frac{\partial^2 v}{\partial x \partial y} - i \frac{\partial^2 u}{\partial x \partial y} = -\frac{\partial^2 u}{\partial y^2} - i \frac{\partial^2 v}{\partial y^2},$$

and so the continuity of f'' implies that all second-order partial derivatives of u and v are continuous at the points where f is analytic. Continuing with this process we obtain the following theorem.

Theorem 17. If $f = u + iv$ is analytic in a domain D, then all partial derivatives of u and v exist and are continuous in D.

(This theorem validates the argument given in Sec. 2.5 that the real and imaginary parts of an analytic function $f = u + iv$ are harmonic[†]. Applying this argument to the functions f', f'', ..., we see that all partials of u and v are harmonic as well.)

Another way to phrase the results on the analyticity of derivatives is to say that whenever a given function f has an antiderivative in a domain D, then f must itself be analytic in D.[‡] Now by Theorem 7 of Sec. 4.3 the existence of an antiderivative for a continuous function is equivalent to the property that all loop integrals vanish. Hence we deduce the following result, known as *Morera's theorem*.

Theorem 18. If f is continuous in a domain D and if

$$\int_{\Gamma} f(z)\, dz = 0$$

for every closed contour Γ in D, then f is analytic in D.

Observe that in establishing equations (7) and (11) we actually verify that for certain types of integrands the process of differentiation with respect to z can be interchanged with the process of integration with respect to ζ. In fact, starting with Cauchy's integral formula, it can be shown inductively that repeated differentiations with respect to z under the integral sign yields valid formulas for the successive derivatives of f. Keeping track of the exponents, then, we have the *generalized Cauchy integral formula*.

[†]Specifically, in Sec. 2.5 we *assumed* that the second partials of u and v were continuous.

[‡] Of course the antiderivative is analytic, since it has a derivative!

Theorem 19. If f is analytic inside and on the simple closed positively oriented contour Γ and if z is any point inside Γ, then

$$f^{(n)}(z) = \frac{n!}{2\pi i} \int_\Gamma \frac{f(\zeta)}{(\zeta - z)^{n+1}} \, d\zeta \quad (n = 1, 2, 3, \ldots). \qquad (13)$$

For purposes of application it is convenient to write Eq. (13) in the equivalent form

$$\int_\Gamma \frac{f(z)}{(z - z_0)^m} \, dz = \frac{2\pi i f^{(m-1)}(z_0)}{(m - 1)!} \quad (z_0 \text{ inside } \Gamma, \ m = 1, 2, \ldots). \qquad (14)$$

Example 4

Compute $\int_\Gamma e^{5z}/z^3 \, dz$, where Γ is the circle $|z| = 1$ traversed once counterclockwise.

Solution.   Observe that $f(z) = e^{5z}$ is analytic inside and on Γ. Therefore, from Eq. (14) with $z_0 = 0$ and $m = 3$ we have

$$\int_\Gamma \frac{e^{5z}}{z^3} \, dz = \frac{2\pi i f''(0)}{2!} = 25\pi i. \quad \blacksquare$$

Example 5

Compute

$$\int_C \frac{2z + 1}{z(z - 1)^2} \, dz$$

along the figure-eight contour C sketched in Fig. 4.54.

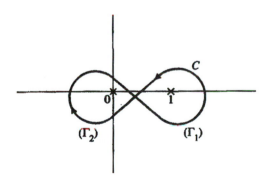

Figure 4.54 Contour for Example 5.

Solution.   Notice that integration along C is equivalent to integrating once around the positively oriented right lobe Γ_1 and then integrating once around the negatively oriented left lobe Γ_2; i.e.,

$$\int_C \frac{2z + 1}{z(z - 1)^2} \, dz = \int_{\Gamma_1} \frac{(2z + 1)/z}{(z - 1)^2} \, dz + \int_{\Gamma_2} \frac{(2z + 1)/(z - 1)^2}{z} \, dz,$$

where we have written the integrand in each term so as to display the relevant singularity. These integrals along Γ_1 and Γ_2 can be evaluated by using Eq. (14) with $m = 2$ and $m = 1$; the desired value is

$$\frac{2\pi i}{1!} \frac{d}{dz} \left(\frac{2z+1}{z} \right) \Big|_{z=1} - 2\pi i \frac{2z+1}{(z-1)^2} \Big|_{z=0} = -2\pi i - 2\pi i = -4\pi i. \quad \blacksquare$$

EXERCISES 4.5

1. Let f be analytic inside and on the simple closed contour Γ. What is the value of

$$\frac{1}{2\pi i} \int_\Gamma \frac{f(z)}{z - z_0} \, dz$$

   when z_0 lies outside Γ?

2. Let f and g be analytic inside and on the simple loop Γ. Prove that if $f(z) = g(z)$ for all z on Γ, then $f(z) = g(z)$ for all z inside Γ.

3. Let C be the circle $|z| = 2$ traversed once in the positive sense. Compute each of the following integrals.

   (a) $\displaystyle \int_C \frac{\sin 3z}{z - \frac{\pi}{2}} \, dz$ (b) $\displaystyle \int_C \frac{ze^z}{2z - 3} \, dz$ (c) $\displaystyle \int_C \frac{\cos z}{z^3 + 9z} \, dz$

   (d) $\displaystyle \int_C \frac{5z^2 + 2z + 1}{(z - i)^3} \, dz$ (e) $\displaystyle \int_C \frac{e^{-z}}{(z + 1)^2} \, dz$ (f) $\displaystyle \int_C \frac{\sin z}{z^2(z - 4)} \, dz$

4. Compute

$$\int_C \frac{z + i}{z^3 + 2z^2} \, dz,$$

   where C is

   (a) the circle $|z| = 1$ traversed once counterclockwise.

   (b) the circle $|z + 2 - i| = 2$ traversed once counterclockwise.

   (c) the circle $|z - 2i| = 1$ traversed once counterclockwise.

5. Let C be the ellipse $x^2/4 + y^2/9 = 1$ traversed once in the positive direction, and define

$$G(z) := \int_C \frac{\zeta^2 - \zeta + 2}{\zeta - z} \, d\zeta \qquad (z \text{ inside } C).$$

   Find $G(1)$, $G'(i)$, and $G''(-i)$.

6. Evaluate

$$\int_\Gamma \frac{e^{iz}}{\left(z^2 + 1\right)^2} \, dz,$$

   where Γ is the circle $|z| = 3$ traversed once counterclockwise. [HINT: Show that the integral can be written as the sum of two integrals around small circles centered at the singularities.]

7. Compute

$$\int_{\Gamma} \frac{\cos z}{z^2(z-3)}\, dz$$

along the contour indicated in Fig. 4.55.

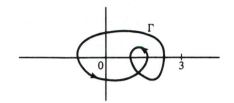

Figure 4.55 Contour for Prob. 7.

8. Use Cauchy's formula to show that if f is analytic inside and on the circle $|z - z_0| = r$, then

$$f(z_0) = \frac{1}{2\pi} \int_0^{2\pi} f(z_0 + re^{i\theta})d\theta.$$

Prove more generally that

$$f^{(n)}(z_0) = \frac{n!}{2\pi r^n} \int_0^{2\pi} f(z_0 + re^{i\theta})e^{-in\theta}\, d\theta.$$

9. Suppose that f is analytic inside and on the unit circle $|z| = 1$. Prove that if $|f(z)| \leq M$ for $|z| = 1$, then $|f(0)| \leq M$ and $\left|f'(0)\right| \leq M$. What estimate can you give for $\left|f^{(n)}(0)\right|$?

10. Let f be analytic inside and on the simple closed contour Γ. Verify from the theorems in this section that

$$\int_{\Gamma} \frac{f'(z)}{z - z_0}\, dz = \int_{\Gamma} \frac{f(z)}{(z - z_0)^2}\, dz$$

for all z_0 not on Γ.

11. Let $f = u + iv$ be analytic in a domain D. Explain why $\partial^2 u/\partial x^2$ is harmonic in D.

12. Prove that the function H defined in Eq. (10) is analytic inside Γ and that its derivative is given by formula (11).

13. According to Theorem 15, when Γ is a simple closed contour the function G defined by

$$G(z) := \frac{1}{2\pi i} \int_{\Gamma} \frac{g(\zeta)}{\zeta - z}\, d\zeta$$

is analytic in the domain enclosed by Γ (assuming only that g is continuous on Γ). Show that the limiting values of $G(z)$ as z approaches Γ need not coincide with the values of g, by considering the situation where Γ is the positively oriented circle $|z| = 1$ and $g(z) = 1/z$. Why doesn't this violate Cauchy's formula? [HINT: Use partial fractions to evaluate G.]

14. Let Γ be a simple closed positively oriented contour that passes through the point $2 + 3i$. Set

$$G(z) := \frac{1}{2\pi i} \int_\Gamma \frac{\cos \zeta}{\zeta - z} \, d\zeta.$$

Find the following limits:

   (a) $\displaystyle\lim_{z \to 2+3i} G(z)$, where z approaches $2 + 3i$ from inside Γ.

   (b) $\displaystyle\lim_{z \to 2+3i} G(z)$, where z approaches $2 + 3i$ from outside Γ.

[The curious reader may speculate on the interpretation of the integral for points z, such as $2 + 3i$, which actually lie on Γ. The theory of Sokhotskyi and Plemelj, which states that in such a case $G(z)$ equals the *average* of the interior and exterior limiting values, is developed in Sec. 8.5.]

15. Suppose that $f(z)$ is analytic at each point of the closed disk $|z| \le 1$ and that $f(0) = 0$. Prove that the function

$$F(z) := \begin{cases} f(z)/z & z \ne 0, \\ f'(0) & z = 0, \end{cases}$$

is analytic on $|z| \le 1$. [HINT: To show that F is analytic at $z = 0$ note that by Theorem 15 the function

$$G(z) := \frac{1}{2\pi i} \oint_{|\zeta|=1} \frac{f(\zeta)/\zeta}{\zeta - z} \, d\zeta$$

is analytic at this point. Using partial fractions deduce that $G(z) = F(z)$ for $|z| < 1$.]

16. Below is an outline of a proof of the fact that *for any analytic function $f(z)$ that is never zero in a simply connected domain D there exists a single-valued branch of $\log f(z)$ analytic in D.* Justify each step in the proof.

   (a) $f'(z)/f(z)$ is analytic in D.

   (b) $f'(z)/f(z)$ has an (analytic) antiderivative in D, say $H(z)$.

   (c) The function $f(z)e^{-H(z)}$ is constant in D, so that $f(z) = ce^{H(z)}$.

   (d) Letting α be a value of $\log c$, the function $H(z) + \alpha$ is a branch of $\log f(z)$ analytic in D.

17. Use the result of Prob. 16 to prove that there exists a single-valued analytic branch of $(z^3 - 1)^{1/2}$ in the unit disk $|z| < 1$.

4.6 Bounds for Analytic Functions

Many interesting facts about analytic functions are uncovered when one considers upper bounds on their moduli. We already have one result in this direction, namely, the integral estimate Theorem 5 of Sec. 4.2. When this is judiciously applied to the Cauchy integral formulas we obtain the *Cauchy estimates* for the derivatives of an analytic function.

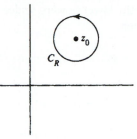

Figure 4.56 Circle for Cauchy estimates.

Theorem 20. Let f be analytic inside and on a circle C_R of radius R centered about z_0. If $|f(z)| \leq M$ for all z on C_R, then the derivatives of f at z_0 satisfy

$$\left| f^{(n)}(z_0) \right| \leq \frac{n!M}{R^n} \qquad (n = 1, 2, 3, \ldots). \qquad (1)$$

Proof. Giving C_R a positive orientation in Fig. 4.56, we have, by the generalized Cauchy formula (Theorem 19),

$$f^{(n)}(z_0) = \frac{n!}{2\pi i} \int_{C_R} \frac{f(\zeta)}{(\zeta - z_0)^{n+1}} \, d\zeta.$$

For ζ on C_R, the integrand is bounded by M/R^{n+1}; the length of C_R is $2\pi R$. Thus from Theorem 5 in Sec. 4.2 there follows

$$\left| f^{(n)}(z_0) \right| \leq \frac{n!}{2\pi} \frac{M}{R^{n+1}} 2\pi R,$$

which reduces to (1). ■

This innocuous-looking theorem actually places rather severe restrictions on the behavior of analytic functions. Suppose, for instance, that f is analytic and bounded by some number M over the whole plane $\mathbf{C}$. Then the conditions of the theorem hold for any z_0 and for any R. Taking $n = 1$ in (1) and letting $R \to \infty$, we conclude that f' vanishes everywhere; i.e., f must be constant. This startling result is known as *Liouville's theorem*.

Theorem 21. The only bounded entire functions are the constant functions.

Clearly, nonconstant *polynomials* are unbounded (over the whole plane). Loosely speaking, we expect a polynomial of degree n to behave like z^n for large $|z|$, because the leading term will dominate the lower powers. Indeed, if $P(z) = a_n z^n + a_{n-1} z^{n-1} + \cdots + a_1 z + a_0$ with $a_n \neq 0$, then

$$P(z) = z^n \left(a_n + a_{n-1}/z + \cdots + a_1/z^{n-1} + a_0/z^n \right)$$

and we see that

$$P(z)/z^n \to a_n \quad \text{as } |z| \to \infty. \tag{2}$$

This observation and Liouville's theorem enable us to prove the *Fundamental Theorem of Algebra*, as we promised in Sec. 3.1.

Theorem 22. Every nonconstant polynomial with complex coefficients has at least one zero.

Proof. Suppose $P(z) = a_n z^n + a_{n-1} z^{n-1} + \cdots + a_1 z + a_0$, with $a_n \neq 0$, has no zeros. It follows immediately that $f(z) = 1/P(z)$ is entire; we shall also see that it is bounded over the whole plane.

(i) From (2) we know that $|P(z)/z^n| \geq |a_n|/2$ for $|z|$ sufficiently large - say, $|z| \geq \rho$. Hence

$$|f(z)| = \frac{1}{|P(z)|} \leq \frac{2}{|z^n| \, |a_n|} \leq \frac{2}{\rho^n |a_n|} \quad (|z| \geq \rho).$$

(ii) For $|z| \leq \rho$, we have the case of a continuous function, $|f(z)|$, on a closed disk. Under such circumstances it is known from calculus that the function must be bounded there (indeed, it achieves its maximum).

But if $1/P(z)$ is bounded and entire it must be constant. Thus $P(z)$, itself, is constant (and its degree n is zero). In other words, the only polynomials that have no zeros are the constants. ∎

Let us now return to the Cauchy formula for the function f, analytic inside and on the circle C_R of radius R around z_0. We have

$$f(z_0) = \frac{1}{2\pi i} \oint_{C_R} \frac{f(z)}{z - z_0} \, dz. \tag{3}$$

Parametrizing C_R by $z = z_0 + Re^{it}, 0 \leq t \leq 2\pi$, we write Eq. (3) as

$$f(z_0) = \frac{1}{2\pi i} \int_0^{2\pi} \frac{f(z_0 + Re^{it})}{Re^{it}} \, iRe^{it} \, dt,$$

or

$$f(z_0) = \frac{1}{2\pi} \int_0^{2\pi} f(z_0 + Re^{it}) \, dt. \tag{4}$$

Equation (4), which is known as the *mean-value property*, displays $f(z_0)$ as the average of its values around the circle C_R. Clearly, if $|f(z)| \le M$ on C_R, then $|f(z_0)|$ is bounded by M also (this verifies the case $n = 0$ of Theorem 20, with $0! = 1$). But more importantly, we can utilize Eq. (4) to establish the following.

Lemma 1. Suppose that f is analytic in a disk centered at z_0 and that the maximum value of $|f(z)|$ over this disk is $|f(z_0)|$. Then $|f(z)|$ is constant in the disk.

Proof.   Assume to the contrary that $|f(z)|$ is not constant. Then there must exist a point z_1 in the disk such that $|f(z_0)| > |f(z_1)|$. Let C_R denote the circle centered at z_0 which passes through z_1. Then by hypothesis $|f(z_0)| \ge |f(z)|$ for all z on C_R. Moreover, by the continuity of f, the strict inequality $|f(z_0)| > |f(z)|$ must hold for z on a portion of C_R containing z_1. This leads to a contradiction of Eq. (4), because the portion containing z_1 would contribute "less than its share" to the average in Eq. (4), and the "deficit" cannot be made up anywhere else on C_R. (We invite the reader to provide a rigorous version of this argument in Prob. 9.)   ∎

Observe that the lemma says that the modulus of an analytic function cannot achieve its maximum at the center of the disk unless $|f|$ is constant. We shall use the lemma to extend this idea in the following version of the *maximum modulus principle.*

Theorem 23. If f is analytic in a domain D and $|f(z)|$ achieves its maximum value at a point z_0 in D, then f is constant in D.

Proof.   We shall prove that $|f|$ is constant in D; by Prob. 12 in Exercises 2.5, we can then conclude that f, itself, is constant.

So for the moment let us suppose that $|f(z)|$ is not constant. Then there must be a point z_1 in D such that $|f(z_1)| < |f(z_0)|$. Let γ be a path in D running from z_0 to z_1, as in Fig. 4.57. Now we consider the values of $|f(z)|$ for z on γ, starting at z_0. Intuitively, we expect to encounter a point w where $|f(z)|$ first starts to decrease on γ. That is, there should be a point w on γ with the following properties:

(i) $|f(z)| = |f(z_0)|$ for all z preceeding w on γ.

(ii) There are points z on γ, arbitrarily close to w, where $|f(z)| < |f(z_0)|$.

(It is possible that w may coincide with z_0.) The existence of w is a fact that can be rigorously demonstrated from the axioms of the real number system, but we shall omit the details here. Naturally, from property (i) and the continuity of f, we have $|f(w)| = |f(z_0)|$.

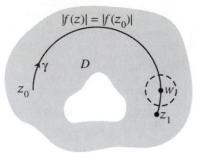

Figure 4.57 Geometry for proof of Theorem 23.

Now since every point of a domain is an interior point, there must be a disk centered at w that lies in D. But Lemma 1 applies and says that $|f|$ is constant in this disk, contradicting property (ii) above. We are forced to conclude, therefore, that our initial supposition about the existence of z_1 is erroneous. Consequently $|f|$, and hence f itself, is constant in D. ∎

Next, consider a function $f(z)$ analytic in a *bounded* domain D and continuous on D and its boundary. From calculus we know that the continuous function $|f(z)|$ must achieve its maximum on this closed bounded region; on the other hand, Theorem 23 says that the maximum cannot occur at an interior point unless f is constant. In any case, we can conservatively state the following modification of the maximum modulus principle.

Theorem 24. A function analytic in a bounded domain and continuous up to and including its boundary attains its maximum modulus on the boundary.

Example 1

Find the maximum value of $\left|z^2 + 3z - 1\right|$ in the disk $|z| \leq 1$.

Solution. The triangle inequality immediately gives us

$$\left|z^2 + 3z - 1\right| \leq \left|z^2\right| + 3|z| + 1 \leq 5 \quad \text{(for } |z| \leq 1\text{).} \tag{5}$$

However, the maximum is actually smaller than this, as the following analysis shows.

The maximum of $\left|z^2 + 3z - 1\right|$ must occur on the boundary of the disk ($|z| = 1$). The latter can be parametrized $z = e^{it}$, $0 \leq t \leq 2\pi$; whence

$$\left|z^2 + 3z - 1\right|^2 = \left(e^{i2t} + 3e^{it} - 1\right)\left(e^{-i2t} + 3e^{-it} - 1\right).$$

Expanding and gathering terms reduces this to $(11 - 2\cos 2t)$. Thus the maximum of $\left|z^2 + 3z - 1\right|$ is $\sqrt{13}$, which occurs at $z = \pm i$. A sketch of the graph of the function $\left|z^2 + 3z - 1\right|$ appears in Fig. 4.58. ∎

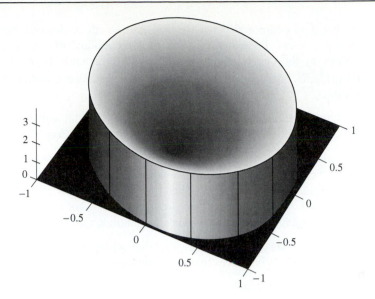

Figure 4.58 Graph of $|z^2 + 3z - 1|$ for $|z| < 1$.

EXERCISES 4.6

1. Let $f(z) = 1\big/(1 - z)^2$, and let $0 < R < 1$. Verify that $\max\limits_{|z|=R} |f(z)| = 1\big/(1 - R)^2$, and also show $f^{(n)}(0) = (n + 1)!$, so that by the Cauchy estimates

$$(n + 1)! \leq \frac{n!}{R^n (1 - R)^2}.$$

2. Suppose that f is analytic in $|z| < 1$ and that $|f(z)| < 1/(1 - |z|)$. Prove that

$$\left| f^{(n)}(0) \right| \leq \frac{n!}{R^n (1 - R)} \qquad (0 < R < 1),$$

and show that the upper bound is smallest when $R = n/(n + 1)$.

3. Let f be analytic and bounded by M in $|z| \leq r$. Prove that

$$\left| f^{(n)}(z) \right| \leq \frac{n! M}{(r - |z|)^n} \qquad (|z| < r).$$

4. If $p(z) = a_0 + a_1 z + \cdots + a_n z^n$ is a polynomial and $\max |p(z)| = M$ for $|z| = 1$, show that each coefficient a_k is bounded by M.

5. Let f be entire and suppose that $\operatorname{Re} f(z) \leq M$ for all z. Prove that f must be a constant function. [HINT: Apply Liouville's theorem to the function e^f.]

6. Let f be entire and suppose that $f^{(5)}$ is bounded in the whole plane. Prove that f must be a polynomial of degree at most 5.

7. Suppose that f is entire and that $|f(z)| \leq |z|^2$ for all sufficiently large values of $|z|$, say $|z| > r_0$. Prove that f must be a polynomial of degree at most 2. [HINT: Show by the Cauchy estimates that for R sufficiently large $\left|f^{(3)}(z_0)\right| \leq 3! \, (R + |z_0|)^2 / R^3$, and thereby conclude that $f^{(3)}$ vanishes everywhere.] Generalize this result.

8. If f is analytic in the annulus $1 \leq |z| \leq 2$ and $|f(z)| \leq 3$ on $|z| = 1$ and $|f(z)| \leq 12$ on $|z| = 2$, prove that $|f(z)| \leq 3|z|^2$ for $1 \leq |z| \leq 2$. [HINT: Consider $f(z)/3z^2$.]

9. Using formula (4) show that if the analytic function f satisfies $|f(z_0)| \geq |f(z)|$ for all z on $C_R : z_0 + Re^{it}$, $0 \leq t \leq 2\pi$, then there is no point z_1 on C_R for which $|f(z_0)| > |f(z_1)|$. [HINT: If z_1 exists, then by the continuity of f there is an $\epsilon > 0$ such that $\left|f(z_0 + Re^{it})\right| \leq |f(z_0)| - \epsilon$ over some interval of t. Using this interval divide up the integration in Eq. (4) to reach the contradiction $|f(z_0)| < |f(z_0)|$.]

10. Find all functions f analytic in $D : |z| < R$ that satisfy $f(0) = i$ and $|f(z)| \leq 1$ for all z in D. [HINT: Where does the maximum modulus occur?]

11. Suppose that f is analytic inside and on the simple closed curve C and that $|f(z) - 1| < 1$ for all z on C. Prove that f has no zeros inside C. [HINT: Suppose $f(z_0) = 0$ for some z_0 inside C and consider the function $g(z) := f(z) - 1$.]

12. It is proved in Chapter 6 that every analytic function which is nonconstant on domains maps open sets onto open sets. Using this fact give another proof of the maximum modulus principle (Theorem 23).

13. Let f and g be analytic in the bounded domain D and continuous up to and including its boundary B. Suppose that g never vanishes. Prove that if the inequality $|f(z)| \leq |g(z)|$ holds for all z on B, then it must hold for all z in D. [HINT: Consider the function $f(z)/g(z)$.]

14. Prove the *minimum modulus principle*: Let f be analytic in a bounded domain D and continuous up to and including its boundary. Then *if f is nonzero in D*, the modulus $|f(z)|$ attains its minimum value on the boundary of D. [HINT: Consider the function $1/f(z)$.] Give an example to show why the italicized condition is essential.

15. Let the nonconstant function f be analytic in the bounded domain D and continuous up to and including its boundary B. Prove that if $|f(z)|$ is constant on B, then f must have at least one zero in D.

16. Show that $\max_{|z| \leq 1} |az^n + b| = |a| + |b|$.

17. Find $\max_{|z| \leq 1} |(z - 1)(z + 1/2)|$. [HINT: Use calculus.]

18. Let P be a polynomial which has no zeros on the simple closed positively oriented contour Γ. Prove that the number of zeros of P (counting multiplicity) that lie *inside* Γ is given by the integral

$$\frac{1}{2\pi i} \int_\Gamma \frac{P'(z)}{P(z)} \, dz.$$

[HINT: See Exercises 3.1, Prob. 17.]

19. Prove that for any polynomial P of the form $P(z) = z^n + a_{n-1}z^{n-1} + \cdots + a_1 z + a_0$, we have $\max_{|z|=1} |P(z)| \geq 1$. [HINT: Consider the polynomial $Q(z) = z^n P(1/z)$, and note that $Q(0) = 1$ and that

$$\max_{|z|=1} |Q(z)| = \max_{|z|=1} |P(z)|.]$$

4.7   *Applications to Harmonic Functions

In the last part of Chapter 2 we discussed the harmonic functions, which are twice-continuously differentiable solutions of Laplace's equation

$$\frac{\partial^2 \phi}{\partial x^2} + \frac{\partial^2 \phi}{\partial y^2} = 0.$$

In particular, we showed that the real and imaginary parts of analytic functions are harmonic, subject to an assumption about their differentiability. This assumption has now been vindicated by Theorem 17. Conversely, we showed how a given harmonic function can be regarded as the real (or imaginary) part of an analytic function by giving a method that constructs the harmonic conjugate, at least in certain elementary domains such as disks. We shall now exploit this interpretation to derive some more facts about harmonic functions.

Our first step will be to extend the harmonic-analytic dualism to simply connected domains, via the next theorem.

Theorem 25. Let ϕ be a function harmonic on a simply connected domain D. Then there is an analytic function f such that $\phi = \operatorname{Re} f$ on D.

Proof.   For motivation, suppose that we *had* such an analytic function, say, $f = \phi + i\psi$. Then one expression for $f'(z)$ would be given by $f' = \partial\phi/\partial x - i\partial\phi/\partial y$ (using the Cauchy-Riemann equations), and f would be an antiderivative of this analytic function.

Accordingly, we begin the proof by defining $g(z) := \partial\phi/\partial x - i\partial\phi/\partial y$. We now claim that g satisfies the Cauchy-Riemann equations in D:

$$\frac{\partial}{\partial x}\left(\frac{\partial\phi}{\partial x}\right) = \frac{\partial}{\partial y}\left(-\frac{\partial\phi}{\partial y}\right)$$

because ϕ is harmonic, and

$$\frac{\partial}{\partial y}\left(\frac{\partial\phi}{\partial x}\right) = -\frac{\partial}{\partial x}\left(-\frac{\partial\phi}{\partial y}\right)$$

because of the equality of mixed second partial derivatives. Of course, these partials are continuous since ϕ is harmonic. Hence $g(z)$ is analytic, and by Theorem 10 (or 13) in Sec. 4.4 it has an analytic antiderivative $G = u + iv$ in the simply connected domain D. Since $G' = g$, we can write

$$\frac{\partial u}{\partial x} - i \frac{\partial u}{\partial y} = \frac{\partial \phi}{\partial x} - i \frac{\partial \phi}{\partial y},$$

showing that u and ϕ have identical first partial derivatives in D. Thus by Theorem 1 of Sec. 1.6, we conclude that $\phi - u$ is constant in D; i.e., $\phi = u + c$. It follows that $f(z) := G(z) + c$ is an analytic function of the kind predicted by the theorem. ∎

With Theorem 25 in hand we are fully equipped to study harmonic functions in simply connected domains, using the theory of analytic functions. Let $\phi(x, y)$ be a harmonic function and $f = \phi + i\psi$ be an "analytic completion" for ϕ in the simply connected domain D. Now observe the following about the function $e^{f(z)}$:

$$|e^f| = |e^{\phi+i\psi}| = |e^\phi||e^{i\psi}| = e^\phi. \tag{1}$$

Because the exponential is a monotonically increasing function of a real variable, Eq. (1) implies that the maximum points of ϕ coincide with the maximum points of the modulus of the analytic function e^f. Thus we immediately have a maximum principle for harmonic functions! Furthermore, since the minimum points of ϕ are the same as the maximum points of the harmonic function $-\phi$, we can state the following versions of the *maximum-minimum principle for harmonic functions*.

Theorem 26. If ϕ is harmonic in a simply connected domain D and $\phi(z)$ achieves its maximum or minimum value at some point z_0 in D, then ϕ is constant in D.

Theorem 27. A function harmonic in a bounded simply connected domain and continuous up to and including the boundary attains its maximum and minimum on the boundary.

Actually, these principles can easily be extended to multiply connected domains by appropriately modifying the proof of Theorem 23; see Prob. 3. We shall utilize this extended form hereafter.

An important problem that arises in electromagnetism, fluid mechanics, and heat transfer is the following.

Dirichlet Problem Find a function $\phi(x, y)$ continuous on a domain D and its boundary, harmonic in D, and taking specified values on the boundary of D.

The function ϕ can be interpreted as electric potential, velocity potential, or steady-state temperature. In studying the Dirichlet problem we are concerned with two main questions: Does a solution exist, and, if so, is it uniquely determined by the given boundary values? For the case of bounded domains the question of uniqueness is answered by the next theorem.

> **Theorem 28.** Let $\phi_1(x, y)$ and $\phi_2(x, y)$ each be harmonic in a bounded domain D and continuous on D and its boundary. Furthermore, suppose that $\phi_1 = \phi_2$ on the boundary of D. Then $\phi_1 = \phi_2$ throughout D.

Proof.   Consider the harmonic function $\phi := \phi_1 - \phi_2$. It must attain its maximum and minimum on the boundary, but it vanishes there! Hence $\phi = 0$ throughout D. ∎

A solution to the Dirichlet problem could be expressed by a formula giving ϕ inside D in terms of its (specified) values on the boundary. Surely this suggests experimenting with the Cauchy integral formula; it expresses the combination $f = \phi + i\psi$ inside D in terms of its values on the boundary (here f is the analytic completion of ϕ). So if we could "uncouple" the real and imaginary parts of f in the Cauchy integral formula, we would solve the Dirchlet problem.

We are, in fact, able to solve the Dirichlet problems for the disk and the half-plane by this technique. Leaving the latter as an exercise for the reader, we proceed with the former.

For simplicity we consider the disk that is bounded by the positively oriented circle $C_R : |z| = R$. The Cauchy integral formula gives the values of an analytic function f inside the disk in terms of its values on the circle:

$$f(z) = \frac{1}{2\pi i} \int_{C_R} \frac{f(\zeta)}{\zeta - z} \, d\zeta \qquad (|z| < R) \qquad (2)$$

(assuming the domain of analyticity includes the circle C_R as well as its interior). We wish to transform Eq. (2) into a formula that involves only the real part of f. To this end, we observe that for fixed z, with $|z| < R$, the function

$$\frac{f(\zeta)\bar{z}}{R^2 - \zeta\bar{z}}$$

is an *analytic function of* ζ inside and on C_R (think about this; the denominator does not vanish). Hence by the Cauchy theorem

$$\frac{1}{2\pi i} \int_{C_R} \frac{f(\zeta)\bar{z}}{R^2 - \zeta\bar{z}} \, d\zeta = 0.$$

The utility of this relationship will become apparent when we add it to Eq. (2):

$$f(z) = \frac{1}{2\pi i} \int_{C_R} \left(\frac{1}{\zeta - z} + \frac{\bar{z}}{R^2 - \zeta\bar{z}} \right) f(\zeta) \, d\zeta$$

$$= \frac{1}{2\pi i} \int_{C_R} \frac{R^2 - |z|^2}{(\zeta - z)(R^2 - \zeta\bar{z})} f(\zeta) \, d\zeta. \qquad (3)$$

If we parametrize C_R by $\zeta = Re^{it}$, $0 \le t \le 2\pi$, Eq. (3) becomes

$$f(z) = \frac{1}{2\pi i} \int_0^{2\pi} \frac{R^2 - |z|^2}{\left(Re^{it} - z\right)\left(R^2 - Re^{it}\bar{z}\right)} f(Re^{it}) R i e^{it} \, dt$$

$$= \frac{R^2 - |z|^2}{2\pi} \int_0^{2\pi} \frac{f(Re^{it})}{\left(Re^{it} - z\right)\left(Re^{-it} - \bar{z}\right)} \, dt$$

$$= \frac{R^2 - |z|^2}{2\pi} \int_0^{2\pi} \frac{f(Re^{it})}{\left|Re^{it} - z\right|^2} \, dt.$$

Finally, by taking the real part of this equation, identifying Re f as the harmonic function ϕ, and writing z in the polar form $z = re^{i\theta}$, we arrive at *the Poisson integral formula*, which we state as follows.

Theorem 29. Let ϕ be harmonic in a domain containing the disk $|z| \le R$. Then for $0 \le r < R$, we have

$$\phi\left(re^{i\theta}\right) = \frac{R^2 - r^2}{2\pi} \int_0^{2\pi} \frac{\phi(Re^{it})}{R^2 + r^2 - 2rR\cos(t - \theta)} \, dt. \qquad (4)$$

Actually, Poisson's formula is more general than is indicated in this statement. We direct the interested reader to Ref. [9] or [10] for a proof of the next theorem.

Theorem 30. Let U be a real-valued function defined on the circle $C_R : |z| = R$ and continuous there except for a finite number of jump discontinuities. Then the function

$$u(re^{i\theta}) := \frac{R^2 - r^2}{2\pi} \int_0^{2\pi} \frac{U(Re^{it})}{R^2 + r^2 - 2rR\cos(t - \theta)} \, dt \qquad (5)$$

is harmonic inside C_R, and as $re^{i\theta}$ approaches any point on C_R where U is continuous, $u\left(re^{i\theta}\right)$ approaches the value of U at that point.

Naturally at the points of discontinuity of U the behavior is more complicated. As an example, consider the harmonic function Arg$(z + 1)$, which is the imaginary part of Log$(z + 1)$, in the domain depicted in Fig. 4.59. Clearly, Arg$(z + 1)$ approaches its boundary values in a reasonable manner except at $z = -1$, where it is erratic (the boundary value jumps from $\pi/2$ to $-\pi/2$ there).

Thus Poisson's formula solves the Dirichlet problem for the disk under very general circumstances. (We solved the simplest case, where U is constant, in Exercises 3.4, Prob. 6.)

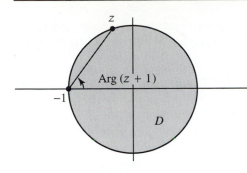

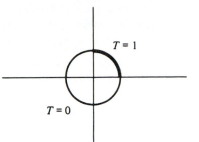

Figure 4.59 Arg(z + 1).

Figure 4.60 Find the temperature inside the disk.

Example 1

Find the steady-state temperature T at each point inside the unit disk if the temperature is prescribed to be $+1$ on the part of the rim in the first quadrant, and 0 elsewhere (Fig. 4.60).

Solution.   The temperature is given by the harmonic function taking the prescribed boundary values. Here formula (5) becomes

$$T(re^{i\theta}) = \frac{1 - r^2}{2\pi} \int_0^{\pi/2} \frac{1}{1 + r^2 - 2r\cos(t - \theta)} \, dt.$$

For example, at the center,

$$T(0) = \frac{1}{2\pi} \int_0^{\pi/2} 1 \, dt = \frac{1}{4},$$

which is the average of its values on the rim. ■

EXERCISES 4.7

1. Find all functions ϕ harmonic in the unit disk $D : |z| < 1$ that satisfy $\phi(i/2) = -5$ and $\phi(z) \geq -5$ for all z in D.

2. Show by example that a harmonic function need not have an analytic completion in a multiply connected domain. [HINT: Consider Log $|z|$.]

3. Prove the maximum-minimum principle for harmonic functions in an arbitrary domain. [HINT: Theorem 26 can be used to establish a lemma analogous to Lemma 1 in Sec. 4.6. Then argue as in Theorem 23.]

4. What is the physical interpretation of the maximum-minimum principle in steady-state heat flow?

5. Show by example that the solution to the Dirichlet problem need not be unique for unbounded domains. [HINT: Construct two functions that are harmonic in the upper half-plane, each vanishing on the x-axis.]

6. Prove the *circumferential mean-value theorem* for harmonic functions: If ϕ is harmonic in a domain containing the disk $|z| \leq \rho$, then

$$\phi(0) = \frac{1}{2\pi} \int_0^{2\pi} \phi(\rho e^{it}) \, dt.$$

7. Prove the following version of the *solid mean-value theorem* for harmonic functions: If ϕ is harmonic in a domain containing the closed disk $D : |z| \leq R$, then

$$\phi(0) = \frac{1}{\pi R^2} \iint_D \phi \, dx \, dy.$$

[HINT: Multiply the equation in Prob. 6 by $\rho \, d\rho$ and integrate.]

8. Without doing any computations, explain why

$$\frac{R^2 - r^2}{2\pi} \int_0^{2\pi} \frac{1}{R^2 + r^2 - 2rR\cos(t - \theta)} \, dt = 1 \quad \text{for } 0 \leq r < R.$$

9. Prove *Harnack's inequality*: If $\phi(z)$ is harmonic and nonnegative in a domain containing the disk $|z| \leq R$, then for $0 \leq r < R$

$$\phi(0)\frac{R - r}{R + r} \leq \phi(re^{i\theta}) \leq \phi(0)\frac{R + r}{R - r}.$$

[HINT: Use Poisson's formula and the mean-value property of Prob. 6, observing also that
$$(R - r)^2 \leq R^2 + r^2 - 2rR\cos(t - \theta) \leq (R + r)^2.]$$

10. Prove *Liouville's theorem* for harmonic functions: If ϕ is harmonic in the whole plane and bounded from above or below there, then ϕ is constant. [HINT: Modify ϕ so that Harnack's inequality (Prob. 9) can be applied.]

11. The temperature of the rim of the unit disk is maintained at the levels indicated in Fig. 4.61. What is the temperature at the center?

12. (*Schwarz Integral Formula*) Let $f = u + iv$ be analytic on the disk $|z| \leq R$. Then Poisson's integral formula expresses the values of the real part u inside the disk in terms of the values of u on the boundary of the disk. In this problem you will derive an expression for the imaginary part v in terms of the boundary values of u.

(a) Show that the Poisson integral formula (4) can be written

$$u(z) = \frac{1}{2\pi} \int_0^{2\pi} P(Re^{it}, z) \, u(Re^{it}) \, dt,$$

where the *Poisson kernel* P is given by

$$P(\zeta, z) := \frac{|\zeta|^2 - |z|^2}{|\zeta - z|^2}.$$

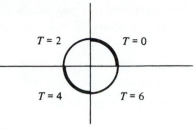

Figure 4.61 Find the temperature at the center.

(b) Show that

$$\frac{\zeta + z}{\zeta - z} = P(\zeta, z) + 2i \frac{\operatorname{Im} z\bar{\zeta}}{|\zeta - z|^2};$$

in other words, $P(\zeta, z)$ is the real part of $(\zeta + z)/(\zeta - z)$.

(c) Utilizing Theorem 15, argue that

$$H(z) := \frac{1}{2\pi i} \oint_{|\zeta|=R} \frac{\zeta + z}{\zeta - z} \frac{u(\zeta)}{\zeta} d\zeta$$

defines an analytic function of z for $|z| < R$.

(d) Insert the parametrization $\zeta = Re^{it}$ into the integral for $H(z)$ and derive

$$\operatorname{Re} H(z) = \frac{1}{2\pi} \int_0^{2\pi} P(Re^{it}, z) u(Re^{it}) dt$$

which, by the Poisson integral formula, equals $u(z)$.

(e) Since $H(z)$ and $f(z)$ are two analytic functions whose real parts coincide, use the theory developed in Sec. 2.5 to argue that they can differ only by an imaginary constant,

$$f(z) = H(z) + iC.$$

Insert the value $z = 0$ into this identity and use the circumferential mean-value theorem (Prob. 6) to demonstrate that C must be $v(0)$.

Assemble the results to obtain the *Schwarz integral formula*

$$\boxed{f(z) = \frac{1}{2\pi i} \oint_{|\zeta|=R} \frac{\zeta + z}{\zeta - z} \frac{u(\zeta)}{\zeta} d\zeta + iv(0), \quad \text{for } |z| < R.}$$

(f) Equate the imaginary parts in the Schwarz integral formula to derive the representation for v in terms of the boundary values of u:

$$v(z) = \frac{1}{2\pi} \int_0^{2\pi} Q(Re^{it}, z) u(Re^{it}) dt + v(0), \quad \text{for } |z| < R,$$

where

$$Q(\zeta, z) := 2 \frac{\operatorname{Im} z\bar{\zeta}}{|\zeta - z|^2}.$$

13. Let f be an entire function whose real part satisfies $|\operatorname{Re} f(z)| \le M|z|^2$ for all sufficiently large values of $|z|$, where M is a constant. Show that f must be a polynomial by arguing as follows.

(a) Use the Schwarz formula [Prob. 12(e)] with $R = 2|z|$ to show that $|f(z)|$ is bounded for large $|z|$ by some multiple of $|z|^2$.

(b) Use the result of Prob. 7 of Exercises 4.6 to conclude that f is a polynomial of degree at most 2.

14. (*Poisson Integral Formula for the Half-Plane*) If $f = \phi + i\psi$ is analytic in a domain containing the x-axis and the upper half-plane and $|f(z)| \le K$ in this domain, then the values of the harmonic function ϕ in the upper half-plane are given in terms of its values on the x-axis by

$$\phi(x, y) = \frac{y}{\pi} \int_{-\infty}^{\infty} \frac{\phi(\xi, 0)\, d\xi}{(\xi - x)^2 + y^2} \qquad (y > 0).$$

Here is an outline of the derivation; justify the steps.

(a) For the situation depicted in Fig. 4.62,

$$f(z) = \frac{1}{2\pi i} \int_{\Gamma_R} \frac{f(\zeta)}{\zeta - z}\, d\zeta.$$

(b) For the same situation

$$0 = \frac{1}{2\pi i} \int_{\Gamma_R} \frac{f(\zeta)}{\zeta - \bar{z}}\, d\zeta.$$

(c) Subtract these two equations to conclude that

$$f(z) = \frac{1}{2\pi i} \int_{-R}^{R} f(\xi) \frac{2i \operatorname{Im} z}{|\xi - z|^2}\, d\xi$$
$$+ \frac{1}{2\pi i} \int_{C_R^+} f(\zeta) \frac{2i \operatorname{Im} z}{(\zeta - z)(\zeta - \bar{z})}\, d\zeta,$$

where C_R^+ is the semicircular portion of Γ_R.

(d) Show that the integral along C_R^+ is bounded by

$$\frac{K}{\pi} \frac{\operatorname{Im} z}{(R - |z|)^2} \pi R.$$

(e) Let $R \to \infty$ in the last equation and take the real part.

15. The Poisson integral formula in Prob. 14 admits a generalization; for suitable functions $U(\xi)$ the integral

$$\phi(x, y) := \frac{y}{\pi} \int_{-\infty}^{\infty} \frac{U(\xi)}{(\xi - x)^2 + y^2}\, d\xi$$

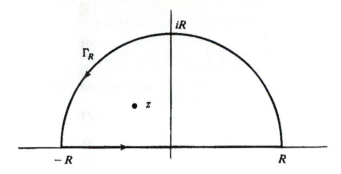

Figure 4.62 Poisson integral formula.

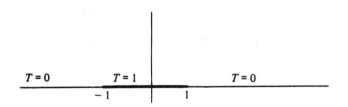

Figure 4.63 Find the temperature in the upper half-plane.

will define a harmonic function in either half-plane taking the limiting value $U(x)$ as z approaches a point of continuity x of U. Use the formula to find the temperature in the upper half-plane if the temperature on the x-axis is maintained as in Fig. 4.63. (This problem was solved by elementary methods in Example 3, Sec. 3.4.)

16. (*Schwarz Integral Formula for the Half-Plane*) If $f = u + iv$ is an analytic function in the closed upper half-plane and $|f|$ is bounded there, then the Poisson integral formula expresses the values of $u(x, y)$ for $y > 0$ in terms of the values of u on the real axis (Prob. 14). If f satisfies a slightly stronger condition, then the derivation in Prob. 14 can be modified to express also the values of the imaginary part $v(x, y)$ for $y > 0$ in terms of these boundary values of u.

   (a) Retrace parts (a) and (b) of Prob. 14, but in part (c) add the two integrals to obtain the identity

   $$f(z) = \frac{1}{2\pi i} \int_{-R}^{R} f(\xi) \frac{2(\xi - x)}{|\xi - z|^2}\, d\xi + \frac{1}{2\pi i} \int_{C_R^+} f(\zeta) \frac{2(\zeta - x)}{(\zeta - z)(\zeta - \bar{z})}\, d\zeta.$$

   (b) Show that the integral along C_R^+ goes to zero as $R \to \infty$ if f satisfies an inequality of the form $|f(z)| \leq K / |z|^\alpha$ with $\alpha > 0$, for sufficiently large $|z|$ in the closed upper half-plane. Thus, for such an f,

   $$f(z) = \frac{1}{\pi i} \int_{-\infty}^{\infty} f(\xi) \frac{(\xi - x)}{|\xi - z|^2}\, d\xi, \qquad \text{for Im } z > 0.$$

(c) Equate imaginary parts in the last equation to derive

$$v(z) = -\frac{1}{\pi} \int_{-\infty}^{\infty} u(\xi, 0) \frac{(\xi - x)}{|\xi - z|^2} \, d\xi, \qquad \text{Im } z > 0.$$

(d) Combine this formula with the Poisson integral formula to obtain the *Schwarz integral formula* for the upper half-plane

$$\boxed{f(z) = \frac{1}{\pi i} \int_{-\infty}^{\infty} \frac{u(\xi, 0)}{\xi - z} \, d\xi, \qquad \text{Im } z > 0.}$$

17. With the assumptions of Prob. 16, the Schwarz integral formula also defines a function in the *lower* half-plane; call this function $\tilde{f}(z)$.

(a) Show that $\tilde{f}(z) = -\overline{f(\overline{z})}$, for Im $z < 0$, and conclude from this that $\tilde{f}(z)$ is analytic in the lower half-plane.

(b) Use the equation in part (a) to show that as z crosses the real axis from above, the change in

$$\frac{1}{\pi i} \int_{-\infty}^{\infty} \frac{u(\xi, 0)}{\xi - z} \, d\xi$$

is the real number $-2u(x, 0)$.

18. Use the theory developed in Probs. 16 and 17 to argue that

$$\frac{1}{z + i} = \frac{1}{\pi i} \int_{-\infty}^{\infty} \frac{\xi}{\left(\xi^2 + 1\right)(\xi - z)} \, d\xi \qquad \text{for Im } z > 0,$$

and that

$$\frac{-1}{z - i} = \frac{1}{\pi i} \int_{-\infty}^{\infty} \frac{\xi}{\left(\xi^2 + 1\right)(\xi - z)} \, d\xi \qquad \text{for Im } z < 0.$$

(These values will be confirmed directly in Prob. 12, Exercises 6.3.) Verify that the jump in the value of the integral as z crosses the real axis equals $-2x / (x^2 + 1)$.

SUMMARY

Integration in the complex plane takes place along contours, which are continuous chains of directed smooth curves. The definite integral over each smooth part is defined by imitating the Riemann sum definition used in calculus, and the contour integral $\int_\Gamma f(z) \, dz$ is the sum of the integrals over the smooth components of Γ.

A "brute-force" technique for computing an integral along a contour Γ involves finding a parametrization $z = z(t)$, $a \le t \le b$, for Γ; then the contour integral can be obtained by performing an integration with respect to the real variable t, in accordance with the formula

$$\int_\Gamma f(z) \, dz = \int_a^b f(z(t)) z'(t) \, dt.$$

When the integrand $f(z)$ is analytic in a domain containing Γ, the following considerations may be useful in evaluating $\int_\Gamma f(z) \, dz$:

1. (*Fundamental Theorem of Calculus*) If f has an antiderivative F in a domain containing Γ, then

$$\int_\Gamma f(z)\,dz = F(z_T) - F(z_I).$$

   where z_I and z_T are the initial and terminal points of Γ.

2. (*Cauchy's Theorem*) If Γ is a simple closed contour (i.e., $z_I = z_T$ but no other self-intersections occur) and f is analytic inside and on Γ, then $\int_\Gamma f(z)\,dz = 0$.

3. (*Cauchy's Integral Formula*) If Γ is a simple closed positively oriented contour and f has the form $f(z) = g(z)/(z - z_0)$, with g analytic inside and on Γ and z_0 lying inside Γ, then

$$\int_\Gamma f(z)\,dz = \int_\Gamma \frac{g(z)}{z - z_0}\,dz = 2\pi i g(z_0).$$

   More generally,

$$\int_\Gamma \frac{g(z)\,dz}{(z - z_0)^m} = \frac{2\pi i g^{(m-1)}(z_0)}{(m-1)!}.$$

4. If f is of the form $f(z) = g(z)/[(z - z_1)^{m_1}(z - z_2)^{m_2}]$, with g as in case (3) and the points z_1 and z_2 lying inside the simple closed contour Γ, then $\int_\Gamma f(z)\,dz$ can be reduced to case (3) by using partial fractions.

5. (*Deformation Invariance Theorem*) If the closed contour Γ can be continuously deformed to another closed contour Γ' without passing through any singularities of f, then

$$\int_\Gamma f(z)\,dz = \int_{\Gamma'} f(z)\,dz.$$

   When the domain of analyticity of f is simply connected (i.e., has no "holes"), then f has an antiderivative and integrals of f are independent of path.

   The Cauchy integral formula has many consequences for analytic functions. Some of these are the infinite differentiability of analytic functions (see 3), Liouville's theorem (bounded entire functions are constant), the Fundamental Theorem of Algebra (every nonconstant polynomial has a zero), and the maximum modulus theorem (the maximum of $|f|$ is attained on the boundary of a bounded domain).

   In simply connected domains harmonic functions can be identified as real parts of analytic functions. Hence these results have analogues for harmonic functions, such as infinite differentiability, maximum principles, and Poisson's formula (the analogue of Cauchy's integral formula).

Suggested Reading

The following references will be helpful to the reader interested in seeing a more detailed treatment of the topics in this chapter:

Theory of Curves, Arc Length

[1] Apostol, T.M. *Mathematical Analysis*, 2nd ed.   Addison-Wesley Publishing Company, Inc., Reading, MA, 1974.

[2] Nevanlinna, R., and Paatero, V. *Introduction to Complex Analysis*.   Addison-Wesley Publishing Company, Inc., Reading, MA, 1969.

Jordan Curve Theorem

[3] Pederson, R.N. "The Jordan Curve Theorem for Piecewise Smooth Curves." *American Mathematical Monthly* **76** (1969), 605–610.

Riemann Sums, Integration

Reference 2.

Continuous Deformation of Curves, Cauchy's Theorem

[4] Conway, J.B. *Functions of One Complex Variable*, 2nd ed.   Springer-Verlag, New York, 1978.

[5] Flatto, L., and Shisha, O. "A Proof of Cauchy's Integral Theorem." *Journal of Approximation Theory* **7** (1973), 386–390.

[6] Courant, R., and Hurwitz, A., *Vorlesungen uber Allgemeine Funktionentheorie und Elliptische Funktionen*, Springer, Berlin, 1964.

Leibniz's Rule for Integrals

[6] Taylor, A.E., and Mann, W.R. *Advanced Calculus*, 3rd ed.   Wiley, New York, 1983.

Green's Theorem

[7] Borisenko, A.I., and Tarapov, THAT IS *Vector and Tensor Analysis with Applications* (trans. from Russian by R.A. Silverman). Dover, New York, 1979.

[8] Davis, H., and Snider, A.D. *Introduction to Vector Analysis*, 7th ed.   Quant Systems, Charleston, SC, 1994.

Poisson's Formula

[9] Fisher, S.D. *Function Theory on Planar Domains*. Wiley-Interscience, New York, 1983.

[10] Hille, E. *Analytic Function Theory*, Vol. 2, 2nd ed. Chelsea, New York, 1973.

Additional Problems

[11] Spiegel, M.R. *Theory and Problems of Complex Variables*. Schaum's Outline Series, McGraw-Hill Book Company, New York, 1964.

Chapter 5

Series Representations
for Analytic Functions

5.1 Sequences and Series

In Chapter 2 we defined what is meant by convergence of a sequence of complex numbers; recall that the sequence $\{A_n\}_{n=1}^{\infty}$ has A as a limit if $|A - A_n|$ can be made arbitrarily small by taking n large enough. For computational convenience it is often advantageous to use an element A_n of the sequence as an approximation to A. Indeed, when we calculate the area of a circle we usually use an element of the sequence $3.14, 3.141, 3.1415, 3.14159, \ldots$ as an approximation to π. The use of sequences, and in particular the kind of sequences associated with *series*, is an important tool in both the theory and applications of analytic functions, and the present chapter is devoted to the development of this subject.

The possibility of summing an infinite string of numbers must have occurred to anyone who has toyed with adding

$$\frac{1}{2} + \frac{1}{4} = \frac{3}{4}, \qquad \frac{1}{2} + \frac{1}{4} + \frac{1}{8} = \frac{7}{8}, \qquad \frac{1}{2} + \frac{1}{4} + \frac{1}{8} + \frac{1}{16} = \frac{15}{16}, \quad \text{etc.}$$

The sequence of sums thus derived obviously has 1 as a limit, and it seems sensible to say $\frac{1}{2} + \frac{1}{4} + \frac{1}{8} + \cdots = 1$. We are motivated to generalize this by saying that an *infinite series* of the form $c_0 + c_1 + c_2 + \cdots$ has the sum S if the sums of the first n terms approach S as a limit as n goes to infinity. The customary nomenclature is summarized in the following definition.

Definition 1. A **series** is a formal expression of the form $c_0 + c_1 + c_2 + \cdots$, or equivalently $\sum_{j=0}^{\infty} c_j$, where the **terms** c_j are complex numbers. The **nth partial sum** of the series, usually denoted S_n, is the sum of the first $n + 1$ terms, that is, $S_n := \sum_{j=0}^{n} c_j$. If the sequence of partial sums $\{S_n\}_{n=0}^{\infty}$ has a limit S, the series is said to **converge**, or **sum**, to S, and we write $S = \sum_{j=0}^{\infty} c_j$. A series that does not converge is said to **diverge**.

Notice that the notion of convergence for a series has been defined in terms of convergence for a sequence. As an illustration, observe that π is the sum of the series $3 + .1 + .04 + .001 + .0005 + .00009 + \cdots$.

Clearly one way to demonstrate that a series converges to S is to show that the *remainder* after summing the first $n + 1$ terms, $S - \sum_{j=0}^{n} c_j$, goes to zero as $n \to \infty$. We use this technique in describing the convergence of the simple but extremely useful *geometric series* $\sum_{j=0}^{\infty} c^j$.

Lemma 1. The series $\sum_{j=0}^{\infty} c^j$ converges to $1/(1-c)$ if $|c| < 1$; that is,

$$1 + c + c^2 + c^3 + \cdots = \frac{1}{1-c} \qquad \text{if } |c| < 1. \tag{1}$$

(In Prob. 6 we shall see that such a series diverges if $|c| \geq 1$.)

Proof.   Observe that

$$(1-c)\left(1 + c + c^2 + \cdots + c^{n-1} + c^n\right)$$
$$= 1 + c + c^2 + \cdots + c^{n-1} + c^n - c - c^2 - \cdots - c^{n-1} - c^n - c^{n+1}$$
$$= 1 - c^{n+1}.$$

Rearranging this yields

$$\frac{1}{1-c} - \left(1 + c + c^2 + \cdots + c^{n-1} + c^n\right) = \frac{c^{n+1}}{1-c}. \tag{2}$$

Since $|c| < 1$, the lemma follows; Eq. (2) displays the remainder as $c^{n+1}/(1-c)$, which certainly goes to zero as $n \to \infty$ (cf. Prob. 6 in Exercises 2.2).   ∎

Another important way to establish the convergence of a series involves comparing it with another series whose convergence is known. The following theorem, which generalizes a result from calculus, seems so transparent that we shall spare our trusting readers the proof and refer the skeptics to Sec. 5.4.

Theorem 1. (*Comparison Test*) Suppose that the terms c_j satisfy the inequality

$$\left|c_j\right| \leq M_j$$

for all integers j larger than some number J. Then if the series $\sum_{j=0}^{\infty} M_j$ converges, so does $\sum_{j=0}^{\infty} c_j$.

Example 1

Show that the series $\sum_{j=0}^{\infty}(3+2i)/(j+1)^j$ converges.

> **Solution.**   We shall compare the series
>
> $$\sum_{j=0}^{\infty}\frac{3+2i}{(j+1)^j}=(3+2i)+\frac{(3+2i)}{2}+\frac{(3+2i)}{9}+\frac{(3+2i)}{64}+\cdots \qquad (3)$$
>
> with the *convergent* geometric series
>
> $$\sum_{j=0}^{\infty}\frac{1}{2^j}=1+\frac{1}{2}+\frac{1}{4}+\frac{1}{8}+\cdots . \qquad (4)$$
>
> Since $|3+2i|=\sqrt{13}<4$, the reader can easily verify that
>
> $$\left|\frac{3+2i}{(j+1)^j}\right|<\frac{4}{(j+1)^j},$$
>
> and that this is less than $1/2^j$ for $j\geq 3$. Thus the terms of (4) dominate those of the series in (3) and, hence, (3) converges.   ∎

Sometimes the *ratio test* can be applied to a series to establish convergence.

Theorem 2. (*Ratio Test*) Suppose that the terms of the series $\sum_{j=0}^{\infty}c_j$ have the property that the ratios $|c_{j+1}/c_j|$ approach a limit L as $j\to\infty$. Then the series converges if $L<1$ and diverges if $L>1$.

The proof of this theorem involves comparing the given series with a series obtained by judiciously modifying the geometric series $\sum_{j=0}^{\infty}L^j$. See Prob. 15.

Example 2

Show that the series $\sum_{j=0}^{\infty}4^j/j!$ converges.

> **Solution.**   We have
>
> $$\left|\frac{c_{j+1}}{c_j}\right|=\frac{4^{j+1}}{(j+1)!}\frac{j!}{4^j}=\frac{4}{j+1}.$$
>
> This ratio approaches zero as $j\to\infty$; thus the series converges.   ∎

We remark that a series $\sum_{j=0}^{\infty}c_j$ is said to be *absolutely convergent* if the series $\sum_{j=0}^{\infty}|c_j|$ converges. Any absolutely convergent series is convergent, by a trivial application of the comparison test.

The kinds of sequences and series that often arise in complex analysis are those where the terms are functions of a complex variable z. Thus if we have a sequence of functions $F_1(z)$, $F_2(z)$, $F_3(z)$, ..., we must consider the possibility that for some values of z the sequence converges, while for others it diverges. As an example, the sequence $(z/2i)^n$, $n = 1, 2, 3, \ldots$, approaches zero for $|z| < |2i| = 2$, approaches 1 for $z = 2i$ (obviously!), and has no limit otherwise (see Prob. 4). Similarly, a *series* of complex functions $\sum_{j=0}^{\infty} f_j(z)$ may converge for some values of z and diverge for others.

Example 3

If z_0 ($\neq 0$) is fixed, show that the series $\sum_{j=0}^{\infty} (z/z_0)^j$ [which is quite distinct from the *sequence* $(z/z_0)^j$] converges for $|z| < |z_0|$.

Solution. This is merely a thinly disguised resurrection of Lemma 1. In fact, setting $c = z/z_0$ in Eq. (2) yields

$$\frac{1}{1 - z/z_0} - \left[1 + \frac{z}{z_0} + \left(\frac{z}{z_0} \right)^2 + \cdots + \left(\frac{z}{z_0} \right)^n \right] = \frac{(z/z_0)^{n+1}}{1 - z/z_0}. \tag{5}$$

We conclude, as before, that for $|z| < |z_0|$ the series sums to the function $1/(1 - z/z_0)$. ■

In applying this theory to analytic functions we need a somewhat stronger notion of convergence. By way of illustration, consider the sequence of real functions $F_n(x) = x^n$, depicted over the half-open interval $0 \leq x < 1$ in Fig. 5.1. Clearly, on this set the sequence $\{F_n(x)\}_{n=1}^{\infty}$ converges to the function $F(x) \equiv 0$; that is, for any *given* x the powers x^n become minuscule, for sufficiently large n. But none of the *curves* $y = x^n$ ($0 \leq x < 1$) would be regarded as good approximations to the *curve* $y = 0$, since each of the former has points near its right edge that generate (relatively) large values of x^n. We say that the convergence is *pointwise*, but not *uniform*. Thus we formulate the following.

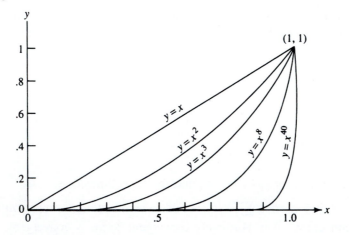

Figure 5.1 The functions $F_n(x) = x^n$ converge to zero pointwise but not uniformly in $[0, 1)$.

> **Definition 2.** The sequence $\{F_n(z)\}_{n=1}^{\infty}$ is said to **converge uniformly to** $F(z)$ **on the set** T if for any $\varepsilon > 0$ there exists an integer N such that when $n > N$,
>
> $$|F(z) - F_n(z)| < \varepsilon \qquad \text{for all } z \text{ in } T.$$
>
> Accordingly, the series $\sum_{j=0}^{\infty} f_j(z)$ converges uniformly to $f(z)$ on T if the sequence of its partial sums converges uniformly to $f(z)$ there.

The essential feature of uniform convergence is that for a given $\varepsilon > 0$, one must be able to find an integer N *that is independent of z in T* such that the error $|F(z) - F_n(z)|$ is less than ε for $n > N$. In contrast, for pointwise convergence, N can depend upon z. (See Prob. 17.) Of course, uniform convergence on T implies pointwise convergence on T.

Example 4

Show that the series $\sum_{j=0}^{\infty}(z/z_0)^j$ of Example 3 is uniformly convergent in every closed disk $|z| \le r$, if $r < |z_0|$.

Solution. Given $\varepsilon > 0$, we have to show that the remainder after $n + 1$ terms will be less than ε for all z in the disk, when n is large enough. This is easy; from Eq. (5) we can find an upper bound, independent of z, for the remainder:

$$\left| \frac{(z/z_0)^{n+1}}{1 - z/z_0} \right| \le \frac{(r/|z_0|)^{n+1}}{1 - r/|z_0|}, \qquad \text{for } |z| \le r.$$

This can be made arbitrarily small since $r < |z_0|$. ■

Combining Examples 3 and 4 we see that the series $\sum_{j=0}^{\infty}(z/z_0)^j$ converges pointwise in the open disk $|z| < |z_0|$ and uniformly on any closed subdisk $|z| \le r < |z_0|$.

EXERCISES 5.1

1. Find the sum of the following convergent series.

(a) $\displaystyle\sum_{j=0}^{\infty} \left(\frac{i}{3} \right)^j$

(b) $\displaystyle\sum_{k=0}^{\infty} \frac{3}{(1+i)^k}$

(c) $\displaystyle\sum_{j=0}^{\infty} (-1)^j \left(\frac{2}{3} \right)^j$

(d) $\displaystyle\sum_{k=14}^{\infty} \left(\frac{1}{2i} \right)^k$

(e) $\displaystyle\sum_{j=0}^{\infty} \left(\frac{1}{3} \right)^{2j}$

(f) $\displaystyle\sum_{j=0}^{\infty} \left[\frac{1}{j+2} - \frac{1}{j+1} \right]$

2. Using the ratio test, show that the following series converge.

(a) $\displaystyle\sum_{j=1}^{\infty} \frac{1}{j!}$

(b) $\displaystyle\sum_{k=1}^{\infty} \frac{(3+i)^k}{k!}$

(c) $\displaystyle\sum_{j=0}^{\infty} \frac{j^2}{4^j}$

(d) $\displaystyle\sum_{k=1}^{\infty} \frac{k!}{k^k}$

3. Prove that if the sequence $\{z_n\}_{n=1}^{\infty}$ converges, then $(z_n - z_{n-1}) \to 0$ as $n \to \infty$.

4. Let $z_0 \neq 0$. Prove that the sequence $(z/z_0)^n$, $n = 1, 2, \ldots$, diverges if $|z| \geq |z_0|$, $z \neq z_0$. [HINT: For $|z| = |z_0|$, observe that

$$\left| \left(\frac{z}{z_0} \right)^n - \left(\frac{z}{z_0} \right)^{n-1} \right| = \left| \frac{z}{z_0} - 1 \right| > 0$$

and use the result of Prob. 3.]

5. Prove that if the series $\sum_{j=0}^{\infty} c_j$ converges, then $c_j \to 0$ as $j \to \infty$. [HINT: Consider the difference $S_n - S_{n-1}$ of consecutive partial sums.]

6. Prove that the series $\sum_{j=0}^{\infty} c^j$ diverges if $|c| \geq 1$. [HINT: See Prob. 5.]

7. For each of the following determine if the given series converges or diverges.

(a) $\displaystyle\sum_{k=0}^{\infty} \left(\frac{1+2i}{1-i} \right)^k$ (b) $\displaystyle\sum_{j=1}^{\infty} \frac{1}{j^2 3^j}$ (c) $\displaystyle\sum_{n=1}^{\infty} \frac{n i^n}{2n+1}$

(d) $\displaystyle\sum_{j=1}^{\infty} \frac{j!}{5^j}$ (e) $\displaystyle\sum_{k=1}^{\infty} \frac{(-1)^k k^3}{(1+i)^k}$ (f) $\displaystyle\sum_{k=1}^{\infty} \left(i^k - \frac{1}{k^2} \right)$

8. Prove the following statements.

(a) If $\sum_{j=0}^{\infty} c_j$ sums to S, then $\sum_{j=0}^{\infty} \bar{c}_j$ sums to $\bar{S}$.

(b) If $\sum_{j=0}^{\infty} c_j$ sums to S and λ is any complex number, then $\sum_{j=0}^{\infty} \lambda c_j$ sums to λS.

(c) If $\sum_{j=0}^{\infty} c_j$ sums to S and $\sum_{j=0}^{\infty} d_j$ sums to T, then $\sum_{j=0}^{\infty} (c_j + d_j)$ sums to $S + T$.

9. Prove that the series $\sum_{j=0}^{\infty} z_j$ converges if and only if both of the series $\sum_{j=0}^{\infty} \text{Re}(z_j)$ and $\sum_{j=0}^{\infty} \text{Im}(z_j)$ converge.

10. Show that the sequence of functions $F_n(z) = z^n / (z^n - 3^n)$, $n = 1, 2, \ldots$, converges to zero for $|z| < 3$ and to 1 for $|z| > 3$.

11. Using the ratio test, find a domain in which convergence holds for each of the following series of functions.

(a) $\displaystyle\sum_{j=1}^{\infty} j z^j$ (b) $\displaystyle\sum_{k=0}^{\infty} \frac{(z-i)^k}{2^k}$

(c) $\displaystyle\sum_{j=0}^{\infty} \frac{z^j}{j!}$ (d) $\displaystyle\sum_{k=0}^{\infty} (z+5i)^{2k} (k+1)^2$

12. Let $F_n(z) = [nz/(n+1)] + (3/n)$, $n = 1, 2, \ldots$. Prove that the sequence $\{F_n(z)\}_1^{\infty}$ converges uniformly to $F(z) = z$ on every closed disk $|z| \leq R$.

13. Prove that $\sum_{j=1}^{\infty} 1/j^p$ converges if $p > 1$. [HINT: Interpret the integral $\int_1^N (1/x^p)\,dx$ as an area; then interpret $\sum_{j=2}^{N} 1/j^p$ as an area and compare.]

14. Using the comparison test and the result of Prob. 13, show that the following series converge.

(a) $\displaystyle\sum_{j=1}^{\infty} \frac{1}{j(j+i)}$

(b) $\displaystyle\sum_{k=1}^{\infty} \frac{\sin\left(k^2\right)}{k^{3/2}}$

(c) $\displaystyle\sum_{k=1}^{\infty} \frac{k^2 i^k}{k^4 + 1}$

(d) $\displaystyle\sum_{k=2}^{\infty} (-1)^k \left(\frac{5k+8}{k^3-1}\right)$

15. Prove the ratio test (Theorem 2). [HINT: If $L < 1$, choose $\varepsilon > 0$ and J so that $|c_{j+1}/c_j| < L + \varepsilon < 1$ for $j \geq J$. Then show that $|c_k| \leq |c_J|(L+\varepsilon)^{k-J}$ for $k > J$ and use the comparison test.]

16. Prove that the sequence $\{z_n\}_1^{\infty}$ converges if and only if the series $\sum_{k=1}^{\infty} (z_{k+1} - z_k)$ converges.

17. Consider the sequence of functions $F_n(x) = x^n$ $(n = 0, 1, 2, \ldots)$ on the real interval $T = (0, 1)$, which converges pointwise to $F(x) = 0$ on T. Show that, for $0 < x < 1$,

$$|F_n(x) - F(x)| < \frac{1}{2}$$

when and only when $n > N_x$, where

$$N_x := \frac{\text{Log } 2}{\text{Log } \left(x^{-1}\right)}.$$

(Observe that $N_x \to \infty$ as $x \to 1^-$, so that it is not possible to fulfill Definition 2 with an N *independent* of x in T when $\varepsilon = \frac{1}{2}$. This proves that this sequence does not converge uniformly on T.)

18. Assume that the sequence of functions $\{F_n(z)\}_1^{\infty}$ converges uniformly to $F(z)$ on a set T. Prove that if $|F(z)| \geq \rho > 0$ for all z in T, then there exists an integer N such that for $n > N$ the inequality $|F_n(z)| > \rho/2$ holds for all z in T. [HINT: Take $\varepsilon = \rho/2$ in Definition 2 and apply the triangle inequality.]

19. It will be shown in the next section that the series $\sum_{k=0}^{\infty} z^k/k!$ converges uniformly to e^z on every disk $|z| \leq R$. Accepting this fact, prove that, for n sufficiently large, none of the polynomials $S_n(z) = \sum_{k=0}^{n} z^k/k!$ has zeros on $|z| \leq 5$. [HINT: Use Prob. 18.] (There is nothing special about the disk $|z| \leq 5$; the same assertion holds on any bounded set.)

20. Prove that the series $\sum_{j=0}^{\infty} z^j$ does not converge *uniformly* to $1/(1-z)$ on the open disk $|z| < 1$.

21. An increasingly lazy and asymmetric frog leaps one meter (from $z = 0$ to $z = 1$) on his first jump, 1/2 meter on his second jump, 1/4 meter on his third jump, 1/8 meter on his fourth jump, and so on, each time turning exactly an angle α to the left of his preceding flight path. Use complex algebra to demonstrate that the limiting location of the frog will inevitably be someplace on the circle $|z - 4/3| = 2/3$ *regardless* of the value of α.

5.2 Taylor Series

Suppose that we wish to find a polynomial $p_n(z)$ of degree at most n that approximates an analytic function $f(z)$ in a neighborhood of a point z_0. Naturally there are differing criteria as to how well the polynomial approximates the function. We shall construct a polynomial that "looks like" $f(z)$ at the point z_0 in the sense that its derivatives match those of f at z_0, insofar as possible:

$$p_n(z_0) = f(z_0)$$
$$p_n'(z_0) = f'(z_0)$$
$$\vdots$$
$$p_n^{(n)}(z_0) = f^{(n)}(z_0).$$

[Of course, the $(n + 1)$st derivative of any polynomial of degree n must equal *zero*.]

As we saw in Sec. 3.1, the preceding equations determine the Taylor form of $p_n(z)$, and according to Eq. (2) of that section *the nth-degree polynomial that matches* $f, f', f'', \ldots, f^{(n)}$ *at z_0 is*

$$p_n(z) = f(z_0) + f'(z_0)(z - z_0) + \frac{f''(z_0)}{2!}(z - z_0)^2 + \cdots + \frac{f^{(n)}(z_0)}{n!}(z - z_0)^n . ^\dagger \quad (1)$$

Naturally we conjecture that as n tends to infinity, $p_n(z)$ becomes a better and better approximation to $f(z)$ near z_0. In fact, the astute reader may have noticed that $p_n(z)$ looks like a partial sum of a series, and so might anticipate that this series converges to $f(z)$. The precise state of affairs is given in the following definition and theorem.

Definition 3. If f is analytic at z_0, then the series

$$f(z_0) + f'(z_0)(z - z_0) + \frac{f''(z_0)}{2!}(z - z_0)^2 + \cdots = \sum_{j=0}^{\infty} \frac{f^{(j)}(z_0)}{j!}(z - z_0)^j \quad (2)$$

is called the **Taylor series** for f around z_0. When $z_0 = 0$, it is also known as the **Maclaurin series** for f.

---

† Actually $p_n(z)$ will have degree less than n if $f^{(n)}(z_0) = 0$.

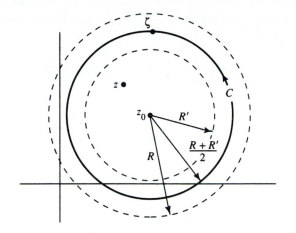

Figure 5.2 Uniform convergence in closed subdisks.

Theorem 3. If f is analytic in the disk $|z - z_0| < R$, then the Taylor series (2) converges to $f(z)$ for all z in this disk. Furthermore, the convergence of the series is uniform in any closed subdisk $|z - z_0| \le R' < R$.

Proof.   Notice that if we prove uniform convergence in *every* closed subdisk $|z - z_0| \le R' < R$, we will have pointwise convergence for each z in the open disk $|z - z_0| < R$ (why?); thus we deal only with the closed subdisk statement. Let C be the circle $|z - z_0| = (R + R')/2$, positively oriented (see Fig. 5.2). Then by Cauchy's integral formula we have, for any z within the closed subdisk,

$$f(z) = \frac{1}{2\pi i} \int_C \frac{f(\zeta)}{\zeta - z} \, d\zeta.$$

We rewrite the integrand by manipulating $(\zeta - z)^{-1}$ into a form suggesting the geometric series in powers of $(z - z_0)/(\zeta - z_0)$, which converges since $|z - z_0|/|\zeta - z_0| < 1$ in Fig. 5.2:

$$\frac{1}{\zeta - z} = \frac{1}{(\zeta - z_0) - (z - z_0)} = \frac{1}{1 - \dfrac{z - z_0}{\zeta - z_0}} \cdot \frac{1}{\zeta - z_0}$$

$$= \left[1 + \frac{z - z_0}{\zeta - z_0} + \frac{(z - z_0)^2}{(\zeta - z_0)^2} + \cdots + \frac{(z - z_0)^n}{(\zeta - z_0)^n} + \frac{\dfrac{(z - z_0)^{n+1}}{(\zeta - z_0)^{n+1}}}{1 - \dfrac{z - z_0}{\zeta - z_0}} \right] \frac{1}{\zeta - z_0},$$

$$\tag{3}$$

using identity (2) of Sec. 5.1 with $c = (z - z_0)/(\zeta - z_0)$. Putting this into the Cauchy integral formula we find

$$f(z) = \frac{1}{2\pi i} \int_C \frac{f(\zeta)}{\zeta - z_0} d\zeta + \frac{z - z_0}{2\pi i} \int_C \frac{f(\zeta)}{(\zeta - z_0)^2} d\zeta$$

$$+ \frac{(z - z_0)^2}{2\pi i} \int_C \frac{f(\zeta)}{(\zeta - z_0)^3} d\zeta + \cdots + \frac{(z - z_0)^n}{2\pi i} \int_C \frac{f(\zeta)}{(\zeta - z_0)^{n+1}} d\zeta + T_n(z),$$

$$(4)$$

where $T_n(z)$ can be expressed (with a little algebra) as

$$T_n(z) = \frac{1}{2\pi i} \int_C \frac{f(\zeta)}{(\zeta - z)} \frac{(z - z_0)^{n+1}}{(\zeta - z_0)^{n+1}} d\zeta. \tag{5}$$

Now by Cauchy's integral formula for derivatives

$$\frac{1}{2\pi i} \int_C \frac{f(\zeta)}{\zeta - z_0} d\zeta = f(z_0), \qquad \frac{1}{2\pi i} \int_C \frac{f(\zeta)}{(\zeta - z_0)^2} d\zeta = f'(z_0),$$

$$\frac{1}{2\pi i} \int_C \frac{f(\zeta)}{(\zeta - z_0)^3} d\zeta = \frac{f''(z_0)}{2!}, \quad \text{etc.,}$$

and so the first $n + 1$ terms on the right-hand side of Eq. (4) yield the nth partial sum of the series in (2). Thus $T_n(z)$ is the remainder, and we must show that it can be made "uniformly small" for all z in the subdisk, by taking n sufficiently large.

This is simply a matter of applying the integral inequality of Chapter 4 (Theorem 5, Sec. 4.2) and being more specific about the ratio $|z - z_0|/|\zeta - z_0|$. For the terms in the integrand of Eq. (5) we have

$$|z - z_0| \le R', \qquad |\zeta - z_0| = \frac{R + R'}{2}, \qquad \frac{|z - z_0|}{|\zeta - z_0|} \le \frac{2R'}{R + R'},$$

and, from Fig. 5.2,

$$|\zeta - z| \ge \frac{R + R'}{2} - R' = \frac{R - R'}{2}.$$

The length of C is $2\pi (R + R')/2$. Thus for z in the subdisk

$$|T_n(z)| \le \frac{1}{2\pi} \cdot \max_{\zeta \text{ on } C} |f(\zeta)| \frac{2}{R - R'} \left(\frac{2R'}{R + R'} \right)^{n+1} 2\pi \left(\frac{R + R'}{2} \right).$$

The right-hand side is independent of z, and since $2R' < R + R'$, it can be made less than any positive ε by taking n large enough. ∎

Notice that the theorem implies that *the Taylor series will converge to $f(z)$ everywhere inside the largest open disk, centered at z_0, over which f is analytic.*

Example 1

Compute and state the convergence properties of the Taylor series for **(a)** Log z around $z_0 = 1$, **(b)** $1/(1 - z)$ around $z_0 = 0$, and **(c)** e^z around $z_0 = 0$.

Solution. **(a)** The consecutive derivatives of Log z are Log z, z^{-1}, $-z^{-2}$, $2z^{-3}$, $-3 \cdot 2z^{-4}$, etc.; in general

$$\frac{d^j \operatorname{Log} z}{dz^j} = (-1)^{j+1}(j-1)!z^{-j} \qquad (j = 1, 2, \ldots).$$

Evaluating these at $z = 1$ we find

$$\operatorname{Log} z = 0 + (z-1) - \frac{(z-1)^2}{2!} + 2!\frac{(z-1)^3}{3!} - 3!\frac{(z-1)^4}{4!} + \cdots$$

$$= \sum_{j=1}^{\infty} \frac{(-1)^{j+1}(z-1)^j}{j}. \tag{6}$$

This is valid for $|z - 1| < 1$, the largest open disk centered at $+1$ over which Log z is analytic.

(b) The consecutive derivatives of $(1 - z)^{-1}$ are $(1 - z)^{-1}$, $(1 - z)^{-2}$, $2(1 - z)^{-3}$, $3 \cdot 2(1 - z)^{-4}$, and, in general,

$$\frac{d^j}{dz^j}(1-z)^{-1} = j!(1-z)^{-j-1}.$$

Evaluating these at $z = 0$ gives the Taylor series

$$\frac{1}{1-z} = 1 + z + \frac{2!z^2}{2!} + \frac{3!z^3}{3!} + \cdots = \sum_{j=0}^{\infty} z^j, \tag{7}$$

valid for $|z| < 1$ (why?). Not surprisingly, this is just the geometric series considered in Lemma 1.

(c) Since

$$\frac{d^j e^z}{dz^j} = e^z$$

for all $j = 0, 1, \ldots$, the common value of these derivatives at $z = 0$ is 1. Thus

$$e^z = 1 + z + \frac{z^2}{2!} + \frac{z^3}{3!} + \cdots = \sum_{j=0}^{\infty} \frac{z^j}{j!},$$

which is valid for all z because e^z is entire. ∎

Let's experiment with Eq. (6), the expansion for Log z. If we differentiate the *series* term by term, we get

$$1 - (z-1) + (z-1)^2 - (z-1)^3 + \cdots = \sum_{j=0}^{\infty} (1-z)^j, \tag{8}$$

properly incorporating the negative signs. But the series (8) has the same form as the series in Eq. (7), with $(1 - z)$ in place of z. Thus (8) actually converges to the function $1/[1 - (1 - z)] = 1/z$; in other words, by differentiating the *series* for Log z we obtained the *series* for $1/z$, which is, in fact, the derivative of Log z. We are led to conjecture the following.

Theorem 4. If f is analytic at z_0, the Taylor series for f' around z_0 can be obtained by termwise differentiation of the Taylor series for f around z_0 and converges in the same disk as the series for f.

Proof.   The jth derivative of f' is, of course, the $(j + 1)$st derivative of f. Thus the Taylor series for f' is given by

$$f'(z_0) + f''(z_0)\,(z - z_0) + \frac{f'''(z_0)}{2!}\,(z - z_0)^2 + \cdots. \tag{9}$$

On the other hand, termwise differentiation of the Taylor series (2) yields

$$0 + 1 \cdot f'(z_0) + \frac{2}{2!} \cdot f''(z_0)\,(z - z_0) + \frac{3}{3!} \cdot f'''(z_0)\,(z - z_0)^2 + \cdots,$$

which is the same as (9). Furthermore, application of Theorem 3 to the function $f'(z)$ establishes that (9) converges in the largest open disk around z_0 over which f' is analytic. But according to Theorem 16 of Sec. 4.5, f' is analytic wherever f is analytic. This completes the proof.   ■

Example 2

Find the Maclaurin series for $\sin z$ and $\cos z$.

Solution.   We expand $\sin z$ as usual. The sequence of the derivatives is $\sin z$, $\cos z$, $-\sin z$, $-\cos z$, $\sin z$, ... Evaluating at the origin yields

$$\sin z = z - \frac{z^3}{3!} + \frac{z^5}{5!} - \frac{z^7}{7!} + \cdots, \tag{10}$$

which holds for all z since $\sin z$ is entire. To get $\cos z$, we differentiate Eq. (10):

$$\cos z = 1 - \frac{z^2}{2!} + \frac{z^4}{4!} - \frac{z^6}{6!} + \cdots. \tag{11}$$

The reader should by now be able to predict what would result if we differentiate Eq. (11).   ■

The next two theorems may sometimes simplify the computation of a Taylor series.

Theorem 5. Let f and g be analytic functions with Taylor series $f(z) = \sum_{j=0}^{\infty} a_j(z - z_0)^j$ and $g(z) = \sum_{j=0}^{\infty} b_j(z - z_0)^j$ around the point z_0 [that is, $a_j = f^{(j)}(z_0)/j!$ and $b_j = g^{(j)}(z_0)/j!$]. Then

   (i) the Taylor series for $cf(z)$, c a constant, is $\sum_{j=0}^{\infty} ca_j(z - z_0)^j$;

   (ii) the Taylor series for $f(z) \pm g(z)$ is $\sum_{j=0}^{\infty} (a_j \pm b_j)(z - z_0)^j$.

The proof is left as an easy exercise. The disk of convergence for $f \pm g$ is, of course, at least as big as the smaller of the convergence disks for f and g.

Example 3

Find the Maclaurin series for $\cos z + i \sin z$.

Solution.   Using the expansions (10) and (11) we find

$$\cos z + i \sin z = 1 + iz - \frac{z^2}{2!} - \frac{iz^3}{3!} + \frac{z^4}{4!} + \frac{iz^5}{5!} - \cdots = e^{iz}$$

for all z. (This validates a computation made in Example 1, Sec. 1.4.)   ■

Theorem 5 naturally leads us to cogitate the corresponding statement for products. First we must find a sensible way of multiplying two Taylor series. The *Cauchy product* of two Taylor series around a point z_0 is defined in the manner suggested by applying the distributive law and then grouping the terms in powers of $(z - z_0)$. Thus, if $z_0 = 0$, we find for the Cauchy product

$$\left[a_0 + a_1 z + a_2 z^2 + a_3 z^3 + \cdots \right] \cdot \left[b_0 + b_1 z + b_2 z^2 + b_3 z^3 + \cdots \right]$$
$$= a_0 b_0 + (a_1 b_0 + a_0 b_1)z + (a_2 b_0 + a_1 b_1 + a_0 b_2)z^2$$
$$+ (a_3 b_0 + a_2 b_1 + a_1 b_2 + a_0 b_3)z^3 + \cdots . \quad (12)$$

The coefficient, c_j, of z^j is therefore given by

$$c_j = a_j b_0 + a_{j-1} b_1 + a_{j-2} b_2 + \cdots + a_1 b_{j-1} + a_0 b_j = \sum_{\ell=0}^{j} a_{j-\ell} b_\ell. \quad (13)$$

Definition 4. The **Cauchy product** of two Taylor series $\sum_{j=0}^{\infty} a_j(z - z_0)^j$ and $\sum_{j=0}^{\infty} b_j(z - z_0)^j$ is defined to be the (formal) series $\sum_{j=0}^{\infty} c_j(z - z_0)^j$, where c_j is given by formula (13).

> **Theorem 6.** Let f and g be analytic functions with Taylor series $f(z) = \sum_{j=0}^{\infty} a_j (z - z_0)^j$ and $g(z) = \sum_{j=0}^{\infty} b_j (z - z_0)^j$ around the point z_0. Then the Taylor series for the product fg around z_0 is given by the Cauchy product of these two series.

Actually, we anticipated this result in electing to write the Cauchy product in (12) as if it were an ordinary product. As in Theorem 5, the Taylor series for fg converges at least in the smaller of the convergence disks for f and g.

Proof of Theorem 6.   We compute the consecutive derivatives of the product fg:

$$(fg)' = f'g + fg',$$
$$(fg)'' = f''g + 2f'g' + fg'',$$
$$(fg)''' = f'''g + 3f''g' + 3f'g'' + fg''',$$

and, in general, we have *Leibniz's formula* for the jth derivative of fg:

$$(fg)^{(j)} = \sum_{\ell=0}^{j} j! \frac{f^{(j-\ell)}}{(j-\ell)!} \cdot \frac{g^{(\ell)}}{\ell!}, \tag{14}$$

which we invite the reader to prove in Prob. 10.

On the other hand, if we identify the constants a_k and b_k in Eq. (13) with their expressions in terms of derivatives of f and g [e.g., $a_k = f^{(k)}(z_0) / k!$], we see from Eq. (14) that $(fg)^{(j)} / j!$ evaluated at z_0 is precisely c_j. This completes the proof. ∎

Example 4

Use the Cauchy product to find the Maclaurin series for $\sin z \cdot \cos z$.

Solution.   We have

$$\left(z - \frac{z^3}{3!} + \frac{z^5}{5!} - \frac{z^7}{7!} + \cdots \right) \cdot \left(1 - \frac{z^2}{2!} + \frac{z^4}{4!} - \frac{z^6}{6!} + \cdots \right)$$

$$= z - \left(\frac{1}{3!} + \frac{1}{2!} \right) z^3 + \left(\frac{1}{5!} + \frac{1}{3!}\frac{1}{2!} + \frac{1}{4!} \right) z^5$$

$$- \left(\frac{1}{7!} + \frac{1}{5!}\frac{1}{2!} + \frac{1}{3!}\frac{1}{4!} + \frac{1}{6!} \right) z^7 + \cdots .$$

It is amusing to try to simplify the coefficients; the reader can verify that

$$\sin z \cdot \cos z = z - \frac{4}{3!} z^3 + \frac{16}{5!} z^5 - \frac{64}{7!} z^7 + \cdots ,$$

which, when rewritten as

$$\frac{1}{2} \left[(2z) - \frac{(2z)^3}{3!} + \frac{(2z)^5}{5!} - \frac{(2z)^7}{7!} + \cdots \right],$$

will be recognized as the Taylor series for $\frac{1}{2}\sin w$, with $w = 2z$. We have reproduced a well-known trigonometric identity!   ■

Example 5

Find the first few terms of the Maclaurin series for $\tan z$.

Solution.   The expressions for the higher derivatives of $\tan z$ are cumbersome, so let's try to use the Cauchy product. First observe that $\cos z \cdot \tan z = \sin z$. Now set $\tan z = \sum_{j=0}^{\infty} a_j z^j$ for $|z| < \pi/2$. (Why $\pi/2$?) The product $\cos z \cdot \tan z$ then becomes

$$\left(1 - \frac{z^2}{2!} + \frac{z^4}{4!} - \cdots\right) \cdot \left(a_0 + a_1 z + a_2 z^2 + a_3 z^3 + a_4 z^4 + a_5 z^5 + \cdots\right)$$

$$= a_0 + a_1 z + \left(a_2 - \frac{a_0}{2!}\right) z^2 + \left(a_3 - \frac{a_1}{2!}\right) z^3 + \left(a_4 - \frac{a_2}{2!} + \frac{a_0}{4!}\right) z^4$$

$$+ \left(a_5 - \frac{a_3}{2!} + \frac{a_1}{4!}\right) z^5 + \cdots .$$

Identifying this with $\sin z = z - z^3/3! + z^5/5! - \cdots$, we solve recursively and find

$$a_0 = 0, \quad a_1 = 1, \quad a_2 = 0, \quad a_3 = \frac{1}{3}, \quad a_4 = 0, \quad a_5 = \frac{2}{15}, \quad \text{etc.}$$

Thus

$$\tan z = z + \frac{z^3}{3} + \frac{2z^5}{15} + \cdots .$$

The shrewd reader will observe that we have actually uncovered an indirect method of dividing Taylor series!   ■

   In closing this section we would like to point out that the proof of the validity of the Taylor expansion substantiates the claim, made in Sec. 2.3, that any *analytic* function can be displayed with a formula involving z alone, and not $\bar{z}$, x, or y.

EXERCISES 5.2

1. Using Definition 3, verify each of the following Taylor expansions by finding a general formula for $f^{(j)}(z_0)$.

   (a) $e^{-z} = \sum_{j=0}^{\infty} \frac{(-z)^j}{j!} = 1 - z + \frac{z^2}{2!} - \frac{z^3}{3!} + \cdots, \qquad z_0 = 0$

   (b) $\cosh z = \sum_{j=0}^{\infty} \frac{z^{2j}}{(2j)!} = 1 + \frac{z^2}{2!} + \frac{z^4}{4!} + \cdots, \qquad z_0 = 0$

   (c) $\sinh z = \sum_{j=0}^{\infty} \frac{z^{2j+1}}{(2j+1)!} = z + \frac{z^3}{3!} + \frac{z^5}{5!} + \cdots, \qquad z_0 = 0$

(d) $\dfrac{1}{1-z} = \displaystyle\sum_{j=0}^{\infty} \dfrac{(z-i)^j}{(1-i)^{j+1}}, \qquad z_0 = i$

(e) $\text{Log}(1-z) = \displaystyle\sum_{j=1}^{\infty} \dfrac{-z^j}{j}, \qquad z_0 = 0$

(f) $z^3 = 1 + 3(z-1) + 3(z-1)^2 + (z-1)^3, \qquad z_0 = 1$

2. Determine the disks over which the Taylor expansions in Prob. 1 are valid.

3. Let $f(z) = \sum_{j=0}^{\infty} a_j z^j$ be the Maclaurin expansion of a function $f(z)$ analytic at the origin. Prove each of the following statements.

   (a) $\displaystyle\sum_{j=0}^{\infty} a_j z^{2j}$ is the Maclaurin expansion of $g(z) := f\left(z^2\right)$.

   (b) $\displaystyle\sum_{j=0}^{\infty} a_j c^j z^j$ is the Maclaurin expansion of $h(z) := f(cz)$.

   (c) $\displaystyle\sum_{j=0}^{\infty} a_j z^{m+j}$ is the Maclaurin expansion of $H(z) := z^m f(z)$.

   (d) $\displaystyle\sum_{j=0}^{\infty} a_j (z - z_0)^j$ is the Taylor expansion of $G(z) := f(z - z_0)$ around z_0.

4. Let α be a complex number. Show that if $(1+z)^\alpha$ is taken as $e^{\alpha \, \text{Log}(1+z)}$, then for $|z| < 1$

$$(1+z)^\alpha = 1 + \frac{\alpha}{1} z + \frac{\alpha(\alpha-1)}{1\cdot 2} z^2 + \frac{\alpha(\alpha-1)(\alpha-2)}{1\cdot 2\cdot 3} z^3 + \cdots .$$

   [REMARK: This generalizes the binomial theorem.]

5. Find and state the convergence properties of the Taylor series for the following.

   (a) $\dfrac{1}{1+z}$ around $z_0 = 0$ **(b)** e^{-z^2} around $z_0 = 0$

   (c) $z^3 \sin 3z$ around $z_0 = 0$ **(d)** $2\cos z - ie^z$ around $z_0 = 0$

   (e) $\dfrac{1+z}{1-z}$ around $z_0 = i$ **(f)** $\cos z$ around $z_0 = \dfrac{\pi}{4}$

   (g) $\dfrac{z}{(1-z)^2}$ around $z_0 = 0$

6. Prove directly that the Taylor expansion of $1/(\zeta - z)$ around $z_0 (\neq \zeta)$ is given by

$$\frac{1}{\zeta - z} = \sum_{j=0}^{\infty} \frac{(z - z_0)^j}{(\zeta - z_0)^{j+1}} \qquad \text{for } |z - z_0| < |\zeta - z_0| .$$

   [REMARK: The expansion lies at the heart of the proof of Theorem 3.]

7. Verify that the identity

$$\text{Log}\left(\frac{1+z}{1-z}\right) = \text{Log}(1+z) - \text{Log}(1-z)$$

holds when $|z| < 1$. Then, using the Maclaurin expansions of $\text{Log}(1+z)$ and $\text{Log}(1-z)$, find the Maclaurin expansion of $\text{Log}[(1+z)/(1-z)]$.

8. Use Taylor series to verify the following identities.

   (a) $\sin(-z) = -\sin z$

   (b) $\dfrac{de^z}{dz} = e^z$

   (c) $e^{-iz} = \cos z - i \sin z$

   (d) $e^{2z} = e^z \cdot e^z$

9. Prove Theorem 5.

10. Prove Leibniz's formula,[†] Eq. (14). [HINT: Use mathematical induction.]

11. Using Theorem 6 for computing the product of Taylor series, find the first three nonzero terms in the Maclaurin expansion of the following.

   (a) $e^z \cos z$

   (b) $\dfrac{e^z}{z-1}$

   (c) $\sec z = \dfrac{1}{\cos z}$

   (d) $\tanh z = \dfrac{\sinh z}{\cosh z}$

12. Prove that the polynomial $p_n(z)$ of Eq. (1) is the *only* polynomial of degree at most n that matches $f, f', f'', \ldots, f^{(n)}$ at z_0.

13. Find an explicit formula for the analytic function $f(z)$ that has the Maclaurin expansion $\sum_{k=0}^{\infty} k^2 z^k$. [HINT: Starting with the expression $(1-z)^{-1} = \sum_{k=0}^{\infty} z^k$, differentiate, multiply by z, differentiate again, and finally multiply by z.]

14. Let $f(z)$ be analytic in the disk $D : |z - z_0| < R$. Prove that if $f^{(k)}(z_0) = 0$ for every $k = 0, 1, 2, \ldots$, then $f(z)$ is identically zero in D.

15. Let $f(z)$ be analytic in the unit disk $|z| < 1$. Prove that if $f'(0) = f^{(3)}(0) = f^{(5)}(0) = \cdots = 0$, then $f(-z) = f(z)$ for all z in this disk. That is, show that f is an even function.

16. Rewrite the polynomial $p(z) = a_0 + a_1 z + \cdots + a_n z^n$ in powers of $(z-1)$; that is, find the coefficients c_i of the expansion $p_n(z) = c_0 + c_1(z-1) + \cdots + c_n(z-1)^n$ in terms of the a_j. [HINT: Do not rearrange; use Taylor series.]

17. Recall from Exercises 5.1, Prob. 20, that the Taylor series $\sum_{j=0}^{\infty} z^j$ does not converge *uniformly* to $(1-z)^{-1}$ on the open disk $D : |z| < 1$. Why doesn't this contradict Theorem 3?

18. The Taylor series provides a workable method of numerically tabulating the functions of mathematical physics when the remainder term can be estimated. Establish each of the following error estimates: for $|z| \le 1$,

---

[†] Gottfried Wilhelm von Leibniz (1646–1716) practiced law in Mainz, Germany.

(a) $\left| e^z - \sum_{k=0}^{n} \frac{z^k}{k!} \right| \leq \frac{1}{(n+1)!} \cdot \left(1 + \frac{1}{n+1} \right)$

(b) $\left| \sin z - \sum_{k=0}^{n} \frac{(-1)^k z^{2k+1}}{(2k+1)!} \right| \leq \frac{1}{(2n+3)!} \left(\frac{4n^2 + 18n + 20}{4n^2 + 18n + 19} \right)$

[HINT: Write the error as an infinite series, factor out the first term, and then compare with a geometric series.]

19. According to the estimate in Prob. 18(a), how many terms of the expansion $\sum_{k=0}^{\infty} z^k / k!$ are needed to compute e^z to within $\pm 10^{-5}$ for $|z| \leq 1$?

20. (*Hermite Formula*) From the proof of Theorem 3 deduce the Hermite[†] formula for the remainder in the Maclaurin series for a function f analytic on $|z| \leq R$:

$$f(z) - \sum_{j=0}^{n} \frac{f^{(j)}(0)}{j!} z^j = \frac{1}{2\pi i} \oint_{|\zeta|=R} \frac{z^{n+1}}{\zeta^{n+1}} \frac{f(\zeta)}{(\zeta - z)} d\zeta \quad \text{for } |z| < R.$$

5.3 Power Series

A Taylor series for an analytic function appears to be a special instance of a certain general type of series of the form $\sum_{j=0}^{\infty} a_j (z - z_0)^j$. Such series have a name:

Definition 5. A series of the form $\sum_{j=0}^{\infty} a_j (z - z_0)^j$ is called a **power series**. The constants a_j are the **coefficients** of the power series.

Suppose that we are presented with an arbitrary power series, such as

$$\sum_{j=0}^{\infty} \frac{z^j}{(j+1)^2} = 1 + \frac{z}{4} + \frac{z^2}{9} + \frac{z^3}{16} + \cdots . \tag{1}$$

Certain questions then arise naturally. For what values of z does the series converge? Is the sum an analytic function? Is the power series representation of a function unique? In short, is every power series a Taylor series? This section is devoted to answering these questions.

The issue of convergence is settled by the following result, which smacks of the Taylor expansion.

[†]Charles Hermite (1822–1901) published the first proof that e is a transcendental number.

Theorem 7. For any power series $\sum_{j=0}^{\infty} a_j (z - z_0)^j$ there is a real number R between 0 and ∞, inclusive, which depends only on the coefficients $\{a_j\}$, such that

   (i) the series converges for $|z - z_0| < R$,

   (ii) the series converges uniformly in any closed subdisk $|z - z_0| \leq R' < R$,

   (iii) the series diverges for $|z - z_0| > R$.

The number R is called the **radius of convergence** of the power series.

In particular, when $R = 0$ the power series converges only at $z = z_0$, and when $R = \infty$ the series converges for all z. For $0 < R < \infty$, the circle $|z - z_0| = R$ is called the *circle of convergence*, but no general convergence statement can be made for z lying on this circle (see Prob. 1). The situation is depicted in Fig. 5.3.

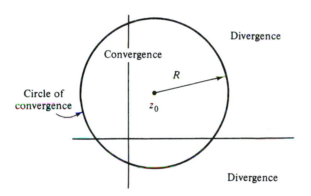

Figure 5.3 Circle of convergence.

Although a rigorous proof of Theorem 7 is deferred to (optional) Sec. 5.4, we can give an informal argument here that shows why the particular format of a power series dictates that its region of convergence has to be a disk. The essential ingredient is the following lemma, which we state for the special case $z_0 = 0$:

Lemma 2. If the power series $\sum_{j=0}^{\infty} a_j z^j$ converges at a point having modulus r, then it converges at every point in the disk $|z| < r$.

Proof of Lemma 2. By hypothesis, there exists a point z_1, with $|z_1| = r$, such that the series $\sum_{j=0}^{\infty} a_j z_1^j$ converges. This implies that the sequence of terms $a_j z_1^j$ is bounded; that is, there exists a constant M such that

$$\left| a_j z_1^j \right| = |a_j| r^j \le M \qquad \text{(for all } j\text{)}.$$

Now for $|z| < r$ we can write

$$\left| a_j z^j \right| = |a_j| r^j \cdot \left(\frac{|z|}{r} \right)^j \le M \left(\frac{|z|}{r} \right)^j.$$

Thus, for $|z|/r < 1$, the terms of the series $\sum_{j=0}^{\infty} a_j z^j$ are dominated by the terms of a convergent geometric series. By the comparison test, therefore, we conclude that the series converges. ∎

To see the existence of the number R in Theorem 7 for the power series $\sum_{j=0}^{\infty} a_j z^j$ we reason informally as follows: Consider the set of all real numbers r such that the series converges at some point having modulus r. Let R be the "largest"[†] of these numbers r. Then, by Lemma 2, the series converges for $|z| < R$, and from the definition of R the series diverges for all z with $|z| > R$.

If z is replaced by $(z - z_0)$ in the preceding argument, we deduce that the region of convergence of the general power series $\sum_{j=0}^{\infty} a_j (z - z_0)^j$ must be a disk with center z_0.

A formula for the radius of convergence R can be given, but we shall postpone it also until Sec. 5.4. However, in Prob. 2, the ratio test is used to show that in the special case when $\left| a_{j+1} / a_j \right|$ has a limit as j goes to infinity, R is the reciprocal of this limit. For example, the coefficients of the power series (1) satisfy

$$\lim_{j \to \infty} \left| \frac{a_{j+1}}{a_j} \right| = \lim_{j \to \infty} \frac{(j + 1)^2}{(j + 2)^2} = 1,$$

so its circle of convergence is $|z| = R = \frac{1}{1} = 1$.

Uniform convergence [cf. (ii) of Theorem 7] is a powerful feature of a sequence, as the next three results show. The first says that the uniform limit of continuous functions is itself continuous.

Lemma 3. Let f_n be a sequence of functions continuous on a set $T \subset \mathbf{C}$ and converging uniformly to f on T. Then f is also continuous on T.

Proof. To prove that f is continuous at a point z_0 of T, we must show that for any $\varepsilon > 0$ there is a $\delta > 0$ such that if z belongs to T and $|z_0 - z| < \delta$, then $|f(z_0) - f(z)| < \varepsilon$. We proceed by first choosing an integer N so large that $|f(z) - f_N(z)| < \varepsilon/3$ for all z in T; this is possible thanks to uniform convergence.

[†]More advanced students will recognize that R is precisely defined as the *least upper bound* of the numbers r.

Now since f_N is continuous, there is a number $\delta' > 0$ such that $|f_N(z_0) - f_N(z)| < \varepsilon/3$ for any z in T satisfying $|z_0 - z| < \delta'$. But then $|f(z_0) - f(z)| < \varepsilon$ for such z, since

$$|f(z_0) - f(z)| = |f(z_0) - f_N(z_0) + f_N(z_0) - f_N(z) + f_N(z) - f(z)|$$
$$\leq |f(z_0) - f_N(z_0)| + |f_N(z_0) - f_N(z)| + |f_N(z) - f(z)|$$
$$< \frac{\varepsilon}{3} + \frac{\varepsilon}{3} + \frac{\varepsilon}{3} = \varepsilon.$$

Thus f is continuous at every point of T.   ∎

Knowing that the uniform limit of a sequence of continuous functions is continuous, we can integrate this limit. In fact the integral of the limit is the limit of integrals.

Theorem 8. Let f_n be a sequence of functions continuous on a set $T \subset \mathbf{C}$ containing the contour Γ, and suppose that f_n converges uniformly to f on T. Then the sequence $\int_\Gamma f_n(z)\, dz$ converges to $\int_\Gamma f(z)\, dz$.

Proof.   This is easy. Let ℓ be the length of Γ, and choose N large enough so that $|f(z) - f_n(z)| < \varepsilon/\ell$ for any $n > N$ and for all z on Γ. Then

$$\left| \int_\Gamma f(z)\, dz - \int_\Gamma f_n(z)\, dz \right| = \left| \int_\Gamma [f(z) - f_n(z)]\, dz \right|$$
$$< \frac{\varepsilon}{\ell} \cdot \ell = \varepsilon.   ∎$$

Combining these results with Morera's theorem (Theorem 18, Chapter 4), we can prove the following.

Theorem 9. Let f_n be a sequence of functions analytic in a simply connected domain D and converging uniformly to f in D. Then f is analytic in D.[†]

Proof.   By Lemma 3, the function f is continuous. Let Γ be any loop contained in D. Then by Theorem 8, the integral $\int_\Gamma f(z)\, dz$ is the limit of $\int_\Gamma f_n(z)\, dz$; but the latter is zero for all n, since each f_n is analytic inside and on Γ. Thus $\int_\Gamma f(z)\, dz = 0$, and Morera's result applies.   ∎

Now we have everything we need to render an account of power series. Since the partial sums of a power series are analytic functions (indeed, polynomials) and since they converge uniformly in any closed subdisk interior to the circle of convergence, Theorem 9 tells us that the limit function is analytic inside every such subdisk. But *any* point within the circle of convergence lies inside such a subdisk, so we can state the following.

[†] In fact this result holds in *any* domain, since a domain is a union of open disks.

Theorem 10. A power series sums to a function that is analytic at every point inside its circle of convergence.

For example, the power series (1) defines an analytic function for $|z| < 1$.

Notice that Theorems 7 and 8 justify integrating a power series termwise, as long as the contour lies inside the circle of convergence. Using this fact, we can identify every power series (with $R > 0$) as a Taylor series, in accordance with the next result.

Theorem 11. If $\sum_{j=0}^{\infty} a_j (z - z_0)^j$ converges to $f(z)$ in some circular neighborhood of z_0 (that is, the radius of its circle of convergence is nonzero), then

$$a_j = \frac{f^{(j)}(z_0)}{j!} \qquad (j = 0, 1, 2, \ldots).$$

Consequently, $\sum_{j=0}^{\infty} a_j (z - z_0)^j$ is the Taylor expansion of $f(z)$ around z_0.

Proof. Let C be a positively oriented circle centered at z_0 and lying inside the circle of convergence. Since the limit $f(z)$ is analytic, we can write the generalized Cauchy integral formula

$$f^{(n)}(z_0) = \frac{n!}{2\pi i} \int_C \frac{f(\zeta)}{(\zeta - z_0)^{n+1}} d\zeta \qquad (n = 0, 1, 2, \ldots). \qquad (2)$$

Now plug in the series $\sum_{j=0}^{\infty} a_j (\zeta - z_0)^j$ for $f(\zeta)$ and integrate termwise; since

$$\int_C \frac{(\zeta - z_0)^j}{(\zeta - z_0)^{n+1}} d\zeta = \begin{cases} 2\pi i & \text{if } j = n, \\ 0 & \text{otherwise}, \end{cases}$$

the only term that survives the integration in (2) is $n! \, a_n$. ∎

At this point we have specified two disks around z_0 in which a power series such as (1) converges. One is the interior of the "circle of convergence" of Theorem 7. The other is the largest disk over whose interior the limit function f is analytic. Are these two disks the same? Observe that the "Taylor disk" cannot extend beyond the circle of convergence because the series is known to *diverge* outside the latter. On the other hand, the limit function is analytic inside the circle of convergence, so the Taylor disk must enclose the interior of this circle. Thus the disks actually do coincide.

Summarizing, we have shown that if a power series converges inside some circle, it is the Taylor series of its (analytic) limit function and can be integrated and differentiated term by term inside this circle; moreover, this limit function must *fail* to be analytic somewhere on the circle of convergence.

Example 1

Find a function f that is analytic and satisfies the differential equation

$$\frac{df(z)}{dz} = 3if(z) \tag{3}$$

in a neighborhood of $z = 0$, taking the value 1 at $z = 0$.

Solution. Since f is analytic at the origin, it must have a Maclaurin series representation. We can find this series by using Eq. (3) to compute the derivatives.

We are given $f(0) = 1$, and from Eq. (3) we have $f'(0) = 3i \cdot 1 = 3i$. By differentiating Eq. (3), we see recursively that

$$f''(0) = 3if'(0) = (3i)^2,$$
$$f'''(0) = 3if''(0) = (3i)^3,$$

and, in general,

$$f^{(j)}(0) = 3if^{(j-1)}(0) = (3i)^2 f^{(j-2)}(0) = \cdots = (3i)^j.$$

Thus we can write the solution as

$$f(z) = 1 + 3iz + \frac{(3i)^2 z^2}{2!} + \cdots = \sum_{j=0}^{\infty} \frac{(3iz)^j}{j!}. \tag{4}$$

Recalling the representation

$$e^w = \sum_{j=0}^{\infty} \frac{w^j}{j!}, \tag{5}$$

we can identify our solution (4) as

$$f(z) = e^{3iz}.$$

Indeed, direct computation quickly verifies that e^{3iz} solves the problem. ∎

The classic initial value problem for the *nth-order linear homogeneous differential equation*

$$\frac{d^n f}{dz^n} + p_{n-1}(z)\frac{d^{n-1} f}{dz^{n-1}} + \cdots + p_2(z)\frac{d^2 f}{dz^2} + p_1(z)\frac{df}{dz} + p_0(z)f = 0$$

is the task of finding a solution that satisfies the initial conditions

$$f(z_0) = a_0, \; f'(z_0) = a_1, \; f''(z_0) = a_2, \ldots, f^{(n-1)}(z_0) = a_{n-1}$$

at some point z_0. In advanced differential equation texts (cf. Refs. [3] and [4]), it is shown that if each of the coefficients $p_j(z)$ is analytic inside a disk centered at z_0, then for arbitrary constants $\{a_0, a_1, \ldots, a_{n-1}\}$, there is one and only one solution of the initial value problem, and it, too, is analytic inside this disk.

Example 2

Let g be a continuous complex-valued function of a real variable on $[0, 2]$, and for each complex number z define

$$F(z) := \int_0^2 e^{zt} g(t) \, dt.$$

Prove that F is entire, and find its power series around the origin.

    Solution.   We first find a power series representation for F. Let z be fixed and define

$$h(t) := e^{zt}, \qquad \text{for all complex numbers } t.$$

Then h is an entire function of t, and so its Maclaurin expansion

$$h(t) = e^{zt} = \sum_{k=0}^{\infty} \frac{z^k}{k!} \cdot t^k = 1 + zt + \frac{z^2 t^2}{2!} + \cdots$$

converges uniformly on every disk $|t| \leq r$; in particular, the convergence is uniform for t in the interval $[0, 2]$. Furthermore, termwise multiplication by the bounded function g preserves this convergence; that is,

$$e^{zt} g(t) = \sum_{k=0}^{\infty} \frac{z^k}{k!} \cdot t^k g(t)$$

uniformly for $0 \leq t \leq 2$ (and fixed z). Now, by Theorem 8, we can integrate term by term with respect to t to obtain

$$F(z) = \int_0^2 e^{zt} g(t) \, dt = \sum_{k=0}^{\infty} \left[\frac{1}{k!} \int_0^2 t^k g(t) \, dt \right] z^k.$$

Since the bracketed quantity in the series is a constant dependent only on k, the preceding expansion is a power series in z; moreover, it converges to $F(z)$ for all z. Hence F is entire by Theorem 10.   ■

EXERCISES 5.3

1.   (a) Prove that the power series $\sum_{j=0}^{\infty} z^j$ converges at no point on its circle of convergence $|z| = 1$.

        (b) Prove that the power series $\sum_{j=1}^{\infty} z^j / j^2$ converges at every point on its circle of convergence $|z| = 1$. [HINT: See Prob. 13 in Exercises 5.1.]

2. Assume for the power series $\sum_{j=0}^{\infty} a_j (z - z_0)^j$ we have $\lim_{j \to \infty} |a_{j+1}/a_j| = L$. Prove, by the ratio test, that the radius of convergence of the power series is given by $R = 1/L$.

3. Using the results of Prob. 2, find the circle of convergence of each of the following power series:

(a) $\displaystyle\sum_{j=0}^{\infty} j^3 z^j$ (b) $\displaystyle\sum_{k=0}^{\infty} 2^k (z-1)^k$ (c) $\displaystyle\sum_{j=0}^{\infty} j! z^j$

(d) $\displaystyle\sum_{k=0}^{\infty} \frac{(-1)^k k}{3^k}(z-i)^k$ (e) $\displaystyle\sum_{k=1}^{\infty} \frac{(3-i)^k}{k^2}(z+2)^k$ (f) $\displaystyle\sum_{j=0}^{\infty} \frac{z^{2j}}{4^j}$

4. Does there exist a power series $\sum_{j=0}^{\infty} a_j z^j$ that converges at $z = 2+3i$ and diverges at $z = 3 - i$?

5. Let $f(z) = \displaystyle\sum_{k=0}^{\infty} \left(k^3/3^k\right) z^k$. Compute each of the following.

(a) $f^{(6)}(0)$ (b) $\displaystyle\oint_{|z|=1} \frac{f(z)}{z^4}\, dz$

(c) $\displaystyle\oint_{|z|=1} e^z f(z)\, dz$ (d) $\displaystyle\oint_{|z|=1} \frac{f(z)\sin z}{z^2}\, dz$

6. Define

$$f(z) := \begin{cases} \dfrac{\sin z}{z} & \text{for } z \neq 0, \\[2mm] 1 & \text{for } z = 0. \end{cases}$$

(a) Using the Maclaurin expansion for $\sin z$, show that for *all* z

$$f(z) = 1 - \frac{z^2}{3!} + \frac{z^4}{5!} - \frac{z^6}{7!} + \cdots .$$

(b) Explain why $f(z)$ is analytic at the origin.

(c) Find $f^{(3)}(0)$ and $f^{(4)}(0)$.

7. Find the first three nonzero terms in the Maclaurin expansion of $f(z) := \int_0^z e^{\zeta^2}\, d\zeta$. [HINT: First expand e^{ζ^2}.]

8. Assume that $f(z)$ is analytic at the origin and that $f(0) = f'(0) = 0$. Prove that $f(z)$ can be written in the form $f(z) = z^2 g(z)$, where $g(z)$ is analytic at $z = 0$.

9. Suppose that g is continuous on the circle $C : |z| = 1$, and that there exists a sequence of polynomials that converges uniformly to g on C. Prove that

$$\oint_C g(z)\, dz = 0.$$

10. Explain why the two power series $\sum_{k=0}^{\infty} a_k z^k$ and $\sum_{k=1}^{\infty} k a_k z^{k-1}$ have the same radius of convergence.

11. Let $\sum_{k=0}^{\infty} a_k z^k$ and $\sum_{k=0}^{\infty} b_k z^k$ be two power series having a positive radius of convergence.

    (a) Show that if $\sum_{k=0}^{\infty} a_k z^k = \sum_{k=0}^{\infty} b_k z^k$ in some circular neighborhood of the origin, then $a_k = b_k$ for all k.

    (b) Show, more generally, that if $\sum_{k=0}^{\infty} a_k x^k = \sum_{k=0}^{\infty} b_k x^k$ for all real x in some open interval containing the origin, then $a_k = b_k$ for all k.

12. Prove by means of power series that the only solution of the initial-value problem

$$\begin{cases} \dfrac{df}{dz} = f \\[2ex] f(0) = 1 \end{cases}$$

that is analytic at $z = 0$ is $f(z) = e^z$.

13. Each of the following initial-value problems has a unique solution that is analytic at the origin. Find the power series expansion $\sum_{j=0}^{\infty} a_j z^j$ of the solution by determining a recurrence relation for the coefficients a_j.

    (a) $\begin{cases} \dfrac{d^2 f}{dz^2} - z\dfrac{df}{dz} - f = 0 \\[2ex] f(0) = 1, \qquad f'(0) = 0 \end{cases}$
    (b) $\begin{cases} \dfrac{d^2 f}{dz^2} + 4f = 0 \\[2ex] f(0) = 1, \qquad f'(0) = 1 \end{cases}$

    (c) $\begin{cases} (1 - z^2)\dfrac{d^2 f}{dz^2} - 6z\dfrac{df}{dz} - 4f = 0 \\[2ex] f(0) = 1, \qquad f'(0) = 0 \end{cases}$

[HINT: The technique demonstrated in Example 1 becomes laborious for more complicated equations such as (c). It is more efficient to substitute the power series expression for $f(z)$ into the equation and collect like powers of z.]

14. Prove by means of power series that the only solution of the initial-value problem

$$\begin{cases} \dfrac{d^2 f}{dz^2} + f = 0 \\[2ex] f(0) = 0, \qquad f'(0) = 1 \end{cases}$$

that is analytic at $z = 0$ is $f(z) = \sin z$.

15. Let g be continuous on the real interval $[-1, 2]$, and define

$$F(z) := \int_{-1}^{2} g(t) \sin(zt)\, dt.$$

    (a) Prove that F is entire and find its power series expansion around the origin.

(b) Prove that for all z

$$F'(z) = \int_{-1}^{2} tg(t)\cos(zt)\,dt.$$

16. Let g be continuous on the real interval $[0, 1]$ and define

$$H(z) := \int_{0}^{1} \frac{g(t)}{1 - zt^2}\,dt \qquad (|z| < 1).$$

Prove that H is analytic in the open disk $|z| < 1$.

17. Define

$$(a)_j := a(a+1)\cdots(a+j-1), \quad j \geq 1, \quad (a)_0 := 1,$$

for any complex number a. The *Gaussian hypergeometric series* $_2F_1(b, c; d; z)$ is defined by

$$_2F_1(b, c; d; z) := \sum_{j=0}^{\infty} \frac{(b)_j (c)_j}{(d)_j} \cdot \frac{z^j}{j!},$$

and the *confluent hypergeometric series* $_1F_1(c; d; z)$ is given by

$$_1F_1(c; d; z) := \sum_{j=0}^{\infty} \frac{(c)_j}{(d)_j} \cdot \frac{z^j}{j!}.$$

(a) Verify that

$$\sum_{j=0}^{\infty} \frac{z^j}{j+1} = {}_2F_1(1, 1; 2; z),$$

$$e^z - \sum_{k=0}^{n-1} \frac{z^k}{k!} = \frac{z^n}{n!}\, {}_1F_1(1; n+1; z) \quad (n = 1, 2, \ldots).$$

(b) Prove that if $d \neq 0, -1, -2, \ldots$, then the series $_2F_1(b, c; d; z)$ converges for $|z| < 1$ and satisfies the differential equation

$$z(1-z)\frac{d^2 f}{dz^2} + [d - (b + c + 1)z]\frac{df}{dz} - bcf = 0.$$

(c) Prove that if $d \neq 0, -1, -2, \ldots$, then the series $_1F_1(c; d; z)$ converges for all z and satisfies the differential equation

$$z\frac{d^2 f}{dz^2} + (d - z)\frac{df}{dz} - cf = 0.$$

18. (*Generalized L'Hôpital's Rule*). Use power series to prove that if f, g are both analytic at z_0 and

$$f(z_0) = g(z_0) = f'(z_0) = g'(z_0) = \cdots = f^{(m-1)}(z_0) = g^{(m-1)}(z_0) = 0$$

but $g^{(m)}(z_0) \neq 0$, then

$$\lim_{z \to z_0} \frac{f(z)}{g(z)} = \frac{f^{(m)}(z_0)}{g^{(m)}(z_0)}.$$

19. (*Area Integral Formula*). By completing the steps below, prove the following area integral representation formula for any function f analytic on the closed unit disk $\overline{\mathbf{D}} = \{z : |z| \leq 1\}$. For $|\zeta| < 1$,

$$f(\zeta) = \frac{1}{\pi} \iint_{\overline{\mathbf{D}}} \frac{f(z)}{(1 - \bar{z}\zeta)^2} dx\, dy, \quad z = x + iy.$$

(a) Show that

$$\iint_{\overline{\mathbf{D}}} z^j \bar{z}^k dx\, dy = \begin{cases} \pi/(k+1) & \text{if } j = k, \\ \\ 0 & \text{if } j \neq k. \end{cases}$$

[HINT: $dx\, dy = r\, dr\, d\theta$.]

(b) Use part (a) to verify that the Maclaurin series coefficients $a_k = f^{(k)}(0)/k!$ of the function f satisfy

$$a_k = \frac{k+1}{\pi} \iint_{\overline{\mathbf{D}}} f(z) \bar{z}^k dx\, dy, \quad k = 0, 1, \ldots.$$

(c) Use part (b) to prove the area integral representation formula given above for $f(z)$. [HINT: $(1 - \bar{z}\zeta)^{-2} = \sum_{k=0}^{\infty}(k+1)\bar{z}^k \zeta^k$, $|\zeta| < 1$, $|z| \leq 1$.]

5.4 *Mathematical Theory of Convergence

In this section we shall backtrack somewhat and provide the mathematical details of the unproved theorems of this chapter. Applications-oriented students may wish to skip ahead to Sec. 5.5.

So far all the conditions we have seen for convergence of a sequence involve the limit explicitly. However, there is a way of testing whether or not a sequence is convergent without mentioning a limit at all. It is known as the *Cauchy criterion* for convergence.

Theorem 12. A necessary and sufficient condition for the sequence of complex numbers $\{A_n\}_{n=1}^{\infty}$ to converge is the following: For any $\varepsilon > 0$ there exists an integer N such that $|A_n - A_m| < \varepsilon$ for every pair of integers m and n satisfying $m > N, n > N$.

Proof. (*Necessity*). If the sequence does converge, say to A, we choose N so that each A_ℓ is within $\varepsilon/2$ of A for $\ell > N$. Then any two such A_ℓ must lie within ε of each other.

The proof that the Cauchy criterion is sufficient for convergence requires a rigorous axiomatization for the real number system; indeed, the criterion can be used to *define* the concept of an irrational real number. We shall not explore this here. ∎

A sequence that satisfies the Cauchy criterion is often called a *Cauchy sequence*. By Theorem 12, every convergent sequence is a Cauchy sequence and vice versa.

Corollary 1. If $\{A_n\}_{n=1}^{\infty}$ is a Cauchy sequence and N is chosen so that $|A_n - A_m| < \varepsilon$ for every m and n greater than N, then each A_n with $n > N$ is within ε of the limit.

Proof. Let $m \to \infty$ in the inequality $|A_n - A_m| < \varepsilon$. The result is

$$|A_n - A| \leq \varepsilon. \qquad \blacksquare$$

The Cauchy criterion, applied to the sequence of partial sums of a series, reads as follows:

Corollary 2. A necessary and sufficient condition for the series $\sum_{j=0}^{\infty} c_j$ to converge is the following: For any $\varepsilon > 0$ there exists an N such that $|\sum_{j=n+1}^{m} c_j| < \varepsilon$ for every pair of integers m and n satisfying $m > n > N$.

The proof is immediate. Such a series is (naturally) called a *Cauchy series*. Corollary 2 justifies the following, almost obvious, result: If $\sum_{j=0}^{\infty} c_j$ converges, then $c_j \to 0$ as $j \to \infty$.

With the Cauchy criterion in hand we can give a proof of the comparison test, Theorem 1 of this chapter. However, it takes only a little more effort to prove the following, more general, theorem, known as the *Weierstrass M-test*:

Theorem 13 (*M-test*). Suppose $\sum_{j=0}^{\infty} M_j$ is a convergent series with real non-negative terms and suppose, for all z in some set T and for all j greater than some number J, that $|f_j(z)| \leq M_j$. Then the series $\sum_{j=0}^{\infty} f_j(z)$ converges uniformly on T.

Proof.   Since $\sum_{j=0}^{\infty} M_j$ is a Cauchy series, we can choose $N > J$ so that for any m and n satisfying $m > n > N$ we have $\sum_{j=n+1}^{m} M_j < \varepsilon$. But then for z in T, the series $\sum_{j=0}^{\infty} f_j(z)$ is a Cauchy series also, because

$$\left| \sum_{j=n+1}^{m} f_j(z) \right| \leq \sum_{j=n+1}^{m} |f_j(z)| \leq \sum_{j=n+1}^{m} M_j < \varepsilon. \qquad (1)$$

Hence $\sum_{j=0}^{\infty} f_j(z)$ converges for each z in T, say to the function $F(z)$. It is easy to see that the convergence is uniform; observe that the inequality (1) can be rewritten in terms of the partial sums as

$$\left| \sum_{j=0}^{m} f_j(z) - \sum_{j=0}^{n} f_j(z) \right| < \varepsilon, \quad \text{for all } z \text{ in } T, \text{ and } m > n > N.$$

Therefore, by Corollary 1,

$$\left| F(z) - \sum_{j=0}^{n} f_j(z) \right| \leq \varepsilon, \quad \text{for all } z \text{ in } T, \text{ and } n > N.$$

This proves uniform convergence. ∎

The comparison test can be regarded as a special case of the M-test wherein each $f_j(z)$ is a constant function.

Now we are ready to analyze Theorem 7 of the previous section, specifying the convergence properties of power series. For convenience, we restate the theorem here.

Theorem 7. For any power series $\sum_{j=0}^{\infty} a_j (z - z_0)^j$ there is a real number R between 0 and ∞, inclusive, which depends only on the coefficients $\{a_j\}$, such that

   (i)   the series converges for $|z - z_0| < R$,

   (ii) the series converges uniformly in any closed subdisk $|z - z_0| \leq R' < R$,

   (iii) the series diverges for $|z - z_0| > R$.

To specify this number R, we must introduce the concept of the *upper limit* of an infinite sequence of real numbers; it generalizes the notion of limit. For motivation, let us first consider a *convergent* sequence of real numbers $\{x_n\}$, with limit x. Then for any $\varepsilon > 0$ there is an N such that all the elements x_n for $n > N$ will lie within ε of x. So, in particular, x has the following property: Given any $\varepsilon > 0$, for only a finite number of values of n does x_n exceed $x + \varepsilon$. Moreover, no number less than x has this property. We now extend this notion to arbitrary sequences.

Definition 6. The **upper limit** of a sequence of real numbers $\{x_n\}_{n=1}^{\infty}$, abbreviated $\limsup x_n$, is defined to be the smallest real number ℓ with the property that for any $\varepsilon > 0$ there are only a finite number of values of n such that x_n exceeds $\ell + \varepsilon$; if there are no such numbers with this property, we set $\limsup x_n := \infty$; if all real numbers have this property we set $\limsup x_n := -\infty$.

As we indicated before, if $\{x_n\}$ converges to x, then $\limsup x_n = x$. Other examples are $\limsup (-1)^n = 1$, $\limsup (n) = \infty$, and $\limsup (-n) = -\infty$.

The number R in Theorem 7 is now specified as follows. From the set of coefficients $\{a_j\}$ we form the sequence $\sqrt[n]{|a_n|}$. Then R is equal to the reciprocal of the upper limit of this sequence,

$$R = \frac{1}{\limsup \sqrt[n]{|a_n|}}, \tag{2}$$

with the usual conventions $1/0 = \infty$, $1/\infty = 0$. Equation (2) is known as the *Cauchy-Hadamard formula*. Starting with this formula, let's address Theorem 7.

Proof of Theorem 7. First we consider the convergence statements (i) and (ii). If $R = 0$, there is nothing to prove. When $R > 0$, the convergence for $|z - z_0| < R$ follows from the uniform convergence in all closed subdisks $|z - z_0| \leq R' < R$, so we attack the latter problem.

Choose a number k in the interval

$$\frac{1}{R} < k < \frac{1}{R'}.$$

Then because of Eq. (2), all but a finite number of the a_j will satisfy $\sqrt[j]{|a_j|} < k$. Consequently, if z lies in the closed subdisk $|z - z_0| \leq R'$, we have the inequality

$$\left| a_j (z - z_0)^j \right| = \left(\sqrt[j]{|a_j|} \, |z - z_0| \right)^j < (kR')^j , \tag{3}$$

valid for j sufficiently large. But inequality (3) tells us that the M-test (Theorem 13) is satisfied when we compare $\sum_{j=0}^{\infty} a_j (z - z_0)^j$ to the geometric series $\sum_{j=0}^{\infty} (kR')^j$, which converges since $kR' < 1$. Accordingly, $\sum_{j=0}^{\infty} a_j (z - z_0)^j$ is uniformly convergent in the closed subdisk and statement (ii) is proved.

To prove divergence when $|z - z_0| > R$, we choose k in the interval

$$\frac{1}{|z - z_0|} < k < \frac{1}{R}.$$

Then it follows from the definition of lim sup and Eq. (2) that there must be an infinite number of a_j satisfying $\sqrt[j]{|a_j|} > k$ (remember that the lim sup is the *smallest* number such that so-and-so). For such a_j

$$\left| a_j (z - z_0)^j \right| = \left(\sqrt[j]{|a_j|} \, |z - z_0| \right)^j > (k \, |z - z_0|)^j > 1;$$

that is, an infinite number of the terms of $\sum_{j=0}^{\infty} a_j (z - z_0)^j$ exceed 1 in modulus. This is clearly incompatible with the Cauchy criterion, so the series must diverge. ∎

In the next example we shall make use of the following lemma.

Lemma 4. $\lim_{n \to \infty} \sqrt[n]{n!} = \infty.$

Proof of Lemma 4. It is not hard to see that, given any positive integer B, the factorial $n!$ exceeds B^n when n is sufficiently large; we simply presume $n > 2B$ and examine the ratio

$$\frac{n!}{B^n} = \frac{(2B)!}{B^{2B}} \frac{(2B+1)}{B} \frac{(2B+2)}{B} \cdots \frac{n}{B} > \frac{(2B)!}{B^{2B}} 2^{n-2B},$$

which exceeds 1 whenever 2^n exceeds $2^{2B} B^{2B}/[(2B)!]$. Taking nth roots we conclude that $\sqrt[n]{n!}$ approaches infinity, because it eventually exceeds *any* integer B. ∎

Example 1

Show that the *Bessel function of the first kind of order zero*[†]

$$J_0(z) := \sum_{j=0}^{\infty} \frac{(-1)^j z^{2j}}{2^{2j} (j!)^2} \tag{4}$$

is entire.

Solution.   Our goal is to prove that $R = \infty$. Keeping in mind that a_j is the coefficient of z^j, we have

$$a_j = \begin{cases} 0 & \text{if } j \text{ is odd,} \\[2ex] \dfrac{(-1)^{j/2}}{2^j[(j/2)!]^2} & \text{if } j \text{ is even.} \end{cases}$$

Obviously $\sqrt[j]{|a_j|} = 0$ for odd j. For even j,

$$\sqrt[j]{|a_j|} = \frac{1}{2[(j/2)!]^{2/j}},$$

and Lemma 4 shows that this goes to zero. Hence $\limsup \sqrt[j]{|a_j|} = 0$, and $R = \infty$. Consequently $J_0(z)$ is an entire function. ∎

The reader should verify (Prob. 7) that $J_0(z)$ satisfies *Bessel's equation of order zero*:

$$\frac{d^2 f}{dz^2} + \frac{1}{z} \frac{df}{dz} + f = 0.$$

EXERCISES 5.4

1. Find the upper limit of each of the following sequences $\{x_n\}_{n=1}^{\infty}$.

(a) $x_n = (-1)^n \left(\dfrac{2n}{n+1} \right)$ (b) $x_n = (-1)^n n$

(c) $x_n = \dfrac{1}{n^2}$ (d) $x_n = n \sin \left(\dfrac{n\pi}{2} \right)$

[†]Friedrich Wilhelm Bessel (1784–1846) worked principally as an astronomer in Konigsberg.

2. Prove that for any sequence $\{x_n\}$ of positive real numbers

$$\limsup \sqrt[n]{x_n} \leq \limsup \frac{x_{n+1}}{x_n}.$$

3. Find the radius of convergence of each of the following power series.

(a) $\displaystyle\sum_{j=1}^{\infty} \frac{2^j}{3^j + 4^j} z^j$ (b) $\displaystyle\sum_{j=0}^{\infty} 2^j z^{j^2}$ (c) $\displaystyle\sum_{j=0}^{\infty} \left[2 + (-1)^j\right]^j z^j$

(d) $\displaystyle\sum_{j=1}^{\infty} \frac{j!}{j^j} z^j$ (e) $\displaystyle\sum_{j=1}^{\infty} \frac{2}{3j} z^{2j}$ (f) $\displaystyle\sum_{j=0}^{\infty} z^{j!}$

4. By considering the series $\sum_{j=1}^{\infty} z^j/j^2$, $\sum_{j=1}^{\infty} z^j/j$, and $\sum_{j=0}^{\infty} z^j$, show that a power series may converge on all, some, or none of the points on its circle of convergence.

5. If the radius of convergence for the series $\sum_{j=0}^{\infty} a_j z^j$ is R, find the radius of convergence for the following.

(a) $\displaystyle\sum_{j=0}^{\infty} j^3 a_j z^j$ (b) $\displaystyle\sum_{j=0}^{\infty} a_j^4 z^j$ (c) $\displaystyle\sum_{j=0}^{\infty} a_j z^{2j}$

(d) $\displaystyle\sum_{j=0}^{\infty} a_j z^{j+7}$ (e) $\displaystyle\sum_{j=1}^{\infty} j^{-j} a_j z^j$

6. Prove that if the radius of convergence for the series $\sum_{j=0}^{\infty} a_j z^j$ is R, then the radius of convergence for the series $\sum_{j=0}^{\infty} \mathrm{Re}(a_j) z^j$ is greater than or equal to R.

7. Show that the Bessel function $J_0(z)$ satisfies Bessel's equation of order zero (as claimed after Example 1).

8. Bessel's equation of order n is

$$\frac{d^2 f(z)}{dz^2} + \frac{1}{z} \frac{df(z)}{dz} + \left(1 - \frac{n^2}{z^2}\right) f(z) = 0.$$

Show that, for integers $n > 0$, the *Bessel function of the first kind of order n*

$$J_n(z) = \sum_{j=0}^{\infty} \frac{(-1)^j}{j!(n+j)!} \cdot \left(\frac{z}{2}\right)^{2j+n}$$

is entire and satisfies the Bessel equation. (Bessel functions arise in the study of two-dimensional wave propagation in radial directions.)

9. Use power series to solve the *functional equation*

$$f(z) = z + f(z^2)$$

for the analytic function f.

10. The *Fibonacci sequence*[†] $1, 1, 2, 3, 5, 8, 13, \ldots$ arises with surprising frequency in natural phenomena. The defining relations for the terms are

$$a_0 = a_1 = 1,$$
$$a_n = a_{n-1} + a_{n-2} \qquad \text{(for } n \geq 2\text{)}.$$

Show that

$$f(z) := a_0 + a_1 z + a_2 z^2 + \cdots$$

defines an analytic function satisfying the equation

$$f(z) = 1 + zf(z) + z^2 f(z).$$

Solve for $f(z)$ and compute the Maclaurin series to derive the expression

$$a_j = \frac{1}{\sqrt{5}} \left[\left(\frac{1 + \sqrt{5}}{2} \right)^{j+1} - \left(\frac{1 - \sqrt{5}}{2} \right)^{j+1} \right].$$

11. The *Legendre polynomials*[‡] $P_j(\zeta)$ are the coefficients of z^j in the Maclaurin series for

$$\left(1 - 2\zeta z + z^2 \right)^{-1/2} = \sum_{j=0}^{\infty} P_j(\zeta) z^j$$

(regarding ζ as a parameter). Show that $P_j(\zeta)$ is a polynomial in ζ of degree j, and compute P_0, P_1, P_2, and P_3. (These polynomials arise in three-dimensional potential theory.)

12. The *Riemann zeta function* has important applications to number theory. It is defined by

$$\zeta(z) := \sum_{j=1}^{\infty} \frac{1}{j^z} \qquad (\text{Re } z > 1)$$

where $j^z := \exp(z \operatorname{Log} j)$. Prove that $\zeta(z)$ is analytic for $\text{Re } z > 1$. [HINT: Let $\text{Re } z \geq \lambda > 1$, and show that $\left| 1 / j^z \right| \leq j^{-\lambda}$. Then use the Weierstrass M-test.] (One of the most famous problems whose solution still eludes mathematicians is the *Riemann hypothesis*. It asserts that all the nonreal zeros of the analytic continuation (cf. Sec. 5.8) of $\zeta(z)$ lie on the vertical line $\text{Re } z = 1/2$.)

13. (*Abel's Limit Theorem*) Let $\sum_{j=0}^{\infty} a_j z^j$ be the power series expansion for a function f analytic in $|z| < 1$ (so that $f(z) = \sum_{j=0}^{\infty} a_j z^j$ for $|z| < 1$). Suppose that $\lim_{r \to 1^-} f(r)$ (with r real) exists and equals A. Prove that if $\sum_{j=0}^{\infty} a_j$ converges, then

$$A = \sum_{j=0}^{\infty} a_j.$$

[†]Fibonacci was the nickname of Leonardo Pisano (1170–1250), who introduced the Hindu-Arabic place-valued decimal system to Europe.

[‡]Adrien-Marie Legendre (1752–1833) discovered these polynomials while analyzing the gravitational attraction of ellipsoids.

[HINT: Set $M_n := \max_{j \geq n} \left| \sum_{k=j}^{\infty} a_k \right|$. Then $M_n \to 0$ as $n \to \infty$ and for $0 \leq r < 1$

$$\left| f(r) - \sum_{j=0}^{n} a_j r^j \right| = \left| \sum_{j=n+1}^{\infty} a_j r^j \right| = \left| \sum_{j=n+1}^{\infty} \left(\sum_{k=j}^{\infty} a_k - \sum_{k=j+1}^{\infty} a_k \right) r^j \right|$$

$$= \left| \left(\sum_{k=n+1}^{\infty} a_k \right) r^{n+1} + \sum_{j=n+2}^{\infty} \left(\sum_{k=j}^{\infty} a_k \right) r^{j-1}(r-1) \right|$$

$$\leq M_{n+1} + M_{n+2}.$$

Now let $r \to 1^-$ and then $n \to \infty$.]

14. Use Abel's limit theorem (Prob. 13) to prove that

$$\text{Log } 2 = 1 - \frac{1}{2} + \frac{1}{3} - \frac{1}{4} + \cdots .$$

5.5 Laurent Series

We now wish to investigate the possibility of a series representation of a function f near a *singularity*.[†] After all, if (for example) the occurrence of a singularity is merely due to a vanishing denominator, might it not be possible to express the function as something like $A/(z - z_0)^P + g(z)$, where g is analytic and has a Taylor series around z_0? To be sure, not all singularities are of this type (recall Log z at $z_0 = 0$). However, if the function is analytic in an annulus surrounding one or more of its singularities (note that Log z does not have this property, due to its branch cut), we can display its "singular part" according to the following theorem.

Theorem 14. Let f be analytic in the annulus[‡] $r < |z - z_0| < R$. Then f can be expressed there as the sum of two series

$$f(z) = \sum_{j=0}^{\infty} a_j (z - z_0)^j + \sum_{j=1}^{\infty} a_{-j} (z - z_0)^{-j} ,$$

both series converging in the annulus, and converging uniformly in any closed subannulus $r < \rho_1 \leq |z - z_0| \leq \rho_2 < R$. The coefficients a_j are given by

$$a_j = \frac{1}{2\pi i} \oint_C \frac{f(\zeta)}{(\zeta - z_0)^{j+1}} d\zeta \qquad (j = 0, \pm 1, \pm 2, \ldots), \qquad (1)$$

[†]Recall from Sec. 2.3 that a singularity of f is a point z_0 where f is not analytic but that is the limit of points where f is analytic.

[‡]We allow $r = 0$, in which case the "annulus" becomes a punctured disk.

> where C is any positively oriented simple closed contour lying in the annulus and containing z_0 in its interior.

Such an expansion, containing negative as well as positive powers of $(z - z_0)$, is called the *Laurent series*[†] for f in this annulus. It is usually abbreviated

$$\sum_{j=-\infty}^{\infty} a_j (z - z_0)^j .$$

Notice that if f is analytic throughout the *disk* $|z - z_0| < R$, the coefficients in (1) with negative subscripts are zero by Cauchy's theorem, and the others reproduce the Taylor series for f.

Proof of Theorem 14. It suffices to prove uniform convergence in every closed subannulus, for this implies (pointwise) convergence in the open annulus.

First we show that for any z satisfying $r < \rho_1 \leq |z - z_0| \leq \rho_2 < R$ we have the representation

$$f(z) = \frac{1}{2\pi i} \oint_{C_1} \frac{f(\zeta)}{\zeta - z} d\zeta + \frac{1}{2\pi i} \oint_{C_2} \frac{f(\zeta)}{\zeta - z} d\zeta, \tag{2}$$

where C_1 is the negatively oriented circle around z_0 of radius $R_1 = (r + \rho_1)/2$, and C_2 is the positively oriented circle around z_0 of radius $R_2 = (R + \rho_2)/2$; see Fig. 5.4. Indeed, Eq. (2) is just a slight variation of the Cauchy integral formula in this case, as the following argument shows. Consider the contour Γ of Fig. 5.5(a); it is simple, closed, positively oriented, and contains z in its interior. Therefore,

$$f(z) = \frac{1}{2\pi i} \int_{\Gamma} \frac{f(\zeta)}{\zeta - z} d\zeta. \tag{3}$$

Let's think of Γ as a doughnut with a bite taken out of it, and let Γ' denote the "bite," as in Fig. 5.5(b). Observe that

$$\frac{1}{2\pi i} \int_{\Gamma'} \frac{f(\zeta)}{\zeta - z} d\zeta = 0$$

because the integrand is analytic inside and on Γ'. Consequently we can put the "bite" back into the doughnut and modify Eq. (3) to read

$$f(z) = \frac{1}{2\pi i} \int_{\Gamma + \Gamma'} \frac{f(\zeta)}{\zeta - z} d\zeta,$$

where $\Gamma + \Gamma'$ is the path indicated in Fig. 5.5(c). But the integrals along the line segments in Fig. 5.5(c) cancel and, keeping track of the orientation, we arrive at Eq. (2).[‡] Now we are ready to proceed with the derivation of the Laurent expansion.

[†]Pierre Alphonse Laurent (1813–1854) was an engineer responsible for the port of Le Havre.

[‡]Some readers may be able to use a deformation-of-contour argument to derive Eq. (2), but the doughnut analogy is probably easier to digest.

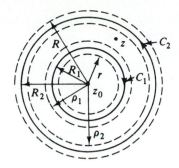

Figure 5.4 Circles of integration for Eq. (2).

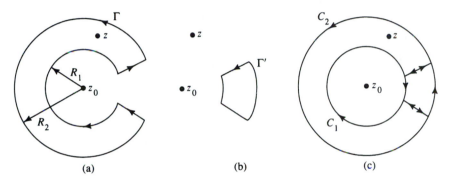

(a) (b) (c)

Figure 5.5 The hole equals the sum of its parts.

Since z lies inside C_2, the integral over C_2 appearing in Eq. (2) is exactly like the integral that arose in the Taylor series theorem (Theorem 3, Sec. 5.2); we treat it the same way, and find

$$\frac{1}{2\pi i} \oint_{C_2} \frac{f(\zeta)}{\zeta - z} \, d\zeta = \sum_{j=0}^{n} a_j \, (z - z_0)^j + T_n(z),$$

where $T_n(z) \to 0$ uniformly as $n \to \infty$ for $|z - z_0| \leq \rho_2$ and a_j is given by

$$a_j = \frac{1}{2\pi i} \oint_{C_2} \frac{f(\zeta)}{(\zeta - z_0)^{j+1}} \, d\zeta \qquad (j = 0, 1, 2, \ldots). \tag{4}$$

Hence

$$\frac{1}{2\pi i} \oint_{C_2} \frac{f(\zeta)}{\zeta - z} \, d\zeta = \sum_{j=0}^{\infty} a_j \, (z - z_0)^j \qquad (|z - z_0| \leq \rho_2).$$

Note, however, that a_j in (4) can no longer be identified as $f^{(j)}(z_0)/j!$, as was the case in the Taylor theorem. Indeed, our hypotheses contain no guarantee that f is differentiable at z_0.

We now turn to the integral around C_1 in Eq. (2). Since z lies outside C_1, we seek an expression for $1/(\zeta - z)$ in powers of $(\zeta - z_0)/(z - z_0)$, whose magnitude is less than 1; accordingly, we write

$$\frac{1}{\zeta - z} = \frac{1}{(\zeta - z_0) - (z - z_0)} = -\frac{1}{(z - z_0)} \frac{1}{1 - \frac{\zeta - z_0}{z - z_0}}$$

$$= -\frac{1}{z - z_0} \left[1 + \frac{\zeta - z_0}{z - z_0} + \frac{(\zeta - z_0)^2}{(z - z_0)^2} + \cdots + \frac{(\zeta - z_0)^m}{(z - z_0)^m} + \frac{\frac{(\zeta - z_0)^{m+1}}{(z - z_0)^{m+1}}}{1 - \frac{\zeta - z_0}{z - z_0}} \right].$$

Inserting this into the integral, we find

$$\frac{1}{2\pi i} \oint_{C_1} \frac{f(\zeta)}{\zeta - z} \, d\zeta = \sum_{j=1}^{m+1} a_{-j} (z - z_0)^{-j} + \mathcal{T}_m(z),$$

where we identify

$$a_{-j} = -\frac{1}{2\pi i} \oint_{C_1} \frac{f(\zeta)}{(\zeta - z_0)^{-j+1}} \, d\zeta \qquad (j = 1, 2, 3, \ldots) \tag{5}$$

(observe the exponent with care) and

$$\mathcal{T}_m(z) = \frac{1}{2\pi i} \oint_{C_1} \frac{f(\zeta)}{(\zeta - z)} \frac{(\zeta - z_0)^{m+1}}{(z - z_0)^{m+1}} \, d\zeta.$$

Now for ζ on C_1 we have $|\zeta - z| \geq \rho_1 - R_1$, $|\zeta - z_0| = R_1$, and $|z - z_0| \geq \rho_1$ (see Fig. 5.4). Thus

$$|\mathcal{T}_m(z)| \leq \frac{1}{2\pi} \cdot \max_{\zeta \text{ on } C_1} |f(\zeta)| \frac{1}{\rho_1 - R_1} \left(\frac{R_1}{\rho_1} \right)^{m+1} 2\pi R_1.$$

Since $R_1/\rho_1 < 1$, $\mathcal{T}_m(z) \to 0$ uniformly for $|z - z_0| \geq \rho_1$, and so

$$\frac{1}{2\pi i} \oint_{C_1} \frac{f(\zeta)}{\zeta - z} \, d\zeta = \sum_{j=1}^{\infty} a_{-j} (z - z_0)^{-j} \qquad (|z - z_0| \geq \rho_1).$$

We have thus expressed both integrals in Eq. (2) as uniformly convergent series of the form mentioned in the theorem with the common region of convergence $\rho_1 \leq |z - z_0| \leq \rho_2$; we still have to verify formula (1) for the coefficients. If j is nonnegative, formula (4) applies, but the analysis of Chapter 4 justifies replacing the integral over C_2 by the integral over the contour C mentioned in the theorem, since the intervening region contains no singularities of $f(\zeta)/(\zeta - z_0)^{j+1}$; hence Eq. (1) holds for $j \geq 0$. Similarly, the integral over C_1 in formula (5) can be changed into an integral over C, incorporating the minus sign to account for the change in orientation. Therefore (1) is verified for every j. ∎

Replacing $(z - z_0)$ with $1/(z - z_0)$ in Theorem 7, one easily sees that any formal series of the form $\sum_{j=1}^{\infty} c_{-j} (z - z_0)^{-j}$ will converge *outside* some "circle of convergence" $|z - z_0| = r$ whose radius depends on the coefficients, with uniform convergence holding in each region $|z - z_0| \geq r' > r$. Thus termwise integration is justified by Theorem 8, and proceeding in a manner analogous to that of Sec. 5.3 we can prove the following.

Theorem 15. Let $\sum_{j=0}^{\infty} c_j (z - z_0)^j$ and $\sum_{j=1}^{\infty} c_{-j} (z - z_0)^{-j}$ be any two series with the following properties:

(i) $\displaystyle\sum_{j=0}^{\infty} c_j (z - z_0)^j$ converges for $|z - z_0| < R$,

(ii) $\displaystyle\sum_{j=1}^{\infty} c_{-j} (z - z_0)^{-j}$ converges for $|z - z_0| > r$, and

(iii) $r < R$.

Then there is a function $f(z)$, analytic for $r < |z - z_0| < R$, whose Laurent series in this annulus is given by $\sum_{j=-\infty}^{\infty} c_j (z - z_0)^j$.

The proof is left to the exercises (Prob. 8).

This theorem, like Theorem 7, justifies the use of other methods of finding Laurent series since it implies that *any* convergent series of the form $\sum_{j=-\infty}^{\infty} c_j (z - z_0)^j$, however obtained, must be the Laurent series of its sum function. In Examples 1 and 2 we shall derive Laurent expansions by making judicious use of the fact that the geometric series $\sum_{j=0}^{\infty} w^j$ converges to $(1 - w)^{-1}$ when $|w| < 1$.

Example 1

Find the Laurent series for the function $(z^2 - 2z + 3)/(z - 2)$ in the region $|z - 1| > 1$.

Solution. Notice that this region is centered at $z_0 = 1$ and excludes the singularity at $z = 2$. First we manipulate $1/(z - 2)$ so that we can apply the geometric series result in the specified region:

$$\frac{1}{z - 2} = \frac{1}{(z - 1) - 1} = \frac{1}{z - 1} \cdot \frac{1}{1 - 1/(z - 1)}.$$

Thus for $|z - 1| > 1$,

$$\frac{1}{z - 2} = \frac{1}{z - 1} \cdot \sum_{j=0}^{\infty} \frac{1}{(z - 1)^j}$$

$$= \frac{1}{z - 1} + \frac{1}{(z - 1)^2} + \frac{1}{(z - 1)^3} + \cdots .$$

Using the methods of Sec. 3.1 we express the numerator $z^2 - 2z + 3$ in powers of $z - 1$:

$$z^2 - 2z + 3 = (z - 1)^2 + 0 \cdot (z - 1) + 2 = (z - 1)^2 + 2.$$

Therefore,

$$
\begin{aligned}
\frac{z^2 - 2z + 3}{z - 2} &= \left[(z - 1)^2 + 2 \right] \cdot \left[\frac{1}{z - 1} + \frac{1}{(z - 1)^2} + \frac{1}{(z - 1)^3} + \cdots \right] \\
&= \left[(z - 1) + 1 + \frac{1}{(z - 1)} + \frac{1}{(z - 1)^2} + \cdots \right] \\
&\quad + \left[\frac{2}{(z - 1)} + \frac{2}{(z - 1)^2} + \cdots \right] \\
&= (z - 1) + 1 + \sum_{j=1}^{\infty} \frac{3}{(z - 1)^j}. \quad \blacksquare
\end{aligned}
$$

Example 2

For the function

$$\frac{1}{(z - 1)(z - 2)},$$

find the Laurent series expansion in

   (a) the region $|z| < 1$,

   (b) the region $1 < |z| < 2$,

   (c) the region $|z| > 2$.

   Solution.   Again using partial fractions, write

$$\frac{1}{(z - 1)(z - 2)} = \frac{1}{z - 2} - \frac{1}{z - 1}.$$

Now we proceed differently in each region in order to derive convergent series.

   (a) For $|z| < 1$,

$$\frac{1}{z - 2} = -\frac{1}{2} \frac{1}{1 - z/2} = -\frac{1}{2} \sum_{j=0}^{\infty} \left(\frac{z}{2} \right)^j = -\sum_{j=0}^{\infty} \frac{z^j}{2^{j+1}} \tag{6}$$

and

$$\frac{1}{z - 1} = -\frac{1}{1 - z} = -\sum_{j=0}^{\infty} z^j. \tag{7}$$

Subtracting Eq. (7) from Eq. (6) gives

$$\frac{1}{(z - 1)(z - 2)} = \sum_{j=0}^{\infty} \left(-\frac{1}{2^{j+1}} + 1 \right) z^j = \frac{1}{2} + \frac{3}{4} z + \frac{7}{8} z^2 + \cdots.$$

(b) For $1 < |z| < 2$, Eq. (6) is still valid, but we have

$$\frac{1}{z-1} = \frac{1}{z}\frac{1}{1-1/z} = \frac{1}{z}\sum_{j=0}^{\infty}\frac{1}{z^j} = \sum_{j=0}^{\infty}\frac{1}{z^{j+1}}. \tag{8}$$

Thus

$$\frac{1}{(z-1)(z-2)} = -\sum_{j=0}^{\infty}\frac{z^j}{2^{j+1}} - \sum_{j=0}^{\infty}\frac{1}{z^{j+1}} = \cdots - \frac{1}{z^2} - \frac{1}{z} - \frac{1}{2} - \frac{z}{4} - \cdots .$$

(c) For $|z| > 2$, Eq. (8) is still valid and

$$\frac{1}{z-2} = \frac{1}{z}\frac{1}{1-2/z} = \frac{1}{z}\sum_{j=0}^{\infty}\left(\frac{2}{z}\right)^j = \sum_{j=0}^{\infty}\frac{2^j}{z^{j+1}}.$$

Hence

$$\frac{1}{(z-1)(z-2)} = \sum_{j=0}^{\infty}\frac{2^j - 1}{z^{j+1}} = \frac{1}{z^2} + \frac{3}{z^3} + \frac{7}{z^4} + \cdots . \quad \blacksquare$$

Example 3

Expand $e^{1/z}$ in a Laurent series around $z = 0$.

    Solution.   As we already know,

$$e^w = 1 + w + \frac{w^2}{2!} + \frac{w^3}{3!} + \cdots$$

for all (finite) w. Thus if $z \neq 0$, we let $w = 1/z$ and find

$$e^{1/z} = 1 + \frac{1}{z} + \frac{1}{2!z^2} + \frac{1}{3!z^3} + \cdots . \quad \blacksquare \tag{9}$$

Example 4

What is the Laurent expansion around $z = 0$ for the function

$$f(z) = \begin{cases} \sin z & \text{if } z \neq 0, \\ 5 & \text{if } z = 0? \end{cases}$$

    Solution.   The alert reader, upon seeing this example, will undoubtedly accuse the authors of toying with trifles—and with good reason! The function f is simply defined "incorrectly" at $z = 0$ in order to make a point. We ask, however, that the audience bear with us, because this example will be useful in the next section.

    Observe that f satisfies the hypothesis of Theorem 14, so it does have a Laurent series, valid for $|z| > 0$. But for such z we have [recall Eq. (10), Sec. 5.2]

$$f(z) = \sin z = z - \frac{z^3}{3!} + \frac{z^5}{5!} - \cdots \quad (|z| > 0). \tag{10}$$

Therefore, Eq. (10) must be the Laurent expansion for f. $\blacksquare$

EXERCISES 5.5

1. Find the Laurent series for the function $1/(z+z^2)$ in each of the following domains:

   (a) $0 < |z| < 1$ $\qquad\qquad\qquad$ **(b)** $1 < |z|$
   (c) $0 < |z + 1| < 1$ $\qquad\qquad$ **(d)** $1 < |z + 1|$

2. Does the principal branch $\sqrt{z}$ have a Laurent series expansion in the domain $\mathbf{C} \setminus \{0\}$?

3. Find the Laurent series for the function $\dfrac{z}{(z+1)(z-2)}$ in each of the following domains.

   (a) $|z| < 1$ $\qquad\qquad$ **(b)** $1 < |z| < 2$ $\qquad\qquad$ **(c)** $2 < |z|$

4. Find the Laurent series for $(\sin 2z)/z^3$ in $|z| > 0$.

5. Find the Laurent series for $\dfrac{(z+1)}{z(z-4)^3}$ in $0 < |z - 4| < 4$.

6. Find the Laurent series for $z^2 \cos\left(\dfrac{1}{3z}\right)$ in $|z| > 0$.

7. Obtain the first few terms of the Laurent series for each of the following functions in the specified domains.

   (a) $\dfrac{e^{1/z}}{z^2 - 1}$ for $|z| > 1$ $\qquad\qquad$ **(b)** $\dfrac{1}{e^z - 1}$ for $0 < |z| < 2\pi$

   (c) $\csc z$ for $0 < |z| < \pi$ $\qquad\qquad$ **(d)** $1 / e^{(1-z)}$ for $1 < |z|$

8. Give a proof of Theorem 15.

9. Determine the annulus of convergence of the Laurent series

$$\sum_{j=-\infty}^{\infty} \frac{z^j}{2^{|j|}}.$$

10. Prove that the Laurent series expansion of the function

$$f(z) = \exp\left[\frac{\lambda}{2}\left(z - \frac{1}{z}\right)\right]$$

   in $|z| > 0$ is given by

$$\sum_{k=-\infty}^{\infty} J_k(\lambda) z^k,$$

   where

$$J_k(\lambda) = (-1)^k J_{-k}(\lambda) := \frac{1}{2\pi} \int_0^{2\pi} \cos(k\theta - \lambda \sin\theta)\, d\theta.$$

The $J_k(\lambda)$ are known as *Bessel functions* of the first kind of order k.[†] [HINT: Use the integral formula (1) with $C : |z| = 1$.]

11. Obtain a general formula for the Laurent expansion of

$$f_n(z) = \frac{1}{(z - \alpha)^n} \qquad (n = 1, 2, \ldots)$$

that is valid for $|z| > |\alpha|$.

12. Prove that if $f(z)$ has a Laurent series expansion of the form $\sum_{j=0}^{\infty} a_j z^j$ in $0 < |z| < \rho$, then $\lim_{z \to 0} f(z)$ exists.

13. Let $f(z)$ be analytic in the annulus $r < |z - z_0| < R$ and bounded by M there. Prove that the coefficients a_j of the Laurent expansion of $f(z)$ in the annulus satisfy

$$|a_j| \leq \frac{M}{R^j}, \qquad |a_{-j}| \leq M r^j \qquad \text{(for } j = 0, 1, 2, \ldots).$$

5.6   Zeros and Singularities

In this section we shall use the Laurent expansion to classify, in general terms, the behavior of an analytic function near its zeros and isolated singularities. A *zero* of a function f is a point z_0 where f is analytic and $f(z_0) = 0$. An *isolated singularity* of f is a point z_0 such that f is analytic in some punctured disk $0 < |z - z_0| < R$ but not analytic at z_0 itself. For example, $\tan(\pi z/2)$ has a zero at each even integer and an isolated singularity at each odd integer.

We shall begin by examining the zeros of f.

Definition 7. A point z_0 is called a **zero of order m** for the function f if f is analytic at z_0 and f and its first $m - 1$ derivatives vanish at z_0, but $f^{(m)}(z_0) \neq 0$.

In other words, we have

$$f(z_0) = f'(z_0) = f''(z_0) = \cdots = f^{(m-1)}(z_0) = 0 \neq f^{(m)}(z_0).$$

In this case the Taylor series for f around z_0 takes the form

$$f(z) = a_m (z - z_0)^m + a_{m+1} (z - z_0)^{m+1} + a_{m+2} (z - z_0)^{m+2} + \cdots$$

or

$$f(z) = (z - z_0)^m \left[a_m + a_{m+1} (z - z_0) + a_{m+2} (z - z_0)^2 + \cdots \right], \qquad (1)$$

where $a_m = f^{(m)}(z_0)/m! \neq 0$. The bracketed series in Eq. (1) clearly converges wherever the series for f does (at any particular point one is just a multiple of the other); hence it defines a function $g(z)$ analytic in a neighborhood of z_0, with $g(z_0) \neq 0$. Conversely, any function with a representation like Eq. (1) must have a zero of order m, so we deduce the following.

[†]It can be shown that this integral formula is equivalent to the series representation in Prob. 8, Exercises 5.4.

Theorem 16. Let f be analytic at z_0. Then f has a zero of order m at z_0 if and only if f can be written as

$$f(z) = (z - z_0)^m g(z),$$

where g is analytic at z_0 and $g(z_0) \neq 0$.

A zero of order 1 is sometimes called a *simple zero*. For instance, the zeros of the function $\sin z$, which occur (as we saw in Chapter 3) at integer multiples of π, are all simple (at such points the first derivative, $\cos z$, is nonzero).

An easy consequence of Theorem 16 is the following result, which asserts that zeros of nonconstant analytic functions are isolated.

Corollary 3. If f is an analytic function such that $f(z_0) = 0$, then either f is identically zero in a neighborhood of z_0 or there is a punctured disk about z_0 in which f has no zeros.

Proof. Let $\sum_{k=0}^{\infty} a_k (z - z_0)^k$ be the Taylor series for f about z_0 (so that $a_k = f^{(k)}(z_0)/k!$). Then, as we know from Theorem 3, this series converges to $f(z)$ in some circular neighborhood of z_0. So if all the Taylor coefficients a_k are zero, then $f(z)$ must be identically zero in this neighborhood. Otherwise, let $m(\geq 1)$ be the smallest subscript such that $a_m \neq 0$. Then, by Definition 7, f has a zero of order m at z_0, and so the representation $f(z) = (z - z_0)^m g(z)$ of Theorem 16 is valid. Since $g(z_0) \neq 0$ and g is continuous at z_0 (indeed, it is analytic there), there exists a disk $|z - z_0| < \delta$ throughout which g is nonzero. Consequently, $f(z) \neq 0$ for $0 < |z - z_0| < \delta$. ∎

Notice that if f is nonconstant, analytic, and zero at z_0, the order of the zero must be a whole number; the condition of analyticity at z_0 in the analysis of Theorem 16 dictates that m be an integer. The function $z^{1/2}$ could be said to have a zero of order $\frac{1}{2}$ at $z = 0$, but of course it is not analytic there.

We now turn to the isolated singularities of f. We know that f has a Laurent expansion around any isolated singularity z_0;

$$f(z) = \sum_{j=-\infty}^{\infty} a_j (z - z_0)^j, \tag{2}$$

for, say, $0 < |z - z_0| < R$. (The "r" of Theorem 14 is zero for an isolated singularity.) We can classify z_0 into one of the following three categories.

Definition 8. Let f have an isolated singularity at z_0, and let (2) be the Laurent expansion of f in $0 < |z - z_0| < R$. Then

---

(i) If $a_j = 0$ for all $j < 0$, we say that z_0 is a **removable singularity** of f;

(ii) If $a_{-m} \neq 0$ for some positive integer m but $a_j = 0$ for all $j < -m$, we say that z_0 is a **pole of order m** for f;

(iii) If $a_j \neq 0$ for an infinite number of negative values of j, we say that z_0 is an **essential singularity** of f.

By examining separately each of these three types of isolated singularities we shall show that they can be distinguished by the qualitative behavior of $f(z)$ near the singularity (that is, without working out the Laurent expansion). The resulting characterizations are summarized in the final theorem of this section.

When f has a removable singularity at z_0, its Laurent series takes the form

$$f(z) = a_0 + a_1(z - z_0) + a_2(z - z_0)^2 + \cdots \qquad (0 < |z - z_0| < R). \qquad (3)$$

Example 4 of the previous section provides an illustration of this. Other examples of functions having removable singularities are

$$\frac{\sin z}{z} = \frac{1}{z}\left(z - \frac{z^3}{3!} + \frac{z^5}{5!} - \cdots\right) = 1 - \frac{z^2}{3!} + \frac{z^4}{5!} - \cdots \qquad (z_0 = 0),$$

$$\frac{\cos z - 1}{z} = \frac{1}{z}\left[\left(1 - \frac{z^2}{2!} + \frac{z^4}{4!} - \cdots\right) - 1\right] = -\frac{z}{2!} + \frac{z^3}{4!} - \cdots \qquad (z_0 = 0),$$

$$\frac{z^2 - 1}{z - 1} = z + 1 = 2 + (z - 1) + 0 + 0 + \cdots \qquad (z_0 = 1).$$

From (3) we can see that, except for the point z_0 itself, $f(z)$ is equal to a function $h(z)$, which is analytic at z_0. In other words, the only reason for the singularity is that $f(z)$ is undefined or defined "peculiarly" at z_0. Since the function $h(z)$ is analytic at z_0, it is obviously bounded[†] in some neighborhood of z_0, and so we have established the following lemma.

Lemma 5. If f has a removable singularity at z_0, then

(i) $f(z)$ is bounded in some punctured circular neighborhood of z_0,

(ii) $f(z)$ has a (finite) limit as z approaches z_0, and

(iii) $f(z)$ can be redefined at z_0 so that the new function is analytic at z_0.

[†]That is, there exists a neighborhood of z_0 and a constant M such that $|h(z)| \leq M$ for all z in this neighborhood.

Conversely, if a function is bounded in some punctured neighborhood of an isolated singularity, that singularity is removable; see Prob. 13 for a direct proof.

Clearly, removable singularities are not too important in the theory of analytic functions. But as we shall see in Lemmas 6 and 8, the concept is occasionally helpful in providing compact descriptions of the other kinds of singularities.

The Laurent series for a function with a pole of order m looks like

$$f(z) = \frac{a_{-m}}{(z - z_0)^m} + \frac{a_{-(m-1)}}{(z - z_0)^{m-1}} + \cdots + \frac{a_{-1}}{z - z_0}$$
$$+ a_0 + a_1 (z - z_0) + a_2 (z - z_0)^2 + \cdots \qquad (a_{-m} \neq 0), \tag{4}$$

valid in some punctured neighborhood of z_0. For example,

$$\frac{e^z}{z^2} = \frac{1}{z^2} \left(1 + z + \frac{z^2}{2!} + \cdots \right) = \frac{1}{z^2} + \frac{1}{z} + \frac{1}{2!} + \frac{z}{3!} + \cdots$$

has a pole of order 2, and

$$\frac{\sin z}{z^5} = \frac{1}{z^5} \left(z - \frac{z^3}{3!} + \frac{z^5}{5!} - \cdots \right) = \frac{1}{z^4} - \frac{1}{3! z^2} + \frac{1}{5!} - \frac{z^2}{7!} + \cdots$$

has a pole of order 4, at $z = 0$.

A pole of order 1 is called a *simple pole*. For example, $z = 0$ is a simple pole of the function $(\sin z) / z^2$.

From Eq. (4), we can deduce the following characterization of a pole:

Lemma 6. If the function f has a pole of order m at z_0, then $| (z - z_0)^\ell f(z)| \to \infty$ as $z \to z_0$ for all integers $\ell < m$, while $(z - z_0)^m f(z)$ has a removable singularity at z_0. In particular, $|f(z)| \to \infty$ as z approaches a pole.[†]

Proof of Lemma 6. Equation (4) implies that in some punctured neighborhood of z_0 we have

$$(z - z_0)^m f(z) = a_{-m} + a_{-m+1} (z - z_0) + \cdots , \tag{5}$$

and since there are no negative powers, the singularity of $(z - z_0)^m f(z)$ at z_0 is removable. Furthermore, $(z - z_0)^m f(z) \to a_{-m} \neq 0$ as $z \to z_0$. Thus for any integer $\ell < m$,

$$\left| (z - z_0)^\ell f(z) \right| = \left| \frac{1}{(z - z_0)^{m-\ell}} (z - z_0)^m f(z) \right| \to \infty \text{ as } z \to z_0, \tag{6}$$

because $(z - z_0)^{m-\ell} \to 0$ and $a_{-m} \neq 0$. ∎

[†]We remind the reader that the notation "$|h(z)| \to \infty$ as $z \to z_0$" means that $|h(z)|$ exceeds any given number for all z sufficiently near z_0.

Lemma 7. A function f has a pole of order m at z_0 if and only if in some punctured neighborhood of z_0

$$f(z) = \frac{g(z)}{(z - z_0)^m},\tag{7}$$

where g is analytic at z_0 and $g(z_0) \neq 0$.

Proof.   If f has a pole of order m at z_0, then it follows from the representation (5) that in some punctured neighborhood of z_0 we have $(z - z_0)^m f(z) = g(z)$, where

$$g(z) := a_{-m} + a_{-m+1}(z - z_0) + \cdots.$$

Setting $g(z_0) := a_{-m} \neq 0$, we see that g is analytic and nonzero at z_0, so (7) follows. Now suppose that the representation (7) holds, and write the Taylor series for $g(z)$:

$$g(z) = b_0 + b_1(z - z_0) + b_2(z - z_0)^2 + \cdots.$$

Then the Laurent series for f near z_0 must be

$$f(z) = \frac{g(z)}{(z - z_0)^m} = \frac{b_0}{(z - z_0)^m} + \frac{b_1}{(z - z_0)^{m-1}} + \cdots.$$

Since $b_0 = g(z_0) \neq 0$, the expansion displays the predicted pole for f.   ∎

Example 1

Classify the singularity at $z = 1$ of the function $(\sin z)/(z^2 - 1)^2$.

Solution.   Since

$$\frac{\sin z}{(z^2 - 1)^2} = \frac{(\sin z)/(z + 1)^2}{(z - 1)^2}$$

and the numerator is analytic and nonzero at $z = 1$, Lemma 7 implies that the function has a pole of order 2.   ∎

Example 2

Show that the only singularities of rational functions are removable singularities or poles.

Solution.   Recall that a rational function is $P(z)/Q(z)$, the ratio of two polynomials, and is analytic everywhere except at the zeros of $Q(z)$. If $Q(z)$ has a zero, say of order m, at z_0, then $Q(z) = (z - z_0)^m q(z)$, where $q(z)$ is a polynomial and $q(z_0) \neq 0$.
If $P(z_0) \neq 0$, we apply Lemma 7 to the expression

$$\frac{P(z)}{Q(z)} = \frac{1}{(z - z_0)^m} \frac{P(z)}{q(z)}$$

to deduce that $P(z)/Q(z)$ has a pole of order m. If, on the other hand, $P(z_0) = 0$, we can write $P(z) = (z - z_0)^n \, p(z)$, where n is the order of the zero at z_0 [we ignore the trivial case $P(z) \equiv 0$]; thus

$$\frac{P(z)}{Q(z)} = \frac{(z - z_0)^n}{(z - z_0)^m} \frac{p(z)}{q(z)},$$

and clearly $P(z)/Q(z)$ will have a pole if $n < m$ or a removable singularity if $n \geq m$. ∎

The following lemma relating zeros and poles is easily derived using the preceding methods of analysis, so we simply state it here for reference purposes and assign the proof to the reader (Prob. 4).

Lemma 8. If f has a zero of order m at z_0, then $1/f$ has a pole of order m at z_0. Conversely, if f has a pole of order m at z_0, then $1/f$ has a removable singularity at z_0, and if we define $(1/f)(z_0) = 0$, then $1/f$ has a zero of order m at z_0.

Some students may have felt that it is obvious that $|f(z)| \to \infty$ as z approaches a pole and that our painstaking analysis was unnecessary. They will probably be surprised to learn that such behavior does not occur as z approaches an essential singularity; instead we have the following.

Theorem 17 (*Picard's Theorem*). A function with an essential singularity assumes every complex number, with possibly one exception, as a value in any neighborhood of this singularity.

The proof of this theorem is beyond our text, but we invite the student to prove a somewhat weaker result, the *Casorati-Weierstrass theorem*, in Prob. 14. We illustrate the Picard[†] theorem in the next example.

Example 3
Verify Picard's result for $e^{1/z}$ near $z = 0$.

Solution.   (Observe first of all that $z = 0$ is an essential singularity; see Example 3, Sec. 5.5.) Obviously $e^{1/z}$ is never zero. However, if $c \neq 0$, we can show that $e^{1/z}$ achieves the value c for $|z|$ less than any positive ε. To this end, recall that

$$\log c = \operatorname{Log} |c| + i \operatorname{Arg} c + 2n\pi i \qquad (n = 0, \pm 1, \pm 2, \ldots)$$

[†]Charles Emile Picard (1856–1941) trained over 10,000 engineers during his career at Ecole Centrale in Paris.

(Sec. 3.3). By picking n sufficiently large, we can find a value w of $\log c$ such that $|w| > 1/\varepsilon$. Then let $z = 1/w$. We will have $|z| < \varepsilon$, and

$$e^{1/z} = e^w = e^{\log c} = c.$$

(To gain further insight into the exotic behavior of $e^{1/z}$ near its essential singularity, the reader is invited (Prob. 16) to sketch the curves $\left|e^{1/z}\right| = s$, where $s = 1, \frac{1}{2}, 2, \frac{1}{3}, 3, \ldots$.)   ∎

From the preceding results we observe that the three different kinds of isolated singularities produce qualitatively different behaviors near these points. *Thus boundedness indicates a removable singularity, approaching ∞ indicates a pole, and anything else must indicate an essential singularity.* These characterizations are often useful in determining the nature of a singularity when it is inconvenient to find the Laurent expansion, as illustrated in the next example.

Example 4

Classify the zeros and singularities of the function $\sin\left(1 - z^{-1}\right)$.

Solution.   Since the zeros of $\sin w$ occur only when w is an integer multiple of π, the function $\sin\left(1 - z^{-1}\right)$ has zeros when

$$1 - z^{-1} = n\pi,$$

that is, at

$$z = \frac{1}{1 - n\pi} \qquad (n = 0, \pm1, \pm2, \ldots).$$

Furthermore, the zeros are simple because the derivative at these points is

$$\frac{d}{dz}\sin\left(1 - z^{-1}\right)\Bigg|_{z=(1-n\pi)^{-1}} = \frac{1}{z^2}\cos\left(1 - z^{-1}\right)\Bigg|_{z=(1-n\pi)^{-1}}$$

$$= (1 - n\pi)^2 \cos n\pi \neq 0.$$

The only singularity of $\sin\left(1 - z^{-1}\right)$ appears at $z = 0$. If we let z approach 0 through positive values, then $\sin\left(1 - z^{-1}\right)$ oscillates between ±1. Such behavior can only characterize an essential singularity.   ∎

Example 5

Classify the zeros and poles of the function $f(z) = (\tan z)/z$.

Solution.   Since $(\tan z)/z = (\sin z)/(z \cos z)$, the only possible zeros are those of $\sin z$; that is, $z = n\pi$ $(n = 0, \pm1, \pm2, \ldots)$. However $z = 0$ is, in fact, a singularity. Furthermore, the points $z = (n + \frac{1}{2})\pi$, which are the zeros of $\cos z$, are also singularities. We shall investigate these in turn.

If n is a nonzero integer, the reader should have no trouble showing that $z = n\pi$ is a *simple* zero for the given function.

Near the point $z = 0$ we can write

$$\frac{\tan z}{z} = \frac{\sin z}{z \cos z}$$

$$= \frac{1}{z \cos z}\left(z - \frac{z^3}{3!} + \frac{z^5}{5!} - \cdots\right)$$

$$= \frac{1}{\cos z}\left(1 - \frac{z^2}{3!} + \frac{z^4}{5!} - \cdots\right),$$

and we see that $(\tan z)/z \to 1$ as $z \to 0$. Hence the origin is a *removable singularity*.

Finally, since $\cos z$ has simple zeros at $z = (n + \frac{1}{2})\pi$ for $n = 0, \pm 1, \pm 2, \ldots$, it is easy to see that $f(z)$ has simple poles at these points. ■

Theorem 18 summarizes the various equivalent characterizations of the three types of isolated singularities. For economy of notation we employ the logician's symbol "$\Leftrightarrow$" to denote logical equivalence; it can be translated "if and only if."

Theorem 18. If f has an isolated singularity at z_0, then the following equivalences hold:

(i) z_0 is a removable singularity $\Leftrightarrow$ $|f|$ is bounded near z_0 $\Leftrightarrow$ $f(z)$ has a limit as $z \to z_0$ $\Leftrightarrow$ f can be redefined at z_0 so that f is analytic at z_0.

(ii) z_0 is a pole $\Leftrightarrow$ $|f(z)| \to \infty$ as $z \to z_0$ $\Leftrightarrow$ f can be written $f(z) = g(z)/(z - z_0)^m$ for some integer $m > 0$ and some function g analytic at z_0 with $g(z_0) \neq 0$.

(iii) z_0 is an essential singularity $\Leftrightarrow$ $|f(z)|$ neither is bounded near z_0 nor goes to infinity as $z \to z_0$ $\Leftrightarrow$ $f(z)$ assumes every complex number, with possibly one exception, as a value in every neighborhood of z_0.

In closing, we make a few some general observations. Earlier we saw that the seemingly innocent-looking property of analyticity for a function f at a point z_0 places enormous restrictions on f; in particular, it must be infinitely differentiable, and expressible by its Taylor series in a neighborhood of z_0. Now we find that if f is merely presumed to be defined, and analytic, in a *punctured* neighborhood of z_0 (like $0 < |z - z_0| < r$), then it is still strongly restricted. One can characterize its behavior near z_0 by asking how many powers of $(z - z_0)$ would it take to "civilize" $f(z)$, in the sense that $(z - z_0)^m f(z)$ would have a finite, *nonzero* limiting value as $z \to z_0$. If the answer (m) is a positive integer, then f has a pole of order m at z_0 and it can be written as $g(z)/(z - z_0)^m$ with g analytic and nonzero at z_0. If m is a negative integer, then f can be written as $g(z)(z - z_0)^{|m|}$ with g, again, analytic and nonzero at z_0; the latter form exhibits a zero of order $|m|$ at z_0. If m is zero, then f has a removable singularity at z_0.

The only other possibility is that no such m exists, that is, no power of $(z - z_0)$ can endow $(z - z_0)^m f(z)$ with a nonzero limit at z_0. Then unless f is identically zero (and not worth "civilizing"), it has an essential singularity at z_0, taking *all* complex numbers as values in any neighborhood of z_0 (with, possibly, one exception).

EXERCISES 5.6

1. Find and classify the isolated singularities of each of the following functions.

   (a) $\dfrac{z^3 + 1}{z^2(z + 1)}$ (b) $z^3 e^{1/z}$ (c) $\dfrac{\cos z}{z^2 + 1} + 4z$ (d) $\dfrac{1}{e^z - 1}$

   (e) $\tan z$ (f) $\cos\left(1 - \dfrac{1}{z}\right)$ (g) $\dfrac{\sin(3z)}{z^2} - \dfrac{3}{z}$ (h) $\cot\left(\dfrac{1}{z}\right)$

2. What is the order of the pole of

   $$f(z) = \frac{1}{\left(2 \cos z - 2 + z^2\right)^2}$$

   at $z = 0$? [HINT: Work with $1/f(z)$.]

3. For each of the following, construct a function f, analytic in the plane except for isolated singularities, that satisfies the given conditions.

   (a) f has a zero of order 2 at $z = i$ and a pole of order 5 at $z = 2 - 3i$.

   (b) f has a simple zero at $z = 0$ and an essential singularity at $z = 1$.

   (c) f has a removable singularity at $z = 0$, a pole of order 6 at $z = 1$, and an essential singularity at $z = i$.

   (d) f has a pole of order 2 at $z = 1 + i$ and essential singularities at $z = 0$ and $z = 1$.

4. Give a proof of Lemma 8.

5. For each of the following, determine whether the statement made is always true or sometimes false.

   (a) If f and g have a pole at z_0, then $f + g$ has a pole at z_0.

   (b) If f has an essential singularity at z_0 and g has a pole at z_0, then $f + g$ has an essential singularity at z_0.

   (c) If $f(z)$ has a pole of order m at $z = 0$, then $f(z^2)$ has a pole of order $2m$ at $z = 0$.

     (d) If f has a pole at z_0 and g has an essential singularity at z_0, then the product $f \cdot g$ has a pole at z_0.

     (e) If f has a zero of order m at z_0 and g has a pole of order n, $n \leq m$, at z_0, then the product $f \cdot g$ has a removable singularity at z_0.

6. Prove that if $f(z)$ has a pole of order m at z_0, then $f'(z)$ has a pole of order $m + 1$ at z_0.

7. If $f(z)$ is analytic in $D : 0 < |z| \leq 1$, and $z^\ell \cdot f(z)$ is unbounded in D for every integer ℓ, then what kind of singularity does $f(z)$ have at $z = 0$?

8. Verify Picard's theorem for the function $\cos(1/z)$ at $z_0 = 0$.

9. Does there exist a function $f(z)$ having an essential singularity at z_0 that is bounded along some line segment emanating from z_0?

10. If the function $f(z)$ is analytic in a domain D and has zeros at the distinct points $z_1, z_2, \ldots, z_n$ of respective orders $m_1, m_2, \ldots, m_n$, then prove that there exists a function $g(z)$ analytic in D such that

$$f(z) = (z - z_1)^{m_1} (z - z_2)^{m_2} \cdots (z - z_n)^{m_n} g(z).$$

11. If f has a pole at z_0, show that $\operatorname{Re} f$ and $\operatorname{Im} f$ take on arbitrarily large positive as well as negative values in any punctured neighborhood of z_0.

12. Prove that if $f(z)$ has a pole of order m at z_0, then $g(z) := f'(z) / f(z)$ has a simple pole at z_0. What is the coefficient of $(z - z_0)^{-1}$ in the Laurent expansion for $g(z)$?

13. Let $f(z)$ have an isolated singularity at z_0 and suppose that $f(z)$ is bounded in some punctured neighborhood of z_0. Prove directly from the integral formula for the Laurent coefficients that $a_{-j} = 0$ for all $j = 1, 2, \ldots$; that is, $f(z)$ must have a removable singularity at z_0.

14. Without appealing to Picard's theorem, prove the theorem of *Casorati* and *Weierstrass*:[†] If $f(z)$ has an essential singularity at z_0, then in any punctured neighborhood of z_0 the function $f(z)$ comes arbitrarily close to any specified complex number. [HINT: Let the specified number be c and assume to the contrary that $|f(z) - c| \geq \delta > 0$ in every small punctured neighborhood of z_0. Then, using Prob. 13, show that $f(z) - c$ [and hence $f(z)$ itself] must have either a pole or a removable singularity at z_0.]

15. Prove that if $f(z)$ has an essential singularity at z_0, then so does the function $e^{f(z)}$. [HINT: Argue that $e^{f(z)}$ is neither bounded nor tends (in modulus) to infinity as $z \to z_0$.]

16. Sketch the graphs for $s = 1, \frac{1}{2}, 2, \frac{1}{3}, 3, \ldots$ of the level curves $\left| e^{1/z} \right| = s$, and observe that they all converge at the essential singularity $z = 0$ of $e^{1/z}$. [HINT: The level curves are all circles.]

17. By completing each of the following steps, prove Schwarz's lemma.

[†]Felice Casorati (1835–1890), Karl Theodor Wilhelm Weierstrass (1853–1897). In his lectures developing the subject of analysis, Weierstrass established standards of rigor for the future of mathematics.

(*Schwarz's Lemma*) If f is analytic in the unit disk $U : |z| < 1$ and satisfies the conditions

$$f(0) = 0 \text{ and } |f(z)| \leq 1 \text{ for all } z \text{ in } U,$$

then $|f(z)| \leq |z|$ for all z in U.

(a) Define $F(z) := f(z)/z$, for $z \neq 0$, and $F(0) = f'(0)$. Show that F is analytic in U.

(b) Let $\zeta (\neq 0)$ be any fixed point in U, and r be any real number that satisfies $|\zeta| < r < 1$. Show by means of the maximum-modulus principle that if C_r denotes the circle $|z| = r$, then

$$|F(\zeta)| \leq \max_{z \text{ on } C_r} \frac{|f(z)|}{r} \leq \frac{1}{r}.$$

(c) Letting $r \to 1^-$ in part (b), deduce that $|f(\zeta)| \leq |\zeta|$ for all ζ in U.

18. Let f be a function satisfying the conditions of Schwarz's lemma (Prob. 17). Prove that if $|f(z_0)| = |z_0|$ for some nonzero z_0 in U, then f must be a function of the form $f(z) = e^{i\theta} z$ for some real θ. Show also that f must be of this form if $|f'(0)| = 1$.

19. Define the function $h(z)$ by

$$h(z) = \frac{1}{\sin z} - \frac{1}{z} + \frac{2z}{z^2 - \pi^2}.$$

(a) Show that $h(z)$ is analytic in the disk $|z| < 2\pi$, except for removable singularities at $z = 0, \pm\pi$.

(b) Find the first four terms of the Taylor series about $z = 0$ for $h(z)$. What is the radius of convergence of this series?

(c) Use the result of part (b) to obtain the first few coefficients (with positive *and* negative indices) in the Laurent series expansion for $\csc z = 1/\sin z$, valid in the annulus $\pi < |z| < 2\pi$.

5.7   The Point at Infinity

From our discussion of singularities in Sec. 5.6 we know that if a mapping is given by an analytic function possessing a pole, it carries points near that pole to indefinitely distant points. It must have occurred to the reader that one might take the value of f *at* the pole to be ∞. Before taking this plunge, however, we should be aware of all the ramifications. Let us look in detail at the behavior of $1/z$ near $z = 0$.

As $z \to 0$ along the positive real axis, $1/z$ goes to "plus infinity"; along the negative real axis, $1/z$ goes to "minus infinity"; and along the positive y-axis, $1/z$ goes to—what? "Minus i times infinity?" If we are to assign the symbol ∞ to $1/0$, we must realize that we are identifying all these "limits" as a single number; geometrically, we are speaking of *the point at infinity*, which can be reached (in a manner of speaking) by proceeding infinitely far along any direction in the complex plane.

The incorporation of the point at infinity into the complex number system was discussed in considerable detail in Section 1.7. Recall that when "∞" is appended to the set of complex numbers the resulting collection $\widehat{\mathbf{C}} = \mathbf{C} \cup \{\infty\}$, known as the *extended complex plane*, can be visualized (via stereographic projection) as points on the Riemann sphere. A neighborhood of ∞ is just a spherical cap centered at the north pole (see Fig. 1.22), which corresponds to the set of all points z in $\mathbf{C}$ lying outside some circle $|z| = M$, together with ∞. Furthermore, a sequence of points z_n in $\mathbf{C}$ ($n = 1, 2, 3, \ldots$) approaches ∞ if $|z_n|$ can be made arbitrarily large by taking n large.

Consequently, we shall write $f(z_0) = \infty$ when $|f(z)|$ increases without bound[†] as $z \to z_0$ and shall write $f(\infty) = w_0$ when $f(z) \to w_0$ as $z \to \infty$. For example, if

$$f(z) = \frac{2z + 1}{z - 1}, \tag{1}$$

then $f(1) = \infty$ and $f(\infty) = 2$.

Observe that for $h(z) = 2z + 1$ we have $h(\infty) = \infty$.

Now we find it convenient to carry this notion still further and speak of functions that are "analytic at ∞." The analyticity properties of f at ∞ are classified by first performing the mapping $w = 1/z$, which maps the point at infinity to the origin, and then examining the behavior of the composite function $g(w) := f(1/w)$ at the origin $w = 0$. Thus we say

1. $f(z)$ is analytic at ∞ if $f(1/w)$ is analytic (or has a removable singularity) at $w = 0$,

2. $f(z)$ has a pole of order m at ∞ if $f(1/w)$ has a pole of order m at $w = 0$, and

3. $f(z)$ has an essential singularity at ∞ if $f(1/w)$ has an essential singularity at $w = 0$.[‡]

From Theorem 18, we can interpret these conditions for a function analytic outside some disk as follows:

1′. $f(z)$ is analytic at ∞ if $|f(z)|$ is bounded for sufficiently large $|z|$,

---

[†] Technically, $f(z_0) = \infty$ if for any $M > 0$ there is a $\delta > 0$ such that $0 < |z - z_0| < \delta$ implies that $|f(z)| > M$.

[‡] Some authors also allow the possibility of a removable singularity at ∞, but we feel that nothing is gained by this generality. Of course, a zero of order m for $f(z)$ at ∞ corresponds to a zero of order m for $f(1/w)$ at 0.

2′. $f(z)$ has a pole at ∞ if $f(z) \to \infty$ as $z \to \infty$, and

3′. $f(z)$ has an essential singularity at ∞ if $|f(z)|$ neither is bounded for large $|z|$ nor goes to infinity as $z \to \infty$.

Example 1

Classify the behavior at ∞ of the functions $z^2 + 2$, $(iz + 1)/(z - 1)$, and $\sin z$.

Solution.   Obviously $f(z) = z^2 + 2$ has a pole at ∞. The pole is of order 2, because

$$f\left(\frac{1}{w}\right) = \frac{1}{w^2} + 2$$

has a pole of order 2 at $w = 0$.

Since

$$\frac{iz + 1}{z - 1} \to i \ \text{ as } \ z \to \infty,$$

this function is analytic at ∞.

Finally, $\sin z$ has no limit as $z \to \infty$, even for real z (it oscillates). Hence ∞ must be an essential singularity.[†]   ■

Example 2

Find all the functions f that are analytic everywhere in the extended complex plane.

Solution.   Since f is analytic at ∞, it is bounded for, say, $|z| > M$. By continuity, f is also bounded for $|z| \le M$. Consequently, f is a bounded entire function. Hence f is constant, by Liouville's theorem.   ■

Example 3

Classify all the functions that are everywhere analytic in the extended complex plane except for a pole at one point.

Solution.   If $f(z)$ has a pole, say, of order m at some finite point z_0, then the Laurent series for f

$$f(z) = \frac{a_{-m}}{(z - z_0)^m} + \frac{a_{-m+1}}{(z - z_0)^{m-1}} + \cdots + \frac{a_{-1}}{z - z_0} + \sum_{n=0}^{\infty} a_n (z - z_0)^n \qquad (2)$$

converges for all $z \ne z_0$. Moreover, since we are assuming that $z_0 \ne \infty$, the function f must be analytic, and hence bounded, at ∞. From Eq. (2), then, we see that the *entire* function defined by the series $\sum_{n=0}^{\infty} a_n (z - z_0)^n$ is also bounded at ∞; thus it must be constant, that is, equal to a_0. Therefore, the most general form for such a function f is

$$f(z) = \frac{a_{-m}}{(z - z_0)^m} + \frac{a_{-m+1}}{(z - z_0)^{m-1}} + \cdots + \frac{a_{-1}}{z - z_0} + a_0. \qquad (3)$$

[†]Alternatively, this can be seen directly from the Laurent expansion for $\sin(1/w)$ about $w = 0$.

If the pole occurs at $z = \infty$, then $f(1/w)$ has a pole at the origin and can be expressed in the form

$$f(\frac{1}{w}) = \frac{a_{-m}}{w^m} + \frac{a_{-m+1}}{w^{m-1}} + \cdots + \frac{a_{-1}}{w} + \sum_{n=0}^{\infty} a_n w^n. \tag{4}$$

Since $f(z)$ is bounded near $z = 0$, it follows that $f(1/w)$ is bounded for large $|w|$, and, as before, we conclude that $a_n = 0$ for $n > 0$. Hence Eq. (4) becomes

$$f(z) = a_{-m}z^m + a_{-m+1}z^{m-1} + \cdots + a_{-1}z + a_0; \tag{5}$$

that is, $f(z)$ is a *polynomial* in z.

Equations (3) and (5) categorize the totality of all functions possessing one pole in the extended complex plane. ∎

We note in passing that the theory of *Fuchsian equations* is based upon considerations of singularities in the extended complex plane, and these have been extremely helpful in relating many of the so-called "special functions" that arise in mathematical physics; Ref. [3] discusses this application.

EXERCISES 5.7

1. Classify the behavior at ∞ for each of the following functions (if a zero or pole, give its order):

   (a) e^z

   (b) $\cosh z$

   (c) $\dfrac{z-1}{z+1}$

   (d) $\dfrac{z}{z^3+i}$

   (e) $\dfrac{z^3+i}{z}$

   (f) $e^{\sinh z}$

   (g) $\dfrac{\sin z}{z^2}$

   (h) $\dfrac{1}{\sin z}$

   (i) $e^{\tan 1/z}$

2. Prove that if $f(z)$ is analytic at ∞, then it has a series expansion of the form

   $$f(z) = \sum_{n=0}^{\infty} \frac{a_n}{z^n}$$

   converging uniformly outside some disk.

3. Construct the series mentioned in Prob. 2 for the following functions.

   (a) $\dfrac{z-1}{z+1}$

   (b) $\dfrac{z^2}{z^2+1}$

   (c) $\dfrac{1}{z^3-i}$

4. State Picard's theorem (Sec. 5.6) for functions with an essential singularity at ∞. Verify for e^z.

5. What is the order of the zero at ∞ if $f(z)$ is a rational function of the form $\dfrac{P(z)}{Q(z)}$ with deg $P <$ deg Q?

6. Suppose that f is analytic on and *outside* the simple closed *negatively* oriented contour Γ. Assume further that f is analytic at ∞ and $f(\infty) = 0$. Prove that

$$f(z) = \frac{1}{2\pi i} \oint_\Gamma \frac{f(\zeta)}{\zeta - z}\, d\zeta$$

   for all z outside Γ. [HINT: Apply Cauchy's integral formula for z in an annulus and let the outer radius tend to ∞.]

7. Prove that if f is analytic on and outside the simple closed contour Γ and has a zero of order 2 or more at ∞, then

$$\int_\Gamma f(z)\, dz = 0.$$

   Does this integral vanish if we merely assume that f has a simple zero at ∞?

Problems 8–12 discuss the class of positive functions.

8. **(a)** Argue that in some neighborhood of a zero, z_0, of order m for the analytic function f one can express $f(z)$ as $(z - z_0)^m [c + \varepsilon(z)]$, where c is constant, $\varepsilon(z)$ is analytic, and $|\varepsilon(z)| < |c|/100$. (The fraction $\frac{1}{100}$ has no special significance.)

   (b) Argue that in some neighborhood of a pole, z_0, of order m for the analytic function f one can express $f(z)$ as $(z - z_0)^{-m} [c + \eta(z)]$, where c is constant, $\eta(z)$ is analytic, and $|\eta(z)| < |c|/100$.

9. A function $f(z)$ is said to be a *positive function* if $f(z)$ is rational (a ratio of polynomials) and if Re $f(z) > 0$ whenever Re $z > 0$. In 1931 Otto Brune proved that the complex *impedance* of any electric circuit must be a positive function when z is interpreted as the "imaginary frequency" $i\omega$ (Sec. 3.6).

   (a) Show that the complex impedance for the electrical circuit studied in Sec. 3.6,

$$R_{\text{eff}} = \frac{R/i\omega C}{R + 1/i\omega C} + i\omega L,$$

   yields a positive function when z is substituted for $i\omega$ (for positive R, L, and C).

   (b) Show that the complex *admittance* $1/R_{\text{eff}}$ is also a positive function.

   (c) Generalize: Show that the reciprocal of any positive function is also positive.

10. By considering the changes of sign in the factor $(z - z_0)^{\pm m}$, use the results of the previous problems to show that if f is a positive function,

    (a) f has no poles or zeros in the right half-plane;

    (b) the pure imaginary poles and zeros (if any) of f must be simple (order 1) and the corresponding constants c in Prob. 8 must be real and positive.

11. Extend the reasoning of the previous problems to argue that a positive function is either analytic and nonzero at ∞ or has a simple pole or a simple zero there. What does this say about the degrees of its numerator and denominator polynomials?

12. (a) Suppose that f is a positive function that has no poles on the imaginary axis. Use the max-min theory of Sec. 4.7 to deduce that the minimal value of Re f in the closed right half-plane occurs on the imaginary axis (including the point at infinity).

    (b) Using the characterization of the imaginary poles of f as described in Prob. 10, remove the "no-poles" restriction in part (a). [HINT: Use an indented contour.]

    (c) Establish the following: If f is a rational function that is analytic in the right half-plane, has only simple poles with positive "residues" (constants c as in Prob. 8) on the imaginary axis (including the point at infinity), and satisfies Re $f \geq 0$ on the imaginary axis, then f is a positive function.

13. (a) Prove there does not exist a (single-valued) branch of $\left(z^2 - 1\right)^{1/3}$ that is analytic in $|z| > 1$. [HINT: Consider the point at infinity.]

    (b) Prove that there *does* exist a (single-valued) branch of $[(z-1)(z-2)(z-3)]^{1/3}$ analytic in $|z| > 3$.

5.8 *Analytic Continuation

When one is given a formula or algorithm for computing an analytic function f in a domain D, it is often of interest to know if this "domain of analyticity" can be extended—that is, if there is a function F analytic in a larger domain whose values agree with those of $f(z)$ for z in D. In such a case we say that F is an *analytic continuation* of f.

This terminology is also used in some situations when the original domain of definition of f is not truly a "domain" in the sense of Chapter 1—and hence the original f is not analytic. We have already encountered some trivial examples of analytic continuation in this sense; for instance, when we extend the real polynomial $x^2 + 1$ to complex numbers as $z^2 + 1$, we have analytically continued the function $f(x) = x^2 + 1$ from the x-axis to the entire plane $\mathbf{C}$. The functions e^z, $\sin z$, and $\log z$ can also be interpreted as analytic continuations.

Analytic continuation arises in the analysis of many engineering systems, wherein one is confronted with a function $f(\omega)$ describing the system response to an excitation at the frequency ω (recall Sec. 3.6). The analytic continuation of f to "complex frequency values" can often be most instructive. In fact, we shall use this approach to relate the Fourier and Laplace transforms in Chapter 8.

Analytic continuation can be a subtle process. True, the continuation of $x^2 + 1$ to $z^2 + 1$ is only a matter of extending the formula; but the identity (recall Sec. 4.4)

$$\oint_C \frac{d\zeta}{\zeta - z} = \begin{cases} 2\pi i & \text{if } z \text{ lies inside } C, \\ 0 & \text{if } z \text{ lies outside } C, \end{cases}$$

demonstrates that the "formula" $\int_C d\zeta / (\zeta - z)$ for the function $f(z) := 2\pi i$ (= constant), valid *inside* C, does not provide the correct analytic continuation of f outside C.

The interesting questions about analytic continuations are these. Is analytic continuation across the boundary of D—or part of the boundary—always possible? Is the continued function F unique? Are there any computational rules for obtaining F? We will try to address these issues at an elementary level in this section.

One tool for studying analytic continuations is the Taylor series. Let us postulate a situation where an analyst needs to investigate the function $f(z)$ defined by the series

$$f(z) := \sum_{j=1}^{\infty} \frac{(-1)^{j-1}(z-1)^j}{j}. \tag{1}$$

For instance, she might have obtained (1) as a result of using the power series method of Example 1, Sec. 5.3, on the differential equation

$$y'' + \frac{y'}{z} = 0 \tag{2}$$

with the initial conditions

$$y(1) = 0, \qquad y'(1) = 1. \tag{3}$$

[In fact, $f(z)$ is a well-known function, but for the purposes of exposition we choose not to identify it for the moment.]

Since the coefficients in the differential equation (2) fail to be analytic only at $z = 0$, the initial-value theorem quoted in Sec. 5.3 guarantees that the circle of convergence for (1) extends at least to this point. Therefore, we can say that the analyst knows the values of this analytic function f inside $C : |z - 1| = 1$ (or, more realistically, she can compute the values of $f(z)$, and its derivatives, to arbitrary accuracy inside C). In particular, she "knows" the function and its derivatives at the point z_1 in Fig. 5.6.

From these data she can construct the power series

$$\sum_{j=0}^{\infty} \frac{f^{(j)}(z_1)}{j!} (z - z_1)^j. \tag{4}$$

Of course, this is a Taylor series for the solution f centered at z_1, and it is guaranteed to converge to f at least inside the small circle C' of Fig. 5.7, wherein f is known to be analytic. The initial-value theorem, however, tells us more—namely, that (4)

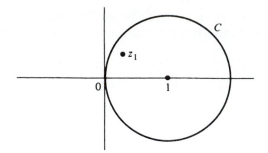

Figure 5.6 Circle of convergence for series (1).

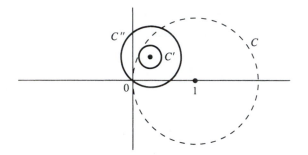

Figure 5.7 Proposed circles of convergence.

converges inside the larger circle C'', extending from z_1 to the origin (again, because the coefficients in the differential equation are analytic inside C''). Thus by employing (4) the analyst has analytically continued the function f, extending its domain of analyticity outside the circle C.

The function defined by the series (1) is, in fact, $\mathrm{Log}\, z$. The reader can directly verify that (1) is a Taylor series for $\mathrm{Log}\, z$ and that $\mathrm{Log}\, z$ satisfies the differential equation (2) and the initial conditions (3). This solution does, indeed, fail to be analytic at $z = 0$, and the circles C and C'' are true circles of convergence.

To continue our study we formalize some terminology covering the particular situation just analyzed.

Definition 9. Suppose that f is analytic in a domain D_1, and that g is analytic in a domain D_2. Then we say that g is a **direct analytic continuation** of f to D_2 if $D_1 \cap D_2$ is nonempty and $f(z) = g(z)$ for all z in $D_1 \cap D_2$.

In our example the first domain D_1 is the interior of C, the second domain D_2 is the interior of C'', and g is the sum of the power series (4). We have not actually proved that $f = g$ over the whole lens-shaped domain $D_1 \cap D_2$; equality has been established only inside C'. So we shall use the following theorem to show that (4) is a bona fide direct analytic continuation of $f(z)$ to D_2.

Theorem 19. If F is analytic in a domain D and vanishes on some open disk contained in D, then it vanishes throughout D.

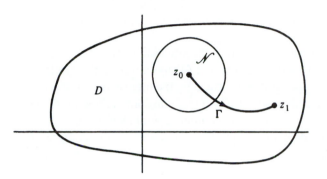

Figure 5.8 Geometry for Theorem 19.

Proof.   By hypothesis, D contains a disk $\mathcal{N}$, say with center z_0, such that $F(z) = 0$ for all z in $\mathcal{N}$. Now let's assume, contrary to the conclusion of the theorem, that there is some point z_1 in D such that $F(z_1) \neq 0$, and let Γ be a path in D joining z_0 to z_1. (See Fig. 5.8.) As we move along Γ from z_0, at first we only observe points where $F(z) = 0$. Eventually, however, we must encounter a point w with the properties

(i) For all z on Γ preceding w, we have $F(z) = 0$;

(ii) There are points z on Γ arbitrarily close to w such that $F(z) \neq 0$. (Recall a similar situation in the proof of Theorem 23, Sec. 4.6.)

First we observe that condition (i) implies that the derivative of $F(z)$ is zero on the portion of Γ *preceding* w, because at any point z on this part of the curve we can evaluate

$$F'(z) = \lim_{\zeta \to z} \frac{F(\zeta) - F(z)}{\zeta - z}$$

by letting ζ approach z along this portion of Γ, where $F(\zeta) = F(z) = 0$. Hence the limit is zero. Continuing in this fashion, we express

$$F''(z) = \lim_{\substack{\zeta \to z \\ \zeta \text{ on } \Gamma}} \frac{F'(\zeta) - F'(z)}{\zeta - z} = \lim_{\substack{\zeta \to z \\ \zeta \text{ on } \Gamma}} 0 = 0,$$

and so on, to conclude that *all derivatives of $F(z)$ vanish on the portion of Γ preceding w*. By continuity, then, F and all its derivatives also vanish at w, which implies that the Taylor coefficients for F about $z = w$ are all zeros; that is, F must vanish in some *disk* around w. But this contradicts condition (ii), so our assumption that F is not identically zero must be wrong.   ■

Corollary 4. If f and g are analytic in a domain D and $f = g$ in some disk contained in D, then $f = g$ throughout D.

Proof.   Simply apply Theorem 19 to the difference $F := f - g$.   ■

Returning to our discussion of the situation depicted in Fig. 5.7 we can now conclude that because the series (1) and (4) agree inside the circle C', they must agree inside the lens-shaped domain formed by the intersection of the interiors of C and C''. This, then, implies that (4) is a direct analytic continuation of $f(z)$ from the interior of C to the interior of C'', in strict accordance with Definition 9.

It is worthwhile at this point to list some salient observations about direct analytic continuation. The first two are so trivial that we shall delete the proofs:

Theorem 20. If f is analytic in a domain D_1 and g is a direct analytic continuation of f to the domain D_2, then the function

$$F(z) := \begin{cases} f(z) & \text{for } z \text{ in } D_1, \\ g(z) & \text{for } z \text{ in } D_2, \end{cases} \tag{5}$$

is single-valued and analytic on $D_1 \cup D_2$.

Theorem 21. If f is analytic in a domain D_1, and D_2 is a domain such that $D_1 \cap D_2$ is nonempty, then the direct analytic continuation of f to D_2, if it exists, is unique.

Theorem 19 and its corollary can be generalized as follows.

Theorem 22. Suppose that f is analytic in a domain D and that $\{z_n\}$ is an infinite sequence of distinct points converging to a point z_0 in D. Suppose, moreover, that $f(z_n) = 0$ for each $n = 1, 2, \ldots$. Then $f(z) \equiv 0$ throughout D.

Proof.   By continuity, z_0 is a zero of f. However, it is not an isolated zero because every punctured disk about z_0 contains points of the sequence $\{z_n\}$. Consequently, by Corollary 3, Sec. 5.6, f must be identically zero in some neighborhood of z_0. Hence Theorem 19 implies $f(z) \equiv 0$ throughout D.   ■

> **Corollary 5.** If f and g are analytic functions in a domain D and $f(z_n) = g(z_n)$ for an infinite sequence of distinct points $\{z_n\}$ converging to a point z_0 in D, then $f \equiv g$ throughout D.

Proof.   Again, consider $f - g$.   ■

Often this corollary is used to extend a known equality from a curve to a domain, as in the next example.

Example 1

Use Corollary 5 to prove that

$$\sin^2 z + \cos^2 z = 1 \qquad \text{for all } z.$$

Solution.   We know from elementary trigonometry that the equality is true for real z. In other words, the two entire functions $f(z) := \sin^2 z + \cos^2 z$ and $g(z) := 1$ agree on the real axis. Corollary 5 thus extends the equality to the whole plane.   ■

Now let's turn to a related topic, the concept of *analytic continuation along a curve*. The situation is as follows (refer to Fig. 5.9): $f(z)$ is analytic in a domain D, z_1 is a point in D, and γ is some path connecting z_1 to a point z^*.

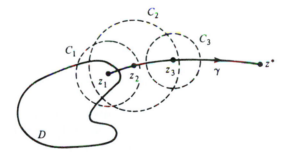

Figure 5.9 Analytic continuation along a curve.

We expand $f(z)$ in a Taylor series around z_1; the resulting power series converges to a function $f_1(z)$ inside, say, the circle C_1. Staying inside C_1, we proceed along γ until we come to some point z_2, and then we expand $f_1(z)$ around z_2. This series converges to some analytic function $f_2(z)$ inside the circle C_2. Next we move further along γ, now staying inside C_2, to some point z_3, and expand around z_3. And so on. If this process eventually produces a circle of convergence, say C_n, that encloses the portion of γ between z_n and z^*, we say that the scheme derived from the sequence of points $\{z_1, z_2, \dots, z_n, z^*\}$ and the corresponding functions $\{f, f_1, f_2, \dots, f_n\}$ constitutes an *analytic continuation of $f(z)$ to z^* along the curve γ*. The value of this analytic continuation at z^* is, of course, $f_n(z^*)$.

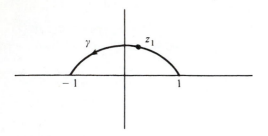

Figure 5.10 Curve for continuation.

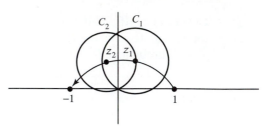

Figure 5.11 Stages of continuation.

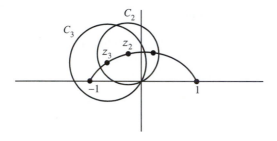

Figure 5.12 Further stages of continuation.

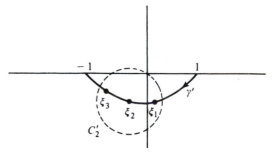

Figure 5.13 Alternative route for continuation.

(We remark that there can be situations in which f_1 is not a *direct* analytic continuation of f; Prob. 6 illustrates this possibility.)

Let's see what would happen if our aforementioned colleague investigating the series $\sum_{j=1}^{\infty}(-1)^{j-1}(z-1)^j/j$ sought to establish an analytic continuation along the curve γ in Fig. 5.10. She expands around z_1 (as before), computing the function $f_1(z)$, which *we* know to equal $\operatorname{Log} z$ inside C_1. Next she expands around z_2, deriving the function $f_2(z)$ inside the circle C_2; see Fig. 5.11.

But now the domain of the analytic function f_2 extends beyond the negative real axis, so we cannot identify f_2 with $\operatorname{Log} z$ over the whole interior of C_2. However, $f_2(z)$ is an analytic function, defined on the interior of C_2, whose value and derivatives coincide with those of $\operatorname{Log} z$ at z_2; hence $f_2(z)$ must agree with some appropriately chosen *branch* of $\log z$ whose branch cut does not intersect C_2. For example, the branch given by $\operatorname{Log}|z| + i \arg z$ with $0 < \arg z \le 2\pi$ matches $\operatorname{Log} z$ at z_2; thus it agrees with $f_2(z)$ inside C_2.

Finally our tireless mathematician expands around the point z_3 and derives the function $f_3(z)$, analytic inside C_3 (see Fig. 5.12). Once again we see that $f_3(z)$ is the branch $\operatorname{Log}|z| + i \arg z$, $0 < \arg z \le 2\pi$, inside C_3; in particular, $f_3(-1) = \pi i$. In short, the mathematician has analytically continued the power series $f(z) =$

$\sum_{j=1}^{\infty}(-1)^{j-1}(z-1)^j/j$ along the curve γ, and the value of the continuation at -1 is πi.

It is instructive to study the result of continuing this same function f along the curve γ' in Fig. 5.13. In this case, the scheme might consist of the points $\{+1, \xi_1, \xi_2, \xi_3, -1\}$ and the functions $\{f, g_1, g_2, g_3\}$. The power series computed around the point ξ_2 would sum to the function $g_2(z)$, whose derivatives would agree with those of $\mathrm{Log}\, z$ at ξ_2 but whose domain of analyticity (the interior of C_2') would enclose a portion of the negative real axis. Thus $g_2(z)$ and, in fact, $g_3(z)$ would agree with the branch

$$\mathcal{L}_{-2\pi}(z) = \mathrm{Log}\,|z| + i \arg z, \quad -2\pi < \arg z \leq 0.$$

In particular, $g_3(-1) = -\pi i$.

Summarizing, we have seen that an analytic continuation of $f(z)$ along γ gives the value πi at $z = -1$, but an analytic continuation along γ' gives the value $-\pi i$. We conclude that, in general, the value obtained by analytic continuation along a curve may depend on the curve itself, not merely on its terminal point.

The alert reader has probably surmised by now that the source of this anomaly is the singularity of all the branches of $\log z$ at the origin. This is in fact the case, and the following result, known as the *Monodromy theorem*, can be considered a vindication of the procedure of analytic continuation along curves.

Theorem 23 (*Monodromy Theorem*). Let $f(z)$ be analytic in a domain D, and suppose that γ and γ' are two directed smooth curves connecting the point z_1 in D to some point z^*. Suppose further that there is some domain D' with the following properties:

  (i) The loop $\Gamma = \{\gamma, -\gamma'\}$ lies in D' and can be continuously deformed[†] to a point in D', and

  (ii) $f(z)$ can be analytically continued along any smooth curve in D'.

Then the value at z^* of the analytic continuation of f along γ agrees with the value of its continuation along γ'.

Thus, for instance, if we continue $\mathrm{Log}\, z$ from $z = +1$ to $z = -1$ along *any* curve lying in the upper half-plane, we shall arrive at the value πi, while the value of the continuation along any curve in the lower half-plane will be $-\pi i$.

The proof of the Monodromy theorem involves some topological constructions with which our readers may be unfamiliar, so we shall delete it (see Ref. [5]). It should be noted, however, that one consequence of the theorem is the fact (which we have tacitly assumed) that the analytic continuation of a function along a particular curve does not depend on the choice of points $\{z_1, z_2, \ldots, z_n, z^*\}$ used in the scheme—again we direct the reader to the references for a rigorous proof.

[†]Compare Sec. 4.4a.

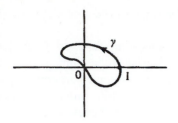

Figure 5.14 Origin lies on γ.

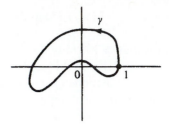

Figure 5.15 Origin lies outside γ.

Example 2

Consider the function $f(z)$ defined for $\operatorname{Re} z > 0$ as the principal branch of $z^{1/2}$, that is, that branch that takes the value $+1$ when $z = 1$. What is the value obtained by continuation of f along a simple closed positively oriented curve γ beginning at $z = 1$ and terminating at the same point, $z = 1$?

Solution. We must consider three possibilities: Either the origin lies on γ, outside γ, or inside γ.

Case 1: The origin lies on γ. Then analytic continuation along γ is not possible; there is no scheme of points and power series that can pass through the "barrier" $z = 0$, since this singularity will be excluded from every circle of convergence of a Taylor series for $z^{1/2}$. See Fig. 5.14.

Case 2: The origin lies outside γ (see Fig. 5.15). Then the conditions of the Monodromy theorem are satisfied with the curve γ' consisting of the single point $z = 1$, and the domain D' given by the entire plane with the origin deleted. The continuation along γ' (and hence γ) results, of course, in the value $+1$.

Case 3: The origin lies inside γ. We can apply the Monodromy theorem, again identifying D' as the punctured plane but now taking γ' to be the unit circle $|z| = 1$, positively oriented (see Fig. 5.16). To get the continuation along γ' we use De Moivre's formula for $z^{1/2}$. Then it is clear that if γ' is parametrized by $z = e^{i\theta}$, $0 \le \theta \le 2\pi$, the value of the continuation of $z^{1/2}$ at $z = e^{i\theta}$ is $e^{i\theta/2}$; this gives $e^{i2\pi/2} = -1$ at the terminal point $z = 1$. Thus we learn that analytic continuation around a *closed* curve may yield functional values that are different from the original ones! ∎

We conclude this section by remarking that there exist functions analytic in a domain D that cannot be analytically continued to *any* point outside D. In such a case the boundary of D is called a *natural boundary*. This situation is illustrated in Prob. 10. *Schwarz reflection*, discussed in Probs. 13 through 15, can provide simple rules for implementing analytic continuations in some cases.

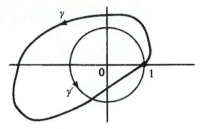

Figure 5.16 Origin lies inside γ.

EXERCISES 5.8

1. Given that $f(z)$ is analytic at $z = 0$ and that $f(1/n) = 1/n^2$, $n = 1, 2, \ldots$, find $f(z)$.

2. Prove that if $f(z)$ is analytic and agrees with a polynomial $\sum_{j=0}^{n} a_j x^j$ for $z = x$ on a segment of the real axis, then $f(z) = \sum_{j=0}^{n} a_j z^j$ everywhere.

3. Does there exist a function $f(z)$, not identically zero, which is analytic in the open disk $D : |z| < 1$ and vanishes at infinitely many points in D?

4. Prove that if f is analytic in a punctured neighborhood of $z = 0$ and if $f(1/n) = 0$ for all $n = \pm 1, \pm 2, \ldots$, then either f is identically zero or f has an essential singularity at $z = 0$.

5. Let $f(z) = \sum_{j=0}^{\infty} z^j$ for $|z| < 1$. For what values of α ($|\alpha| < 1$) does the Taylor expansion of $f(z)$ about $z = \alpha$ yield a direct analytic continuation of $f(z)$ to a disk extending outside $|z| < 1$?

6. Show that when a function f is analytically continued along a curve (as depicted in Fig. 5.9, the first function $f_1(z)$ generated by the power series expansion around the initial point z_1 of γ need not be a direct analytic continuation of f. [HINT: Take $f(z) = \operatorname{Log} z$, with D and z_1 as depicted in Fig. 5.17.]

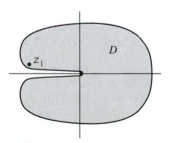

Figure 5.17 Example for Prob. 6.

7. By summing the series, show that $-\sum_{j=0}^{\infty}(2-z)^j$ is an analytic continuation, along some curve, of $\sum_{j=0}^{\infty} z^j$.

8. The *Gamma function* $\Gamma(z)$ is defined for Re $z > 0$ by the integral

$$\Gamma(z) := \int_0^\infty e^{-t} t^{z-1}\, dt.$$

(a) Show that $\Gamma(z + 1) = z\Gamma(z)$ for Re $z > 0$.

(b) In most advanced texts it is shown that $\Gamma(z)$ is analytic in the right half-plane. Assuming this, argue that the functional equation in part (a) can be used to analytically continue $\Gamma(z)$ to the entire plane, except for the nonpositive integers $z = -n$, $n = 0, 1, 2, \ldots$.

(c) Show that $\Gamma(n) = (n-1)!$ for positive integers n.

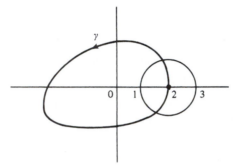

Figure 5.18 Continuation path for Prob. 8.

9. For each of the following functions, choose a branch that is analytic in the circle $|z - 2| < 1$. Then analytically continue this branch along the curve γ indicated in Fig. 5.18. Do the new functional values agree with the old?

(a) $3z^{2/3}$ (b) $\sin 5z$ (c) $(e^z)^{1/3}$

(d) $\sin\left(z^{1/2}\right)$ (e) $\left(z^{1/2}\right)^2$ (f) $\left(z^2\right)^{1/2}$

10. Show that the unit circle $|z| = 1$ is a natural boundary for the function $f(z) = \sum_{j=1}^\infty z^{j!}$, $|z| < 1$. [HINT: Argue that $|f(re^{ip\pi})| \to \infty$ as $r \to 1^-$ for any rational number p.]

11. Show that $|z| = 1$ is a natural boundary for the function $g(z) = \sum_{j=1}^\infty z^{j!}/j!$, although the series converges for $|z| = 1$. [HINT: Relate $g(z)$ to the function $f(z)$ of Prob. 10.]

12. Show that $|z| = 1$ is a natural boundary for $\sum_{j=0}^\infty z^{2^j}$.

13. The *Schwarz reflection principle* provides a formula for analytic continuation across a straight-line segment under certain circumstances. Its simplest form is stated as follows: Suppose that $f(z) = u(x, y) + iv(x, y)$ is analytic in a simply connected domain D that lies in the upper half-plane and which has a segment γ of the real axis as part of its boundary (see Fig. 5.19). Suppose furthermore that $v(x, y) \to 0$

as (x, y) approaches γ and that $u(x, y)$ also takes continuous limiting values on γ, denoted by $U(x)$ [so that $U(x)$ is continuous on γ]. Then the function f can be analytically continued across γ into the domain D', which is the reflection of D in the real axis. Specifically, the function $F(z)$ defined by

$$
F(z) = \begin{cases} u(x, y) + iv(x, y) & \text{for } z = x + iy \text{ in } D, \\ U(x) & \text{for } z = x \text{ on } \gamma, \\ u(x, -y) - iv(x, -y) & \text{for } z = x + iy \text{ in } D' \end{cases}
$$

is analytic in the domain $D \cup \gamma \cup D'$. Justify this principle based upon the following observations.

(a) $F(z)$ satisfies the Cauchy-Riemann equations in D'.

(b) $F(z)$ is continuous in $D \cup \gamma \cup D'$.

(c) Morera's theorem can be applied to $F(z)$, if the contour of integration Γ is decomposed as illustrated in Fig. 5.20.

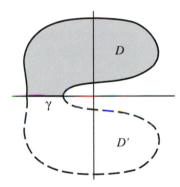

Figure 5.19 Domain for Prob. 13.

14. State and prove a generalization of the Schwarz reflection principle to the case where the boundary of D contains an arbitrary line segment upon which the limiting values of $f(z)$ all lie on some straight line (as would be the case, for example, if these limiting values were all real).

15. State and prove two reflection principles for harmonic functions $\phi(x, y)$, based upon the Schwarz reflection principle. One should cover the case when $\phi(x, y) \to 0$ on the real axis, and the other should apply when $\partial \phi / \partial y \to 0$ on the real axis.

16. Prove the following generalization of Theorem 23 in Sec. 4.6: If f is analytic in a domain D and $|f(z)|$ achieves a *local* maximum at a point z_0 in D, the f is constant in D.

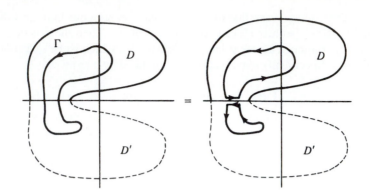

Figure 5.20 The integral of $F(z)$ along Γ is zero.

SUMMARY

The principal achievement of this chapter was to establish an equivalence (roughly speaking) between analytic functions and convergent power series. The equivalence is as follows: Any function f can be expressed as a power series around any point z_0 at which f is analytic, and the series converges uniformly in every closed disk (centered at z_0) that excludes the singularities of f. This series is known as the Taylor series and has the form

$$\sum_{j=0}^{\infty} \frac{f^{(j)}(z_0)}{j!} \, (z - z_0)^j \, .$$

On the other hand, any power series $\sum_{j=0}^{\infty} a_j \, (z - z_0)^j$ converging in a disk $|z - z_0| < R$ sums to an analytic function and, in fact, is the Taylor series of this function.

Power series can be added, integrated, and differentiated termwise, as can any uniformly convergent sequence of analytic functions. Moreover, power series can be multiplied like polynomials.

If a function f fails to be analytic at certain points but is analytic in an annulus surrounding or excluding these points, it can be expanded in a Laurent series,

$$f(z) = \sum_{j=-\infty}^{\infty} a_j \, (z - z_0)^j \, ,$$

converging uniformly in any closed subannulus; the nonanalyticity is reflected in the appearance of negative exponents in the expansion. In fact, if z_0 is an isolated singularity of the function f, the Laurent series can be used to classify z_0 into one of three categories: a removable singularity (f bounded near z_0), a pole ($|f| \to \infty$ at z_0), or an essential singularity (neither of the above). Similarly, the Taylor series allows one to classify the order of an isolated zero of f.

The proof of the Taylor and Laurent expansions, and the actual computation of many Taylor and Laurent series, is made easier by the use of the geometric series

$\sum_{j=0}^{\infty} w^j$, which converges uniformly to $(1 - w)^{-1}$ on any closed disk of the form $|w| \leq \rho < 1$.

Finally, we have seen how the power series expansions lead one to the possibility of extending the domain of definition of f as an analytic function. Analytic continuation, however, is a subtle process in that it may result in multiple-valuedness.

Suggested Reading

More detailed treatments of some of the topics of this chapter can be found in the following references:

Theory of Series

[1] Dienes, P. *The Taylor Series*. Dover Publications, Inc., New York, 1957.

[2] Knopp, K. *Infinite Sequences and Series*. Dover Publications, Inc., New York, 1956.

Differential Equations

[3] Birkhoff, G., and Rota, G. C. *Ordinary Differential Equations*, 4th ed. John Wiley & Sons, New York, 1989.

[4] Rainville, E. D. *Intermediate Differential Equations*, 2nd ed. Chelsea, New York, 1972.

Analytic Continuation and the Reflection Principle

[5] Hille, E. *Analytic Function Theory*, Vol. II, 2nd ed. Chelsea, New York, 1973.

[6] Nehari, Z. *Conformal Mapping*. Dover, New York, 1975.

Positive Functions and Circuit Analysis

[7] Van Valkenburg, M. E. *Introduction to Modern Network Synthesis*. John Wiley & Sons, New York, 1960.

[8] Levinson, N., and Redheffer, R. M. *Complex Variables*. Holden-Day, Inc., San Francisco, 1970.

Applications of Bessel and Legendre Functions

[9] Snider, A.D., *Partial Differential Equations: Sources and Solutions*. Prentice-Hall, Inc., Upper Saddle River NJ, 1999.

Chapter 6

Residue Theory

We have already seen how the theory of contour integration lends great insight into the properties of analytic functions. In this chapter we shall explore another dividend of this theory, namely, its usefulness in evaluating certain *real* integrals. We shall begin by presenting a technique for evaluating contour integrals that is known as *residue theory*.

6.1   The Residue Theorem

Let us consider the problem of evaluating the integral

$$\int_\Gamma f(z)\,dz,$$

where Γ is a simple closed positively oriented contour and $f(z)$ is analytic on and inside Γ *except* for a single isolated singularity, z_0, lying interior to Γ. As we know, the function $f(z)$ has a Laurent series expansion

$$f(z) = \sum_{j=-\infty}^{\infty} a_j\,(z - z_0)^j, \tag{1}$$

converging in some punctured circular neighborhood of z_0; in particular Eq. (1) is valid for all z on the small positively oriented circle C indicated in Fig. 6.1. By the methods of Sec. 4.4, integration over Γ can be converted to integration over C without changing the integral:

$$\int_\Gamma f(z)\,dz = \int_C f(z)\,dz.$$

This last integral can be computed by termwise integration of the series (1) along C. For all $j \neq -1$ the integral is zero, and for $j = -1$ we obtain the value $2\pi i\,a_{-1}$. Consequently we have

$$\int_\Gamma f(z)\,dz = 2\pi i\,a_{-1}. \tag{2}$$

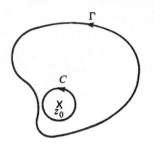

Figure 6.1 Contours for integration.

[Compare this with the formula for a_{-1} given in Theorem 14 of Chapter 5.]

Thus the constant a_{-1} plays an important role in contour integration. Accordingly, we adopt the following terminology.

Definition 1. If f has an isolated singularity at the point z_0, then the coefficient a_{-1} of $1/(z - z_0)$ in the Laurent expansion for f around z_0 is called the **residue of f at z_0** and is denoted by

$$\mathrm{Res}\,(f; z_0) \quad \text{or} \quad \mathrm{Res}\,(z_0).$$

Example 1

Find the residue at $z = 0$ of the function $f(z) = ze^{3/z}$ and compute

$$\oint_{|z|=4} ze^{3/z}\, dz.$$

Solution. Since e^w has the Taylor expansion

$$e^w = \sum_{j=0}^{\infty} \frac{w^j}{j!} \qquad \text{(for all } w\text{)},$$

the Laurent expansion for $ze^{3/z}$ around $z = 0$ is given by

$$ze^{3/z} = z\sum_{j=0}^{\infty} \frac{1}{j!}\left(\frac{3}{z}\right)^j = z + 3 + \frac{3^2}{2!z} + \frac{3^3}{3!z^2} + \cdots .$$

Hence

$$\mathrm{Res}(0) = \frac{3^2}{2!} = \frac{9}{2},$$

and since $z = 0$ is the only singularity inside $|z| = 4$, we have, by formula (2),

$$\oint_{|z|=4} ze^{3/z}\, dz = 2\pi i \cdot \frac{9}{2} = 9\pi i. \quad \blacksquare$$

Now if f has a *removable* singularity at z_0, all the coefficients of the negative powers of $(z - z_0)$ in its Laurent expansion are zero, and so, in particular, the residue at z_0 is zero. Furthermore, if f has a *pole* at z_0, we shall see that its residue there can be computed from a formula. Suppose first that z_0 is a simple pole, that is, a pole of order 1. Then for z near z_0, we have

$$f(z) = \frac{a_{-1}}{z - z_0} + a_0 + a_1 (z - z_0) + a_2 (z - z_0)^2 + \cdots ,$$

and so

$$(z - z_0) f(z) = a_{-1} + (z - z_0) \left[a_0 + a_1 (z - z_0) + a_2 (z - z_0)^2 + \cdots \right].$$

By taking the limit as $z \to z_0$ we deduce that

$$\lim_{z \to z_0} (z - z_0) f(z) = a_{-1} + 0.$$

Hence *at a simple pole*

$$\operatorname{Res} (f; z_0) = \lim_{z \to z_0} (z - z_0) f(z). \tag{3}$$

For example, the function $f(z) = \dfrac{e^z}{z(z + 1)}$ has simple poles at $z = 0$ and $z = -1$; therefore,

$$\operatorname{Res}(f; 0) = \lim_{z \to 0} z \, f(z) = \lim_{z \to 0} \frac{e^z}{z + 1} = 1,$$

and

$$\operatorname{Res}(f; -1) = \lim_{z \to -1} (z + 1) f(z) = \lim_{z \to -1} \frac{e^z}{z} = -e^{-1}.$$

Another consequence of formula (3) is illustrated in the next example.

Example 2

Let $f(z) = P(z)/Q(z)$, where the functions $P(z)$ and $Q(z)$ are both analytic at z_0, and Q has a simple zero at z_0, while $P(z_0) \neq 0$. Prove that

$$\operatorname{Res} (f; z_0) = \frac{P(z_0)}{Q'(z_0)}.$$

Solution.   Obviously f has a simple pole at z_0 (see Sec. 5.6), so we can apply formula (3). Using the fact that $Q(z_0) = 0$, we see directly that

$$\operatorname{Res} (f; z_0) = \lim_{z \to z_0} (z - z_0) \frac{P(z)}{Q(z)} = \lim_{z \to z_0} \frac{P(z)}{\frac{Q(z) - Q(z_0)}{z - z_0}} = \frac{P(z_0)}{Q'(z_0)}. \quad \blacksquare$$

Example 3

Compute the residue at each singularity of $f(z) = \cot z$.

Solution. Since $\cot z = \cos z / \sin z$, the singularities of this function are simple poles occurring at the points $z = n\pi$, $n = 0, \pm 1, \pm 2, \ldots$. Utilizing Example 2 with $P(z) = \cos z$, $Q(z) = \sin z$, the residues at these points are given by

$$\text{Res}(\cot z; n\pi) = \left.\frac{\cos z}{(\sin z)'}\right|_{z=n\pi} = \frac{\cos n\pi}{\cos n\pi} = 1. \quad \blacksquare$$

To obtain the general formula for the residue at a pole of order m we need some method of picking out the coefficient a_{-1} from the Laurent expansion. The reader should encounter no difficulty in following the derivation of the next formula, which was obtained for rational functions in Sec. 3.1 (see Eq. (21)).

Theorem 1. If f has a pole of order m at z_0, then

$$\text{Res}(f; z_0) = \lim_{z \to z_0} \frac{1}{(m-1)!} \frac{d^{m-1}}{dz^{m-1}} \left[(z - z_0)^m f(z)\right]. \qquad (4)$$

Proof. Starting with the Laurent expansion for f around z_0,

$$f(z) = \frac{a_{-m}}{(z - z_0)^m} + \cdots + \frac{a_{-2}}{(z - z_0)^2} + \frac{a_{-1}}{z - z_0} + a_0 + a_1(z - z_0) + \cdots,$$

we multiply by $(z - z_0)^m$,

$$(z - z_0)^m f(z) = a_{-m} + \cdots + a_{-2}(z - z_0)^{m-2} + a_{-1}(z - z_0)^{m-1}$$
$$+ a_0(z - z_0)^m + a_1(z - z_0)^{m+1} + \cdots.$$

and differentiate $m - 1$ times to derive

$$\frac{d^{m-1}}{dz^{m-1}}\left[(z - z_0)^m f(z)\right]$$
$$= (m-1)!\, a_{-1} + m!\, a_0(z - z_0) + \frac{(m+1)!}{2} a_1(z - z_0)^2 + \cdots.$$

Hence

$$\lim_{z \to z_0} \frac{d^{m-1}}{dz^{m-1}}\left[(z - z_0)^m f(z)\right] = (m-1)!a_{-1} + 0,$$

which is equivalent to Eq. (4). $\blacksquare$

Example 4

Compute the residues at the singularities of

$$f(z) = \frac{\cos z}{z^2(z-\pi)^3}.$$

Solution.   This function has a pole of order 2 at $z = 0$ and a pole of order 3 at $z = \pi$. Applying formula (4) we find

$$\operatorname{Res}(0) = \lim_{z \to 0} \frac{1}{1!} \frac{d}{dz} \left[z^2 f(z) \right] = \lim_{z \to 0} \frac{d}{dz} \left[\frac{\cos z}{(z-\pi)^3} \right]$$

$$= \lim_{z \to 0} \left[\frac{-(z-\pi)\sin z - 3\cos z}{(z-\pi)^4} \right] = \frac{-3}{\pi^4},$$

$$\operatorname{Res}(\pi) = \lim_{z \to \pi} \frac{1}{2!} \frac{d^2}{dz^2} \left[(z-\pi)^3 f(z) \right] = \lim_{z \to \pi} \frac{1}{2} \frac{d^2}{dz^2} \left[\frac{\cos z}{z^2} \right]$$

$$= \lim_{z \to \pi} \frac{1}{2} \left[\frac{(6 - z^2)\cos z + 4z\sin z}{z^4} \right] = \frac{-(6 - \pi^2)}{2\pi^4}. \quad \blacksquare$$

We have already seen how to compute the integral $\int_\Gamma f(z)\, dz$ when $f(z)$ has only one singularity inside Γ. Let's now turn to the more general case where Γ is a simple closed positively oriented contour and $f(z)$ is analytic inside and on Γ except for a finite number of isolated singularities at the points $z_1, z_2, \ldots, z_n$ inside Γ (see Fig. 6.2). Notice that by the methods of Sec. 4.4 we can express the integral along Γ in terms of the integrals around the circles C_j in Fig. 6.3:

$$\int_\Gamma f(z)\, dz = \sum_{j=1}^{n} \int_{C_j} f(z)\, dz.$$

However, because z_j is the only singularity of f inside C_j, we know that

$$\int_{C_j} f(z)\, dz = 2\pi i \ \operatorname{Res}(z_j).$$

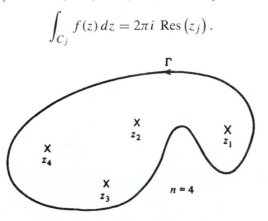

Figure 6.2 Isolated singularities inside contour.

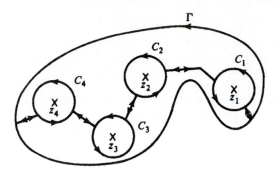

Figure 6.3 Equivalent contours for integration.

Hence we have established the following important result.

Theorem 2 (*Cauchy's Residue Theorem*). If Γ is a simple closed positively oriented contour and f is analytic inside and on Γ except at the points $z_1, z_2, \ldots, z_n$ inside Γ, then

$$\int_\Gamma f(z)\,dz = 2\pi i \sum_{j=1}^{n} \text{Res}\,(z_j). \tag{5}$$

Example 5

Evaluate

$$\oint_{|z|=2} \frac{1-2z}{z(z-1)(z-3)}\,dz.$$

Solution. The integrand $f(z) = (1-2z)/[z(z-1)(z-3)]$ has simple poles at $z = 0$, $z = 1$, and $z = 3$. However, only the first two of these points lie inside $\Gamma : |z| = 2$. Thus by the residue theorem

$$\oint_{|z|=2} f(z)\,dz = 2\pi i [\text{Res}(0) + \text{Res}(1)],$$

and since

$$\text{Res}(0) = \lim_{z\to 0} z\,f(z) = \lim_{z\to 0} \frac{1-2z}{(z-1)(z-3)} = \frac{1}{3},$$

$$\text{Res}(1) = \lim_{z\to 1}(z-1)f(z) = \lim_{z\to 1} \frac{1-2z}{z(z-3)} = \frac{1}{2},$$

we obtain

$$\oint_{|z|=2} f(z)\,dz = 2\pi i \left(\frac{1}{3} + \frac{1}{2}\right) = \frac{5\pi i}{3}. \quad\blacksquare$$

Example 6
Compute

$$\oint_{|z|=5} \left[ze^{3/z} + \frac{\cos z}{z^2(z-\pi)^3} \right] dz.$$

Solution.   The given integral can obviously be expressed as the sum

$$\oint_{|z|=5} ze^{3/z}\, dz + \oint_{|z|=5} \frac{\cos z}{z^2(z-\pi)}\, dz,$$

which, by the residue theorem, equals

$$2\pi i \left[\text{Res}\left(ze^{3/z}; 0 \right) + \text{Res}\left(\frac{\cos z}{z^2(z-\pi)^3}; 0 \right) + \text{Res}\left(\frac{\cos z}{z^2(z-\pi)^3}; \pi \right) \right].$$

These residues were computed in Examples 1 and 4; the desired answer is therefore

$$2\pi i \left[\frac{9}{2} - \frac{3}{\pi^4} - \frac{(6-\pi^2)}{2\pi^4} \right]. \quad \blacksquare$$

EXERCISES 6.1

1. Determine all the isolated singularities of each of the following functions and compute the residue at each singularity.

   (a) $\dfrac{e^{3z}}{z-2}$

   (b) $\dfrac{z+1}{z^2-3z+2}$

   (c) $\dfrac{\cos z}{z^2}$

   (d) $\left(\dfrac{z-1}{z+1}\right)^3$

   (e) $\dfrac{e^z}{z(z+1)^3}$

   (f) $\sin\left(\dfrac{1}{3z}\right)$

   (g) $\tan z$

   (h) $\dfrac{z-1}{\sin z}$

   (i) $z^2/(1-\sqrt{z})$, where $\sqrt{z}$ denotes the principal branch.

2. Explain why Cauchy's integral formula can be regarded as a special case of the residue theorem.

3. Evaluate each of the following integrals by means of the Cauchy residue theorem.

   (a) $\displaystyle\oint_{|z|=5} \frac{\sin z}{z^2-4}\, dz$

   (b) $\displaystyle\oint_{|z|=3} \frac{e^z}{z(z-2)^3}\, dz$

   (c) $\displaystyle\oint_{|z|=2\pi} \tan z\, dz$

   (d) $\displaystyle\oint_{|z|=3} \frac{e^{iz}}{z^2(z-2)(z+5i)}\, dz$

   (e) $\displaystyle\oint_{|z|=1} \frac{1}{z^2 \sin z}\, dz$

   (f) $\displaystyle\oint_{|z|=3} \frac{3z+2}{z^4+1}\, dz$

   (g) $\displaystyle\oint_{|z|=8} \frac{1}{z^2+z+1}\, dz$

4. Let f have an isolated singularity at z_0 (f analytic in a punctured neighborhood of z_0). Show that the residue of the derivative f' at z_0 is equal to zero.

5. Is there a function f having a simple pole at z_0 with Res $(f; z_0) = 0$? How about a function with a pole of order 2 at z_0 and Res$(f; z_0) = 0$?

6. Suppose that f is analytic and has a zero of order m at the point z_0. Show that the function $g(z) = f'(z)/f(z)$ has a simple pole at z_0 with Res$(g; z_0) = m$.

7. Evaluate

$$\oint_{|z|=1} e^{1/z} \sin(1/z)\, dz.$$

6.2   Trigonometric Integrals over $[0, 2\pi]$

Our goal in this section is to apply the residue theorem to evaluate real integrals of the form

$$\int_0^{2\pi} U(\cos\theta, \sin\theta)\, d\theta, \tag{1}$$

where $U(\cos\theta, \sin\theta)$ is a rational function (with real coefficients) of $\cos\theta$ and $\sin\theta$ and is finite over $[0, 2\pi]$. An example of such an integral is

$$\int_0^{2\pi} \frac{\sin^2\theta}{5 + 4\cos\theta}\, d\theta \ .$$

We shall show that (1) can be identified as the parametrized form of a contour integral, $\int_C F(z)\, dz$, of some complex function F around the positively oriented unit circle $C : |z| = 1$. To establish this identification we parametrize C by

$$z = e^{i\theta} \qquad (0 \le \theta \le 2\pi).$$

For such z we have

$$\frac{1}{z} = \frac{1}{e^{i\theta}} = e^{-i\theta},$$

and since

$$\cos\theta = \frac{e^{i\theta} + e^{-i\theta}}{2}, \qquad \sin\theta = \frac{e^{i\theta} - e^{-i\theta}}{2i},$$

we have the identities[†]

$$\cos\theta = \frac{1}{2}\left(z + \frac{1}{z}\right), \qquad \sin\theta = \frac{1}{2i}\left(z - \frac{1}{z}\right). \tag{2}$$

Furthermore, when integrating along C,

$$dz = ie^{i\theta}\, d\theta = iz\, d\theta,$$

[†]Of course we could use $\bar{z}$ instead of $1/z$, but this would forfeit analyticity.

so that

$$d\theta = \frac{dz}{iz}. \tag{3}$$

Making the substitutions (2) and (3) in the integral (1), we see that

$$\int_0^{2\pi} U(\cos\theta, \sin\theta)\, d\theta = \int_C F(z)\, dz, \tag{4}$$

where the new integrand F is

$$F(z) := U\left[\frac{1}{2}\left(z + \frac{1}{z}\right), \frac{1}{2i}\left(z - \frac{1}{z}\right)\right] \cdot \frac{1}{iz};$$

integration over $[0, 2\pi]$ has thus been replaced by integration around C.

Because of the form of U, the function F must be a rational function of z. Hence it has only removable singularities (which can be ignored in evaluating integrals) or poles. Consequently, by the residue theorem, our trigonometric integral equals $2\pi i$ times the sum of the residues at those poles of F that lie inside C.

The procedure is illustrated in the next example.

Example 1

Evaluate

$$I = \int_0^{2\pi} \frac{\sin^2\theta}{5 + 4\cos\theta}\, d\theta.$$

Solution.   First observe that the denominator, $5 + 4\cos\theta$, is never zero, so the integrand is finite over $[0, 2\pi]$. Performing the substitutions (2) and (3) for $\cos\theta$, $\sin\theta$, and $d\theta$, we obtain

$$I = \int_C \frac{\left[\frac{1}{2i}\left(z - \frac{1}{z}\right)\right]^2}{5 + 4\left[\frac{1}{2}\left(z + \frac{1}{z}\right)\right]}\, \frac{dz}{iz},$$

which after some algebra reduces to

$$I = -\frac{1}{4i}\int_C \frac{\left(z^2 - 1\right)^2}{z^2\left(2z^2 + 5z + 2\right)}\, dz.$$

Clearly the integrand

$$g(z) := \frac{\left(z^2 - 1\right)^2}{z^2\left(2z^2 + 5z + 2\right)} = \frac{\left(z^2 - 1\right)^2}{2z^2\left(z + \frac{1}{2}\right)(z + 2)}$$

has simple poles at $z = -\frac{1}{2}$ and $z = -2$ and has a pole of order 2 at the origin. However, only $-\frac{1}{2}$ and 0 lie inside the unit circle C, so that

$$I = -\frac{1}{4i} \cdot 2\pi i \left[\text{Res}\left(g; -\frac{1}{2}\right) + \text{Res}(g; 0)\right].$$

Utilizing the formulas of the preceding section we find

$$\text{Res}\left(g; -\frac{1}{2}\right) = \lim_{z \to -1/2}\left(z + \frac{1}{2}\right)g(z) = \lim_{z \to -1/2}\frac{(z^2 - 1)^2}{2z^2(z + 2)} = \frac{3}{4},$$

and

$$\text{Res}(g; 0) = \lim_{z \to 0}\frac{1}{1!}\frac{d}{dz}\left[z^2 g(z)\right] = \lim_{z \to 0}\frac{d}{dz}\left[\frac{(z^2 - 1)^2}{2z^2 + 5z + 2}\right]$$

$$= \frac{(2z^2 + 5z + 2) \cdot 2\left(z^2 - 1\right)2z - \left(z^2 - 1\right)^2 (4z + 5)}{\left(2z^2 + 5z + 2\right)^2}\Bigg|_{z=0}$$

$$= \frac{-5}{4}.$$

Hence

$$I = \frac{-1}{4i}\, 2\pi i\left[\frac{3}{4} - \frac{5}{4}\right] = \frac{\pi}{4}. \quad \blacksquare$$

As a rough check on our calculations we observe that the integrand of Example 1 is real and nonnegative, so I must be a positive real number, which is consistent with our answer of $\pi/4$.

Example 2

Evaluate

$$I = \int_0^\pi \frac{d\theta}{2 - \cos\theta}.$$

Solution.   The catch here is that the integral is taken over $[0, \pi]$ instead of $[0, 2\pi]$. However it is easy to see that, since $\cos\theta = \cos(2\pi - \theta)$,

$$\int_0^\pi \frac{d\theta}{2 - \cos\theta} = \int_\pi^{2\pi} \frac{d\theta}{2 - \cos\theta},$$

and, therefore,

$$\int_0^{2\pi} \frac{d\theta}{2 - \cos\theta} = 2I.$$

Substituting for $\cos\theta$ and $d\theta$ we have

$$2I = \int_C \frac{1}{2 - \frac{1}{2}\left(z + \frac{1}{z}\right)} \cdot \frac{dz}{iz} = -\frac{2}{i}\int_C \frac{dz}{z^2 - 4z + 1}. \tag{5}$$

By the quadratic formula the zeros of the denominator are

$$z_1 := 2 - \sqrt{3} \quad \text{and} \quad z_2 := 2 + \sqrt{3},$$

and so the integrand

$$g(z) := \frac{1}{z^2 - 4z + 1} = \frac{1}{(z - z_1)(z - z_2)}$$

has simple poles at these points. But only z_1 lies inside C, and the residue there is given by

$$\text{Res}\,(g; z_1) = \lim_{z \to z_1}\,(z - z_1)\,g(z) = \lim_{z \to z_1} \frac{1}{(z - z_2)}$$

$$= \frac{1}{z_1 - z_2} = -\frac{1}{2\sqrt{3}}.$$

Hence from Eq. (5)

$$2I = -\frac{2}{i} \cdot 2\pi i \left(-\frac{1}{2\sqrt{3}}\right) = \frac{2\pi}{\sqrt{3}},$$

or

$$I = \frac{\pi}{\sqrt{3}}. \quad \blacksquare$$

EXERCISES 6.2

Using the method of residues, verify each of the following.

1. $\displaystyle\int_0^{2\pi} \frac{d\theta}{2 + \sin\theta} = \frac{2\pi}{\sqrt{3}}$

2. $\displaystyle\int_0^{\pi} \frac{8\,d\theta}{5 + 2\cos\theta} = \frac{8\pi}{\sqrt{21}}$

3. $\displaystyle\int_0^{\pi} \frac{d\theta}{(3 + 2\cos\theta)^2} = \frac{3\pi\sqrt{5}}{25}$

4. $\displaystyle\int_{-\pi}^{\pi} \frac{d\theta}{1 + \sin^2\theta} = \pi\sqrt{2}$

5. $\displaystyle\int_0^{2\pi} \frac{d\theta}{1 + a\cos\theta} = \frac{2\pi}{\sqrt{1 - a^2}}, \quad a^2 < 1$

6. $\displaystyle\int_0^{2\pi} \frac{\sin^2\theta}{a + b\cos\theta}\,d\theta = \frac{2\pi}{b^2}\left(a - \sqrt{a^2 - b^2}\right), \quad a > |b| > 0$

7. $\displaystyle\int_0^{\pi} \frac{d\theta}{(a + \sin^2\theta)^2} = \frac{\pi(2a + 1)}{2\sqrt{(a^2 + a)^3}}, \quad a > 0$

8. $\displaystyle\int_0^{2\pi} \frac{d\theta}{a^2\sin^2\theta + b^2\cos^2\theta} = \frac{2\pi}{ab}, \quad a, b > 0$

9. $\displaystyle\int_0^{2\pi} (\cos\theta)^{2n}\,d\theta = \frac{\pi\cdot(2n)!}{2^{2n-1}(n!)^2},\quad n = 1, 2, \ldots$

10. $\displaystyle\int_0^{2\pi} e^{\cos\theta}\cos(n\theta - \sin\theta)\,d\theta = \frac{2\pi}{n!},\quad n = 1, 2, \ldots$

11. $\displaystyle\int_0^{\pi} \tan(\theta + ia)\,d\theta = \pi i\cdot\operatorname{sign} a,\quad a$ real and nonzero

6.3   Improper Integrals of Certain Functions over $(-\infty, \infty)$

If $f(x)$ is a function continuous on the nonnegative real axis $0 \le x < \infty$, then the improper integral of f over $[0, \infty)$ is defined by

$$\int_0^{\infty} f(x)\,dx := \lim_{b\to\infty}\int_0^b f(x)\,dx, \tag{1}$$

provided that this limit exists.[†] For example,

$$\int_0^{\infty} e^{-2x}\,dx = \lim_{b\to\infty}\int_0^b e^{-2x}\,dx = \lim_{b\to\infty}\frac{-e^{-2x}}{2}\Big|_0^b$$

$$= \lim_{b\to\infty}\left[\frac{-e^{-2b}}{2} + \frac{1}{2}\right] = \frac{1}{2}.$$

Similarly, when $f(x)$ is continuous on $(-\infty, 0]$, we set

$$\int_{-\infty}^0 f(x)\,dx := \lim_{c\to-\infty}\int_c^0 f(x)\,dx. \tag{2}$$

If it turns out that both of the limits (1) and (2) exist for a function f continuous on the whole real line, then we say that the *improper integral of f over $(-\infty, \infty)$ exists* and we write

$$\int_{-\infty}^{\infty} f(x)\,dx := \lim_{c\to-\infty}\int_c^0 f(x)\,dx + \lim_{b\to\infty}\int_0^b f(x)\,dx$$

$$= \int_{-\infty}^0 f(x)\,dx + \int_0^{\infty} f(x)\,dx.$$

In such a case the value of the improper integral over $(-\infty, \infty)$ can be computed by taking a single limit, namely,

$$\int_{-\infty}^{\infty} f(x)\,dx = \lim_{\rho\to\infty}\int_{-\rho}^{\rho} f(x)\,dx.$$

[†]More generally, $\int_a^{\infty} f(x)\,dx := \lim\limits_{b\to\infty}\int_a^b f(x)\,dx$ if this limit exists.

However, we caution the reader that this last limit may exist for certain functions f even though its improper integral over $(-\infty, \infty)$ does not. Indeed, consider $f(x) = x$. Its improper integral over $(-\infty, \infty)$ does not exist because the limit

$$\lim_{b \to \infty} \int_0^b x \, dx = \lim_{b \to \infty} \frac{x^2}{2} \bigg|_0^b = \lim_{b \to \infty} \frac{b^2}{2}$$

does not exist (as a finite number). However,

$$\lim_{\rho \to \infty} \int_{-\rho}^{\rho} x \, dx = \lim_{\rho \to \infty} \frac{x^2}{2} \bigg|_{-\rho}^{\rho} = \lim_{\rho \to \infty} 0 = 0.$$

For this reason we introduce the following terminology: Given *any* function f continuous on $(-\infty, \infty)$ the limit

$$\lim_{\rho \to \infty} \int_{-\rho}^{\rho} f(x) \, dx$$

(if it exists) is called the *Cauchy principal value* of the integral of f over $(-\infty, \infty)$, and we write

$$\text{p.v.} \int_{-\infty}^{\infty} f(x) \, dx := \lim_{\rho \to \infty} \int_{-\rho}^{\rho} f(x) \, dx.$$

For example,

$$\text{p.v.} \int_{-\infty}^{\infty} x \, dx = 0.$$

We reiterate that whenever the improper integral $\int_{-\infty}^{\infty} f(x) \, dx$ exists, it must equal its principal value (p.v.).

We shall now show how the theory of residues can be used to compute p.v. integrals for certain functions f.

Example 1

Evaluate

$$I = \text{p.v.} \int_{-\infty}^{\infty} \frac{dx}{x^4 + 4} \left(= \lim_{\rho \to \infty} \int_{-\rho}^{\rho} \frac{dx}{x^4 + 4} \right).$$

Solution.   As a first step, we recognize that the integral I_ρ defined by

$$I_\rho := \int_{-\rho}^{\rho} \frac{dx}{x^4 + 4}$$

can be interpreted as a contour integral of an analytic function; in fact

$$I_\rho = \int_{\gamma_\rho} \frac{dz}{z^4 + 4},$$

where γ_ρ is the directed segment of the real axis from $-\rho$ to $+\rho$. Now the key to using residue theory to find I lies in constructing (for each sufficiently large value of ρ) a simple *closed* contour Γ_ρ such that γ_ρ is one of its components, that is, $\Gamma_\rho = (\gamma_\rho, \gamma'_\rho)$, and such that the integral of $1/(z^4 + 4)$ along the other component γ'_ρ is somehow known. For then we will have

$$\int_{\Gamma_\rho} \frac{dz}{z^4 + 4} = I_\rho + \int_{\gamma'_\rho} \frac{dz}{z^4 + 4},$$

and if Γ_ρ is positively oriented, the residue theorem yields

$$2\pi i \cdot \sum (\text{residues inside } \Gamma_\rho) = I_\rho + \int_{\gamma'_\rho} \frac{dz}{z^4 + 4}.$$

Consequently, the integral I is evaluated by

$$I = \lim_{\rho \to \infty} I_\rho = \lim_{\rho \to \infty} 2\pi i \sum (\text{residues inside } \Gamma_\rho) - \lim_{\rho \to \infty} \int_{\gamma'_\rho} \frac{dz}{z^4 + 4}, \qquad (3)$$

provided the limits on the right exist. Actually, we see from Eq. (3) that it is only the *limiting value* of the integrals over γ'_ρ that must be known in order to apply the residue theory.

Now among the many curves that "close the contour γ_ρ," that is, that start at $z = \rho$ and terminate at $z = -\rho$, how are we to find a suitable curve γ'_ρ? Observe that the integrand, $1/(z^4 + 4)$, is quite small in modulus when $|z|$ is large. This suggests that if we choose our curves γ'_ρ far enough away from the origin, the integrals over them might well be negligible, that is, approach zero as $\rho \to \infty$. Thus an obvious thing to try for γ'_ρ is the half-circle C^+_ρ parametrized by

$$C^+_\rho: \quad z = \rho e^{it} \qquad (0 \le t \le \pi) \qquad (4)$$

(see Fig. 6.4). To see if this works, we note that $|z| = \rho$ on C^+_ρ, so that by the triangle inequality

$$\left| \frac{1}{z^4 + 4} \right| \le \frac{1}{|z|^4 - 4} = \frac{1}{\rho^4 - 4} \qquad \left(\text{for } \rho^4 > 4 \right),$$

and hence

$$\left| \int_{C^+_\rho} \frac{dz}{z^4 + 4} \right| \le \frac{1}{\rho^4 - 4} \cdot \pi \rho,$$

which certainly does go to zero as $\rho \to \infty$.

So now all we have to do in Eq. (3) is to evaluate the appropriate residues. First we locate the singularities of $1/(z^4 + 4)$. These occur at the zeros of $z^4 + 4$, that is, at

$$z_1 = 1 + i, \qquad z_2 = -1 + i, \qquad z_3 = -1 - i, \qquad z_4 = 1 - i,$$

and the function

$$\frac{1}{z^4 + 4} = \frac{1}{(z - z_1)(z - z_2)(z - z_3)(z - z_4)}$$

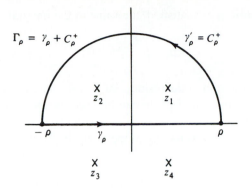

Figure 6.4 Closing the contour.

has simple poles at these points. Since z_3 and z_4 lie in the lower half-plane, they are always excluded from the interior of the semicircular contour Γ_ρ of Fig. 6.4, but z_1 and z_2 lie inside Γ_ρ for every $\rho > \sqrt{2}$. Thus, for such ρ,

$$\int_{\Gamma_\rho} \frac{dz}{z^4 + 4} = 2\pi i \,[\text{Res}\,(z_1) + \text{Res}\,(z_2)]$$

$$= 2\pi i \left(\lim_{z \to z_1} \frac{z - z_1}{z^4 + 4} + \lim_{z \to z_2} \frac{z - z_2}{z^4 + 4} \right)$$

$$= 2\pi i \left[\frac{1}{(z_1 - z_2)(z_1 - z_3)(z_1 - z_4)} + \frac{1}{(z_2 - z_1)(z_2 - z_3)(z_2 - z_4)} \right]$$

$$= 2\pi i \left[\frac{1}{2(2 + 2i)2i} + \frac{1}{(-2)(2i)(-2 + 2i)} \right]$$

$$= 2\pi i \left[\frac{-1 - i}{16} + \frac{1 - i}{16} \right] = \frac{\pi}{4}.$$

Putting this all together in Eq. (3) (with $\gamma'_\rho = C^+_\rho$) we have

$$I = \lim_{\rho \to \infty} \frac{\pi}{4} - \lim_{\rho \to \infty} \int_{C^+_\rho} \frac{dz}{z^4 + 4} = \frac{\pi}{4} - 0 = \frac{\pi}{4}. \quad \blacksquare$$

The technique of using expanding semicircular contours Γ_ρ can readily be applied to a general class of integrands f. Indeed, the success of the procedure illustrated in Example 1 depends only on the following two conditions:

(i) f is analytic on and above the real axis except for a *finite* number of isolated singularities in the open upper half-plane Im $z > 0$ (this ensures that for ρ sufficiently large, all the singularities in the upper half-plane will lie inside the contour Γ_ρ of Fig. 6.4), and

(ii) $\lim\limits_{\rho \to \infty} \int_{C^+_\rho} f(z)\,dz = 0$.

Whenever these conditions are satisfied, the value of the integral

$$\text{p.v.} \int_{-\infty}^{\infty} f(x)\, dx$$

is given by $2\pi i$ times the sum of the residues of f at the singularities in the upper half-plane. (Of course, the lower half-plane can be used whenever analogous conditions hold there; see Prob. 8.)

A class of rational functions having property (ii) is given in the next lemma.

Lemma 1. If $f(z) = P(z)/Q(z)$ is the quotient of two polynomials such that

$$\text{degree } Q \geq 2 + \text{degree } P, \tag{5}$$

then

$$\lim_{\rho \to \infty} \int_{C_\rho^+} f(z)\, dz = 0, \tag{6}$$

where C_ρ^+ is the upper half-circle of radius ρ defined in Eq. (4).

Proof. Estimate $f(z)$ by writing

$$|f(z)| = \frac{|P(z)|}{|Q(z)|} = \frac{|a_0 + a_1 z + a_2 z^2 + \cdots + a_m z^m|}{|b_0 + b_1 z + b_2 z^2 + \cdots + b_n z^n|}$$

$$= \frac{|a_0/z^m + a_1/z^{m-1} + a_2/z^{m-2} + \cdots + a_m|}{|b_0/z^n + b_1/z^{n-1} + b_2/z^{n-2} + \cdots + b_n|} \frac{|z^m|}{|z^n|}.$$

As $|z| \to \infty$ the first term approaches $|a_m|/|b_n|$, so for sufficiently large $|z|$ it will certainly be less than, say, $|a_m|/|b_n| + 1$. Consequently when $n \geq 2 + m$ and ρ is sufficiently large,

$$\left| \int_{C_\rho^+} f(z)\, dz \right| \leq \left(\frac{|a_m|}{|b_n|} + 1 \right) \rho^{m-n} \cdot \pi \rho = \left(\frac{|a_m|}{|b_n|} + 1 \right) \pi \rho^{1+m-n} \to 0$$

as $\rho \to \infty$. ∎

We emphasize that the same proof shows that Eq. (6) remains valid if integration along C_ρ^+ is replaced by integration along the lower half-circle $C_\rho^- : z = \rho e^{-it}$ $(0 \leq t \leq \pi)$.

Example 2

Compute

$$\text{p.v.} \int_{-\infty}^{\infty} \frac{x^2}{\left(x^2 + 1\right)^2}\, dx.$$

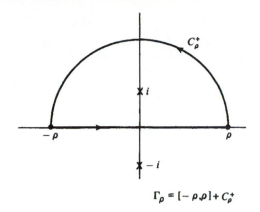

$$\Gamma_\rho = [-\rho,\rho] + C_\rho^+$$

Figure 6.5 Closed contour for Example 2.

Solution. Since the integrand has no singularities on the real axis and has numerator degree 2 and denominator degree 4, the expanding semicircular contour method is justified thanks to Lemma 1. On writing

$$f(z) := \frac{z^2}{\left(z^2 + 1\right)^2} = \frac{z^2}{(z - i)^2(z + i)^2},$$

we see that poles occur at $z = \pm i$. Thus, for any $\rho > 1$, the integral along the closed contour Γ_ρ in Fig. 6.5 is given by

$$\int_{\Gamma_\rho} f(z)\, dz = 2\pi i \operatorname{Res}(f; +i).$$

Since $+i$ is a second-order pole, we have from the residue formula (Theorem 1)

$$\operatorname{Res}(f; +i) = \lim_{z \to i} \frac{1}{1!} \frac{d}{dz}\left[(z - i)^2 f(z)\right] = \lim_{z \to i} \frac{d}{dz}\left[\frac{z^2}{(z + i)^2}\right]$$

$$= \lim_{z \to i}\left[\frac{(z + i)2z - 2z^2}{(z + i)^3}\right] = \frac{1}{4i}.$$

Therefore,

$$\int_{\Gamma_\rho} f(z)\, dz = 2\pi i \cdot \frac{1}{4i} = \frac{\pi}{2} \qquad \text{(for all } \rho > 1\text{).} \tag{7}$$

On the other hand,

$$\int_{\Gamma_\rho} f(z)\, dz = \int_{-\rho}^{\rho} f(x)\, dx + \int_{C_\rho^+} f(z)\, dz,$$

and so on taking the limit as $\rho \to \infty$ in this last equation we deduce from Eq. (7) and Lemma 1 that

$$\frac{\pi}{2} = \lim_{\rho \to \infty} \int_{-\rho}^{\rho} f(x)\, dx + 0.$$

Hence

$$\frac{\pi}{2} = \text{p.v.} \int_{-\infty}^{\infty} \frac{x^2}{(x^2 + 1)^2} \, dx. \quad \blacksquare$$

As we shall see in the next section, semicircular contours are also useful in evaluating certain integrals involving trigonometric functions. The following example is atypical of the integrals that arise in many applications, because the closing contour is *not* semicircular.

Example 3

Compute

$$\text{p.v.} \int_{-\infty}^{\infty} \frac{e^{ax}}{1 + e^x} \, dx, \qquad \text{for } 0 < a < 1.$$

Solution. Observe that the function $e^{ax}/(1 + e^x)$ has an infinite number of singularities in both the upper and lower half-planes; these occur at the points

$$z = (2n + 1)\pi i \qquad (n = 0, \pm 1, \pm 2, \ldots).$$

Hence if we employ expanding semicircles, the contribution due to the residues will result in an infinite series, which is undesirable. Moreover, there is no obvious way to estimate the contribution due to the semicircles themselves!

A better "return path" is revealed through careful examination of the integrand. The denominator of the function $e^{ax}/(1 + e^x)$ is unchanged if z is shifted by $2\pi i$, whereas the numerator changes by a factor of $e^{2\pi ai}$. Thus if we consider the rectangular contour Γ_ρ in Fig. 6.6, the contribution from γ_3 is easy to assess; it's merely $-e^{2\pi ai}$ times the contribution from γ_1 (negative because the path runs from right to left). Therefore

$$\int_{\gamma_3} \frac{e^{az}}{1 + e^z} \, dz = -e^{2\pi ai} \int_{\gamma_1} \frac{e^{az}}{1 + e^z} \, dz.$$

For $\gamma_2 : z = \rho + it, 0 \le t \le 2\pi$, we have

$$\left| \int_{\gamma_2} \frac{e^{az}}{1 + e^z} \, dz \right| = \left| \int_0^{2\pi} \frac{e^{a(\rho + it)}}{1 + e^{\rho + it}} \, i \, dt \right|$$

$$\le \frac{e^{a\rho}}{e^\rho - 1} \cdot 2\pi,$$

which goes to zero as $\rho \to \infty$ since $a < 1$.

Similarly, on $\gamma_4 : z = -\rho + i(2\pi - t), 0 \le t \le 2\pi$, we have

$$\left| \int_{\gamma_4} \frac{e^{az}}{1 + e^z} \, dz \right| = \left| \int_0^{2\pi} \frac{e^{a[-\rho + i(2\pi - t)]}}{1 + e^{-\rho + i(2\pi - t)}} \, (-i) \, dt \right|$$

$$\le \frac{e^{-a\rho}}{1 - e^{-\rho}} \cdot 2\pi,$$

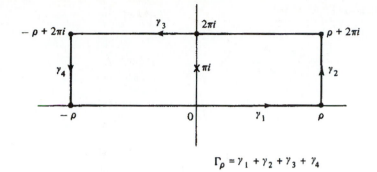

Figure 6.6 Closed contour for Example 3.

again approaching zero as $\rho \to \infty$ since $a > 0$.

As a result, on taking the limit as $\rho \to \infty$ we have

$$\lim_{\rho \to \infty} \int_{\Gamma_\rho} \frac{e^{az}}{1 + e^z}\, dz = \left(1 - e^{a2\pi i}\right)\, \text{p.v.} \int_{-\infty}^{\infty} \frac{e^{ax}}{1 + e^x}\, dx. \tag{8}$$

Now we use residue theory to evaluate the contour integral in Eq. (8). For each $\rho > 0$, the function $e^{az}/(1 + e^z)$ is analytic inside and on Γ_ρ except for a simple pole at $z = \pi i$, the residue there being given by

$$\text{Res}(\pi i) = \frac{e^{az}}{\dfrac{d}{dz}(1 + e^z)}\Bigg|_{z = \pi i} = \frac{e^{a\pi i}}{e^{\pi i}} = -e^{a\pi i} \tag{9}$$

(recall Example 2, Sec. 6.1). Consequently, putting Eqs. (8) and (9) together we obtain

$$\text{p.v.} \int_{-\infty}^{\infty} \frac{e^{ax}}{1 + e^x}\, dx = \frac{1}{1 - e^{a2\pi i}} \cdot (2\pi i)\left(-e^{a\pi i}\right)$$

$$= \frac{-2\pi i}{e^{-a\pi i} - e^{a\pi i}}$$

$$= \frac{\pi}{\sin a\pi}. \quad \blacksquare$$

EXERCISES 6.3

Verify the integral formulas in Problems 1–7 with the aid of residues.

1. p.v. $\displaystyle\int_{-\infty}^{\infty} \frac{dx}{x^2 + 2x + 2} = \pi$

2. p.v. $\displaystyle\int_{-\infty}^{\infty} \frac{x^2}{\left(x^2 + 9\right)^2}\, dx = \frac{\pi}{6}$

3. $\displaystyle\int_0^\infty \frac{x^2+1}{x^4+1}\,dx = \frac{\pi}{\sqrt{2}}$

4. p.v. $\displaystyle\int_{-\infty}^\infty \frac{dx}{(x^2+1)(x^2+4)} = \frac{\pi}{6}$

5. p.v. $\displaystyle\int_{-\infty}^\infty \frac{x}{(x^2+4x+13)^2}\,dx = -\frac{\pi}{27}$

6. $\displaystyle\int_0^\infty \frac{x^2}{(x^2+1)(x^2+4)}\,dx = \frac{\pi}{6}$

7. $\displaystyle\int_0^\infty \frac{x^6}{(x^4+1)^2}\,dx = \frac{3\pi\sqrt{2}}{16}$

8. Show that if $f(z) = P(z)/Q(z)$ is the quotient of two polynomials such that $\deg Q \ge 2 + \deg P$, where Q has no real zeros, then p.v. $\int_{-\infty}^\infty f(x)\,dx$ equals

$$-2\pi i \cdot \sum [\text{residues of } f(z) \text{ at the poles in the } \textit{lower} \text{ half-plane}].$$

9. Show that

$$\text{p.v.} \int_{-\infty}^\infty \frac{e^{2x}}{\cosh(\pi x)}\,dx = \sec 1$$

by integrating $e^{2z}/\cosh(\pi z)$ around rectangles with vertices at $z = \pm\rho,\ \rho + i,\ -\rho + i$.

10. Given that

$$\int_0^\infty e^{-x^2}\,dx = \frac{\sqrt{\pi}}{2},$$

integrate e^{-z^2} around a rectangle with vertices at $z = 0,\ \rho,\ \rho + \lambda i$, and λi (with $\lambda > 0$) and let $\rho \to \infty$ to derive

(a) $\displaystyle\int_0^\infty e^{-x^2}\cos(2\lambda x)\,dx = \frac{\sqrt{\pi}}{2}e^{-\lambda^2}$

(b) $\displaystyle\int_0^\infty e^{-x^2}\sin(2\lambda x)\,dx = e^{-\lambda^2}\int_0^\lambda e^{y^2}\,dy$

(The right-hand side of (b), as a function of λ, is known as the *Dawson integral* and is tabulated by Abramowitz and Stegun in Ref. [5].)

11. Show that

$$\int_0^\infty \frac{dx}{x^3+1} = \frac{2\pi\sqrt{3}}{9}$$

by integrating $1/(z^3+1)$ around the boundary of the circular sector $S_\rho : \{z = re^{i\theta} : 0 \le \theta \le 2\pi/3,\ 0 \le r \le \rho\}$ and letting $\rho \to \infty$.

12. Confirm the values of the integrals discussed in Prob. 18, Exercises 4.7.

13. Show that

$$\int_{-\infty}^{\infty} \frac{1}{\left(1+x^2\right)^{n+1}}\, dx = \frac{\pi (2n)!}{2^{2n} (n!)^2}, \qquad \text{for } n = 0, 1, 2, \ldots.$$

Summation of Series

14. Let $f(z)$ be a rational function of the form $P(z)/Q(z)$, where $\deg Q \geq 2 + \deg P$. Assume that no poles of $f(z)$ occur at the integer points $z = 0, \pm 1, \pm 2, \ldots.$ Complete each of the following steps to establish the summation formula

$$\lim_{N \to +\infty} \sum_{k=-N}^{N} f(k) = -\{\text{sum of the residues of } \pi f(z) \cot(\pi z) \text{ at the poles of } f(z)\}.$$

$$(10)$$

(a) Show that for the function $g(z) := \pi f(z) \cot(\pi z)$, we have

$$\text{Res}(g; k) = f(k), \quad k = 0, \pm 1, \pm 2, \ldots.$$

(b) Let Γ_N be the boundary of the square with vertices at $(N + \frac{1}{2})(1 + i)$, $(N + \frac{1}{2})(-1 + i)$, $(N + \frac{1}{2})(-1 - i)$, $(N + \frac{1}{2})(1 - i)$, taken in that order, where N is a positive integer. Show that there is a constant M independent of N such that $|\pi \cot(\pi z)| \leq M$ for all z on Γ_N.

(c) Prove that

$$\lim_{N \to +\infty} \int_{\Gamma_N} \pi f(z) \cot(\pi z)\, dz = 0,$$

where Γ_N is defined previously.

(d) Use the residue theorem and parts (a) and (c) to derive (10).

15. Using the summation formula in Prob. 14 verify that

(a) $\displaystyle \sum_{k=-\infty}^{\infty} \frac{1}{k^2 + 1} = \pi \coth(\pi)$ [HINT: Take $f(z) = 1/(z^2 + 1)$.]

(b) $\displaystyle \sum_{k=-\infty}^{\infty} \frac{1}{\left(k - \frac{1}{2}\right)^2} = \pi^2$

(c) $\displaystyle \sum_{k=1}^{\infty} \frac{1}{k^2} = \frac{\pi^2}{6}$ [HINT: The formula in Prob. 14 needs to be modified to compensate for the pole of $f(z) = 1/z^2$ at $z = 0$.]

16. Show that for n a positive integer,

$$\sum_{k=1}^{\infty} \frac{1}{k^{2n}} = (-1)^{n-1} \pi^{2n} \frac{2^{2n-1}}{(2n)!} B_{2n},$$

where the constants B_{2n} are the *Bernoulli numbers*, which are defined by the power series expansion

$$\frac{z}{e^z - 1} = \sum_{k=0}^{\infty} \frac{B_k}{k!} z^k.$$

[Compare Prob. 15(c).] [HINT: To determine the required residue at $z = 0$ when $f(z) = 1/z^{2n}$, show that

$$\pi z \cot(\pi z) = \sum_{k=0}^{\infty} (-1)^k \frac{B_{2k}}{(2k)!} (2\pi z)^{2k} .]$$

17. Show that if a is real and noninteger and $0 < r < 1$,

(a) $\displaystyle\sum_{k=-\infty}^{\infty} \frac{1}{(k+a)^2} = \pi^2 \csc^2 \pi a$

(b) $\displaystyle\sum_{k=-\infty}^{\infty} \frac{1}{k^2 + a^2} = \frac{\pi}{a} \coth \pi a$

(c) $\displaystyle\sum_{k=-\infty}^{\infty} \frac{k^2 - a^2}{\left(k^2 + a^2\right)^2} = -\pi^2 \text{csch}^2 \pi a$

(d) $\displaystyle\sum_{k=-\infty}^{\infty} \frac{1}{(k-r)^2 + a^2} = \frac{\pi}{2a} \frac{\sinh 2a\pi}{\sin^2 \pi r + \sinh^2 \pi a}$

(e) $\displaystyle\sum_{k=-\infty}^{\infty} \frac{(k-r)^2 - a^2}{\left[(k-r)^2 + a^2\right]^2} = \frac{\pi^2}{2} \frac{1 - \cos 2\pi r \cosh 2\pi a}{\left(\sin^2 \pi r + \sinh^2 \pi a\right)^2}$

(f) For which *complex* values of a are the preceding identities valid?

18. To evaluate sums of the form $\sum_{k=-\infty}^{\infty} (-1)^k f(k)$ involving a sign alternation, we modify the approach of Prob. 14 by replacing $\pi f(z) \cot(\pi z)$ by $\pi f(z) \csc(\pi z)$. Again assuming that $f(z)$ is a rational function of the form P/Q, with $\deg Q \geq 2 + \deg P$ and that f has no poles at the integer points, derive the formula

$$\sum_{k=-\infty}^{\infty} (-1)^k f(k) = -\{\text{sum of residues of } \pi f(z) \csc(\pi z) \text{ at the poles of } f\}.$$

19. Use the formula of Prob. 18 to verify that

$$\sum_{k=1}^{\infty} \frac{(-1)^k}{k^2} = -\frac{\pi^2}{12}.$$

6.4 Improper Integrals Involving Trigonometric Functions

Our purpose in this section is to use residue theory to evaluate integrals of the general forms

$$\text{p.v.} \int_{-\infty}^{\infty} \frac{P(x)}{Q(x)} \cos mx \, dx, \qquad \text{p.v.} \int_{-\infty}^{\infty} \frac{P(x)}{Q(x)} \sin mx \, dx,$$

where m is real and $P(x)/Q(x)$ denotes a certain rational function continuous on $(-\infty, \infty)$. As we shall show in the following example, the semicircular contour technique of the previous section can be applied, but some modifications are necessary.

Example 1

Compute

$$I = \text{p.v.} \int_{-\infty}^{\infty} \frac{\cos 3x}{x^2 + 4}\, dx.$$

Solution.   In utilizing semicircular contours our first inclination is to deal with the complex function

$$\frac{\cos 3z}{z^2 + 4}. \tag{1}$$

However, with this choice for $f(z)$ we are doomed to failure because the modulus of (1) does not go to zero in either the upper or lower half-plane. Indeed, when $z = \pm \rho i$ we have

$$\left| \frac{\cos 3z}{z^2 + 4} \right| = \left| \frac{e^{i3z} + e^{-i3z}}{2(z^2 + 4)} \right| = \frac{e^{-3\rho} + e^{3\rho}}{2\left| -\rho^2 + 4 \right|}, \tag{2}$$

which becomes infinite as $\rho \to \infty$.

Nonetheless, (2) contains a clue as to how to circumvent this difficulty. The term $e^{i3z} = e^{i3x}e^{-3y}$ is bounded in the upper half-plane, and $e^{-i3z} = e^{-i3x}e^{3y}$ is bounded in the *lower* half-plane. So we write

$$I = I_1 + I_2 = \text{p.v.} \int_{-\infty}^{\infty} \frac{e^{i3x}}{2(x^2 + 4)}\, dx + \text{p.v.} \int_{-\infty}^{\infty} \frac{e^{-i3x}}{2(x^2 + 4)}\, dx \tag{3}$$

and close the contour for I_1 with semicircles in the upper half-plane, but use semicircles in the lower half-plane for I_2. Specifically, for I_1 we deal with the function

$$f_1(z) := \frac{e^{3iz}}{2(z^2 + 4)}.$$

We encounter singularities at $z = \pm 2i$, and because

$$|f_1(z)| = |f_1(x + iy)| = \frac{\left| e^{3ix} \cdot e^{-3y} \right|}{2\left| z^2 + 4 \right|} = \frac{e^{-3y}}{2\left| z^2 + 4 \right|},$$

we have in the *upper* half-plane ($y \geq 0$)

$$|f_1(z)| \leq \frac{1}{2\left| z^2 + 4 \right|}.$$

Thus for any $\rho > 2$, the integral over the upper half-circle C_ρ^+ in Fig. 6.7 is bounded by

$$\left| \int_{C_\rho^+} f_1(z)\, dz \right| \leq \frac{\pi \rho}{2(\rho^2 - 4)},$$

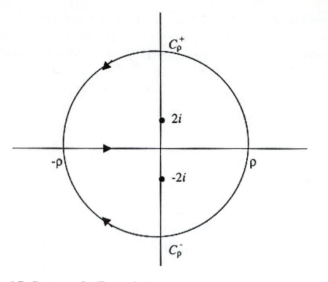

Figure 6.7 Contours for Example 1.

and so it goes to zero as $\rho \to \infty$. Furthermore, since $+2i$ is the only singularity in the upper half-plane, we have for $\rho > 2$

$$\int_{-\rho}^{\rho} f_1(x)\,dx + \int_{C_\rho^+} f(z)\,dz = 2\pi i\,\mathrm{Res}(f_1; 2i).$$

Hence on taking the limit as $\rho \to \infty$ we get

$$\mathrm{p.v.} \int_{-\infty}^{\infty} \frac{e^{3ix}}{2(x^2+4)}\,dx + 0 = 2\pi i\,\mathrm{Res}(f_1; 2i).$$

But

$$\mathrm{Res}(f_1; 2i) = \lim_{z\to 2i}(z-2i)f_1(z) = \lim_{z\to 2i}\frac{e^{3iz}}{2(z+2i)} = \frac{e^{-6}}{8i}.$$

Thus

$$I_1 = 2\pi i \cdot \frac{e^{-6}}{8i} = \frac{\pi}{4e^6}.$$

The integral I_2 is very similar: we employ the function

$$f_2(z) := \frac{e^{-3iz}}{2(z^2+4)},$$

whose singularities coincide with those of f_1, and whose integral over the *lower* half-circle C_ρ^- in Fig. 6.7 also satisfies

$$\left| \int_{C_\rho^-} f_2(z)\,dz \right| \le \frac{\pi\rho}{2(\rho^2-4)}.$$

However, the combination

$$\int_{-\rho}^{\rho} f_2(x)\, dx + \int_{C_\rho^-} f_2(z)\, dz$$

equals *minus* $2\pi i \operatorname{Res}(f_2; -2i)$, because the contour enclosing the singularity $-2i$ is *counterclockwise* in this case! Consequently, taking the limit as $\rho \to \infty$ we find

$$\text{p.v.} \int_{-\infty}^{\infty} \frac{e^{-3ix}}{2(x^2+4)}\, dx + 0 = -2\pi i \lim_{z \to -2i} (z+2i) f_2(z)$$

$$= -2\pi i \lim_{z \to -2i} \frac{e^{-3iz}}{2(z-2i)} = \frac{\pi}{4e^6}.$$

Thus $I = I_1 + I_2 = \pi/(2e^6).^\dagger$ ∎

The technique of Example 1 can be used to evaluate any integral of the form

$$\text{p.v.} \int_{-\infty}^{\infty} e^{imx} \frac{P(x)}{Q(x)}\, dx \qquad (m > 0), \tag{4}$$

where P and Q are polynomials, Q has no real zeros, and the degree of Q exceeds that of P by at least 2. Indeed, the function $e^{imz} P(z)/Q(z)$ has only a finite number of singularities and, for large ρ, its integral over C_ρ^+ is bounded by $1 \cdot (K/\rho^2) \cdot \pi\rho$, which goes to zero as $\rho \to \infty$. Of course, if $m < 0$ we simply use C_ρ^-. However, in applications it is sometimes necessary to evaluate integrals such as (4) where the degree of Q is just *one* higher than that of P. For example, consider

$$\text{p.v.} \int_{-\infty}^{\infty} \frac{e^{ix} x}{1+x^2}\, dx. \tag{5}$$

If we estimate the integral of $e^{iz} z/(1+z^2)$ over C_ρ^+ as before, we find that it is bounded by $1 \cdot (K/\rho) \cdot \pi\rho = K\pi$, for some constant K. Since this does not go to zero, it is by no means obvious that the semicircular contour method will work in evaluating (5).

Surprisingly, it turns out that the integrals of such a function over C_ρ^+ *do* go to zero as $\rho \to \infty$; however, a much finer integral estimate is needed to show this. *Jordan's lemma* fills this need. First, we establish a rather obvious inequality.

Lemma 2. Suppose that $f(t)$ and $M(t)$ are continuous functions on the real interval $a \le t \le b$, with f complex and M real-valued. If $|f(t)| \le M(t)$ on this interval then

$$\left| \int_a^b f(t)\, dt \right| \le \int_a^b M(t)\, dt. \tag{6}$$

---

† See remark at end of this section for an alternative approach to computing I.

Proof.   For a careful proof of (6) we must revert to the integration theory presented in Chapter 4. Choose an arbitrary subdivision of the interval $[a, b]$, say,

$$a = \tau_0 < \tau_1 < \cdots < \tau_n = b,$$

and form a Riemann sum for f:

$$\sum_{k=1}^{n} f(c_k) \Delta \tau_k.$$

Since $|f(t)| \le M(t)$ for all t, we have by the triangle inequality

$$\left| \sum_{k=1}^{n} f(c_k) \Delta \tau_k \right| \le \sum_{k=1}^{n} M(c_k) \Delta \tau_k,$$

and we note that the right-hand side is a Riemann sum for $M(t)$ over $[a, b]$. Because the inequality holds for every partition of $[a, b]$, inequality (6) follows by letting the mesh tend to zero.   ■

As an immediate consequence of this lemma, we deduce that for any complex-valued continuous function $f(t)$,

$$\left| \int_a^b f(t) \, dt \right| \le \int_a^b |f(t)| \, dt. \tag{7}$$

Lemma 3 (Jordan's Lemma). If $m > 0$ and P/Q is the quotient of two polynomials such that

$$\text{degree } Q \ge 1 + \text{degree } P, \tag{8}$$

then

$$\lim_{\rho \to +\infty} \int_{C_\rho^+} e^{imz} \frac{P(z)}{Q(z)} \, dz = 0, \tag{9}$$

where C_ρ^+ is the upper half-circle of radius ρ.

Proof.   Parametrizing C_ρ^+ we have

$$\int_{C_\rho^+} e^{imz} \frac{P(z)}{Q(z)} \, dz = \int_0^{\pi} g(t) \, dt,$$

where

$$g(t) := e^{im(\rho e^{it})} \frac{P\left(\rho e^{it}\right)}{Q\left(\rho e^{it}\right)} \rho i e^{it}.$$

Now

$$\left| e^{im(\rho e^{it})} \right| = \left| e^{im\rho \cos t - m\rho \sin t} \right| = e^{-m\rho \sin t}.$$

Furthermore, from (8) we know that there is some constant K such that

$$\left| \frac{P\left(\rho e^{it}\right)}{Q\left(\rho e^{it}\right)} \right| \leq \frac{K}{\rho} \qquad \text{(for ρ large)}.$$

Thus

$$|g(t)| \leq e^{-m\rho \sin t} \cdot \frac{K}{\rho} \cdot \rho = K e^{-m\rho \sin t},$$

and so, by Lemma 2,

$$\left| \int_{C_\rho^+} e^{imz} \frac{P(z)}{Q(z)}\, dz \right| = \left| \int_0^\pi g(t)\, dt \right| \leq K \int_0^\pi e^{-m\rho \sin t}\, dt. \qquad (10)$$

To estimate the right-hand integral we first note that the function $e^{-m\rho \sin t}$ on $[0, \pi]$ is symmetric about $t = \pi/2$. Consequently,

$$\int_0^\pi e^{-m\rho \sin t}\, dt = 2 \int_0^{\pi/2} e^{-m\rho \sin t}\, dt.$$

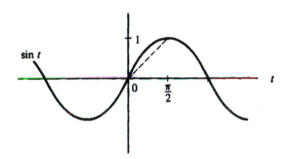

Figure 6.8 Graph of $\sin t$.

Furthermore, if we consider the graph of $\sin t$, we notice that it is concave downward on $[0, \pi/2]$; thus it lies above the dashed line in Fig. 6.8. In other words,

$$\sin t \geq \frac{2}{\pi} t, \qquad \text{for } 0 \leq t \leq \frac{\pi}{2}.$$

When $m\rho > 0$, this last inequality implies that

$$e^{-m\rho \sin t} \leq e^{-m\rho 2t/\pi} \qquad \text{on} \quad [0, \pi/2],$$

and so

$$\int_0^\pi e^{-m\rho \sin t}\, dt = 2 \int_0^{\pi/2} e^{-m\rho \sin t}\, dt \leq 2 \int_0^{\pi/2} e^{-m\rho 2t/\pi}\, dt$$

$$= 2 \left(\frac{-\pi}{m\rho 2} \right) \left[e^{-m\rho} - 1 \right] < \frac{\pi}{m\rho}.$$

Hence from (10) we see that

$$\left| \int_{C_\rho^+} e^{imz} \frac{P(z)}{Q(z)} \, dz \right| \le K \cdot \frac{\pi}{m\rho};$$

consequently, the integral goes to zero as $\rho \to \infty$. ∎

Clearly when m is *negative* the function $e^{imz} P(z)/Q(z)$ is not bounded in the upper half-plane, but under assumption (8),

$$\lim_{\rho \to \infty} \int_{C_\rho^-} e^{imz} \frac{P(z)}{Q(z)} \, dz = 0 \qquad (m < 0),$$

where C_ρ^- is the lower half-circle of radius ρ. To prove this result we need only make the change of variable $w = -z$ in (9).

Example 2
Evaluate

$$\text{p.v.} \int_{-\infty}^{\infty} \frac{x \sin x}{1 + x^2} \, dx.$$

Solution.   Since $\sin x = (e^{ix} - e^{-ix})/(2i)$, we first compute the integral

$$I_1 := \text{p.v.} \frac{1}{2i} \int_{-\infty}^{\infty} \frac{x e^{ix}}{1 + x^2} \, dx.$$

Because the hypotheses of Jordan's lemma are satisfied when $m = 1$ and $P/Q = x/(1+x^2)$, the integral I_1 equals $2\pi i$ times the sum of the residues of $ze^{iz}/[2i(1+z^2)]$ in the upper half-plane. Writing

$$\frac{ze^{iz}}{2i(1 + z^2)} = \frac{ze^{iz}}{2i(z + i)(z - i)},$$

we find for the residue at $z = +i$ the value

$$\lim_{z \to i} \frac{ze^{iz}}{2i(z + i)} = \frac{ie^{i^2}}{2i(i + i)} = \frac{-ie^{-1}}{4}.$$

Consequently,

$$I_1 = 2\pi i \cdot \frac{-ie^{-1}}{4} = \frac{\pi e^{-1}}{2}.$$

By similar reasoning we compute

$$I_2 := -\text{p.v.} \frac{1}{2i} \int_{-\infty}^{\infty} \frac{x e^{-ix}}{1 + x^2} \, dx,$$

closing the contour in the lower half-plane. We have

$$I_2 = (-2\pi i) \, \text{Res} \left(\frac{-ze^{-iz}}{2i(z + i)(z - i)}; -i \right) = -2\pi i \frac{ie^{(-i)^2}}{2i(-i - i)} = \frac{\pi e^{-1}}{2}.$$

Consequently, $I = I_1 + I_2 = \pi/e$. ∎

Example 3

Evaluate

$$I = \text{p.v.} \int_{-\infty}^{\infty} \frac{\cos x}{x+i} \, dx.$$

Solution. The substitution

$$\cos x = \frac{e^{ix} + e^{-ix}}{2}$$

leads to the representation

$$I = \text{p.v.} \frac{1}{2} \int_{-\infty}^{\infty} \frac{e^{ix}}{x+i} \, dx + \text{p.v.} \frac{1}{2} \int_{-\infty}^{\infty} \frac{e^{-ix}}{x+i} \, dx, \tag{11}$$

and we deal with each integral separately.

For

$$I_1 := \text{p.v.} \frac{1}{2} \int_{-\infty}^{\infty} \frac{e^{ix}}{x+i} \, dx$$

we close the contour $[-\rho, \rho]$ with the half-circle C_ρ^+ in the upper half-plane. Then, by Jordan's lemma,

$$\lim_{\rho \to \infty} \frac{1}{2} \int_{C_\rho^+} \frac{e^{iz}}{z+i} \, dz = 0,$$

and since the only singularity of the integrand is in the *lower* half-plane at $z = -i$, we deduce that $I_1 = 0$.

The second integral

$$I_2 := \text{p.v.} \frac{1}{2} \int_{-\infty}^{\infty} \frac{e^{-ix}}{x+i} \, dx$$

requires us to close the contour $[-\rho, \rho]$ in the lower half-plane with the semicircle C_ρ^-, enclosing the singularity at $-i$. We obtain

$$I_2 = -2\pi i \, \text{Res} \left(\frac{e^{-iz}}{2(z+i)}; -i \right)$$

$$= \frac{-2\pi i}{2} \lim_{z \to -i} e^{-iz} = -i\pi e^{-1}.$$

Consequently, $I = I_1 + I_2 = -i\pi/e$. ∎

An alternate, slightly quicker, procedure for Examples 1 and 2 is based on recognizing that the integrals therein are expressible as

$$\text{p.v.} \int_{-\infty}^{\infty} \frac{\cos 3x}{x^2+4} \, dx = \text{Re p.v.} \int_{-\infty}^{\infty} \frac{e^{i3x}}{x^2+4} \, dx$$

and

$$\text{p.v.} \int_{-\infty}^{\infty} \frac{x \sin x}{1+x^2} \, dx = \text{Im p.v.} \int_{-\infty}^{\infty} \frac{x e^{ix}}{1+x^2} \, dx,$$

respectively. Then we only have to perform *one* evaluation of residues, in the upper half-plane, and take the real or imaginary parts at the end. However, this shortcut is not valid for Example 3 since

$$\text{p.v.} \int_{-\infty}^{\infty} \frac{\cos x}{x + i} \, dx \neq \text{Re p.v.} \int_{-\infty}^{\infty} \frac{e^{ix}}{x + i} \, dx \ .$$

In fact the left-hand member is pure imaginary, as we have seen.

EXERCISES 6.4

Using the method of residues, verify the integral formulas in Problems 1–3.

1. p.v. $\displaystyle\int_{-\infty}^{\infty} \frac{\cos(2x)}{x^2 + 1} \, dx = \frac{\pi}{e^2}$

2. p.v. $\displaystyle\int_{-\infty}^{\infty} \frac{x \sin x}{x^2 - 2x + 10} \, dx = \frac{\pi}{3e^3}(3\cos 1 + \sin 1)$

3. $\displaystyle\int_{0}^{\infty} \frac{\cos x}{\left(x^2 + 1\right)^2} \, dx = \frac{\pi}{2e}$

Compute each of the integrals in Problems 4–9.

4. p.v. $\displaystyle\int_{-\infty}^{\infty} \frac{e^{3ix}}{x - 2i} \, dx$

5. p.v. $\displaystyle\int_{-\infty}^{\infty} \frac{x \sin(3x)}{x^4 + 4} \, dx$

6. p.v. $\displaystyle\int_{-\infty}^{\infty} \frac{e^{-2ix}}{x^2 + 4} \, dx$

7. p.v. $\displaystyle\int_{-\infty}^{\infty} \frac{\cos x}{\left(x^2 + 1\right)\left(x^2 + 4\right)} \, dx$

8. $\displaystyle\int_{0}^{\infty} \frac{x^3 \sin(2x)}{\left(x^2 + 1\right)^2} \, dx$

9. p.v. $\displaystyle\int_{-\infty}^{\infty} \frac{\cos(2x)}{x - 3i} \, dx$

10. Derive the formula

$$\text{p.v.} \int_{-\infty}^{\infty} \frac{\cos x}{x - w} \, dx = \begin{cases} \pi i e^{iw} & \text{if } \operatorname{Im} w > 0, \\ -\pi i e^{-iw} & \text{if } \operatorname{Im} w < 0. \end{cases}$$

11. Give conditions under which the following formula is valid:

$$\text{p.v.} \int_{-\infty}^{\infty} e^{imx} \frac{P(x)}{Q(x)} \, dx$$

$$= 2\pi i \cdot \sum \left[\text{residues of } e^{imz} P(z)/Q(z) \text{ at poles in the upper half-plane} \right].$$

12. Given that $\int_0^\infty e^{-x^2} \, dx = \sqrt{\pi}/2$, integrate e^{iz^2} around the boundary of the circular sector $S_\rho : \{z = re^{i\theta} : 0 \le \theta \le \pi/4, 0 \le r \le \rho\}$, and let $\rho \to +\infty$ to prove that

$$\int_0^\infty e^{ix^2} \, dx = \frac{\sqrt{2\pi}}{4}(1 + i).$$

6.5   Indented Contours

In the preceding sections the integrands f were assumed to be defined and continuous over the whole interval of integration. We turn now to the problem of evaluating special integrals where $|f(x)| \to \infty$ as x approaches certain finite points. Our first step is to give precise meaning to the integrals of f.

Let $f(x)$ be continuous on $[a, b]$ except at the point c, $a < c < b$. Then the *improper integrals* of f over the intervals $[a, c]$, $[c, b]$, and $[a, b]$ are defined by

$$\int_a^c f(x) \, dx := \lim_{r \to 0^+} \int_a^{c-r} f(x) \, dx,$$

$$\int_c^b f(x) \, dx := \lim_{s \to 0^+} \int_{c+s}^b f(x) \, dx,$$

and

$$\int_a^b f(x) \, dx := \lim_{r \to 0^+} \int_a^{c-r} f(x) \, dx + \lim_{s \to 0^+} \int_{c+s}^b f(x) \, dx, \tag{1}$$

provided the appropriate limit(s) exists. For example,

$$\int_0^1 \frac{1}{\sqrt{x}} \, dx = \lim_{s \to 0^+} \int_s^1 \frac{1}{\sqrt{x}} = \lim_{s \to 0^+} 2\sqrt{x} \Big|_s^1$$

$$= \lim_{s \to 0^+} \left[2 - 2\sqrt{s} \right] = 2,$$

and therefore one can say that the area under the graph in Fig. 6.9 is *finite*, despite the vertical asymptote.

On the other hand, the areas on either side of the vertical asymptote in the graph of $f(x) = 1/(x - 2)$, depicted in Fig. 6.10, are both infinite, because

$$\int_{2+s}^4 \frac{dx}{x - 2} = \text{Log} |x - 2| \Big|_{x=2+s}^{x=4} = \text{Log } 2 - \text{Log } s \to \infty \quad \text{as} \quad s \to 0^+,$$

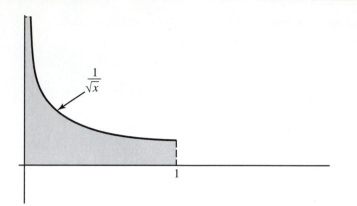

Figure 6.9 Graph of $1/\sqrt{x}$.

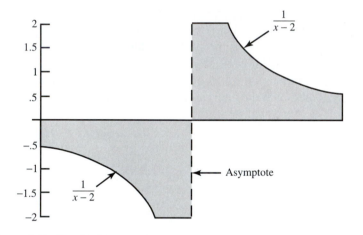

Figure 6.10 Graph of $1/(x-2)$.

and

$$\int_1^{2-r} \frac{dx}{x-2} = \text{Log}\,|x-2|\,\Big|_{x=1}^{x=2-r} = \text{Log}\,r - \text{Log}\,1 \to -\infty \quad \text{as} \quad r \to 0^+.$$

However if we close in on the asymptote from either side *symmetrically*, with $r = s$, the infinities "cancel each other" in the sense that

$$\int_1^{2-r} \frac{dx}{x-2} + \int_{2+r}^4 \frac{dx}{x-2} = \text{Log}\,|x-2|\,\Big|_{x=1}^{x=2-r} + \text{Log}\,|x-2|\,\Big|_{x=2+r}^{x=4}$$
$$= \text{Log}\,r - \text{Log}\,1 + \text{Log}\,2 - \text{Log}\,r$$
$$= \text{Log}\,2.$$

To take advantage of this again we adopt the principal value notation for improper integrals when the limits in (1) are taken symmetrically:

$$\text{p.v.} \int_a^b f(x)\,dx := \lim_{r\to 0^+} \left\{ \int_a^{c-r} f(x)\,dx + \int_{c+r}^b f(x)\,dx \right\}.$$

The improper integral $\int_1^4 dx/(x-2)$ does not exist, but its principal value is Log 2.

When the function $f(x)$ is continuous on the whole real line except at c, the principal value of its integral over $(-\infty, \infty)$ is defined by

$$\text{p.v.} \int_{-\infty}^\infty f(x)\,dx := \lim_{\substack{\rho\to\infty \\ r\to 0^+}} \left[\int_{-\rho}^{c-r} f(x)\,dx + \int_{c+r}^\rho f(x)\,dx \right], \tag{2}$$

provided the limit exists as $\rho \to \infty$ and $r \to 0^+$ independently.[†] In the case of several discontinuities occurring at points $x = c_i$ we extend the definition of the p.v. integral over $(-\infty, \infty)$ in a natural way; namely, we remove a small symmetric interval $(c_i - r_i, c_i + r_i)$ about each c_i and then take the limit of the integral as the variables $r_i \to 0^+$ and $\rho \to \infty$ independently. (See Fig. 6.11.)

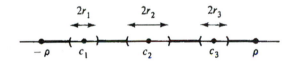

Figure 6.11 Contour for p.v. integrals.

Residue theory is useful in evaluating certain integrals of the form (2) when the integrand, considered as a function of z, has a simple pole at the exceptional point c. Assuming this to be the case, we must consider the integrals of f along $[-\rho, c-r]$ and $[c+r, \rho]$, and to utilize residue theory we must form some closed contour that contains these segments. In the last two sections we discussed suitable ways to join $+\rho$ to $-\rho$. But now we also need to join $c-r$ to $c+r$. In so doing we cannot proceed along the real axis, for such a segment would pass through the singularity at c. Instead, we detour around c by forming, for example, the half-circle S_r indicated in Fig. 6.12. Eventually we will let r tend to zero, and so it will be necessary to determine the limit

$$\lim_{r\to 0^+} \int_{S_r} f(z)\,dz.$$

This is handled by the following lemma, which deals not only with half-circles, but with arbitrary circular arcs.

[†]More precisely, we say that the limit in (2) exists and equals L if for every $\varepsilon > 0$ there exist positive constants M and δ such that for $\rho > M$ and $0 < r < \delta$ the bracketed expression in (2) is within ε of L.

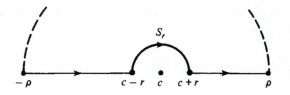

Figure 6.12 Contour with detour.

Lemma 4. If f has a simple pole at $z = c$ and T_r is the circular arc of Fig. 6.13 defined by

$$T_r : \quad z = c + re^{i\theta} \qquad (\theta_1 \le \theta \le \theta_2), \tag{3}$$

then

$$\lim_{r \to 0^+} \int_{T_r} f(z)\, dz = i\,(\theta_2 - \theta_1)\,\mathrm{Res}(f; c). \tag{4}$$

Consequently, for the clockwise oriented half-circle S_r of Fig. 6.12 we have

$$\lim_{r \to 0^+} \int_{S_r} f(z)\, dz = -i\pi\,\mathrm{Res}(f; c). \tag{5}$$

Proof. Since f has a simple pole at c, its Laurent expansion has the form

$$f(z) = \frac{a_{-1}}{z - c} + \sum_{k=0}^{\infty} a_k (z - c)^k,$$

valid in some punctured neighborhood of c, say $0 < |z - c| < R$. Thus if $0 < r < R$, we can write

$$\int_{T_r} f(z)\, dz = a_{-1} \int_{T_r} \frac{dz}{z - c} + \int_{T_r} g(z)\, dz, \tag{6}$$

where

$$g(z) := \sum_{k=0}^{\infty} a_k (z - c)^k.$$

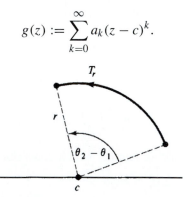

Figure 6.13 Circular arc of Lemma 4.

Now g is analytic at c (why?) and hence is bounded in some neighborhood of this point; that is,

$$|g(z)| \le M, \qquad \text{for } |z - c| < R_1.$$

Consequently, for $0 < r < R_1$, we have

$$\left| \int_{T_r} g(z)\,dz \right| \le M\ell\,(T_r) = M\,(\theta_2 - \theta_1)\,r,$$

and the last term goes to zero as $r \to 0^+$. Therefore,

$$\lim_{r \to 0^+} \int_{T_r} g(z)\,dz = 0.$$

To deal with the integral of $1/(z - c)$ we use the parametrization (3) to derive

$$\int_{T_r} \frac{dz}{z - c} = \int_{\theta_1}^{\theta_2} \frac{1}{re^{i\theta}}\,rie^{i\theta}\,d\theta = i \int_{\theta_1}^{\theta_2} d\theta = i\,(\theta_2 - \theta_1),$$

the value being independent of r. Hence from (6) we obtain

$$\lim_{r \to 0^+} \int_{T_r} f(z)\,dz = a_{-1}i\,(\theta_2 - \theta_1) + 0 = \text{Res}(f; c)\,i\,(\theta_2 - \theta_1),$$

which is the desired limit (4).

In particular, when the T_r are counterclockwise-oriented *half*-circles, we get the limiting value $i\pi\,\text{Res}(f; c)$, and thus for the oppositely oriented half-circles S_r in Fig. 6.12 we get minus this value.   ∎

Example 1

Evaluate

$$I = \text{p.v.} \int_{-\infty}^{\infty} \frac{e^{ix}}{x}\,dx.$$

Solution.   First notice that the integrand is continuous except at $x = 0$. Hence

$$I = \lim_{\substack{\rho \to \infty \\ r \to 0^+}} \left(\int_{-\rho}^{-r} \frac{e^{ix}}{x}\,dx + \int_{r}^{\rho} \frac{e^{ix}}{x}\,dx \right).$$

Now we introduce the complex function

$$f(z) := \frac{e^{iz}}{z},$$

which has a simple pole at the origin but is analytic elsewhere. Next we must form a closed contour containing the segments $[-\rho, -r]$ and $[r, \rho]$. Observing that Jordan's lemma applies to $f(z)$ we join $+\rho$ to $-\rho$ by the half-circle C_ρ^+ in the upper half-plane. In joining $-r$ to r we indent around the origin by using a half-circle S_r. This yields

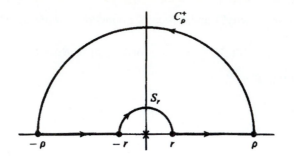

Figure 6.14 Contour for Example 1.

the closed contour of Fig. 6.14. Now since $e^{iz}z$ has no singularities inside the closed contour, we have

$$\left(\int_{-\rho}^{-r} + \int_{S_r} + \int_{r}^{\rho} + \int_{C_\rho^+} \right) \frac{e^{iz}}{z}\, dz = 0;$$

that is,

$$\int_{-\rho}^{-r} \frac{e^{ix}}{x}\, dx + \int_{r}^{\rho} \frac{e^{ix}}{x}\, dx = -\int_{S_r} \frac{e^{iz}}{z}\, dz - \int_{C_\rho^+} \frac{e^{iz}}{z}\, dz. \tag{7}$$

By Jordan's lemma,

$$\lim_{\rho \to \infty} \int_{C_\rho^+} \frac{e^{iz}}{z}\, dz = 0,$$

and by Eq. (5) of Lemma 4

$$\lim_{r \to 0^+} \int_{S_r} \frac{e^{iz}}{z}\, dz = -i\pi \, \mathrm{Res}(0)$$

$$= -i\pi \lim_{z \to 0} z \cdot \frac{e^{iz}}{z} = -i\pi.$$

Thus from Eq. (7) we obtain

$$\text{p.v.} \int_{-\infty}^{\infty} \frac{e^{ix}}{x}\, dx = -(-i\pi) - 0 = i\pi. \qquad \blacksquare \tag{8}$$

Example 2

Find

$$\int_{0}^{\infty} \frac{\sin x}{x}\, dx = \lim_{\substack{\rho \to \infty \\ r \to 0^+}} \int_{r}^{\rho} \frac{\sin x}{x}\, dx.$$

Solution. Observe that the integrand $g(x) := (\sin x)/x$ is an even function of x; that is, $g(-x) = g(x)$ for all x. Hence

$$2 \int_{0}^{\infty} \frac{\sin x}{x}\, dx = \text{p.v.} \int_{-\infty}^{\infty} \frac{\sin x}{x}\, dx.$$

Furthermore, the right-hand integral is the imaginary part of the integral of e^{ix}/x over $(-\infty, \infty)$ and so, by Example 1, it equals $\text{Im}(i\pi) = \pi$. Thus

$$\int_0^\infty \frac{\sin x}{x}\, dx = \frac{\pi}{2}. \quad \blacksquare$$

We remark that as another consequence of Example 1 we have

$$\text{p.v.} \int_{-\infty}^\infty \frac{\cos x}{x}\, dx = \text{Re}(i\pi) = 0,$$

but this is scarcely surprising because the integrand $h(x) := (\cos x)/x$ is an odd function of x; that is, $h(-x) = -h(x)$ for all x.

Example 3

Compute

$$\text{p.v.} \int_{-\infty}^\infty \frac{x e^{2ix}}{x^2 - 1}\, dx.$$

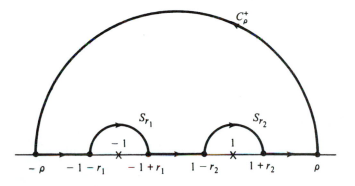

Figure 6.15 Contour for Example 3.

Solution.   Here the integrand is discontinuous at two real points, $x = \pm 1$. Thus we need to find

$$\lim_{\substack{\rho \to \infty \\ r_1, r_2 \to 0^+}} \left(\int_{-\rho}^{-1-r_1} + \int_{-1+r_1}^{1-r_2} + \int_{1+r_2}^{\rho} \right) \frac{x e^{2ix}}{x^2 - 1}\, dx.$$

For this purpose we work with

$$f(z) := \frac{z e^{2iz}}{z^2 - 1}.$$

and indent around each of its simple poles, as indicated in Fig. 6.15. Then since $f(z)$ is analytic inside the closed contour, we obtain

$$\left(\int_{-\rho}^{-1-r_1} + \int_{-1+r_1}^{1-r_2} + \int_{1+r_2}^{\rho} \right) \frac{xe^{2ix}}{x^2-1}\,dx + J_{r_1} + J_{r_2} + J_\rho = 0, \qquad (9)$$

where J_{r_1}, J_{r_2}, J_ρ are the integrals of $f(z)$ over S_{r_1}, S_{r_2}, C_ρ^+, respectively. Now by Jordan's lemma we have

$$\lim_{\rho \to \infty} J_\rho = 0,$$

and from Eq. (5) of Lemma 4,

$$\lim_{r_1 \to 0^+} J_{r_1} = -i\pi \operatorname{Res}(-1) = -i\pi \lim_{z \to -1} (z+1)f(z)$$

$$= -i\pi \lim_{z \to -1} \frac{ze^{2iz}}{z-1} = \frac{-i\pi e^{-2i}}{2},$$

and

$$\lim_{r_2 \to 0^+} J_{r_2} = -i\pi \operatorname{Res}(1) = -i\pi \lim_{z \to 1} (z-1)f(z)$$

$$= -i\pi \lim_{z \to 1} \frac{ze^{2iz}}{z+1} = \frac{-i\pi e^{2i}}{2}.$$

Hence on taking the limits in Eq. (9) we get

$$\text{p.v.} \int_{-\infty}^{\infty} \frac{xe^{2ix}}{x^2-1}\,dx = \frac{i\pi e^{-2i}}{2} + \frac{i\pi e^{2i}}{2} - 0 = i\pi \cos 2. \quad \blacksquare$$

EXERCISES 6.5

1. Compute each of the following limits along the given circular arcs.

   (a) $\lim_{r \to 0^+} \int_{T_r} \dfrac{2z^2+1}{z}\,dz$, where $T_r : z = re^{i\theta}, 0 \le \theta \le \dfrac{\pi}{2}$

   (b) $\lim_{r \to 0^+} \int_{\Gamma_r} \dfrac{e^{3iz}}{z^2-1}\,dz$, where $\Gamma_r : z = 1 + re^{i\theta}, \dfrac{\pi}{4} \le \theta \le \pi$

   (c) $\lim_{r \to 0^+} \int_{\gamma_r} \dfrac{\operatorname{Log} z}{z-1}\,dz$, where $\gamma_r : z = 1 + re^{-i\theta}, \pi \le \theta \le 2\pi$

   (d) $\lim_{r \to 0^+} \int_{S_r} \dfrac{e^z-1}{z^2}\,dz$, where $S_r : z = re^{-i\theta}, \pi \le \theta \le 2\pi$

Using the technique of residues, verify each of the integral formulas in Problems 2–8.

2. p.v. $\displaystyle\int_{-\infty}^{\infty} \frac{e^{2ix}}{x+1}\,dx = \pi i e^{-2i}$

3. p.v. $\displaystyle\int_{-\infty}^{\infty} \frac{e^{ix}}{(x-1)(x-2)}\, dx = \pi i\left(e^{2i} - e^i\right)$

4. $\displaystyle\int_0^{\infty} \frac{\sin(2x)}{x\left(x^2+1\right)^2}\, dx = \pi\left(\frac{1}{2} - \frac{1}{e^2}\right)$

5. $\displaystyle\int_0^{\infty} \frac{\cos x - 1}{x^2}\, dx = -\frac{\pi}{2}$

6. p.v. $\displaystyle\int_{-\infty}^{\infty} \frac{\sin x}{\left(x^2+4\right)(x-1)}\, dx = \frac{\pi}{5}\left[\cos(1) - e^{-2}\right]$

7. p.v. $\displaystyle\int_{-\infty}^{\infty} \frac{x\cos x}{x^2 - 3x + 2}\, dx = \pi\left[\sin(1) - 2\sin(2)\right]$

8. p.v. $\displaystyle\int_{-\infty}^{\infty} \frac{\cos(2x)}{x^3+1}\, dx = \frac{\pi}{3}e^{-\sqrt{3}}\left[\sin(1) + \sqrt{3}\cos(1)\right] + \frac{\pi\sin(2)}{3}$

9. Compute p.v. $\displaystyle\int_{-\infty}^{\infty} \frac{\sin^3 x}{x^3}\, dx.$ $\left[\text{HINT: } \sin^3 x = \text{Im}\left(\frac{3e^{ix}}{4} - \frac{e^{3ix}}{4} - \frac{1}{2}\right).\right]$

10. Verify that

$$\int_0^{\infty} \frac{\sin^2 x}{x^2}\, dx = \frac{\pi}{2}.$$

[HINT: $\sin^2 x = \frac{1}{2}(1 - \cos 2x) = \frac{1}{2}\,\text{Re}\left(1 - e^{2ix}\right).$]

11. Compute p.v. $\displaystyle\int_{-\infty}^{\infty} \frac{e^{ax}}{e^x - 1}\, dx$ for $0 < a < 1$. [HINT: Indent the contour of Fig. 6.6 around the points $z = 0$ and $z = 2\pi i$.]

12. Verify that for $a > 0$ and $b > 0$

$$\int_0^{\infty} \frac{\sin(ax)}{x\left(x^2+b^2\right)}\, dx = \frac{\pi}{2b^2}\left(1 - e^{-ab}\right).$$

6.6 Integrals Involving Multiple-Valued Functions

In attempting to apply residue theory to compute an integral of $f(x)$, it may turn out that the complex function $f(z)$ is multiple-valued. If this happens, we need to modify our procedure by taking into account not only isolated singularities but also branch points and branch cuts. In fact we may find it necessary to integrate along a branch cut, so we turn first to a discussion of this technique.

To be specific, let α denote a real number, but not an integer, and let $f(z)$ be the branch of z^α obtained by restricting the argument of z to lie between 0 and 2π; that is,

$$f(z) = e^{\alpha(\text{Log}\, r + i\theta)}, \qquad \text{where } z = re^{i\theta},\ 0 < \theta < 2\pi. \tag{1}$$

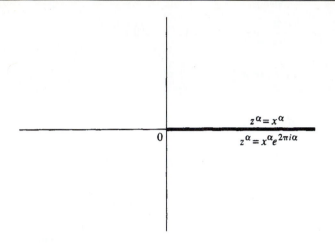

Figure 6.16 Branch cut for z^α.

As shown in Chapter 3 this function is analytic in the plane except along its branch cut, the nonnegative real axis. (See Fig. 6.16.) In fact, as z approaches a point x (> 0) on the cut from the upper half-plane, θ goes to zero, and

$$f(z) \to e^{\alpha \operatorname{Log} x} = x^\alpha \tag{2}$$

(x^α being the principal value as in calculus); while if z approaches x from the lower half-plane, θ tends to 2π, and so

$$f(z) \to e^{\alpha(\operatorname{Log} x + i2\pi)} = x^\alpha \cdot e^{2\pi i \alpha}. \tag{3}$$

In this sense we visualize f as being equal to x^α on the "upper side" of the cut and being equal to $x^\alpha \cdot e^{2\pi i \alpha}$ on the "lower side."

Now if we are to integrate $f(z)$ along the cut, we avoid ambiguity by placing a direction arrow either above or below the cut to indicate which values of f are to be used. For example, the integrals of f along the segments γ_1 and γ_2 of Fig. 6.17 are given by

$$\int_{\gamma_1} f(z)\, dz = \int_\varepsilon^\rho x^\alpha\, dx,$$

$$\int_{\gamma_2} f(z)\, dz = -\int_\varepsilon^\rho x^\alpha \cdot e^{2\pi i \alpha}\, dx = -e^{2\pi i \alpha} \int_{\gamma_1} f(z)\, dz. \tag{4}$$

These same remarks apply to arbitrary functions with branch cuts—the values of the functions on each side of the cut being determined by continuity from that side. We shall now illustrate how the residue theorem applies to such functions.

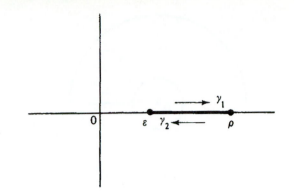

Figure 6.17 Integration along a branch cut.

Example 1

Let α be a real number (not an integer) and let $R(z)$ be a rational function having no poles on the closed contour Γ of Fig. 6.18. Prove that if $f(z)$ is the branch of $z^\alpha R(z)$ obtained by restricting the argument of z to be between 0 and 2π, that is,

$$f(z) = \begin{cases} e^{\alpha(\text{Log}\, r + i\theta)} \cdot R\left(re^{i\theta}\right), & \text{for } z = re^{i\theta},\, 0 < \theta < 2\pi, \\ x^\alpha R(x), & \text{for } z = x \text{ when integrating on } \gamma_1, \\ x^\alpha e^{2\pi i \alpha} R(x), & \text{for } z = x \text{ when integrating on } \gamma_2, \end{cases}$$

then

$$\int_\Gamma f(z)\, dz = 2\pi i \cdot \sum [\text{residues of } f(z) \text{ at the poles inside } \Gamma].^\dagger \qquad (5)$$

Solution.   Notice that the residue theorem cannot be directly applied here because the integrand is multiple-valued on the portion $[\varepsilon, \rho]$ of the branch cut. (Notice also that the integrals along γ_1 and γ_2 do *not* cancel here.) To circumvent this difficulty we introduce a segment, indicated by the dashed line in Fig. 6.19, which joins the inner circle to the outer one and does not pass through any poles of $R(z)$. This creates two positively oriented closed contours Γ_1 and Γ_2 (as shown in Fig. 6.19) such that

$$\int_\Gamma f(z)\, dz = \int_{\Gamma_1} f(z)\, dz + \int_{\Gamma_2} f(z)\, dz \qquad (6)$$

(the integrals along the dashed segment cancel). Now on Γ_1 we *can* apply the residue theorem because $f(z)$ agrees with a function analytic on this contour; indeed, for z on or inside Γ_1

$$f(z) = e^{\alpha \text{Log}\, z} R(z)$$

---

† To be precise, by the *inside* of Γ we mean the set of those points between the two circles but not lying on the segment $[\varepsilon, \rho]$.

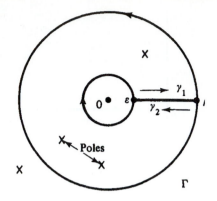

Figure 6.18 Contour for Example 1.

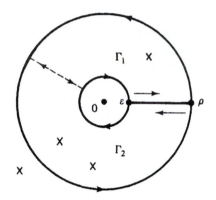

Figure 6.19 Modification of Fig. 6.18.

(the *principal branch* Log z having its branch cut along the negative real axis). Hence

$$\int_{\Gamma_1} f(z)\, dz = 2\pi i \cdot \sum \text{(residues of } f \text{ at poles inside } \Gamma_1).\qquad (7)$$

Similarly, on Γ_2 the function $f(z)$ again agrees with an analytic function; for instance, for $z = re^{i\theta}$ on Γ_2

$$f(z) = e^{\alpha(\operatorname{Log} r + i\theta)} R\left(re^{i\theta}\right), \qquad \frac{\pi}{2} < \theta < \frac{5\pi}{2}$$

(with cut along the positive imaginary axis). Consequently

$$\int_{\Gamma_2} f(z)\, dz = 2\pi i \cdot \sum \text{(residues of } f \text{ at poles inside } \Gamma_2).\qquad (8)$$

Therefore, since every pole inside Γ lies either inside Γ_1 or inside Γ_2, adding Eqs. (7) and (8) gives the desired equation (5). ∎

Armed with this extension of the residue theorem we now can tackle problems of integrating certain functions involving fractional powers of x.

Example 2

Compute

$$I := \int_0^\infty \frac{dx}{\sqrt{x}(x+4)},$$

where $\sqrt{x}$ denotes the principal value for $x > 0$.

Solution.   Observe that we are required here to find

$$I = \lim_{\substack{\rho \to \infty \\ \varepsilon \to 0^+}} \int_\varepsilon^\rho \frac{dx}{\sqrt{x}(x+4)},$$

and for this purpose we take the branch of $z^{1/2}$ defined by

$$\sqrt{z} = e^{(\operatorname{Log} r + i\theta)/2} \quad \text{for } z = re^{i\theta}, \quad 0 < \theta < 2\pi,$$

which has the nonnegative real axis as its branch cut. With this choice of $\sqrt{z}$ we set

$$f(z) := \frac{1}{\sqrt{z}(z+4)}.$$

Then, according to our convention, for $x > 0$ on the upper side of the cut we have

$$f(x) = \frac{1}{\sqrt{x}(x+4)},$$

and, for $x > 0$ on the lower side,

$$f(x) = \frac{1}{\sqrt{x}e^{i2\pi/2}(x+4)} = \frac{-1}{\sqrt{x}(x+4)}.$$

Now we need to form a closed contour containing the segment $[\varepsilon, \rho]$, and in so doing we must take into account the branch point at the origin as well as the pole at $z = -4$. Consider then the closed contour of Fig. 6.20, where ε is small enough and ρ is large enough so that the pole at -4 lies inside the contour. Then for such ε and ρ we have, by Example 1,

$$\left(\int_{\Gamma_\varepsilon} + \int_{C_\rho} + \int_{\gamma_1} + \int_{\gamma_2} \right) f(z)\, dz = 2\pi i \operatorname{Res}(f; -4). \tag{9}$$

As discussed previously,

$$\int_{\gamma_1} f(z)\, dz + \int_{\gamma_2} f(z)\, dz = \int_\varepsilon^\rho \frac{1}{\sqrt{x}(x+4)}\, dx - \int_\varepsilon^\rho \frac{-1}{\sqrt{x}(x+4)}\, dx$$

$$= 2 \int_\varepsilon^\rho \frac{1}{\sqrt{x}(x+4)}\, dx,$$

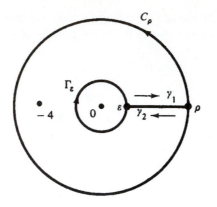

Figure 6.20 Contour for Example 2.

and so we can identify

$$\lim_{\substack{\rho \to \infty \\ \varepsilon \to 0^+}} \left(\int_{\gamma_1} f(z)\,dz + \int_{\gamma_2} f(z)\,dz \right) = 2I.$$

Furthermore, on the circle of radius ρ we have

$$|f(z)| = \frac{1}{\left|\sqrt{z}\right| |z+4|} \le \frac{1}{\sqrt{\rho}(\rho-4)} \quad (\rho > 4),$$

which yields the estimate

$$\left| \int_{C_\rho} f(z)\,dz \right| \le \frac{2\pi\rho}{\sqrt{\rho}(\rho-4)}.$$

Consequently, the integral over C_ρ tends to zero as $\rho \to \infty$. Similarly, on the inner circle of radius ε we have

$$|f(z)| \le \frac{1}{\sqrt{\varepsilon}(4-\varepsilon)} \quad (\varepsilon < 4),$$

which implies that

$$\left| \int_{\Gamma_\varepsilon} f(z)\,dz \right| \le \frac{2\pi\varepsilon}{\sqrt{\varepsilon}(4-\varepsilon)} = \frac{2\pi\sqrt{\varepsilon}}{4-\varepsilon}.$$

As $\varepsilon \to 0^+$ this also goes to zero.

Hence on taking the limit as $\rho \to \infty$ and $\varepsilon \to 0^+$ in Eq. (9), we obtain

$$0 + 0 + 2I = 2\pi i \, \text{Res}(f; -4). \tag{10}$$

Finally, since $z = -4$ is a simple pole of f,

$$\text{Res}(f; -4) = \lim_{z \to -4} (z+4)f(z) = \lim_{\substack{r \to 4 \\ \theta \to \pi}} \frac{1}{\sqrt{z}} = \lim_{\substack{r \to 4 \\ \theta \to \pi}} \frac{1}{e^{(\text{Log}\, r + i\theta)/2}}$$

$$= e^{-(\text{Log}\, 4)/2} e^{-i\pi/2} = \frac{1}{\sqrt{4}}(-i) = \frac{-i}{2}.$$

Therefore from Eq. (10) we get

$$I = \frac{2\pi i}{2}\left(\frac{-i}{2}\right) = \frac{\pi}{2}. \quad \blacksquare$$

A somewhat more complicated situation arises in the following example.

Example 3

Compute

$$I = \text{p.v.} \int_0^\infty \frac{dx}{x^\lambda(x-4)}, \quad \text{where } 0 < \lambda < 1.$$

Solution.   There is a significant difference between this and the preceding example; here we have a singularity at $x = +4$ that lies *on* the interval of integration. Also notice that we have generalized the exponent of x in the denominator. Thus we must compute

$$I = \lim_{\substack{\rho \to \infty \\ \varepsilon, \delta \to 0^+}} \left(\int_\varepsilon^{4-\delta} + \int_{4+\delta}^\rho\right) \frac{dx}{x^\lambda(x-4)}.$$

To do this we modify the approach of the preceding problem by indenting around the singularity. Choosing the branch

$$f(z) = \frac{1}{e^{\lambda(\text{Log}\, r + i\theta)}\left(re^{i\theta} - 4\right)}, \quad \text{for } z = re^{i\theta}, \quad 0 < \theta < 2\pi,$$

we form the contour of Fig. 6.21.

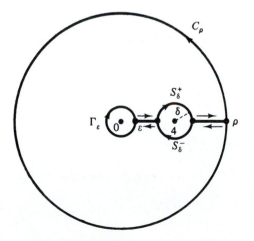

Figure 6.21 Contour for Example 3.

Since $f(x)$ has no singularities "inside" the closed contour, the integral over the latter must be zero. Utilizing different definitions for f on the upper and lower sides of the branch cut, we can write this as

$$\left(1 - e^{-2\pi i\lambda}\right)\left(\int_{\varepsilon}^{4-\delta} + \int_{4+\delta}^{\rho}\right)\frac{dx}{x^{\lambda}(x-4)}$$

$$+ \left(\int_{\Gamma_{\varepsilon}} + \int_{S_{\delta}^{+}} + \int_{S_{\delta}^{-}} + \int_{C_{\rho}}\right)f(z)\,dz = 0, \quad (11)$$

where the contours are as indicated in Fig. 6.21.

Now for $0 < \lambda < 1$, it is easy to extend the estimates used in Example 2 to show that

$$\lim_{\varepsilon \to 0^{+}} \int_{\Gamma_{\varepsilon}} f(z)\,dz = 0 \quad \text{and} \quad \lim_{\rho \to \infty} \int_{C_{\rho}} f(z)\,dz = 0. \quad (12)$$

To compute the limits as $\delta \to 0^{+}$ of the integrals over S_{δ}^{+} and S_{δ}^{-}, we apply the results of the preceding section concerning the behavior of integrals near simple poles. On the upper half-circles around $z = 4$, the function f agrees with the principal branch

$$f_1(z) := \frac{1}{e^{\lambda \operatorname{Log} z}(z-4)},$$

which is *analytic* on the positive real axis except for its simple pole at $z = 4$. Hence by Lemma 4 of Sec. 6.5,

$$\lim_{\delta \to 0^{+}} \int_{S_{\delta}^{+}} f(z)\,dz = -i\pi \operatorname{Res}(f_1; 4) = -i\pi \lim_{z \to 4} e^{-\lambda \operatorname{Log} z} = -i\pi 4^{-\lambda}. \quad (13)$$

However, on the lower half-circles $f(z)$ equals $e^{-2\pi i\lambda}$ *times* $f_1(z)$, and so

$$\lim_{\delta \to 0^{+}} \int_{S_{\delta}^{-}} f(z)\,dz = -i\pi 4^{-\lambda} e^{-2\pi i\lambda}. \quad (14)$$

Finally, on taking the limit as $\rho \to \infty$, $\varepsilon \to 0^{+}$, and $\delta \to 0^{+}$ in Eq. (11), we deduce from (12), (13), and (14) that

$$\left(1 - e^{-2\pi i\lambda}\right)I + 0 - i\pi 4^{-\lambda} - i\pi 4^{-\lambda} e^{-2\pi i\lambda} + 0 = 0,$$

or, equivalently,

$$I = i\pi 4^{-\lambda}\frac{\left(1 + e^{-2\pi i\lambda}\right)}{\left(1 - e^{-2\pi i\lambda}\right)} = i\pi 4^{-\lambda}\frac{e^{i\pi\lambda} + e^{-i\pi\lambda}}{e^{i\pi\lambda} - e^{-i\pi\lambda}} = \pi 4^{-\lambda}\cot(\pi\lambda). \quad \blacksquare$$

Example 4

Compute

$$I = \int_0^\infty \frac{dx}{(x+1)(x^2+2x+2)}.$$

Solution.   Note that the integrand here is *not* an even function of x, so there is no way to relate I to an integral over the whole real line. Motivated by the preceding examples, we shall exploit the features of the log function to attain a suitable closure of the contour. Once again choosing the branch of $\log z$ that is cut along the *nonnegative* x-axis, or $\mathcal{L}_0(z) := \text{Log} |z| + i \arg_0 z$ in the notation of Sec. 3.3, we consider the integral

$$I_{\epsilon,\rho} = \int \frac{\mathcal{L}_0(z)\,dz}{(z+1)(z^2+2z+2)},$$

along the same contour that we used for Example 2 (Fig. 6.20). On the upper side of the cut we have $\mathcal{L}_0(z) = \text{Log}\, x + i \arg_0 x = \text{Log}\, x$, and on the lower side we have $\mathcal{L}_0(z) = \text{Log}\, x + i2\pi$. Integrating back and forth along the cut, then, we observe the cancellation of the real parts containing the unwanted $\text{Log}\, x$ factors, and we obtain

$$\int_\epsilon^\rho \frac{\text{Log}\, x\,dx}{(x+1)(x^2+2x+2)} + \int_\rho^\epsilon \frac{(\text{Log}\, x + i2\pi)\,dx}{(x+1)(x^2+2x+2)}$$

$$= -2\pi i \int_\epsilon^\rho \frac{dx}{(x+1)(x^2+2x+2)}.$$

We estimate the integrals along the circles Γ_ϵ and C_ρ of Fig. 6.20 as usual. For sufficiently small ϵ,

$$\left| \int_{\Gamma_\epsilon} \frac{\mathcal{L}_0(z)\,dz}{(z+1)(z^2+2z+2)} \right| \leq \frac{\sqrt{(\text{Log}\, \varepsilon)^2 + (2\pi)^2}\, 2\pi \varepsilon}{(1-\varepsilon)(2 - 2\varepsilon - \varepsilon^2)},$$

which goes to zero as $\varepsilon \to 0^+$. And for sufficiently large ρ

$$\left| \int_{C_\rho} \frac{\mathcal{L}_0(z)\,dz}{(z+1)(z^2+2z+2)} \right| \leq \frac{\sqrt{(\text{Log}\, \rho)^2 + (2\pi)^2}\, 2\pi \rho}{(\rho - 1)(\rho^2 - 2\rho - 2)}$$

$$= \frac{\sqrt{(\text{Log}\, \rho)^2 + (2\pi)^2}\, 2\pi \rho}{\rho^3(1 - 1/\rho)(1 - 2/\rho - 2/\rho^2)},$$

which goes to zero as $\rho \to \infty$ (since $(\text{Log}\, \rho)/\rho^2 \to 0$). Consequently, in the limit the contour integral $I_{\epsilon,\rho}$ approaches

$$-2\pi i \int_0^\infty \frac{dx}{(x+1)(x^2+2x+2)} = -2\pi i\, I.$$

But, by the residue theorem, $I_{\epsilon,\rho}$ equals $2\pi i$ times the sum of the residues inside the contour. Writing the integrand in $I_{\epsilon,\rho}$ as

$$\frac{\mathcal{L}_0(z)}{[z+1][z-(-1+i)][z-(-1-i)]},$$

we find the sum of the residues to be (for small ϵ and large ρ)

$$\frac{\mathcal{L}_0(-1)}{(-i)i} + \frac{\mathcal{L}_0(-1+i)}{(+i)(2i)} + \frac{\mathcal{L}_0(-1-i)}{(-i)(-2i)}$$

$$= \frac{\pi i}{+1} + \frac{\text{Log}(\sqrt{2}) + 3\pi i/4}{-2} + \frac{\text{Log}(\sqrt{2}) + 5\pi i/4}{-2}$$

$$= -\text{Log}(\sqrt{2}).$$

Therefore the desired integral $I = 2\pi i(-\text{Log}(\sqrt{2})/(-2\pi)i = \text{Log}(\sqrt{2}).$ ■

EXERCISES 6.6

Use residue theory to verify each of the integral formulas in Problems 1–7.

1. $\displaystyle\int_0^\infty \frac{\sqrt{x}}{x^2+1}\,dx = \frac{\pi}{\sqrt{2}}$

2. $\displaystyle\int_0^\infty \frac{x^{\alpha-1}}{x+1}\,dx = \frac{\pi}{\sin(\pi\alpha)},\quad 0 < \alpha < 1$

3. $\displaystyle\int_0^\infty \frac{x^\alpha}{(x+9)^2}\,dx = \frac{9^{\alpha-1}\pi\alpha}{\sin(\pi\alpha)},\quad -1 < \alpha < 1,\ \alpha \neq 0$

4. $\displaystyle\int_0^\infty \frac{x^\alpha}{\left(x^2+1\right)^2}\,dx = \frac{\pi(1-\alpha)}{4\cos(\alpha\pi/2)},\quad -1 < \alpha < 3,\ \ \alpha \neq 1$

5. $\displaystyle\int_0^\infty \frac{x^{\alpha-1}}{x^2+x+1}\,dx = \frac{2\pi}{\sqrt{3}}\cos\left(\frac{2\alpha\pi+\pi}{6}\right)\csc(\alpha\pi),\ 0 < \alpha < 2,\ \alpha \neq 1$

6. p.v. $\displaystyle\int_0^\infty \frac{x^\alpha}{x^2-1}\,dx = \frac{\pi}{2\sin(\pi\alpha)}[1-\cos(\pi\alpha)],\quad -1 < \alpha < 1,\ \ \alpha \neq 0$

7. $\displaystyle\int_0^\infty \frac{x^\alpha}{1+2x\cos\phi+x^2}\,dx = \frac{\pi}{\sin(\pi\alpha)}\frac{\sin(\phi\alpha)}{\sin\phi},\quad -1 < \alpha < 1,$
$\alpha \neq 0,\ -\pi < \phi < \pi,\ \phi \neq 0$

8. Verify that

$$\text{p.v.} \int_{-\infty}^\infty \frac{\text{Log}|x|}{x^2+4}\,dx = \frac{\pi}{2}\text{Log}\,2.$$

[HINT: Integrate $(\text{Log}\,z)/(z^2+4)$ around a *semi*circular contour indented at the origin (see Fig. 6.12) and note that

$$(\rho\,\text{Log}\,\rho)/(\rho^2-4) \to 0 \text{ as } \rho \to +\infty \text{ or as } \rho \to 0^+]$$

9. Using the method of Example 4, compute the following integrals:

$$\text{(a)} \int_0^\infty \frac{x}{(x+1)(x^2+2x+2)}\,dx \quad \text{(b)} \int_0^\infty \frac{1}{x^3+1}\,dx$$

10. Verify that

$$\int_0^\infty \frac{\text{Log}\,x}{x^2+1}\,dx = 0.$$

[HINT: See Prob. 9.]

11. Verify that

$$\int_0^\infty \frac{\text{Log}\,x}{\left(x^2+1\right)^2}\,dx = -\frac{\pi}{4}.$$

[HINT: See Prob. 9.]

12. Verify that

$$\int_0^\infty x^{\alpha-1}\sin x\,dx = \sin\left(\frac{\pi\alpha}{2}\right)\cdot\Gamma(\alpha) \qquad (0 < \alpha < 1),$$

where $\Gamma(\alpha) := \int_0^\infty e^{-x}x^{\alpha-1}dx$ is the *Gamma function*. [HINT: Integrate $e^{-z}z^{\alpha-1}$ around a quarter-circle indented at the origin.]

13. Evaluate the integral

$$\int_0^1 (x^2-x^3)^{-1/3}\,dx$$

by using a barbell-shaped contour with shrinking ends surrounding 0 and 1, together with a large circle containing the barbell. [HINT: Write $(z^2-z^3)^{1/3} = z(z^{-1}-1)^{1/3}$ and select a suitable branch of $(z^{-1}-1)^{1/3}$.]

6.7   The Argument Principle and Rouché's Theorem

In this section we shall use Cauchy's residue theorem to derive two theoretical results that have important practical applications. These results pertain to functions all of whose singularities are poles. Such functions are given a special name in the next definition.

Definition 2. A function f is said to be **meromorphic** in a domain D if at every point of D it is either analytic or has a pole.

In particular, we regard the analytic functions on D as being special cases of meromorphic functions. The rational functions are examples of functions that are meromorphic in the whole plane.

Suppose now that we are given a function f that is analytic and nonzero at each point of a simple closed contour C and is meromorphic inside C. Under these conditions it can be shown that f has at most a *finite* number of poles inside C. The proof of this depends on two facts: first, that the only singularities of f are *isolated* singularities (poles), and, second, that every infinite sequence of points inside C has a subsequence that converges to some point on or inside C. (The last fact is proved in advanced calculus texts under the name Bolzano-Weierstrass theorem.) Hence if f had an infinite number of poles inside C, some subsequence of them would converge to a point that must be a singularity, but not an *isolated* singularity, of f. By contradiction, then, the number of poles must be finite.

By the same token, if f had an infinite number of zeros inside C, a subsequence would converge to some point z_0. But if f were analytic at z_0, then by Corollary 3, Sec. 5.6, the function f would have to be identically zero in a neighborhood of z_0— and thus also on C, by analytic continuation; and if f were *not* analytic at z_0, the deliberations of Sec. 5.6 would require that z_0 be an essential singularity. Both situations contradict our hypotheses; thus the number of zeros inside C is also finite.

In counting the number of zeros or poles of a function it is common practice to include the multiplicity. For example, let $C : |z| = 4$ and take

$$f(z) = \frac{(z-8)^2 z^3}{(z-5)^4 (z+2)^2 (z-1)^5}.$$ (1)

Then the number $N_p(f)$ of poles of f *inside* C is to be interpreted as

$$N_p(f) := \sum_{\text{poles inside } C} (\text{order of each pole})$$

$$= (\text{order of pole at } z = -2) + (\text{order of pole at } z = 1)$$

$$= 2 + 5 = 7,$$

while the number $N_0(f)$ of its zeros inside C is

$$N_0(f) := (\text{order of the zero at } z = 0) = 3.$$

Example 1

For the function in Eq. (1), evaluate $\int_C f'(z)/f(z)\, dz$, where $C : |z| = 4$ is positively oriented (Fig. 6.22).

Solution.   To introduce a new concept we are going to attack this problem via a nonstandard approach. Observe that, formally speaking,

$$\frac{f'(z)}{f(z)} = \frac{d}{dz} \log f(z) = \frac{d}{dz} \{\text{Log } |f(z)| + i \arg f(z)\}.$$

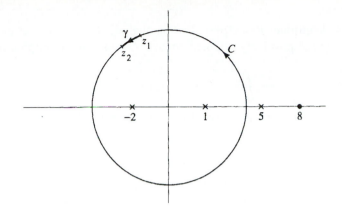

Figure 6.22 Contour for Example 1.

Now let γ be a subarc of C, sufficiently short so that (some branch of) arg $f(z)$ varies by less than 2π along γ. Then there is a branch of log $f(z)$ that is analytic on γ. It follows from Theorem 6, Sec. 4.3, that

$$\int_\gamma \frac{f'(z)}{f(z)}\, dz = \{\text{Log}\,|f(z)| + i \arg f(z)\}\Big|_{z_1}^{z_2} \tag{2}$$

for this branch, where z_1 and z_2 are the endpoints of γ (see Fig. 6.22). If we break C into such subarcs and piece together the contributions (2), we can write the result as

$$\int_C \frac{f'(z)}{f(z)}\, dz = \Delta_C \, \text{Log}\,|f(z)| + i\Delta_C \arg f(z) = i\Delta_C \arg f(z), \tag{3}$$

where we interpret $\Delta_C \arg f(z)$ to be the *net* excursion in arg $f(z)$ as we go around C. (Since Log $|f(z)|$ is single-valued, its net excursion on a closed contour is zero.)

From (1) we see that

$$\arg f(z) = 2\arg(z-8) + 3\arg z - 4\arg(z-5) - 2\arg(z+2) - 5\arg(z-1).$$

The net excursion in arg z, as we go around C, is 2π, since C encircles the origin. Similarly, the net excursion in $\arg(z+2)$ and $\arg(z-1)$ is 2π, since C encircles -2 and $+1$. But the excursion in $\arg(z-8)$ and $\arg(z-5)$ is zero; 8 and 5 lie outside C. Thus

$$\int_C \frac{f'(z)}{f(z)}\, dz = i2\pi(3 - 2 - 5) = -4(2\pi i) = -8\pi i. \quad \blacksquare$$

In the preceding example, every zero of $f(z)$ inside C contributed $2\pi i$ times its multiplicity, and every pole contributed $-2\pi i$ times its multiplicity, to the integral. The *Argument Principle* generalizes this result for all meromorphic functions.

Theorem 3 (Argument Principle). If f is analytic and nonzero at each point of a simple closed positively oriented contour C and is meromorphic inside C, then

$$\frac{1}{2\pi i} \int_C \frac{f'(z)}{f(z)} \, dz = N_0(f) - N_p(f), \tag{4}$$

where $N_0(f)$ and $N_p(f)$ are, respectively, the number of zeros and poles of f inside C (multiplicity included).

Proof.   The strategy is quite straightforward; we locate the singularities and compute the residues of the integrand

$$G(z) := \frac{f'(z)}{f(z)}.$$

Notice that this function is analytic at each point *on* C because f is analytic and nonzero there. Inside C the singularities of G occur at those points where f has a zero or a pole.

Consider first a point z_0 inside C that is a zero of f of order m. Then we know that f can be written in the form

$$f(z) = (z - z_0)^m \cdot h(z),$$

where $h(z)$ is analytic and not zero at $z = z_0$ (recall Theorem 16, Sec. 5.6). Hence in some punctured neighborhood of z_0 we compute that

$$G(z) = \frac{f'(z)}{f(z)} = \frac{m}{z - z_0} + \frac{h'(z)}{h(z)}.$$

Since the function h'/h is analytic at z_0, this representation shows that G has a simple pole at z_0 with residue equal to m.

On the other hand, if f has a pole of order k at z_p, then

$$f(z) = \frac{H(z)}{\left(z - z_p\right)^k},$$

where $H(z)$ is analytic at z_p and $H\left(z_p\right) \neq 0$ (recall Lemma 7, Sec. 5.6). This time we derive that in a punctured neighborhood of z_p

$$G(z) = \frac{f'(z)}{f(z)} = \frac{-k}{z - z_p} + \frac{H'(z)}{H(z)},$$

and since H'/H is analytic at z_p, we find that G has a simple pole at z_p with residue equal to *minus* k.

Finally, by the residue theorem, the integral of G around C must equal $2\pi i$ times the sum of the residues at the singularities inside C. This, from the preceding deliberations, equals $2\pi i$ times the sum of the orders of the zeros of f inside C plus the sum of the negatives of the orders of the poles of f inside C; that is,

$$\int_C G(z)\,dz = \int_C \frac{f'(z)}{f(z)}\,dz = 2\pi i\left[N_0(f) - N_p(f)\right],$$

which is the same as Eq. (4). ∎

Of course, if the function f of Theorem 3 has no poles inside C, then $N_p(f) = 0$, and we have the following.

Corollary 1. If f is analytic inside and on a simple closed positively oriented contour C and if f is nonzero on C, then

$$\frac{1}{2\pi i}\int_C \frac{f'(z)}{f(z)}\,dz = N_0(f),$$

where $N_0(f)$ is the number of zeros of f inside C (multiplicity included).

As we have discussed, the conclusion of the argument principle can be written in the form

$$\frac{1}{2\pi}\Delta_C \arg f(z) = N_0(f) - N_p(f). \tag{5}$$

There is yet another way to express the variation in the argument of f along C; it involves the *image curve* $f(C)$. This is simply the image (in the w-plane) of the curve C under the mapping $w = f(z)$; that is, if C is parametrized by $z = z(t)$, $a \le t \le b$, then $f(C)$ is the curve parametrized by

$$w = f(z(t)) \qquad (a \le t \le b). \tag{6}$$

Obviously the image curve is closed, but unlike C it need not be simple or positively oriented.

Now if we sketch the image curve as in Fig. 6.23, it is easy to follow the net change in the argument of $f(z)$. Every time $f(z(t))$ encircles the origin $w = 0$ in the positive (counterclockwise) direction, arg f increases by 2π, while it decreases by 2π for a negative circuit. Hence, since $f(C)$ is closed, $\Delta_C \arg f(z)$ equals 2π multiplied by the net number of times $f(C)$ winds around $w = 0$ *in the positive sense*, that is, counterclockwise minus clockwise.[†]

As a specific illustration, consider

$$f(z) = z^3 \quad \text{and} \quad C : z = e^{it} \qquad (0 \le t \le 2\pi).$$

Notice that the image of this circle under $w = f(z)$ winds around the origin three times in the positive sense (see Fig. 6.24). Hence the net change in the argument of f is 6π, and Eq. (5) correctly predicts $N_0(f) - N_p(f) = 6\pi/2\pi = 3$.

[†]Some authors call this the *winding number* of $f(C)$ about the origin.

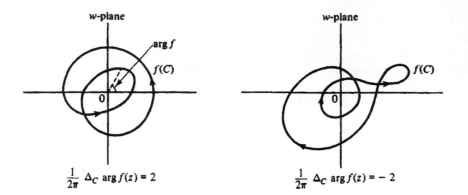

Figure 6.23 Visualization of $\Delta_C \arg f(z)$.

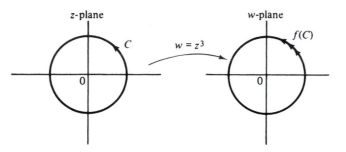

Figure 6.24 $\Delta_C \arg(z^3)$.

Next, suppose we have a function $f(z)$ analytic on C and meromorphic inside, and we know how many times $f(C)$ winds around the origin $w = 0$. Now we are interested in *perturbing* $f(z)$ by some analytic function $h(z)$ to form $g(z) := f(z) + h(z)$. We would like to know how small the perturbation h must be to guarantee that $g(C)$ winds around $w = 0$ the same number of times as $f(C)$.

The problem is a familiar one to anyone who has tried to walk a dog on a leash in a big city. If the pair encounters a lamppost and the leash is long, the canine will inevitably tangle the leash around the post. But if the human continually adjusts the length of the leash as they walk so it never quite extends to the post, then both dog and human will wind around the post an equal number of times and avoid entanglement.

Let us interpret the origin in the w-plane as the lamppost, the contour $f(C)$ as the path of the human, and $g(C)$ as the dog's path. (See Fig. 6.25.) Then $h(z) = g(z) - f(z)$ becomes the leash, and the condition that the leash never extends from the human to the lamppost—and thus that $f(C)$ and $g(C)$ wind around the origin the same number of times—is expressed as

$$|h(z)| < |f(z)|, \quad z \text{ on } C.$$

Rouché's theorem results when we combine these deliberations with the argument principle, for functions f with no poles inside C.

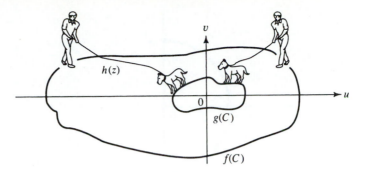

Figure 6.25 M. Rouché and Truffles.

Theorem 4 (Rouché's Theorem). If f and h are each functions that are analytic inside and on a simple closed contour C and if the strict inequality

$$|h(z)| < |f(z)| \tag{7}$$

holds at each point on C, then f and $f + h$ must have the same total number of zeros (counting multiplicities) inside C.

Observe that the inequality (7) need only hold *on* C, not inside, and that (7) prevents f (as well as $g = f + h$) from being zero on C. See Prob. 15 for an extension to the case when f is meromorphic inside C.

One typically uses Rouché's theorem to deduce some information about the location of zeros of a complicated analytic function g by comparing it with an analytic function f whose zeros are known.

Example 2

Prove that all five zeros of the polynomial

$$g(z) = z^5 + 3z + 1$$

lie in the disk $|z| < 2$.

Solution.   We take C as the circle $|z| = 2$, and we regard g as a perturbation of the function $f(z) = z^5$, which clearly has five zeros inside C. To test condition (7) we estimate the perturbation $h(z) = 3z + 1$ on C by

$$|h(z)| = |3z + 1| \le 3|z| + 1 = 3 \cdot 2 + 1 = 7,$$

which sure enough, is strictly less than $|f(z)| = \left|z^5\right| = 2^5 = 32$. Therefore, g also has five zeros inside $|z| < 2$.   ■

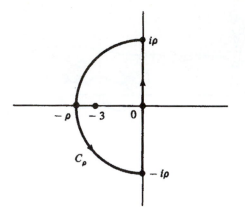

Figure 6.26 Contour for Example 3.

Example 3

Prove that the equation

$$z + 3 + 2e^z = 0$$

has precisely one root in the left half-plane.

Solution.   Since Rouché's theorem refers to domains bounded by contours, we cannot apply it directly to an unbounded set such as a half-plane. But by this time the reader will probably be able to anticipate our strategy; we choose C_ρ as in Fig. 6.26 and regard the function

$$g(z) = z + 3 + 2e^z$$

as a perturbation of

$$f(z) = z + 3,$$

which has exactly one zero in the left half-plane. Then for z on C_ρ we have

$$|h(z)| = |g(z) - f(z)| = |2e^z| = 2e^{\operatorname{Re} z} \le 2e^0 = 2,$$

while $f(z)$ is bounded from below on C_ρ by (see Fig. 6.26)

$$|f(z)| = |z + 3| \ge \begin{cases} 3, & \text{for } z = iy, \\ |z| - 3 = \rho - 3, & \text{for } |z| = \rho. \end{cases}$$

Thus when $\rho > 5$ we have $|h(z)| < |f(z)|$ for all z on C_ρ. This implies that g also has precisely one (simple) zero inside C_ρ, and hence (letting $\rho \to \infty$) in the left half-plane.   ∎

Rouché's theorem can also be used to give an alternative proof of the Fundamental Theorem of Algebra.

Example 4

Prove that every polynomial of degree n has n zeros.

Solution.   Consider the polynomial $g(z)$ of degree n

$$g(z) = a_n z^n + a_{n-1} z^{n-1} + \cdots + a_1 z + a_0 \quad (a_n \neq 0),$$

as a perturbation of

$$f(z) = a_n z^n,$$

which has n zeros (at the origin). The difference

$$h(z) = g(z) - f(z) = a_{n-1} z^{n-1} + \cdots + a_1 z + a_0$$

is then a polynomial of degree at most $n - 1$ and hence does not grow as rapidly as the polynomial f of degree n. In more precise terms, on the circle $C : |z| = R$ we have

$$|f(z)| = |a_n| R^n,$$

and

$$|h(z)| \leq |a_{n-1}| R^{n-1} + \cdots + |a_1| R + |a_0|.$$

Therefore if we choose $R \, (> 1)$ so large that

$$\frac{|a_{n-1}|}{|a_n|} + \cdots + \frac{|a_1|}{|a_n|} + \frac{|a_0|}{|a_n|} < R,$$

then the inequality $|h(z)| < |f(z)|$ will be valid on C. Therefore, $g(z)$ also has n zeros.   ∎

Now suppose $f(z)$ is analytic in some open neighborhood of a point z_0 and $f(z_0) = 0$. Then from Sec. 5.6 we know that unless f is identically zero, there is some circle C, centered at z_0 and lying in the neighborhood, such that f is nonzero on C. Let σ be the minimum value of $|f(z)|$ for z on C, and perturb f by $h(z) = -c$, where c is *any* complex number smaller in magnitude than σ (continuity implies $\sigma > 0$). It follows from Rouché's theorem that $f(z) - c$ also achieves the value 0, and hence f achieves the value c, inside the circle.

In other words, the values taken on by $w = f(z)$ in this neighborhood of z_0 completely cover the open disk $|w| < \sigma$ in the w-plane. So the image of *every* open neighborhood of z_0 contains an open neighborhood of $w_0 = f(z_0) = 0$, unless f is identically zero.

If $f(z_0) = w_0$ is not zero, we can apply this same argument to the function $f(z_0) - w_0$ and conclude that the image of every open neighborhood of z_0 contains an open neighborhood of $f(z_0)$, unless f is constant. This is the *open mapping* property of analytic functions, which we state as follows.

Theorem 5. If f is nonconstant and analytic in a domain D, then its range

$$f(D) := \{w \mid w = f(z) \text{ for some } z \text{ in } D\}$$

is an open set.

EXERCISES 6.7

1. Which of the following functions are meromorphic in the whole plane?

   (a) $2z + z^3$

   (b) $\operatorname{Log} z$

   (c) $\dfrac{\sin z}{z^3 + 1}$

   (d) $e^{1/z}$

   (e) $\tan z$

   (f) $\dfrac{2i}{(z - 3)^2} + \cos z$

2. Let $P(z) = a_n z^n + a_{n-1} z^{n-1} + \cdots + a_1 z + a_0$, where $a_n \neq 0$. Explain why for each sufficiently large value of R

$$\oint_{|z|=R} \frac{P'(z)}{P(z)} \, dz = 2n\pi i.$$

3. Evaluate

$$\frac{1}{2\pi i} \oint_{|z|=3} \frac{f'(z)}{f(z)} \, dz,$$

   where $f(z) = \dfrac{z^2 (z - i)^3 e^z}{3(z + 2)^4 (3z - 18)^5}$.

4. Let $f(z)$ be analytic on the closed disk $|z| \leq \rho$, and suppose that $f(z) \neq w_0$ for all z on the circle $|z| = \rho$. Explain why the value of the integral

$$\frac{1}{2\pi i} \oint_{|z|=\rho} \frac{f'(z)}{f(z) - w_0} \, dz$$

   equals the number of solutions of $f(z) = w_0$ inside the disk.

5. Prove that if $f(z)$ is analytic inside and on a simple closed contour C and is one-to-one on C, then $f(z)$ is one-to-one inside C. [HINT: Consider the image curve $f(C)$.]

6. Use Rouché's theorem to show that the polynomial $z^6 + 4z^2 - 1$ has exactly two zeros in the disk $|z| < 1$.

7. Prove that the equation $z^3 + 9z + 27 = 0$ has no roots in the disk $|z| < 2$.

8. Prove that all the roots of the equation $z^6 - 5z^2 + 10 = 0$ lie in the annulus $1 < |z| < 2$.

9. Find the number of roots of the equation $6z^4 + z^3 - 2z^2 + z - 1 = 0$ in the disk $|z| < 1$.

10. Prove that the equation $z = 2 - e^{-z}$ has exactly one root in the right half-plane. Why must this root be real?

11. Prove that the polynomial $P(z) = z^4 + 2z^3 + 3z^2 + z + 2$ has exactly two zeros in the right half-plane. [HINT: Write $P(iy) = (y^2 - 2)(y^2 - 1) + iy(1 - 2y^2)$, and show that

$$\lim_{R \to \infty} \arg P(iy) \Big|_{-R}^{R} = 0.]$$

12. Suppose that $f(z)$ is analytic on $|z| \le 1$ and satisfies $|f(z)| < 1$ for $|z| = 1$.

    (a) Prove that the equation $f(z) = z$ has exactly one root (counting multiplicity) in $|z| < 1$. (This root is called a *fixed point* of f; see also Sec. 2.7.)

    (b) Prove that if $|z_0| \le 1$, then the sequence z_n defined recursively by $z_n = f(z_{n-1})$, $n = 1, 2, \ldots$, converges to the fixed point of f. (See Prob. 17, Sec. 5.6.)

13. Give an example to show that the conclusion of Rouché's theorem may be false if the strict inequality $|h(z)| < |f(z)|$ is replaced by $|h(z)| \le |f(z)|$ for z on C.

14. State and prove a generalization of Rouché's theorem for meromorphic functions f and h, which concludes that $N_0(f) - N_p(f) = N_0(f + h) - N_p(f + h)$.

15. Prove: If f is analytic and nonzero at each point of a simple closed contour C and is *meromorphic* inside C and if h is *analytic* inside and on C and satisfies $|h(z)| < |f(z)|$ on C, then f and $f + h$ have the same number of zeros inside C (counting multiplicity).

16. Let $\lambda > 0$ be fixed, and let $g(z) = \tan z - \lambda z$. The zeros of g are important in certain problems related to heat flow and transmission line theory. By completing each of the following steps, show that for n large, $g(z)$ has exactly $2n + 1$ zeros inside the square Γ_n with vertices at $n\pi(1 \pm i)$, $n\pi(-1 \pm i)$.

    (a) Show that

    $$\tan(x + iy) = \left[\frac{\sin(2x)}{\cosh(2y) + \cos(2x)} \right] + i \left[\frac{\sinh(2y)}{\cosh(2y) + \cos(2x)} \right].$$

    (b) Prove that for all large integers n, the inequality $|\tan z| \le 2$ holds on the boundary of Γ_n. [HINT: Use the formula in part (a) for the horizontal segments.]

    (c) Show that for all large integers n, the inequality $|g(z) + \lambda z| < \lambda |z|$ holds on the boundary of Γ_n.

    (d) Show that $g(z)$ has exactly $2n$ poles inside Γ_n, $n = 1, 2, \ldots$.

    (e) Conclude from the general form of Rouché's theorem (Prob. 14) that $g(z)$ has $2n + 1$ zeros inside Γ_n for large integers n.

17. Let $f(z)$ be analytic in a domain D, and suppose that $f(z) - f(z_0)$ has a zero of order n at z_0 in D. Prove that for $\varepsilon > 0$ sufficiently small, there exists a $\delta > 0$ such that for all w in $|w - f(z_0)| < \delta$ the equation $f(z) - w = 0$ has exactly n roots in $|z - z_0| < \varepsilon$.

18. Use the open mapping property (Theorem 5) to give a quick proof of the following familiar facts: If f is analytic in a domain D, then f is identically constant in D if any of the following conditions hold.

    (a) Re $f(z)$ is constant in D.

    (b) Im $f(z)$ is constant in D.

(c) $|f(z)|$ is constant in D.

19. Let $f_n(z)$, $n = 1, 2, \ldots$, be a sequence of functions analytic in the disk $D : |z| < R$ which converges uniformly to the analytic function $f(z)$ on each closed subset of D. Prove that if $f(z) \neq 0$ on $|z| = \delta$, $0 < \delta < R$, then for each n sufficiently large $f_n(z)$ has the same number of zeros in $|z| < \delta$ as does $f(z)$.

20. Let $P(z) = a_n z^n + \cdots + a_1 z + a_0$ and let

$$P^*(z) = z^n \overline{P(1/\bar{z})} = \bar{a}_0 z^n + \bar{a}_1 z^{n-1} + \cdots + \bar{a}_n.$$

Prove that if $\left| a_0 / a_n \right| > 1$, then $P(z)$ has the same number of zeros in $|z| < 1$ as does the polynomial $\bar{a}_0 P(z) - a_n P^*(z)$. [HINT: $|P(z)| = |P^*(z)|$ for $|z| = 1$.]

21. To establish the stability of feedback control systems one often has to ensure that a certain meromorphic function of the form $F(z) = 1 + P(z)$ has all its zeros in the left half-plane. The *Nyquist stability criterion* proceeds as follows: First consider the contour Γ_r, shown in Fig. 6.27, and let m equal the net number of times that the image contour $P(\Gamma_r)$ encircles the point $w_0 = -1$ in a counterclockwise direction. Then let n equal the number of poles of $P(z)$ with positive real parts. Argue that if m equals n for all sufficiently large r, then all the zeros of $F(z)$ lie in the left half-plane (and thus the system is stable). [If for r large, $P(\Gamma_r)$ passes through the point $w_0 = -1$, then of course $F(z)$ has a zero on the imaginary axis; in such a case stability cannot be guaranteed.]

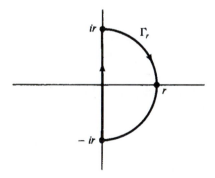

Figure 6.27 Contour for Prob. 21.

22. A stronger version (in that the hypothesis is weaker) of Rouché's theorem was discovered by Glicksberg. With reference to the dog-and-human situation in Fig. 6.25, let τ denote the ray extending from the lamppost in the direction away from the human, as in Fig. 6.28. (Obviously τ turns as the human traverses the path $f(C)$.) Now, if the dog is restricted to stay on one side or the other of τ—and never to cross it—then the leash will not tangle around the lamppost and both dog and human will encircle the post the same number of times.

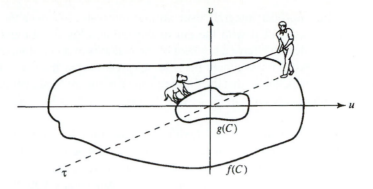

Figure 6.28 M. Glicksberg and Fritz.

(a) From this consideration argue that condition (7) in Rouché's theorem can be replaced by

$$|h(z)| < |f(z)| + |f(z) + h(z)|. \tag{8}$$

[HINT: Inequality (8) is a *strict* triangle inequality, and as such it ensures that the points $h(z)$, $f(z)$, and $f(z) + h(z)$ do not align unfavorably; recall Sec. 1.3.]

(b) Give an alternative derivation based on the observation that (8) implies $f/(f + h)$ is never negative or zero, and thus

$$\int_C \left[\frac{f'}{f} - \frac{(f+h)'}{(f+h)} \right] dz = \int_C \left[\text{Log} \frac{f}{f+h} \right]' dz = 0;$$

now apply Corollary 1.

SUMMARY

A useful way to evaluate certain contour integrals is by means of residues. The residue of a function $f(z)$ at an isolated singularity z_0 is the coefficient a_{-1} of $1/(z - z_0)$ in the Laurent expansion for $f(z)$ about z_0. Simple formulas exist for computing the residues at the poles of $f(z)$. When all the singularities of $f(z)$ are isolated, its integral along a simple closed positively oriented contour is equal to $2\pi i$ times the sum of the residues at the singularities inside the contour.

Residue theory can be employed to evaluate certain integrals that arise in the calculus of functions of a real variable. For example, definite integrals over $[0, 2\pi]$ that involve $\sin \theta$ and $\cos \theta$ can be rewritten as contour integrals around the unit circle $C : |z| = 1$ after making the identification

$$\cos \theta = \frac{1}{2} \left(z + \frac{1}{z} \right), \qquad \sin \theta = \frac{1}{2i} \left(z - \frac{1}{z} \right) \qquad \left(\text{for } z = e^{i\theta} \right).$$

In addition, certain improper integrals over infinite intervals, say over the real axis $(-\infty, +\infty)$, can be computed with the aid of expanding closed contours (such as semicircles or rectangles) that have a segment of the real axis as a component. For this approach to be successful the contours must be selected so that the sum of residues inside is easily computed, and so that the limiting values of the integral over the nonreal portions of the contours are known. Some specific results such as Jordan's lemma are useful in this regard.

Sometimes these contours must be modified by indention to compensate for singularities on the original interval of integration. In the case when the complex version of an integrand is multiple-valued, it might be necessary to integrate along a branch cut. For this purpose the cut is regarded as having two distinct sides, with the integrand being defined differently on each side.

When the only singularities of $f(z)$ are poles, the variation in the argument of $f(z)$ along a simple closed contour is related to the difference in the number of its zeros and poles inside the contour. A consequence of this fact is Rouché's theorem, which provides a comparison technique for counting the number of zeros of an analytic function in a certain domain.

Suggested Reading

Residue Calculus

[1] Conway, J. B. *Functions of One Complex Variable*, 2nd ed. Springer-Verlag Inc., New York, 1978.

[2] Copson, E. T. *An Introduction to the Theory of Functions of a Complex Variable*. Oxford University Press, Inc., New York, 1962.

[3] Eves, H. W. *Functions of a Complex Variable*, Vol. 2. Prindle, Weber & Schmidt, Inc., Boston, 1966.

[4] Henrici, P. *Applied and Computational Analysis*, Vol. 1. John Wiley & Sons, Inc., New York, 1974.

Special Functions

[5] Abramowitz, M. and Stegun, I. A. *Handbook of Mathematical Functions with Formulas, Graphs, and Mathematical Tables*, Dover Publications, New York, 1974.

Stability and Control

[6] Dorf, R. C. *Modern Control Systems*, 9th ed. Prentice-Hall, Upper Saddle River NJ, 2000.

Rouché's Theorem

[7] Glicksberg, I. "A Remark on Rouché's Theorem," *Amer. Math. Mon.* 83 (1976), 186–187.

Chapter 7

Conformal Mapping

In this chapter we shift our point of view somewhat; rather than dealing with the algebraic properties of an analytic function $f(z)$, we are going to regard f as a *mapping* from its domain to its range and consider its *geometric* properties. The ability to map one region onto another via an analytic function proves invaluable in applied mathematics, as we shall see in Sec. 7.1.

7.1 Invariance of Laplace's Equation

One of the most valuable aspects of mappings generated by analytic functions is the persistence of Laplace's equation; roughly, this means that if $\phi(x, y)$ is harmonic in a certain domain D of the xy-plane (so that ϕ satisfies Laplace's equation

$$\frac{\partial^2 \phi}{\partial x^2} + \frac{\partial^2 \phi}{\partial y^2} = 0$$

in D), and if

$$w = f(z) \tag{1}$$

is an analytic function mapping D onto a domain D' in the uv-plane, then ϕ is "carried over" by the mapping to a function that is harmonic in D'.

To express this fact precisely we must elaborate somewhat on the nature of the mapping. We assume that relation (1) provides a *one-to-one* correspondence between the points of D and those of D'.[†] Recall that this means $f(z_1) = f(z_2)$ only if $z_1 = z_2$. We also assume that the derivative df/dz is never zero in D; actually the latter is a consequence of the one-to-one assumption, but we won't prove it here.

Now since the mapping is one-to-one, it has an inverse; that is, with each point of D' there can be associated a point of D, namely, its preimage under f. This relationship, which is the inverse of $w = f(z)$, is suggestively written

$$z = f^{-1}(w). \tag{2}$$

_____

[†]Such a mapping is sometimes called *univalent*, or *schlicht*.

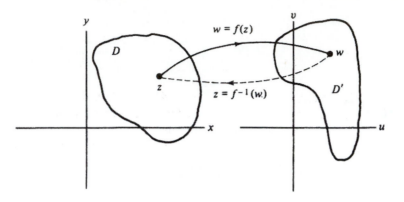

Figure 7.1 Inverse mapping.

Figure 7.1 depicts both the relationships. Observe that f^{-1} is a single-valued function (since only one z is mapped to a particular w). The reader will probably not be surprised to learn that it is, in fact, an *analytic* function, and that its derivative is given by

$$\frac{df^{-1}}{dw}(w) = \frac{1}{\dfrac{df}{dz}(z)} \qquad \text{[where } w = f(z)\text{].} \tag{3}$$

This equation can also be written in the form

$$\frac{dz}{dw} = \frac{1}{\dfrac{dw}{dz}}, \tag{4}$$

keeping in mind the functional relationships. We have already proved a special case of Eq. (3) for the function $w = e^z$ and its inverse $z = \text{Log } w$, taking D to be the strip $|\text{Im } z| < \pi$ and D' to be the entire plane slit along the negative real axis (cf. Sec. 3.3). We shall invite the reader to prove Eq. (3) in general at the end of the next section.

It is sometimes helpful to indicate the mappings (1) and (2) in terms of real variables; thus Eq. (1) becomes

$$u = u(x, y), \qquad v = v(x, y) \tag{5}$$

and its inverse, Eq. (2), becomes

$$x = x(u, v), \qquad y = y(u, v). \tag{6}$$

Now we are prepared to demonstrate the property stated before about the persistence of Laplace's equation. Observe that if $\phi(x, y)$ is a function defined on D, then the domain D' "inherits" ϕ through the one-to-one mapping; that is, the function $\psi(u, v)$ defined by

$$\psi(u, v) := \phi(x(u, v), y(u, v)),$$

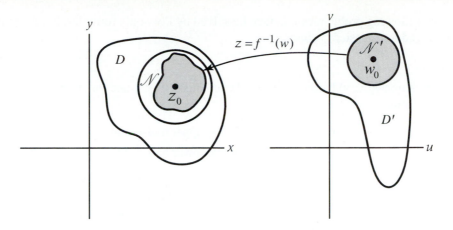

Figure 7.2 Mappings of neighborhoods.

or, equivalently,

$$\psi(w) := \phi\left(f^{-1}(w)\right),$$

agrees with $\phi(x, y)$ at corresponding points. Our claim is that if ϕ is harmonic in D, then ψ is harmonic in D'.

The proof is quite simple. Consider any point w_0 in D', say $w_0 = f(z_0)$. Now D is an open set, so there exists an open disk $\mathcal{N}$ centered at z_0 and lying entirely in D. Since $f^{-1}(w)$ is an analytic function, and thus continuous, there must be a sufficiently small neighborhood $\mathcal{N}'$ around w_0 whose image under f^{-1} lies entirely inside $\mathcal{N}$ (this is depicted in Fig. 7.2). To see that $\psi(u, v)$ is harmonic in the neighborhood $\mathcal{N}'$, recall that in Sec. 2.5 we proved that any harmonic function can be taken as the real part of an analytic function in "nice enough" domains—in particular, this is true for disks. Hence $\phi(x, y)$ is the real part of an analytic function $g(z)$ on $\mathcal{N}$. But then $\psi(u, v)$ is the real part of the composite function $g\left(f^{-1}(w)\right)$ on $\mathcal{N}'$, and since the composition of analytic functions is again analytic, ψ must be harmonic in the neighborhood $\mathcal{N}'$ of w_0. Consequently, as w_0 is an arbitrary point of D', the function ψ must be harmonic everywhere on D'.

Another proof of this fact can be based upon the direct verification of Laplace's equation in the w-plane, using the Cauchy-Riemann equations (see Prob. 2).

One can readily see why the preceding result is so useful in applications. Consider the *Dirichlet problem*, which requires us to find a function harmonic in a domain and taking specified values on its boundary (cf. Sec. 4.7). Once we have solved this problem on a particular domain, we immediately have solutions on all the domains that we can map onto the original one via a one-to-one analytic function, as long as the boundary values correspond. So we select whichever domain renders the problem simplest: usually a washer, wedge, or wall (where the methods of Sec. 3.4 apply), or a disk or upper half-plane (where we can apply Poisson's integral formulas, Sec. 4.7).

The following problem can, in fact, be solved by Poisson's formulas, but the mapping procedure is easier to apply.

Example 1

Find a function $\phi(x, y)$ harmonic inside the unit disk $|z| < 1$ and satisfying the boundary conditions

$$\phi(x, y) \to +1 \qquad \text{on the upper half-circle,}$$
$$\phi(x, y) \to -1 \qquad \text{on the lower half-circle}$$

(see Fig. 7.3). (This function gives the temperature profile inside an infinitely long right circular cylinder whose outer wall is partitioned into sections maintained at different temperatures. Thermal insulation, of course, must be provided at the points of discontinuity $z = \pm 1$, and the temperature is not specified there.)

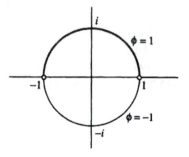

Figure 7.3 Dirichlet problem for Example 1.

Solution.   This complicated-looking problem will be solved by means of a mapping that carries the disk onto a 180° wedge — the right half-plane. In Sec. 7.3 the reader will be instructed on how to construct such mappings, but for now we simply want to illustrate the power of the technique. So we state without proof that the function

$$w = f(z) = \frac{1 + z}{1 - z} \tag{7}$$

maps the unit circle onto the imaginary axis and its interior onto the right half-plane. (The exterior maps to the left half-plane.) The correspondences are illustrated in Fig. 7.4. Observe that $|w|$ increases without bound as $z \to 1$, whereas $z = -1$ maps to $w = 0$. Moreover, the mapping (7) is one-to-one. In fact, the inverse is easily computed by solving for z in terms of w:

$$z = \frac{w - 1}{w + 1}. \tag{8}$$

The pair of functions (7) and (8) are employed in microwave engineering to construct the *Smith chart*, which is discussed in Exercises 7.3. Figure 7.5, from the paper

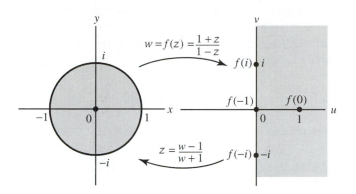

Figure 7.4 Mapping for Example 1.

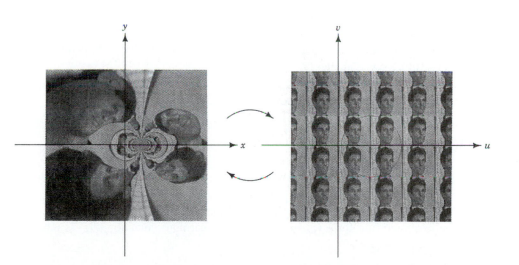

Figure 7.5 Correspondences for mapping Eqs. (7) and (8). (From C. Frederick and E.L. Schwartz, "Conformal Image Warping," *IEEE Computer Graphics and Applications*, March 1990, p. 26. Copyright 1990 IEEE.)

"Conformal Image Warping" by Frederick and Schwartz (see Ref. [9]), offers a some-what whimsical depiction of the mapping.

Notice that the upper half-circle, where the unknown function ϕ equals 1, is mapped to the positive imaginary axis, whereas the lower half-circle (where $\phi = -1$) corresponds to the negative imaginary axis. Consequently, by the methods of Sec. 3.4 we find the solution in the w-plane to be

$$\psi(u, v) = \frac{2}{\pi} \text{Arg}(w).$$

Hence the solution to the original problem is derived from ψ by the mapping (7):

$$\phi(x, y) = \psi(u(x, y), v(x, y)) = \frac{2}{\pi} \text{Arg}(f(z)) = \frac{2}{\pi} \text{Arg}\left(\frac{1+z}{1-z}\right).$$

A little algebra results in the expression

$$\phi(x, y) = \frac{2}{\pi} \tan^{-1} \frac{2y}{1 - x^2 - y^2},$$

where the value of the arctangent is taken to be between $-\pi/2$ and $\pi/2$. Note that $\phi(x, 0) = 0$, as we would expect from symmetry. ∎

With this example as motivation we devote the next few sections to a study of mappings given by analytic functions. The final two sections of the chapter will return us to applications, illustrating the power of this technique in handling many different situations. A table of some of the more useful mappings appears as Appendix II, for the reader's future convenience.

The MATLAB toolbox mentioned in the preface provides an excellent tool for visualizing most of the mappings studied in the chapter.

EXERCISES 7.1

1. Show that the function $w = e^z$ maps the half-strip $x > 0$, $-\pi/2 < y < \pi/2$ onto the portion of the right half w-plane that lies outside the unit circle (see Fig. 7.6). What harmonic function $\psi(w)$ does the w-plane "inherit," via this mapping, from the harmonic function $\phi(z) = x + y$? What harmonic function $\phi(z)$ is inherited from $\psi(w) = u + v$?

2. Suppose that Eqs. (5) and (6) describe a one-to-one analytic mapping. Let $\phi(x, y)$ be a real-valued twice-continuously differentiable function that is carried over in the w-plane to the function

$$\psi(u, v) := \phi(x(u, v), y(u, v)).$$

   (a) The *gradient* of $\phi(x, y)$ is the vector $(\partial\phi/\partial x, \partial\phi/\partial y)$; it corresponds to the complex number (recall Sec. 1.3) $\partial\phi/\partial x + i(\partial\phi/\partial y)$. Similarly, the gradient

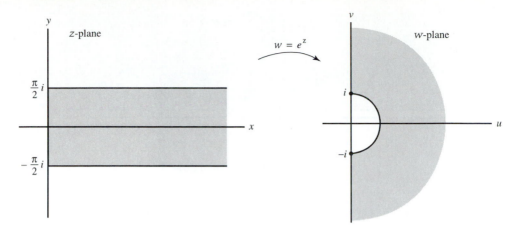

Figure 7.6 Exponential mapping of half-strip.

of ψ corresponds to $\partial\psi/\partial u + i(\partial\psi/\partial v)$. Use the chain rule and the Cauchy-Riemann equations to show that these gradients are related by

$$\frac{\partial\psi}{\partial u} + i\frac{\partial\psi}{\partial v} = \left(\frac{\partial\phi}{\partial x} + i\frac{\partial\phi}{\partial y}\right)\overline{\left(\frac{dz}{dw}\right)}.$$

(b) Show that the *Laplacians* of ψ and ϕ are related by

$$\left\{\frac{\partial^2\psi}{\partial u^2} + \frac{\partial^2\psi}{\partial v^2}\right\} = \left\{\frac{\partial^2\phi}{\partial x^2} + \frac{\partial^2\phi}{\partial y^2}\right\}\left|\frac{dz}{dw}\right|^2.$$

(c) Show that if $\phi(x, y)$ satisfies Laplace's equation in the z-plane, then ψ satisfies Laplace's equation in the w-plane.

(d) Show that if ϕ satisfies *Helmholtz's* equation,

$$\frac{\partial^2\phi}{\partial x^2} + \frac{\partial^2\phi}{\partial y^2} = \Lambda\phi$$

(Λ is a constant), in the z-plane, then ψ satisfies

$$\frac{\partial^2\psi}{\partial u^2} + \frac{\partial^2\psi}{\partial v^2} = \Lambda\left|\frac{dz}{dw}\right|^2\psi$$

in the w-plane. (Helmholtz's equation arises in transient thermal analysis.)

3. Find a function ϕ harmonic in the upper half-plane and taking boundary values as indicated in Fig. 7.7. [HINT: Reread Sec. 3.4.]

4. Consider the problem of finding a function ϕ that is harmonic in the right half-plane and takes the values $\phi(0, y) = y/(1 + y^2)$ on the imaginary axis. Observe that the obvious first guess

$$\phi(z) = \operatorname{Im}\frac{z}{1 - z^2},$$

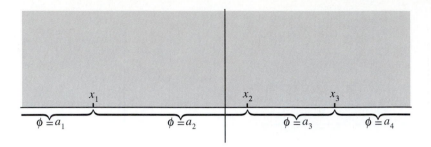

Figure 7.7 Dirichlet problem in Prob. 3.

fails because $z / (1 - z^2)$ is *not* analytic at $z = 1$. However, the following strategy can be used.

(a) According to the text, the mappings (7) and (8) provide a correspondence between the right half-plane and the unit disk. (Of course, one should interchange the roles of z and w in the formulas.) Thus the w-plane inherits from $\phi(z)$ a function $\psi(w)$ harmonic in the unit disk. Show that the values of $\psi(w)$ on the unit circle $w = e^{i\theta}$ must be given by

$$\psi(e^{i\theta}) = \frac{\sin\theta}{2}.$$

(b) Argue that the harmonic function $\psi(w)$ must be given by

$$\psi(w) = \frac{1}{2}\operatorname{Im} w$$

throughout the unit disk.

(c) Use the mappings to carry $\psi(w)$ back to the z-plane, producing the function

$$\phi(z) = \frac{y}{y^2 + (x + 1)^2}$$

as a solution of the problem.

5. Use the strategy of Prob. 4 to find a function ϕ harmonic in the right half-plane such that $\phi(0, y) = 1/(y^2 + 1)$.

6. Suppose that the harmonic function $\phi(x, y)$ in the domain D is carried over to the harmonic function $\psi(u, v)$ in the domain D' via the one-to-one analytic mapping $w = f(z)$. Prove that if the normal derivative $\partial\phi/\partial n$ is zero on a curve Γ in D, then the normal derivative $\partial\psi/\partial n$ is zero on the image curve of Γ under f. (The boundary condition $\partial\phi/\partial n = 0$ is known as a *Neumann* condition.) [HINT: $\partial\phi/\partial n$ is the projection of the gradient $(\partial\phi/\partial x) + i(\partial\phi/\partial y)$ onto the normal, and the gradient is orthogonal to the level curves $\phi(x, y) = $ constant.]

7. Suppose that $f(z)$ is analytic and one-to-one. Then, according to the text, you may presume that f^{-1} is also analytic. If x, y, u, v are as in Eqs. (5) and (6), explain the identities

$$\frac{\partial x}{\partial u} = \frac{\partial y}{\partial v}, \qquad \frac{\partial x}{\partial v} = -\frac{\partial y}{\partial u}.$$

7.2 Geometric Considerations

The geometric aspects of analytic mappings split rather naturally into two categories: *local* properties and *global* properties. Local properties need only hold in sufficiently small neighborhoods, while global properties hold throughout a domain. For example, consider the function e^z. It is one-to-one in any disk of diameter less than 2π, and hence it is locally one-to-one, but since $e^{z_1} = e^{z_2}$ when $z_1 - z_2 = 2\pi i$, the function is not globally one-to-one. On the other hand, sometimes local properties can be extended to global properties; in fact, this is the essence of *analytic continuation* (see Sec. 5.8).

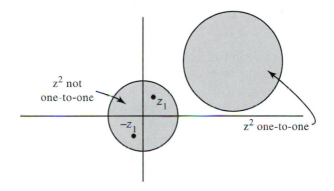

Figure 7.8 Locally one-to-one mapping.

Let us begin our study of local properties by considering "one-to-oneness." As the example e^z shows, a function may be locally one-to-one without being globally one-to-one. (Of course, the opposite situation is impossible.) Furthermore, an analytic function may be locally one-to-one at some points but not at others. Indeed, consider

$$f(z) = z^2.$$

In any open set that contains the origin there will be distinct points z_1 and z_2 such that $z_2 = -z_1$, and hence (since $z_2^2 = z_1^2$) the function f will not be one-to-one. However, around any point other than the origin, we *can* find a neighborhood in which z^2 is one-to-one (any disk that excludes the origin will do; see Fig. 7.8). Thus $f(z) = z^2$ is locally one-to-one at every point other than the origin. An explanation of the exceptional nature of $z = 0$ in this example is provided by the following.

Theorem 1. If f is analytic at z_0 and $f'(z_0) \neq 0$, then there is an open disk D centered at z_0 such that f is one-to-one on D.

Proof. Since f' is also analytic at z_0 and $|f'(z_0)| > 0$, there is an open disk D centered at z_0 such that $|f'(z) - f'(z_0)| \leq |f'(z_0)|/2$ for all z in D. We shall show that, in fact, $|f(z_1) - f(z_2)| \geq |z_1 - z_2| |f'(z_0)/2|$ for z_1, z_2 inside this disk; so f is certainly one-to-one in D.

Let Γ be the segment joining z_1 and z_2. Then we have

$$|f(z_1) - f(z_2)| = \left| \int_\Gamma f'(z)\,dz \right|$$

$$= \left| \int_\Gamma f'(z_0)\,dz - \int_\Gamma [f'(z_0) - f'(z)]\,dz \right|$$

$$\geq |f'(z_0)(z_2 - z_1)| - \left| \frac{f'(z_0)}{2} \right| |z_2 - z_1|$$

$$= \left| \frac{f'(z_0)}{2} \right| |z_2 - z_1|. \quad \blacksquare$$

Theorem 1 says that an analytic function is locally one-to-one at points where its derivative does not vanish. In advanced texts it is shown more generally that if z_0 is a zero of order m for f', then f is locally "$(m+1)$-to-one" around (but excluding) z_0; in other words, each value of f is taken on $m+1$ times. This is reinforced by the observation that $f(z) = z^2$ is two-to-one in any punctured neighborhood of the origin.

The next local property we shall discuss is *conformality*. Consider the following situation: $f(z)$ is analytic and one-to-one in a neighborhood of the point z_0, and γ_1 and γ_2 are two directed smooth curves (in this neighborhood) intersecting at z_0. Under the mapping f the images of these curves, γ_1' and γ_2', are also directed smooth curves, and they will intersect at $w_0 = f(z_0)$. At the point z_0 we construct vectors $\mathbf{v}_1$ and $\mathbf{v}_2$ tangent to γ_1 and γ_2, respectively, and pointing in the directions consistent with the orientations of the curves (see Fig. 7.9). Then the *angle from γ_1 to γ_2* is the angle θ through which $\mathbf{v}_1$ must be rotated counterclockwise in order to lie along $\mathbf{v}_2$. The angle θ' from γ_1' to γ_2' is defined similarly.

Now the mapping f is said to be *conformal at z_0* if these angles are preserved; that is, $\theta = \theta'$ for every pair of directed smooth curves that intersect at z_0. For analytic mappings we have the following theorem.

Theorem 2. An analytic function f is conformal at every point z_0 for which $f'(z_0) \neq 0$.

Proof. By Theorem 1 we know that there is some open disk containing z_0 in which f is one-to-one. We will argue that every directed smooth curve through z_0

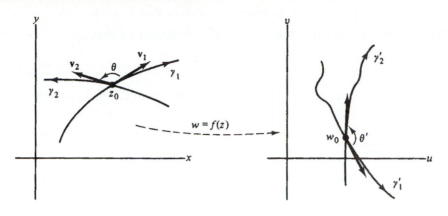

Figure 7.9 Conformality.

has its tangent (at z_0) turned through the same angle, under the mapping $w = f(z)$. Consequently, the angle between any two curves intersecting at z_0 will be preserved.

So let γ be any directed smooth curve through z_0 parametrized, say, by $z = z(t)$ with $z(t_0) = z_0$. The vector $z'(t_0)$ is then tangent to γ at z_0. Under the mapping the image, γ', of γ has the parametrization

$$w = w(t) = f(z(t))$$

with

$$w_0 := f(z_0) = f(z(t_0)),$$

and the vector $w'(t_0)$ (if it is nonzero) is tangent to γ' at w_0. But the chain rule implies

$$w'(t_0) = f'(z_0) z'(t_0), \tag{1}$$

so $w'(t_0)$ is nonzero since $f'(z_0) \neq 0$. Furthermore, we see from Eq. (1) that the angles which the tangent vectors $z'(t_0)$, $w'(t_0)$ make with the horizontal are related by

$$\arg w'(t_0) = \arg f'(z_0) + \arg z'(t_0).$$

Hence every curve through z_0 is rotated through the same angle $\arg f'(z_0)$ which, of course, is a constant independent of the particular curve. ■

The condition that $f'(z_0)$ not vanish in Theorem 2 is crucial; the function $f(z) = z^2$, for which $f'(0) = 0$, does not preserve angles at the origin—it *doubles* them. However, it is common practice to call any mapping generated by a nonconstant analytic function a "conformal map," overlooking the violations occurring at the points where f' is zero. (Incidentally, these exceptional points will be reexamined in Sec. 7.5; they turn out to be quite important.)

Moving now to the global aspects of conformal mapping, we begin with a property that has both local and global ramifications: the *open mapping property*. A function is said to be an *open mapping* if the image of every open set in its domain is, itself, open; that is, the function maps open sets to open sets. For analytic functions we have the following theorem.

Theorem 3. Any analytic function that is nonconstant on domains is an open mapping.

A proof of this theorem was given in Sec. 6.7. Note that the theorem prohibits, for instance, a nonconstant analytic mapping of a disk onto a portion of a line.

It is very useful in investigating conformal maps to exploit the concept of *connectivity*. For example, one can show that any nonconstant analytic function takes domains, that is, open connected sets, to domains. Openness is preserved because of Theorem 3, and as for connectivity, we argue as follows. Let us say f maps the domain D onto the open set $\mathcal{O}$. To join $w_1 = f(z_1)$ to $w_2 = f(z_2)$ by a polygon path in $\mathcal{O}$, first join z_1 to z_2 in D. Then the image of this path is a path joining w_1 to w_2 in $\mathcal{O}$. Of course, the image path need not be polygonal, but any competent topologist can prove that such a path, lying inside an *open* set, can be deformed into a polygonal path without leaving the set. Hence $\mathcal{O}$ is connected. (See Fig. 7.10.)

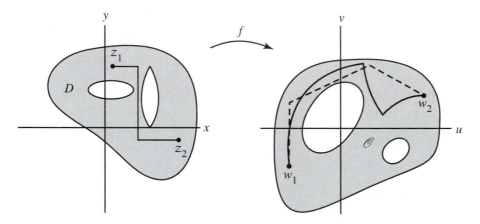

Figure 7.10 Conformal maps preserve connectivity.

The remaining topic from the global theory of analytic functions that we shall consider here is the *Riemann mapping theorem*. Because it is primarily an *existence* theorem, its usefulness in applied mathematics is somewhat limited.

Theorem 4 (Riemann Mapping Theorem). Let D be any simply connected domain in the plane other than the entire plane itself. Then there is a one-to-one analytic function that maps D onto the open unit disk. Moreover, one can prescribe an arbitrary point of D and a direction through that point which are to

> be mapped to the origin and the direction of the positive real axis, respectively. Under such restrictions the mapping is unique.

A direction through a point z_0 is specified, of course, by an angle ϕ, as in Fig. 7.11. The Riemann mapping theorem allows us to specify z_0 and ϕ so that all curves through z_0 with tangent in the direction ϕ are mapped to curves through the origin with tangent along the positive real axis. This yields three "degrees of freedom," or three choices to be made, in fixing the map: the real and the imaginary parts of the point that goes to 0, and the direction.

Again, we appeal to the references for a complete treatment of this theorem. (As an ominous note, we tantalize the reader by pointing out that the theorem makes no predictions about the boundary values of the function.)

From the Riemann mapping theorem we can conclude that any simply connected domain D_1 can be analytically mapped one-to-one onto any other simply connected domain D_2, assuming that neither D_1 nor D_2 is the whole plane. Indeed, let f map D_1 onto the unit disk, and let g map D_2 onto this disk, in accordance with the theorem. Then $g^{-1}(f(z))$ maps D_1 onto D_2 and is one-to-one and analytic. (See Fig. 7.12.)

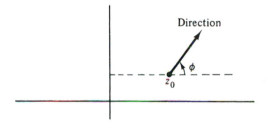

Figure 7.11 Three degrees of freedom.

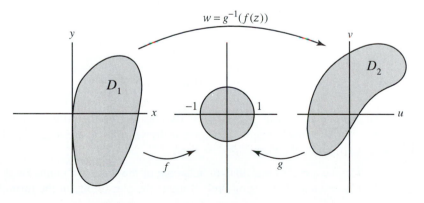

Figure 7.12 Mapping of simply connected domains.

The remainder of this chapter will deal with constructing and applying *specific* conformal mappings.

EXERCISES 7.2

1. For each of the following functions, determine the order m of the zero of the derivative f' at z_0 and show explicitly that the function is not one-to-one in any neighborhood of z_0.

   (a) $f(z) = z^2 + 2z + 1$, $z_0 = -1$
   (b) $f(z) = \cos z$, $z_0 = 0, \pm\pi, \pm 2\pi, \ldots$
   (c) $f(z) = e^{z^3}$, $z_0 = 0$

2. Prove that if $w = f(z)$ is analytic at z_0 and $f'(z_0) \neq 0$, then $z = f^{-1}(w)$ is analytic at $w_0 = f(z_0)$, and

$$\frac{df^{-1}}{dw}(w) = \frac{1}{\dfrac{df}{dz}(z)}$$

   for $w = w_0$, $z = z_0$. [HINT: Theorem 1 guarantees that $f^{-1}(w)$ exists near w_0 and Theorem 3 implies that $f^{-1}(w)$ is continuous. Now generalize the proof in Sec. 3.2.]

3. What happens to angles at the origin under the mapping $f(z) = z^\alpha$ for $\alpha > 1$? For $0 < \alpha < 1$?

4. Use the open mapping theorem to prove the maximum-modulus principle.

5. Find all functions $f(z)$ analytic in $D : |z| < 1$ that assume only pure imaginary values in D.

6. If f is analytic at z_0 and $f'(z_0) \neq 0$, show that the function $g(z) = \overline{f(z)}$ preserves the magnitude, but reverses the orientation, of angles at z_0.

7. Show that the mapping $w = z + 1/z$ maps circles $|z| = \rho$ ($\rho \neq 1$) onto ellipses

$$\frac{u^2}{\left(\rho + \dfrac{1}{\rho}\right)^2} + \frac{v^2}{\left(\rho - \dfrac{1}{\rho}\right)^2} = 1.$$

8. Let f be analytic at z_0 with $f'(z_0) \neq 0$. By considering the difference quotient, argue that "infinitesimal" lengths of segments drawn from z_0 are magnified by the factor $|f'(z_0)|$ under the mapping $w = f(z)$.

9. Let $w = f(z)$ be a one-to-one analytic mapping of the domain D onto the domain D', and let $A' = \text{area}(D')$. Using Prob. 8, argue the plausibility of the formula

$$A' = \iint_D |f'(z)|^2 \, dx \, dy.$$

10. Why is it impossible for D to be the whole plane in the Riemann mapping theorem? [HINT: Appeal to Liouville's theorem.]

11. Describe the image of each of the following domains under the mapping $w = e^z$.

    (a) the strip $0 < \operatorname{Im} z < \pi$

    (b) the slanted strip between the two lines $y = x$ and $y = x + 2\pi$

    (c) the half-strip $\operatorname{Re} z < 0, 0 < \operatorname{Im} z < \pi$

    (d) the half-strip $\operatorname{Re} z > 0, 0 < \operatorname{Im} z < \pi$

    (e) the rectangle $1 < \operatorname{Re} z < 2, 0 < \operatorname{Im} z < \pi$

    (f) the half-planes $\operatorname{Re} z > 0$ and $\operatorname{Re} z < 0$

12. Let $P(z) = (z - \alpha)(z - \beta)$, and let L be any straight line through $(\alpha + \beta)/2$. Prove that P is one-to-one on each of the open half-planes determined by L.

13. Describe the image of each of the following domains under the mapping $w = \cos z = \cos x \cosh y - i \sin x \sinh y$. [HINT: Consider the image of the boundary in each case.]

    (a) the half-strip $0 < \operatorname{Re} z < \pi$, $\operatorname{Im} z < 0$

    (b) the half-strip $0 < \operatorname{Re} z < \dfrac{\pi}{2}$, $\operatorname{Im} z > 0$

    (c) the strip $0 < \operatorname{Re} z < \pi$

    (d) the rectangle $0 < \operatorname{Re} z < \pi, -1 < \operatorname{Im} z < 1$

14. Prove that if f has a simple pole at z_0, then there exists a punctured neighborhood of z_0 on which f is one-to-one.

15. A domain D is said to be *convex* if for any two points z_1, z_2 in D, the line segment joining z_1 and z_2 lies entirely in D. Prove the *Noshiro-Warschawski theorem*: Let f be analytic in a convex domain D. If $\operatorname{Re} f'(z) > 0$ for all z in D, then f is one-to-one in D. [HINT: Write $f(z_2) - f(z_1)$ as an integral of f'.]

16. (For students who have read Sec. 4.4a) Argue that a one-to-one analytic function will map simply connected domains to simply connected domains.

7.3   Möbius Transformations

The problem of finding a one-to-one analytic function that maps one domain onto another can be quite perplexing, so it is worthwhile to investigate a few elementary mappings in order to compile some rules of thumb that we can draw upon. The basic properties of *Möbius transformations*,[†] which we shall investigate in this section, constitute an essential portion of every analyst's bag of tricks. (Some of these mappings were previewed in Exercises 2.1.)

[†] In 1865 August Möbius (1790-1860) described the *Möbius strip*, a piece of paper that has only one side and one edge.

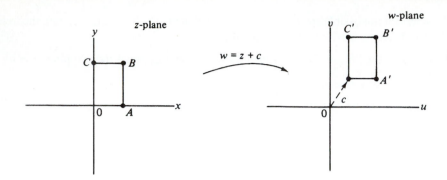

Figure 7.13 Translation.

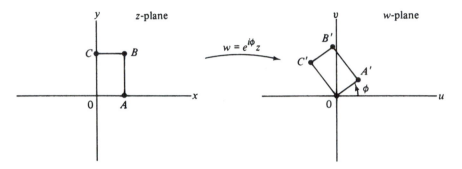

Figure 7.14 Rotation.

First, let's consider the simplest mapping of all, the *translation* defined by the function

$$w = f(z) = z + c, \tag{1}$$

where c is a fixed complex number. Under this mapping every point is shifted by the vector corresponding to c. Its properties are quite apparent: The entire complex plane is mapped one-to-one onto itself, and every geometric object is mapped onto a congruent object. (See Fig. 7.13.)

Rotations are quite simple also. Observe that under the transformation

$$w = f(z) = e^{i\phi}z, \tag{2}$$

with ϕ real, every point is rotated about the origin through the angle ϕ. Such transformations are also one-to-one mappings of the complex plane onto itself and map geometric objects onto congruent objects. (See Fig. 7.14.)

The mapping defined by

$$w = f(z) = \rho z, \tag{3}$$

where ρ is a positive real constant, simply enlarges (or contracts) the distance of every point from the origin by the factor ρ; hence such a transformation is called a *magni-*

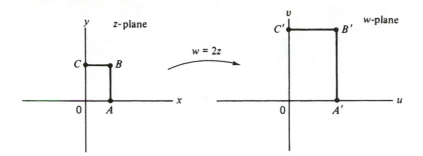

Figure 7.15 Magnification.

fication. Observe that the distance between *any* two points is multiplied by this same constant, since

$$|w_1 - w_2| = |f(z_1) - f(z_2)| = |\rho z_1 - \rho z_2| = \rho |z_1 - z_2|.$$

Magnifications thus rescale distances and (since they are conformal) preserve angles; consequently any geometric object is mapped onto an object that is *similar* to the original. And again, the complex plane is mapped one-to-one onto itself. (See Fig. 7.15.)

A *linear transformation*[†] is any mapping of the form

$$w = f(z) = az + b, \tag{4}$$

where a and b are complex constants with $a \neq 0$. Such a transformation can be considered as the composition of a rotation, a magnification, and a translation (each of which is, of course, a special case of the linear transformation): Writing a in polar form as $a = \rho e^{i\phi}$, we express the linear transformation (4) by the composition of

$$w_1 = e^{i\phi} z,$$
$$w_2 = \rho w_1,$$

and, finally,

$$w = w_2 + b.$$

Hence the linear transformation is, once again, one-to-one in the complex plane, and the image of any object is geometrically similar to the original. (See Fig. 7.16.)

Example 1

Find the linear transformation that rotates the entire complex plane through an angle θ about a given point z_0.

[†]This is, unfortunately, bad terminology, because in other branches of mathematics a "linear transformation" has the property $f(z_1 + z_2) = f(z_1) + f(z_2)$, while this is true of Eq. (4) only when $b = 0$. Worse yet, *Möbius* transformations are called linear transformations by some authors.

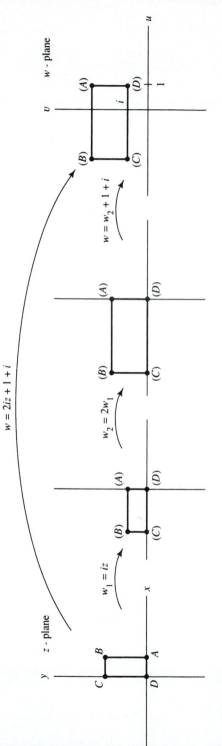

Figure 7.16 Linear transformation $w = 2iz + 1 + i$.

Solution.   We know that the mapping

$$w_1 = e^{i\theta} z$$

rotates the plane through the angle θ about the origin. In particular, the point z_0 is mapped to the point $e^{i\theta} z_0$. If we now shift the whole plane so that the latter point is carried *back to* z_0, the net result will be a rotation of the plane about z_0. (Think about this; every straight line gets rotated through the angle θ, and z_0 is left fixed.) Thus the required answer is

$$w = w_1 + (z_0 - e^{i\theta} z_0) = e^{i\theta} z + (1 - e^{i\theta}) z_0. \quad \blacksquare$$

Example 2

Find a linear transformation that maps the circle $C_1 : |z - 1| = 1$ onto the circle $C_2 : |w - 3i/2| = 2$.

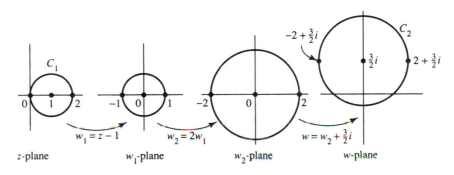

Figure 7.17 Mappings for Example 2.

Solution.   Refer to Fig. 7.17. First we translate by -1 so that C_1 becomes a unit circle centered at the origin. Then we magnify by the factor 2. Finally we translate $3/2$ units up the imaginary axis, bringing us to C_2. The mappings are

$$w_1 = z - 1,$$
$$w_2 = 2w_1 = 2z - 2,$$

and, last,

$$w = w_2 + 3i/2 = 2z - 2 + 3i/2.$$

Moreover, any subsequent rotation about the point $3i/2$ can be permitted. $\quad \blacksquare$

Now we consider the *inversion* transformation defined by

$$w = f(z) = \frac{1}{z}. \tag{5}$$

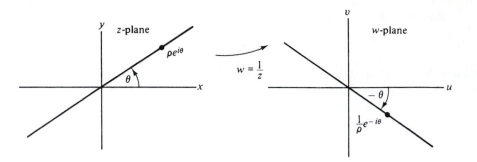

Figure 7.18 Inversion of line through origin.

It is easy to see that the inversion is a one-to-one mapping of the *extended* complex plane onto itself.[†] It "turns the unit circle inside out," mapping the interior to the exterior and vice versa.

Note that the point $z = \rho e^{i\theta}$ is mapped to

$$w = \frac{1}{\rho e^{i\theta}} = \frac{1}{\rho} e^{-i\theta}. \tag{6}$$

If we interpret (6) with the real number ρ varying from $-\infty$ to $+\infty$ while θ is fixed, we see that the line through the origin making an angle θ with the real axis is mapped onto the line through the origin making the angle $-\theta$ with the real axis; the point at ∞ goes to the origin and vice versa. (See Fig. 7.18.) On the other hand, if we interpret (6) with θ varying from 0 to 2π while ρ is fixed, we see that the circle of radius ρ centered at the origin is mapped onto the circle of radius $1/\rho$ centered at the origin (*traced backward*, we might say).

However it is wrong to jump to the conclusion that inversion always maps lines to lines, and circles to circles. But it does the next best thing: the image of a line is always *either a line or a circle*, and so is the image of a circle! So as long as we're treating ∞ as a point, we could regard a line as a "generalized circle" with an infinite radius and say that *inversion maps generalized circles to generalized circles*.

A direct verification of this statement is lengthy, involving the verification of the appropriate equations from analytic geometry. so we relegate it to problems 15–17.

(Note, however, than one can readily derive this statement by considering the stereographic projection interpretation described in Sec. 1.7. First observe that the stereographic projection of generalized circles in the plane are circles on the Riemann sphere (Example 2, Sec. 1.7) and conversely, every circle on the sphere is the image of a generalized circle. Second recall the fact that inversion corresponds to a $180°$ rotation of the Riemann sphere about a diameter (Example 4, Sec. 2.1). Since such a rotation preserves circles on the sphere, the mapping $1/z$ preserves generalized circles in the plane.)

[†]Think about this statement: "The inversion is its own inverse."

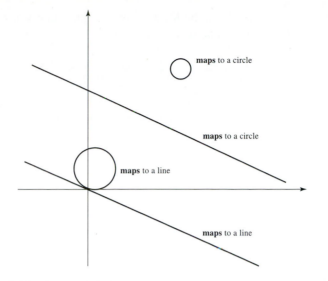

Figure 7.19 Lines and circles.

At any rate, it is always easy to tell whether the image will be a line or circle; if the *original* line or circle passes through the origin, its image will contain the point at infinity, so it must be a line; if the original line or circle misses the origin, its image will be bounded and, hence, a circle. See Fig. 7.19.

To recap, the inversion mapping is one-to-one and carries the class of straight lines and circles into itself, a property shared with translations, rotations, and magnifications.

Now we are ready to define the *Möbius transformations*.

Definition 1. A **Möbius** (or Moebius) **transformation** (sometimes known as a **fractional linear transformation** or **bilinear transformation**) is any function of the form

$$w = f(z) = \frac{az + b}{cz + d} \tag{7}$$

with the restriction that $ad \neq bc$ (so that w is not a constant function).

Notice that since

$$f'(z) = \frac{ad - bc}{(cz + d)^2}$$

does not vanish, the Möbius transformation $f(z)$ is conformal at every point except its pole $z = -d/c$.

It is easy to see that the Möbius transformations include the previous elementary transformations of this section as special cases. More important is the fact that any Möbius transformation can be decomposed into a succession of these elementary transformations. If $c = 0$, we have the linear transformation that was treated earlier. For $c \neq 0$, the decomposition can be seen by writing

$$\frac{az + b}{cz + d} = \frac{\dfrac{a}{c}(cz + d) - \dfrac{ad}{c} + b}{cz + d} = \frac{a}{c} + \frac{b - \dfrac{ad}{c}}{cz + d},$$

which shows that the Möbius transformation can be expressed as a linear transformation (rotation + magnification + translation)

$$w_1 = cz + d, \tag{8}$$

followed by an inversion

$$w_2 = \frac{1}{w_1}, \tag{9}$$

and then another linear transformation

$$w = \left(b - \frac{ad}{c}\right) w_2 + \frac{a}{c}. \tag{10}$$

As a result of this decomposition and of our previous deliberations, we can summarize some properties of Möbius transformations.

Theorem 5. Let f be any Möbius transformation. Then

(i) f can be expressed as the composition of a finite sequence of translations, magnifications, rotations, and inversions.

(ii) f maps the extended complex plane one-to-one onto itself.

(iii) f maps the class of circles and lines to itself.

(iv) f is conformal at every point except its pole.

The possibilities in property (iii) are distinguished as follows. If a line or circle passes through the pole ($z = -d/c$) of the Möbius transformation, it gets mapped to an unbounded figure. Hence its image is a straight line. A line or circle that avoids the pole, then, is mapped to a circle.

Example 3

Find the image of the *interior* of the circle $C : |z - 2| = 2$ under the Möbius transformation

$$w = f(z) = \frac{z}{2z - 8}.$$

Solution.   First we find the image of the circle C. Since f has a pole at $z = 4$ and this point lies on C, the image has to be a straight line. To specify this line all we need is to determine two of its finite points. The points $z = 0$ and $z = 2 + 2i$ which lie on C have, as their images,

$$w = f(0) = 0 \text{ and } w = f(2 + 2i) = \frac{2 + 2i}{2(2 + 2i) - 8} = -\frac{i}{2}.$$

Thus the image of C is the imaginary axis in the w-plane. From our discussion on connectivity in Sec. 7.2, we know that the interior of C is, therefore, mapped either onto the right half-plane $\operatorname{Re} w > 0$ or onto the left half-plane $\operatorname{Re} w < 0$. Since $z = 2$ lies inside C and

$$w = f(2) = \frac{2}{4 - 8} = -\frac{1}{2}$$

lies in the left half-plane, we conclude that the image of the interior of C is the left half-plane. ∎

Now we shall present an example showing how to *construct* a conformal map of one region onto another.

Example 4

Find a conformal map of the unit disk $|z| < 1$ onto the right half-plane $\operatorname{Re} w > 0$.

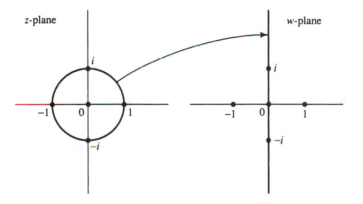

Figure 7.20 Mapping for Example 4.

Solution. We are naturally led to look for a Möbius transformation that maps the circle $|z| = 1$ onto the imaginary axis (Fig. 7.20). The transformation must therefore have a pole on the circle, according to our earlier remarks. Moreover, the origin $w = 0$ must also lie on the image of the circle. As a first step, let's look at

$$w = f_1(z) = \frac{z + 1}{z - 1}, \tag{11}$$

which maps 1 to ∞ and -1 to 0.

From the geometric properties of Möbius transformations that we have learned, we can conclude that (11) maps $|z| = 1$ onto *some* straight line through the origin. To see *which* straight line, we plug in $z = i$ and find that the point

$$w = \frac{i+1}{i-1} = -i$$

also lies on the line. Hence the image of the circle under f_1 must be the imaginary axis.

To see which half-plane is the image of the interior of the circle, we check the point $z = 0$. It is mapped by (11) to the point $w = -1$ in the *left* half-plane. This is not what we want, but it can be corrected by a final rotation of π, yielding

$$w = f(z) = -\frac{z+1}{z-1} = \frac{1+z}{1-z} \tag{12}$$

as an answer to the problem. (Of course, any subsequent vertical translation or magnification can be permitted.) Observe that (12) is precisely the mapping that was introduced in Example 1, Sec. 7.1, to solve a thermal problem, and we have thus verified the claims made there. ∎

EXERCISES 7.3

1. Find a linear transformation mapping the circle $|z| = 1$ onto the circle $|w - 5| = 3$ and taking the point $z = i$ to $w = 2$.

2. What is the image of the strip $0 < \operatorname{Im} z < 1$ under the mapping $w = (z - i)/z$?

3. Discuss the image of the circle $|z - 2| = 1$ and its interior under the following transformations.

   (a) $w = z - 2i$ (b) $w = 3iz$ (c) $w = \dfrac{z-2}{z-1}$

   (d) $w = \dfrac{z-4}{z-3}$ (e) $w = \dfrac{1}{z}$

4. Find a Möbius transformation mapping the lower half-plane to the disk $|w + 1| < 1$. [HINT: Do it in steps.]

5. Find a Möbius transformation mapping the unit disk $|z| < 1$ onto the right half-plane and taking $z = -i$ to the origin.

6. A *fixed point* of a function $f(z)$ is a point z_0 satisfying $f(z_0) = z_0$. Show that a Möbius transformation $f(z)$ can have at most two fixed points in the complex plane unless $f(z) \equiv z$.

7. Find the Möbius transformation that maps $0, 1, \infty$ to the following respective points.

   (a) $0, i, \infty$ (b) $0, 1, 2$ (c) $-i, \infty, 1$ (d) $-1, \infty, 1$

8. What is the image, under the mapping $w = (z + i)/(z - i)$, of the third quadrant?

9. What is the image of the sector $-\pi/4 < \operatorname{Arg} z < \pi/4$ under the mapping $w = z/(z-1)$?

10. Find a conformal map of the semidisk $|z| < 1$, $\operatorname{Im} z > 0$, onto the upper half-plane. [HINT: Combine a Möbius transformation with the mapping $w = z^2$. Make sure you cover the entire upper half-plane.]

11. Map the shaded region in Fig. 7.21 conformally onto the upper half-plane. [HINT: Use a Möbius transformation to map the point 2 to ∞. Argue that the image region will be a *strip*. Then use the exponential map.]

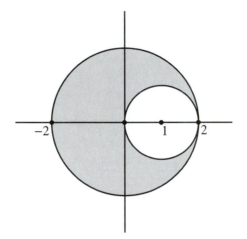

Figure 7.21 Region for Prob. 11.

12. Find a Möbius transformation that takes the half-plane depicted in Fig. 7.22 onto the unit disk $|w| < 1$.

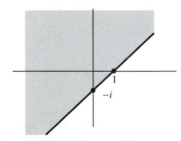

Figure 7.22 Region for Prob. 12.

13. (*Smith Chart*) The *impedance* Z of an electrical circuit oscillating at a frequency ω is a complex number, denoted $Z = R + iB$, which characterizes the voltage-current relationship of the circuit; recall Sec. 3.6. In practice R can take any value from 0 to ∞ and B can take any value from $-\infty$ to ∞. Thus the usual representation of

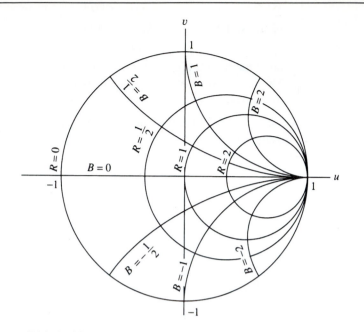

Figure 7.23 Smith chart.

Z as a point in the complex plane becomes unwieldy (inasmuch as the entire right half-plane comes into play). The *Smith chart* provides a more compact graphical description, displaying the entire range of impedances within the unit circle. The impedance Z is depicted as the point

$$W = \frac{Z-1}{Z+1}.$$

This mapping (its inverse, actually) is portrayed in Figs. 7.4 and 7.5. W is also known as the *reflection coefficient* corresponding to Z.

(a) Show that the circles in the Smith chart depicting the lines $\operatorname{Re} Z = R =$ constant, indicating constant-resistance contours, have the equations

$$\left(u - \frac{R}{1+R}\right)^2 + v^2 = \frac{1}{(1+R)^2}.$$

(b) Show that the circles in the Smith chart depicting the lines $\operatorname{Im} Z = B =$ constant, indicating constant-reactance contours, have the equations

$$(u-1)^2 + \left(v - \frac{1}{B}\right)^2 = \frac{1}{B^2}.$$

(See Fig. 7.23.)

14. If a circuit with impedance Z is connected to a length ℓ of *transmission line* with "phase constant" β and a "characteristic impedance" of unity, then the new configuration has a transformed impedance Z' given by

$$Z' = \frac{Z \cos \beta\ell + i \sin \beta\ell}{\cos \beta\ell + iZ \sin \beta\ell}.$$

Show that the Smith chart point depicting Z' can be obtained from the Smith chart point depicting Z by a clockwise rotation of $2\beta\ell$ radians about the origin.[†]

15. Show that the transformation (5) maps lines not passing through the origin onto circles passing through the origin. [HINT: The equation of such a line is $Ax + By = C$, with $C \neq 0$. Solve

$$z = x + iy = 1/w = 1/(u + iv) \tag{13}$$

for x and y in terms of u and v and substitute. Show that the result can be expressed in the form

$$u^2 + v^2 - \frac{A}{C}u + \frac{B}{C}v = 0 .] \tag{14}$$

16. Show that the transformation (5) maps circles passing through the origin onto lines not passing through the origin. [HINT: Use the preceding problem.]

17. Show that the transformation (5) maps circles not passing through the origin onto circles not passing through the origin. [HINT: The equation of such circles is

$$x^2 + y^2 + Ax + By = C, \quad \text{with} \quad C \neq 0.$$

Substitute the expressions for x and y derived from (13) to obtain

$$u^2 + v^2 - \frac{A}{C}u + \frac{B}{C}v = \frac{1}{C} .]$$

7.4   Möbius Transformations, Continued

We shall now explore some additional properties of Möbius transformations that enhance their usefulness as conformal mappings. These are the group properties, the cross-ratio formula, and the symmetry property.

Given any Möbius transformation

$$w = f(z) = \frac{az + b}{cz + d} \quad (ad \neq bc), \tag{1}$$

its inverse $f^{-1}(w)$ can be found by simply solving Eq. (1) for z in terms of w. This computation yields

$$z = f^{-1}(w) = \frac{dw - b}{-cw + a},$$

[†]P. H. Smith patented the Smith chart in the late 1930s. It is the only known conformal mapping to be protected by copyright!

and we see that *the inverse of any Möbius transformation is again a Möbius transformation.* Furthermore, if we take the composition of two Möbius transformations, say

$$w = f_1(z) = \frac{a_1 z + b_1}{c_1 z + d_1} \quad \text{and} \quad \zeta = f_2(w) = \frac{a_2 w + b_2}{c_2 w + d_2},$$

it can be readily shown that

$$\zeta = f_2(f_1(z)) = \frac{(a_2 a_1 + b_2 c_1) z + (a_2 b_1 + b_2 d_1)}{(c_2 a_1 + d_2 c_1) z + (c_2 b_1 + d_2 d_1)}.$$

Hence *the composition of any two Möbius transformations is also a Möbius transformation.* [Rigorously speaking, we should make this claim only after verifying that $f_2(f_1)$ cannot reduce to a constant, but this is obviously true because the one-to-oneness of f_1 and f_2 implies that $f_2(f_1)$ is one-to-one.] Of course when we take $f_1^{-1}(f_1)$ we get the identity function $I(z) \equiv z$, which is certainly a Möbius transformation. These facts are known as the *group properties* of the Möbius transformations (see Prob. 21).

We have already seen that Möbius transformations map the class of circles and lines to itself. Now we turn to the problem of finding a specific transformation that maps a *given* circle (or line) C_z in the z-plane to a *given* circle (or line) C_w in the w-plane. Recall from geometry that any three distinct noncollinear points uniquely determine a circle; if these points are collinear, then, of course, they uniquely determine a line (in particular this will be the case when one of them is ∞). Hence if we choose three points z_1, z_2, z_3 on C_z and three points w_1, w_2, w_3 on C_w and find a Möbius transformation f satisfying

$$f(z_1) = w_1, \qquad f(z_2) = w_2, \qquad f(z_3) = w_3, \tag{2}$$

then f must map C_z onto C_w.

It is not difficult to write down a Möbius transformation that satisfies Eqs. (2) in the case when $w_1 = 0$, $w_2 = 1$, and $w_3 = \infty$; this corresponds to the problem of mapping C_z onto the real axis. If all the points z_1, z_2, z_3 are finite, it is easy to check that

$$T(z) = \frac{(z - z_1)(z_2 - z_3)}{(z - z_3)(z_2 - z_1)} \tag{3}$$

satisfies

$$T(z_1) = 0, \qquad T(z_2) = 1, \qquad T(z_3) = \infty, \tag{4}$$

while if one of the z_i is ∞, the conditions (4) will be satisfied by

$$T(z) = \frac{z_2 - z_3}{z - z_3} \quad (z_1 = \infty), \qquad T(z) = \frac{z - z_1}{z - z_3} \quad (z_2 = \infty),$$

$$\tag{5}$$

$$\text{or} \qquad T(z) = \frac{z - z_1}{z_2 - z_1} \quad (z_3 = \infty).^\dagger$$

---

†Note that the Möbius transformations in (5) can be obtained immediately from (3) by simply deleting the factors involving ∞.

We remark that the right-hand side of Eq. (3) [or Eqs. (5)] is called the *cross-ratio* of the four points z, z_1, z_2, z_3 and is abbreviated by writing (z, z_1, z_2, z_3); that is,

$$(z, z_1, z_2, z_3) := \frac{(z - z_1)(z_2 - z_3)}{(z - z_3)(z_2 - z_1)} \tag{6}$$

in the case of finite points. Notice that the order in which the points are listed is crucial in this notation. For example,

$$(z, 3, 0, i) = \frac{(z - 3)(0 - i)}{(z - i)(0 - 3)} = \frac{-iz + 3i}{-3z + 3i},$$

but

$$(z, i, 3, 0) = \frac{(z - i)(3 - 0)}{(z - 0)(3 - i)} = \frac{3z - 3i}{(3 - i)z}.$$

Now if we wish to solve the general problem of finding a Möbius transformation f that maps

$$z_1 \text{ to } w_1, \qquad z_2 \text{ to } w_2, \qquad z_3 \text{ to } w_3,$$

where w_1, w_2, w_3 are any three distinct points, we can proceed as follows: Let $T(z)$ be the Möbius transformation just discussed, taking

$$z_1 \text{ to } 0, \qquad z_2 \text{ to } 1, \qquad z_3 \text{ to } \infty,$$

and let $S(w)$ be the analogous transformation that maps

$$w_1 \text{ to } 0, \qquad w_2 \text{ to } 1, \qquad w_3 \text{ to } \infty.$$

Then the desired Möbius transformation f is given by the composition

$$w = f(z) = S^{-1}(T(z)), \tag{7}$$

because

$$f(z_1) = S^{-1}(T(z_1)) = S^{-1}(0) = w_1,$$
$$f(z_2) = S^{-1}(T(z_2)) = S^{-1}(1) = w_2,$$
$$f(z_3) = S^{-1}(T(z_3)) = S^{-1}(\infty) = w_3.$$

Notice that Eq. (7) is equivalent to the equation

$$S(w) = T(z);$$

in other words, to map z_1, z_2, z_3 to the respective points w_1, w_2, w_3, we need merely equate the two cross-ratios

$$(w, w_1, w_2, w_3) = (z, z_1, z_2, z_3) \tag{8}$$

and solve for w in terms of z.

Example 1

Find a Möbius transformation that maps 0 to i, 1 to 2, and -1 to 4.

Solution.   The appropriate cross-ratios are given by

$$(z, 0, 1, -1) = \frac{(z - 0)[1 - (-1)]}{[z - (-1)](1 - 0)} = \frac{2z}{z + 1}$$

and

$$(w, i, 2, 4) = \frac{(w - i)(2 - 4)}{(w - 4)(2 - i)} = \frac{-2(w - i)}{(w - 4)(2 - i)}.$$

Hence, solving the equation

$$\frac{-2(w - i)}{(w - 4)(2 - i)} = \frac{2z}{z + 1}$$

for w yields the desired transformation

$$w = \frac{(16 - 6i)z + 2i}{(6 - 2i)z + 2}. \quad \blacksquare$$

It is important to note that the circle or line Γ determined by the three points z_1, z_2, z_3 is also *oriented* by the order of these points. That is, Γ acquires the direction obtained by proceeding through the points z_1, z_2, z_3 in succession. [Notice that lines are regarded as "closed" at ∞ in the present context. Hence they, like circles, require a sequence of *three* points to determine a direction. See Fig. 7.24(d).] This orientation, in turn, uniquely specifies the "left region," the region that lies to the left of an observer traversing Γ (Fig. 7.24). Since Möbius transformations are conformal, it can be shown that a Möbius transformation that takes z_1, z_2, z_3 to the respective points w_1, w_2, w_3 must map the left region of the circle (or line) oriented by z_1, z_2, z_3 onto the left region of the circle (or line) oriented by w_1, w_2, w_3. To see this in the special case depicted in Fig. 7.25 imagine a short directed segment drawn from a point on the circle in the z-plane into the left region. The image of this segment is, by conformality, a curve

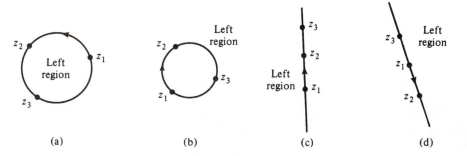

Figure 7.24 Left regions determined by three-point sequence.

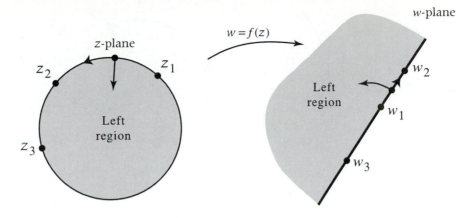

Figure 7.25 Correspondence of left regions.

from the image line into its left region. By connectivity, then, we conclude that the left region is mapped to the left region. The other situations of mapping a circle to a circle, a line to a circle, and a line to a line can be treated in a similar manner.

Thus by judiciously selecting ordered points we can quickly write down an algebraic formula for a mapping of one "circular" region onto another.

Example 2

Find a Möbius transformation that maps the region $D_1 : |z| > 1$ onto the region $D_2 : \operatorname{Re} w < 0$.

Solution. We shall take both D_1 and D_2 to be left regions. This is accomplished for D_1 by choosing any three points on the circle $|z| = 1$ that give it a negative (clockwise) orientation, say

$$z_1 = 1, \qquad z_2 = -i, \qquad z_3 = -1.$$

Similarly the three points

$$w_1 = 0, \qquad w_2 = i, \qquad w_3 = \infty$$

on the imaginary axis make D_2 its left region. Hence a solution to the problem is given by the transformation that takes

$$1 \text{ to } 0, \qquad -i \text{ to } i, \qquad -1 \text{ to } \infty.$$

This we obtain by setting

$$(w, 0, i, \infty) = (z, 1, -i, -1),$$

that is,

$$\frac{w - 0}{i - 0} = \frac{(z - 1)(-i + 1)}{(z + 1)(-i - 1)},$$

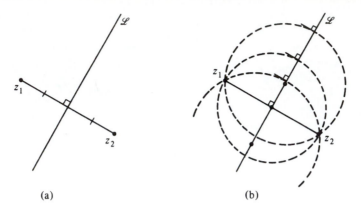

Figure 7.26 z_1, z_2 symmetric with respect to line L.

which yields

$$w = \frac{(z-1)(1+i)}{(z+1)(-i-1)} = \frac{1-z}{1+z}. \quad \blacksquare$$

Another important aspect of Möbius transformations is their symmetry-preserving property. First recall that two points z_1 and z_2 are symmetric with respect to a straight line $\mathcal{L}$ if $\mathcal{L}$ is the perpendicular bisector of the line segment joining z_1 and z_2 [see Fig. 7.26(a)]. From elementary geometry this is equivalent to saying that *every* line or circle through z_1 and z_2 intersects $\mathcal{L}$ orthogonally, that is, at right angles. See Fig. 7.26(b). (Remember that a circle is orthogonal to $\mathcal{L}$ if and only if its center lies on $\mathcal{L}$.)

These considerations suggest the following definition of symmetry with respect to a circle C.

Definition 2. Two points z_1 and z_2 are said to be **symmetric with respect to a circle** C if every straight line or circle passing through z_1 and z_2 intersects C orthogonally (Fig. 7.27).

In particular the center a of the circle C, and the point ∞, are symmetric with respect to C; there are no circles through these two points, and any line containing a (and necessarily ∞) is orthogonal to C, so the condition in Definition 2 holds.

Now we are in a position to state the symmetry-preserving property of Möbius transformations.

Theorem 6 (*Symmetry Principle*). Let C_z be a line or circle in the z-plane, and let $w = f(z)$ be any Möbius transformation. Then two points z_1 and z_2 are symmetric with respect to C_z if and only if their images $w_1 = f(z_1)$, $w_2 = f(z_2)$ are symmetric with respect to the image of C_z under f.

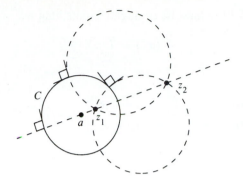

Figure 7.27 z_1, z_2 symmetric with respect to circle C.

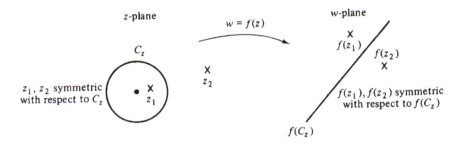

Figure 7.28 Symmetry Principle.

The theorem is illustrated in Fig. 7.28 for the special case when C_z is a circle that maps to a straight line.

Proof. This is easy; think about it. Two points are symmetric with respect to a circle (line) if every circle or line containing the points intersects the given circle (line) orthogonally. But Möbius transformations preserve the class of circles and lines, and they also preserve the orthogonality; hence they preserve the symmetry condition. ∎

Now given a circle C with center a and radius R, and given a point α, it would be convenient to have a formula for the point α^* symmetric to α with respect to C. For this purpose observe that the transformation

$$T(z) = (z, a - R, a + Ri, a + R)$$
$$= \frac{[z - (a - R)](Ri - R)}{[z - (a + R)](Ri + R)} = i\frac{z - (a - R)}{z - (a + R)} \tag{9}$$

maps three points of C, and hence all of C, onto the real axis. Thus, by the symmetry principle α^* is symmetric to α with respect to C if and only if $T(\alpha^*)$ is symmetric to $T(\alpha)$ with respect to the real axis. But the latter condition is clearly equivalent to saying that $T(\alpha^*)$ and $T(\alpha)$ must be conjugate points; that is,

$$T(\alpha^*) = \overline{T(\alpha)},$$

or, using Eq. (9),

$$i\frac{\alpha^* - (a - R)}{\alpha^* - (a + R)} = \overline{\left[i\frac{\alpha - (a - R)}{\alpha - (a + R)}\right]} = -i\frac{\bar{\alpha} - (\bar{a} - R)}{\bar{\alpha} - (\bar{a} + R)}. \tag{10}$$

Solving Eq. (10) for α^* yields the formula

$$\alpha^* = \frac{R^2}{\bar{\alpha} - \bar{a}} + a. \tag{11}$$

(Notice that this also shows that the point symmetric to α with respect to C is *unique*.)
From representation (11) we see that

$$\arg\left(\alpha^* - a\right) = \arg\left(\frac{R^2}{\bar{\alpha} - \bar{a}}\right) = \arg\left[\frac{R^2(\alpha - a)}{|\alpha - a|^2}\right] = \arg(\alpha - a),$$

and

$$\left|\alpha^* - a\right| = \frac{R^2}{|\bar{\alpha} - \bar{a}|} = \frac{R^2}{|\alpha - a|},$$

implying that symmetric points α^* and α lie on the same ray from the center a and that
the product of their distances from the center ($|\alpha^* - a| \cdot |\alpha - a|$) is equal to the radius
squared. Figure 7.29 suggests a construction of symmetric points.

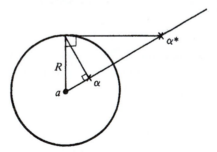

Figure 7.29 Similar triangles give $R/|\alpha - a| = |\alpha^* - a|/R$.

Example 3
Find all Möbius transformations that map $|z| < 1$ onto $|w| < 1$.

Solution.   Let $f(z)$ be any such Möbius transformation. Then f maps the circle
$C_z : |z| = 1$ onto $C_w : |w| = 1$. Furthermore, there must be some point α, $|\alpha| < 1$,

that is mapped to the origin; that is, $f(\alpha) = 0$. According to formula (11) (with $a = 0$, $R = 1$) the point

$$\alpha^* = \frac{1^2}{\bar{\alpha} - 0} + 0 = \frac{1}{\bar{\alpha}}$$

is symmetric to α with respect to C_z. Hence $f(1/\bar{\alpha})$ must be symmetric to $f(\alpha) = 0$ with respect to C_w. But since the origin is the center of C_w, its symmetric point is ∞; that is,

$$f\left(\frac{1}{\bar{\alpha}}\right) = \infty.$$

Consequently, f has a zero at α and a pole at $1/\bar{\alpha}$, so f is of the form

$$f(z) = k \cdot \frac{z - \alpha}{z - \frac{1}{\bar{\alpha}}} = k\bar{\alpha}\frac{z - \alpha}{\bar{\alpha}z - 1}$$

for some constant k. Moreover, since $f(1)$ lies on C_w, we have

$$1 = |f(1)| = |k\bar{\alpha}| \cdot \left|\frac{1 - \alpha}{\bar{\alpha} - 1}\right| = |k\bar{\alpha}|.$$

Thus $k\bar{\alpha} = e^{i\theta}$ for some real θ, and we find

$$f(z) = e^{i\theta} \cdot \frac{z - \alpha}{\bar{\alpha}z - 1} \qquad (|\alpha| < 1). \qquad (12)$$

Conversely, the reader can easily show (Prob. 15) that any transformation of the form (12) maps $|z| < 1$ onto $|w| < 1$. ∎

More generally, it can be shown that the functions in Eq. (12) are the only one-to-one *analytic* mappings of the unit disk onto itself.

EXERCISES 7.4

1. Let $f_1(z) = (z + 2)/(z + 3)$, $f_2(z) = z/(z + 1)$. Find $f_1^{-1}(f_2(z))$.

2. Argue why the Möbius transformation defined by

$$(w, -i, 1, i) = (z, -i, i, 1)$$

maps the unit circle onto itself but maps the interior onto the exterior. [HINT: Consider orientation.]

3. Find the point symmetric to $4 - 3i$ with respect to each of the following circles.

(a) $|z| = 1$ (b) $|z - 1| = 1$ (c) $|z - 1| = 2$

4. Prove that if z_2, z_3, and z_4 are distinct points in the extended complex plane and T is any Möbius transformation, then

$$(z_1, z_2, z_3, z_4) = (T(z_1), T(z_2), T(z_3), T(z_4))$$

for any point z_1 in the extended plane. That is, *the cross-ratio is invariant under Möbius transformations.*

5. Let $w = f(z)$ be the Möbius transformation mapping the points 0, λ, ∞ to $-i$, 1, i, respectively, where λ is real. For what values of λ is the upper half-plane mapped onto $|w| < 1$?

6. Using the cross-ratio notation, write an equation defining a Möbius transformation that maps the half-plane below the line $y = 2x - 3$ onto the interior of the circle $|w - 4| = 2$. Repeat for the exterior of this circle.

7. Does there exist a Möbius transformation f that maps the real axis onto the unit circle $|w| = 1$ and satisfies $f(i) = 2$, $f(-i) = -\frac{1}{2}$?

8. Prove that if z_1, z_2, z_3 are distinct points and w_1, w_2, w_3 are distinct points, then the Möbius transformation T satisfying $T(z_1) = w_1$, $T(z_2) = w_2$, $T(z_3) = w_3$ is *unique*. [HINT: Suppose that S is another such Möbius transformation and consider the fixed points of the composition $T^{-1} \circ S$ (see Prob. 6, Exercises 7.3).]

9. Let f be a Möbius transformation such that $f(1) = \infty$ and f maps the imaginary axis onto the unit circle $|w| = 1$. What is the value of $f(-1)$?

10. By completing the following steps, prove that given a line L and a circle C with no points in common, there always exist two distinct points z_1 and z_2 that are symmetric with respect to L and C *simultaneously.*

    (a) Argue that there exists a Möbius transformation that both maps L onto the real axis and maps C onto a circle of the form $|w - \lambda i| = R$ with λ real and $R < |\lambda|$ (see Fig. 7.30).

    (b) Then show that w_1 and w_2 are symmetric with respect to both the real axis and the circle $|w - \lambda i| = R$ if and only if

$$w_2 = \overline{w_1} \quad \text{and} \quad w_2 = \frac{R^2}{\overline{w_1} + \lambda i} + \lambda i.$$

    Solve this pair of equations to obtain

$$w_1 = i\sqrt{\lambda^2 - R^2}, \qquad w_2 = -i\sqrt{\lambda^2 - R^2}$$

    as the simultaneously symmetric points.

    (c) Use the results of parts (a) and (b) and the symmetry principle to conclude that there are points z_1 and z_2 symmetric in both L and C.

11. Prove that given any two nonintersecting circles C_1 and C_2, there always exist two distinct points z_1 and z_2 that are symmetric with respect to C_1 and C_2 *simultaneously.* [HINT: Argue that there exists a Möbius transformation that maps C_1 onto a line and C_2 onto another circle. Then use the result of the preceding problem.]

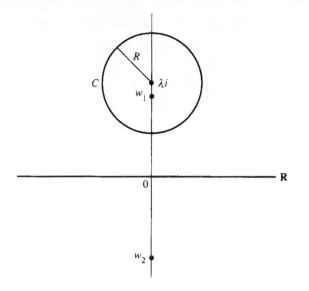

Figure 7.30 w_1, w_2 symmetric in **R** and C.

12. Use the result of Prob. 11 to show that for any two nonintersecting circles C_1 and C_2 there always exists a Möbius transformation that maps C_1 and C_2 onto *concentric circles*. [HINT: Map z_1 to the origin and z_2 to infinity, where z_1, z_2 are points symmetric with respect to both circles.]

13. Let z_1, z_2, and z_3 be three distinct points that lie on a circle (or line) C. Prove that z and z^* are symmetric with respect to C if and only if $\overline{(z^*, z_1, z_2, z_3)} = (z, z_1, z_2, z_3)$.

14. Show that the distinct points w_1, w_2, w_3, and w_4 all lie on the same circle or line if and only if the cross-ratio (w_1, w_2, w_3, w_4) is real. [HINT: Consider the Möbius transformation defined by

$$(w, w_2, w_3, w_4) = (z, 0, 1, \infty)$$

and observe that $(z, 0, 1, \infty) \equiv z$.]

15. Verify that any transformation of the form (12) maps $|z| < 1$ onto $|w| < 1$.

16. Prove that every Möbius transformation mapping the interior of the unit disk to its exterior takes the form (12) with $|\alpha| > 1$.

17. Find a conformal map of the unit disk onto itself, taking the point $i/2$ to the origin.

18. Show that the Möbius transformation taking z_i to w_i $(i = 1, 2, 3)$ can be expressed in determinant form as

$$\begin{vmatrix} 1 & z & w & zw \\ 1 & z_1 & w_1 & z_1 w_1 \\ 1 & z_2 & w_2 & z_2 w_2 \\ 1 & z_3 & w_3 & z_3 w_3 \end{vmatrix} = 0.$$

19. Find all Möbius transformations that map the upper half-plane onto itself.

20. Show that every Möbius transformation that maps the upper half-plane onto the open unit disk must be of the form

$$f(z) = e^{i\theta}\frac{z - z_0}{z - \bar{z}_0}, \qquad \text{where } \text{Im}(z_0) > 0.$$

21. A set $\mathcal{G}$ of mathematical objects (such as numbers or mappings) together with an operation $*$ defined on ordered pairs of objects in $\mathcal{G}$ is called a *group* if it satisfies the following conditions:

  i. $a * b$ is a unique element of $\mathcal{G}$ for every ordered pair (a, b) of elements of $\mathcal{G}$.

  ii. The operation $*$ is associative; that is, for any three elements a, b, c of $\mathcal{G}$,

$$(a * b) * c = a * (b * c).$$

  iii. There exists an element e in $\mathcal{G}$ (called the *identity* element) with the property that

$$e * a = a * e = a \qquad \text{for all } a \text{ in } \mathcal{G}.$$

  iv. For each a in $\mathcal{G}$ there exists an element a^{-1} in $\mathcal{G}$ (called the *inverse* of a) such that

$$a^{-1} * a = a * a^{-1} = e.$$

  (a) Prove that the set $\mathcal{M}$ of Möbius transformations forms a group under the operation of composition $\circ$ of mappings.

  (b) Is the group of Möbius transformations commutative? (That is, is $T \circ S = S \circ T$ for all $S, T \in \mathcal{M}$?)

22. Let $\mathcal{L}$ be the set of all two-by-two (complex number) matrices having determinant 1:

$$\begin{pmatrix} a & b \\ c & d \end{pmatrix}, \qquad ad - bc = 1.$$

  (a) Prove that $\mathcal{L}$ forms a group under ordinary multiplication of matrices (see Prob. 21 for the definition of *group*).

  (b) Show that on multiplying numerator and denominator by a suitable number, any Möbius transformation T can be written in the form

$$T(z) = \frac{\alpha z + \beta}{\gamma z + \delta}$$

  with $\alpha\delta - \beta\gamma = 1$. Thus T can be associated with the element

$$\begin{pmatrix} \alpha & \beta \\ \gamma & \delta \end{pmatrix}$$

  of $\mathcal{L}$.

(c) Show that if the Möbius transformations T_1 and T_2 are associated as in part (b) with the elements

$$S_1 = \begin{pmatrix} \alpha_1 & \beta_1 \\ \gamma_1 & \delta_1 \end{pmatrix} \quad \text{and} \quad S_2 = \begin{pmatrix} \alpha_2 & \beta_2 \\ \gamma_2 & \delta_2 \end{pmatrix}$$

of $\mathcal{L}$, then the composition $T_1 \circ T_2$ is associated with the product matrix $S_1 S_2$.

23. Let z be fixed with $\operatorname{Re} z \geq 0$, and let

$$T_0(w) = \frac{a_0}{z + a_0 + b_1 + w}, \quad T_k(w) = \frac{a_k}{z + b_{k+1} + w} \quad (k = 1, 2, \ldots, n-1)$$

be a sequence of Möbius transformations such that each a_k is real and positive and each b_k is pure imaginary or zero. Prove, by induction, that the composition

$$\zeta = S(w) := T_0 \circ T_1 \circ \cdots \circ T_{n-2} \circ T_{n-1}(w)$$

maps the half-plane $\operatorname{Re} w > 0$ onto a region contained in the disk $|\zeta - \frac{1}{2}| < \frac{1}{2}$.

24. Let $P(z) = z^n + c_1 z^{n-1} + c_2 z^{n-2} + \cdots + c_n$ be a polynomial of degree $n > 0$ with complex coefficients $c_k = p_k + i q_k$, $k = 1, 2, \ldots, n$. Set $Q(z) := p_1 z^{n-1} + i q_2 z^{n-2} + p_3 z^{n-3} + i q_4 z^{n-4} + \cdots$. Prove *Wall's criterion* that if $Q(z)/P(z)$ can be written in the form

$$\frac{Q(z)}{P(z)} = \cfrac{a_0}{z + a_0 + b_1 + \cfrac{a_1}{z + b_2 + \cfrac{a_2}{z + b_3 + \cfrac{\ddots}{\ddots + \cfrac{a_{n-1}}{z + b_n}}}}},$$

where each a_k is real and positive and each b_k is pure imaginary or zero, then all the zeros of $P(z)$ have negative real parts. [HINT: Write $Q(z)/P(z) = T_0 \circ T_1 \circ \cdots \circ T_{n-1}(0)$, where the transformations T_k are defined as in Prob. 23.]

25. Prove that $P(z) = z^3 + 3z^2 + 6z + 6$ has all its zeros in the left half-plane by applying the result of Prob. 24. [HINT: Use ordinary long division to obtain the representation for $Q(z)/P(z)$.]

7.5   The Schwarz-Christoffel Transformation

We have seen that a function $f(z)$ is conformal at every point at which it is analytic and its derivative is nonzero. It is instructive to analyze what happens at certain isolated points where these conditions are not met. For concreteness, let x_1 be a fixed point on the real axis and let $f(z)$ be a function whose *derivative* $f'(z)$ is given by $(z - x_1)^\alpha$ for some real α satisfying $-1 < \alpha < 1$. [To be precise, we shall take the argument of $z - x_1$ to lie between $-\pi/2$ and $3\pi/2$, introducing a branch cut vertically downward from x_1; see Fig. 7.31(a).] We are going to use the equation

$$f'(z) = (z - x_1)^\alpha \tag{1}$$

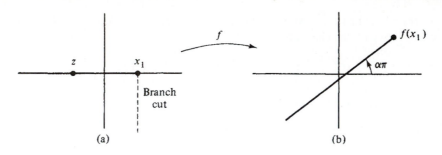

Figure 7.31 Geometry for Eq. (1).

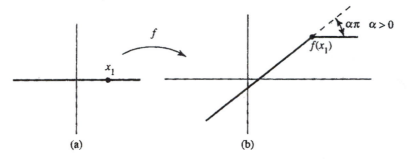

Figure 7.32 Mapping of x-axis.

to determine certain features of the image of the real axis under the mapping f.

If z lies on the x-axis to the left of x_1, as in Fig. 7.31(a), then two observations follow from Eq. (1); namely, f is conformal at z (since f' exists and is nonzero), and the argument of $f'(z)$ is constant for all such z:

$$\arg f'(z) = \arg (z - x_1)^\alpha = \alpha \arg (z - x_1) = \alpha\pi$$

(we ignore multiples of 2π in this derivation). From this we can conclude that f maps the interval $(-\infty, x_1)$ onto a portion of a *straight line* terminating at $f(x_1)$; after all, if we view $(-\infty, x_1)$ as a curve whose tangents are all parallel to the real axis, then according to the discussion of Sec. 7.2, its image must be a curve, all of whose tangents make an angle $\alpha\pi$ with the real axis—that is, a straight line. See Fig. 7.31(b).

For z on the real axis to the right of x_1, we have

$$\arg f'(z) = \alpha \arg (z - x_1) = \alpha \cdot 0 = 0.$$

Hence, by similar reasoning, the interval (x_1, ∞) is mapped to a *horizontal* straight line, and the whole picture looks like Fig. 7.32(b).

For the special case where $f(z) = \frac{2}{3}z^{3/2}$, which has $f'(z) = z^{1/2}$, the mapping (for the branch described earlier) is sketched in Fig. 7.33.

Now we start to generalize this model. If, instead of Eq. (1) we have

$$f'(z) = A (z - x_1)^\alpha \tag{2}$$

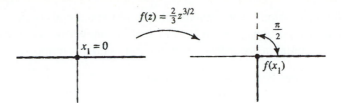

Figure 7.33 Mapping of x-axis by $f(z) = \frac{2}{3}z^{3/2}$.

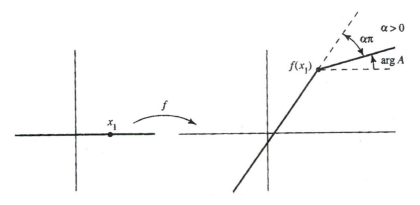

Figure 7.34 Mapping for Eq. (2).

for some complex constant A ($\neq 0$), then

$$\arg f'(z) = \arg A + \alpha \arg (z - x_1),$$

and the mapping can be visualized by rotating Fig. 7.32(b) by an amount $\arg A$; see Fig. 7.34. In particular, the angle made by the image of the interval (x_1, ∞) is now $\arg A$, but the angle of the turn at $f(x_1)$ is unchanged.

The next generalization is to consider a mapping given by a function f with a derivative of the form

$$f'(z) = A \, (z - x_1)^{\alpha_1} \, (z - x_2)^{\alpha_2} \cdots (z - x_n)^{\alpha_n} ; \tag{3}$$

here A ($\neq 0$) is a complex constant, each α_i lies between -1 and $+1$, and the (real) x_i satisfy

$$x_1 < x_2 < \cdots < x_n.$$

(As before we take the argument of each $z - x_i$ to be between $-\pi/2$ and $3\pi/2$.) What does this mapping f do to the real axis?

From the equation

$$\arg f'(z) = \arg A + \alpha_1 \arg (z - x_1) + \alpha_2 \arg (z - x_2) + \cdots + \alpha_n \arg (z - x_n)$$

and the previous discussion we see that the images of the intervals $(-\infty, x_1)$, (x_1, x_2), $\ldots$, (x_n, ∞) are each portions of straight lines, making angles measured counterclockwise from the horizontal in accordance with the following prescription:

$$
\begin{array}{ll}
\textit{Interval} & \textit{Angle of image} \\
(-\infty, x_1) & \arg A + \alpha_1 \pi + \alpha_2 \pi + \cdots + \alpha_n \pi \\
(x_1, x_2) & \arg A + \alpha_2 \pi + \cdots + \alpha_n \pi \\
\quad\vdots & \qquad\vdots \\
(x_{n-1}, x_n) & \arg A + \alpha_n \pi \\
(x_n, \infty) & \arg A.
\end{array}
$$

Hence as z traverses the real axis from left to right $f(z)$ generates a polygonal path whose tangent at the point $f(x_i)$ *makes a right turn through the angle* $\alpha_i \pi$; see Fig. 7.35.

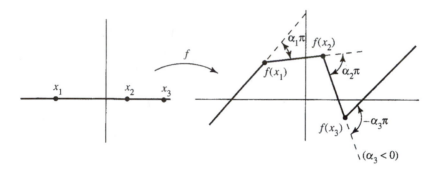

Figure 7.35 Mapping for Eq. (3).

Now if the function $f(z)$ satisfies Eq. (3) it is, a priori, differentiable and hence analytic on the complex plane with the exception of the (downward) branch cuts from the points x_i. So for any z in the upper half-plane we can set

$$
g(z) := \int_\Gamma f'(\zeta)\, d\zeta, \tag{4}
$$

where Γ is, for definiteness, the straight line segment from 0 to z, and conclude then that $f(z) = g(z) + B$ for some constant B. In particular, we can write

$$
f(z) = A \int_0^z (\zeta - x_1)^{\alpha_1} (\zeta - x_2)^{\alpha_2} \cdots (\zeta - x_n)^{\alpha_n}\, d\zeta + B. \tag{5}
$$

Functions of the form (5) are known as *Schwarz-Christoffel transformations*.[†] We have seen that such transformations map the real axis onto a polygonal path. Now one of the most important problems in conformal mapping applications is the construction

---

[†]Hermann Amandus Schwarz (1842-1921), Elwin Bruno Christoffel (1829-1900).

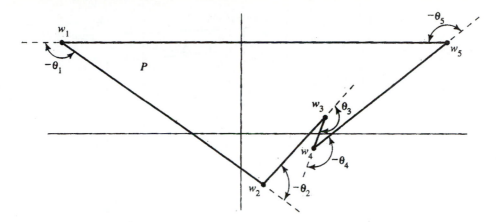

Figure 7.36 Positively oriented polygon (θ_2, θ_3, θ_4, θ_5 are negative).

of a one-to-one analytic function carrying the upper half-plane to the interior of a *given* polygon. We thus turn to the task of tailoring a Schwarz-Christoffel transformation to accomplish this.

To be specific, let the polygon P have vertices at the consecutive points w_1, w_2, ..., w_n taken in *counterclockwise* order, giving P a positive orientation, as in Fig. 7.36. In traversing the polygon we make a right turn at vertex w_i through the angle θ_i. Thus each angle lies between $-\pi$ and π and a negative value of θ_i indicates a left turn. The *net* rotation for a counterclockwise tour must be 2π radians to the left:

$$\theta_1 + \theta_2 + \cdots + \theta_n = -2\pi. \tag{6}$$

To map the x-axis onto P with a Schwarz-Christoffel transformation $w = g(z)$ we begin by picking real points $x_1, x_2, \ldots, x_{n-1}$ as the preimages of the vertices $w_1, w_2, \ldots, w_{n-1}$, and presume that both $x = -\infty$ and $x = \infty$ are the preimages of w_n; see Fig. 7.37. From the discussion of Eq. (5) it follows that the function

$$g(z) := \int_0^z (\zeta - x_1)^{\theta_1/\pi} (\zeta - x_2)^{\theta_2/\pi} \cdots (\zeta - x_{n-1})^{\theta_{n-1}/\pi} \, d\zeta \tag{7}$$

maps the real axis onto *some* polygon P'. Although P' may not be the desired polygon P, it does have the proper right-turn angles $\alpha_i \pi = \theta_i$ at the corners $g(x_i)$ for $i = 1, 2, \ldots, n-1$; and since the initial and final segments intersect at $g(\pm\infty)$, the right turn at this final vertex must match the angle θ_n (because both are given by $-2\pi - \theta_1 - \theta_2 - \cdots - \theta_{n-1}$).

Now because P' has the same angles as P, by adjusting the lengths of the sides of P' we can make it *geometrically similar* to P. And it seems quite plausible that we could accomplish this by adjusting the points $x_1, x_2, \ldots, x_{n-1}$; after all, they determine where the corners of P' lie. Then, with the use of a rotation, a magnification, and a translation—in other words, a linear transformation—we could make these similar polygons coincide.

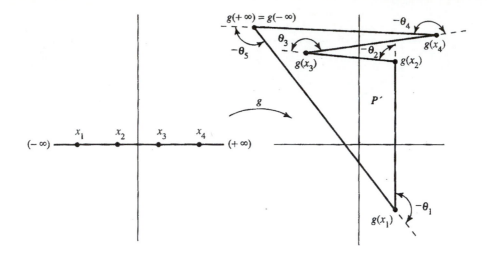

Figure 7.37 Mapping for Eq. (7).

Summarizing, we are led to speculate that with an appropriate choice of the constants we can construct a function

$$f(z) = Ag(z) + B$$

$$= A \int_0^z (\zeta - x_1)^{\theta_1/\pi} (\zeta - x_2)^{\theta_2/\pi} \cdots (\zeta - x_{n-1})^{\theta_{n-1}/\pi} \, d\zeta + B, \tag{8}$$

that is, a Schwarz-Christoffel transformation, which maps the real axis onto the perimeter of a given polygon P, with the correspondences

$$f(x_1) = w_1, \quad f(x_2) = w_2, \quad \ldots, \quad f(x_{n-1}) = w_{n-1}, \quad f(\infty) = w_n. \tag{9}$$

Moreover, if our speculations are valid, we can use conformality and connectivity arguments to show that f maps the upper half-plane to the interior of P, as was requested; for observe that if γ is a segment as indicated in Fig. 7.38, conformality requires that its image, γ', have a tangent that initially points inward as shown, and connectivity completes the argument (assuming one-to-oneness). The whole story about Schwarz-Christoffel transformations is given in Theorem 7, whose proof can be found in the references.

Theorem 7. Let P be a positively oriented polygon having consecutive corners at $w_1, w_2, \ldots, w_n$ with corresponding right-turn angles θ_i ($i = 1, 2, \ldots, n$). Then there exists a function of the form (8) that is a one-to-one conformal map from the upper half-plane onto the interior of P. Furthermore, the correspondences (9) hold.

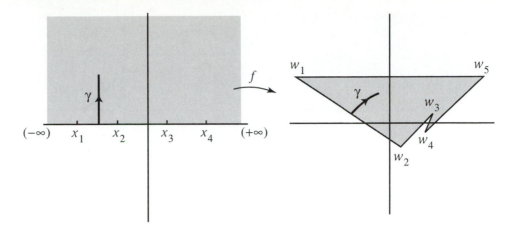

Figure 7.38 The upper half-plane is mapped to the interior of P.

Before we illustrate the technique, we must make two remarks. First, recall that in constructing the map we have three "degrees of freedom" at our disposal (from the Riemann mapping theorem). Thus we can specify three points on the real axis to be the preimages of three of the w_i. However, formula (9) already designates ∞ as the preimage of w_n, so we are free to choose only, say, x_1 and x_2, and the other x_i are then determined.

Second, to get a closed-form expression for the mapping we must be able to compute the integral in Eq. (8). A glance through a standard table of integrals shows that this is hopeless for $n > 4$ and not always possible even for smaller n. Numerical integration, however, is always feasible. In Appendix I, L. N. Trefethen and T. Driscoll discuss how to implement these computations, and provide reference to their readily accessible software package.

Example 1

Derive a Schwarz-Christoffel transformation mapping the upper half-plane onto the triangle in Fig. 7.39.

Solution.   The right turns are through angles $\theta_1 = \theta_2 = -3\pi/4$, $\theta_3 = -\pi/2$. Hence, choosing $x_1 = -1$ and $x_2 = 1$ we have

$$f(z) = A \int_0^z (\zeta + 1)^{-3/4}(\zeta - 1)^{-3/4}\, d\zeta + B$$

$$= A \int_0^z \left(\zeta^2 - 1\right)^{-3/4} d\zeta + B.$$

The integration must be performed numerically. To evaluate the constants we compute

$$f(x_1) = f(-1) = A \int_0^{-1} \left(\zeta^2 - 1\right)^{-3/4} d\zeta + B = A\eta + B,$$

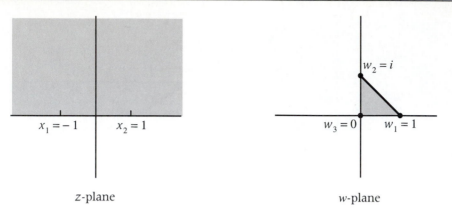

z-plane   *w*-plane

Figure 7.39 Mapping onto a triangle.

where

$$\eta := \int_0^{-1} \left(\zeta^2 - 1\right)^{-3/4} d\zeta \approx 1.85(1 + i)$$

and

$$f(x_2) = f(1) = A \int_0^1 \left(\zeta^2 - 1\right)^{-3/4} d\zeta + B = -A\eta + B.$$

Setting these equal to w_1 and w_2, respectively, we find

$$A\eta + B = 1,$$
$$-A\eta + B = i.$$

Consequently,

$$A = \frac{1 - i}{2\eta}, \qquad B = \frac{1 + i}{2}. \quad \blacksquare$$

Example 2
Determine a Schwarz-Christoffel transformation that maps the upper half-plane onto the semi-infinite strip $|\operatorname{Re} w| < 1$, $\operatorname{Im} w > 0$ (Fig. 7.40).

    Solution. We return to the analysis surrounding Eq. (3) for mapping the real axis onto a polygonal path. To have the upper half-plane map onto the interior of the strip we choose the orientation indicated by the arrows in Fig. 7.40. Left turns of $\pi/2$ radians at w_1 and w_2 can be accommodated by a mapping whose derivative is of the form

$$f'(z) = A (z - x_1)^{-1/2} (z - x_2)^{-1/2}.$$

Choosing $x_1 = -1$ and $x_2 = 1$ again, we compute

$$f(z) = A \int_0^z (\zeta + 1)^{-1/2}(\zeta - 1)^{-1/2} d\zeta + B = \frac{A}{i} \int_0^z \frac{d\zeta}{\sqrt{1 - \zeta^2}} + B$$

$$= \frac{A}{i} \sin^{-1} z + B.$$

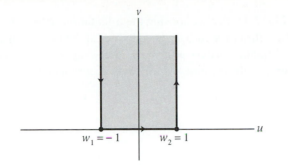

Figure 7.40 Semi-infinite strip for Example 2.

Setting $f(-1) = w_1 = -1$ and $f(1) = w_2 = 1$, we have

$$-iA \sin^{-1}(-1) + B = -1,$$
$$-iA \sin^{-1}(1) + B = 1,$$

which implies that $B = 0$ and $A = 2i/\pi$. Hence

$$f(z) = \frac{2}{\pi} \sin^{-1} z. \quad \blacksquare$$

Example 3

Map the upper half-plane onto the domain consisting of the fourth quadrant plus the strip $0 < v < 1$. (This is a crude model of the continental shelf.)

Solution.   The boundary of this domain consists of the line $v = 1$, the negative u-axis, and the negative v-axis. We shall regard this as the limiting form of the polygonal

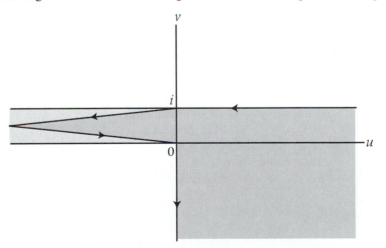

Figure 7.41 Domain for Example 3.

path indicated in Fig. 7.41, again choosing the orientation so that the specified domain lies to the left. A left turn of π radians is called for at the corner "near $w = -\infty$" and a right turn of $\pi/2$ radians occurs at $w = 0$. Selecting $x_1 = -1$ and $x_2 = 1$ as the respective preimages of these points we write, in accordance with Eq. (3),

$$f'(z) = A(z + 1)^{-1}(z - 1)^{1/2}.$$

Using integral tables, with some labor we arrive at

$$f(z) = Ai \left\{ 2\sqrt{1 - z} + \sqrt{2} \log \frac{\sqrt{1 - z} - \sqrt{2}}{\sqrt{1 - z} + \sqrt{2}} \right\} + B.$$

The selection of branches is quite involved in this case, so we shall leave it to the industrious reader (Prob. 6) to verify that with the choice

$$\log \zeta = \mathrm{Log}\, |\zeta| + i \arg \zeta, \qquad -\frac{3}{2}\pi < \arg \zeta \le \frac{\pi}{2},$$

$$\sqrt{\zeta} = e^{(\log \zeta)/2}, \qquad\qquad \log \zeta \text{ as above,}$$

we find that

$$f(z) = \frac{\sqrt{2}}{\pi}\sqrt{1 - z} + \frac{1}{\pi} \log \frac{\sqrt{1 - z} - \sqrt{2}}{\sqrt{1 - z} + \sqrt{2}} + i$$

satisfies the required conditions

$$
\begin{aligned}
&\mathrm{Re}\, f(x) \to +\infty, && \mathrm{Im}\, f(x) \to 1 && \text{as} && x \to -\infty, \\
&\mathrm{Re}\, f(x) \to -\infty, && \mathrm{Im}\, f(x) \to 1 && \text{as} && x \to (-1)^-, \\
&\mathrm{Re}\, f(x) \to -\infty, && \mathrm{Im}\, f(x) \to 0 && \text{as} && x \to (-1)^+, \\
&f(1) = 0 \\
&\mathrm{Re}\, f(x) \to 0, && \mathrm{Im}\, f(x) \to -\infty && \text{as} && x \to +\infty. \quad \blacksquare
\end{aligned}
$$

EXERCISES 7.5

1. Use the techniques in this section to find a conformal map of the upper half-plane onto the whole plane slit along the negative real axis up to the point -1. [HINT: Consider the slit as the limiting form of the wedge indicated in Fig. 7.42.]

2. Use the Schwarz-Christoffel formula to derive the mapping $w = \sqrt{z}$ of the upper half-plane onto the first quadrant.

3. Map the upper half-plane onto the semi-infinite strip $u > 0, 0 < v < 1$, indicated in Fig. 7.43.

4. Show that the transformation

$$w = \int_0^z \frac{d\zeta}{\left(1 - \zeta^2\right)^{2/3}}$$

maps the upper half-plane onto the interior of an equilateral triangle.

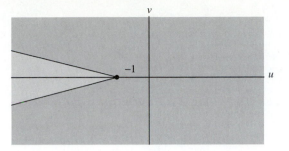

Figure 7.42 Region for Prob. 1.

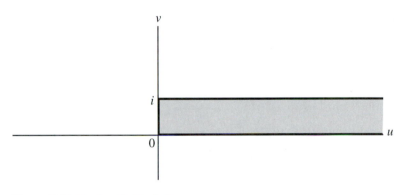

Figure 7.43 Region for Prob. 3.

5. Map the upper half-plane onto the *exterior* of the semi-infinite strip in Fig. 7.40.

6. Verify that the choice of branches indicated in Example 3 yields the appropriate correspondences for $f(-\infty)$, $f(-1)$, $f(1)$, and $f(+\infty)$. [HINT: Argue that if z stays in the upper half-plane, $1 - z$ stays in the lower half-plane, $\sqrt{1 - z}$ stays in the fourth quadrant, and

$$\frac{\sqrt{1 - z} - \sqrt{2}}{\sqrt{1 - z} + \sqrt{2}}$$

stays in the lower half-plane.]

7. Map the upper half-plane onto the shaded region in Fig. 7.44.

8. Derive the expression

$$w = f(z) = \int_0^z \frac{d\zeta}{\sqrt{(1 - \zeta^2)(k^2 - \zeta^2)}}$$

for a conformal map of the upper half-plane onto a rectangle, as indicated in Fig. 7.45. Show that the rectangular dimensions b and c must be related, through k, by

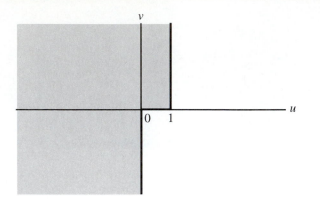

Figure 7.44 Region for Prob. 7.

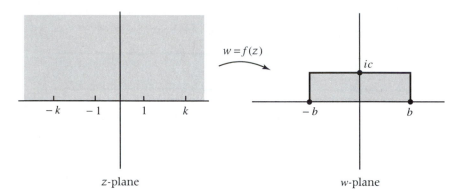

Figure 7.45 Mapping onto a rectangle.

the equations

$$b = \frac{1}{k} \int_0^1 \frac{dx}{\sqrt{(1 - x^2)(1 - x^2/k^2)}},$$

$$c = \frac{1}{k} \int_1^k \frac{dx}{\sqrt{(x^2 - 1)(1 - x^2/k^2)}}.$$

(These are so-called *elliptic integrals*; see Ref. [2].)

9. Map the upper half-plane onto the strip $0 < v < 1$, considered as the limiting form of Fig. 7.46.

10. Argue that a conformal mapping of the unit disk $|z| < 1$ onto the interior of a positively oriented polygon with consecutive corners at $w_1, w_2, \ldots, w_n$ should have the form

$$w = A \int_0^z (\zeta - z_1)^{\theta_1/\pi} (\zeta - z_2)^{\theta_2/\pi} \cdots (\zeta - z_n)^{\theta_n/\pi} \, d\zeta + B,$$

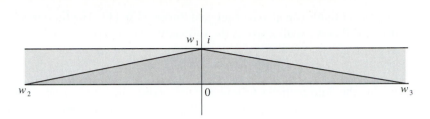

Figure 7.46 Region for Prob. 9.

where the θ_i are the corresponding right-turn angles and the points z_i on the unit circle are the preimages of the corresponding w_i.

11. What does the Schwarz reflection principle (Probs. 13 and 14 in Exercises 5.8) say about the image of the lower half-plane under the Schwarz-Christoffel transformations? (Consider, in turn, Example 2, then Example 1, and then the general case of Fig. 7.38.)

7.6   Applications in Electrostatics, Heat Flow, and Fluid Mechanics

The next two sections in this chapter are devoted to the solution of certain physical problems involving Laplace's equation

$$\frac{\partial^2 \phi}{\partial x^2} + \frac{\partial^2 \phi}{\partial y^2} = 0, \tag{1}$$

using conformal mapping techniques.

We remind the reader that Eq. (1) governs the temperature distribution in two-dimensional steady-state heat flow (Sec. 2.6). The function $\phi(x, y)$ is the temperature, and the curves $\phi = $ constant are the *isotherms*. Usually one assumes that idealized heat sources or heat sinks are used to maintain fixed (specified) values of ϕ on certain parts of the boundary of a domain and that the rest of the boundary is thermally insulated. The latter condition is expressed mathematically by saying that the normal derivative of ϕ is zero; that is, $\partial \phi / \partial n = 0$, where n is a coordinate measured perpendicular to the boundary. The problem, of course, is to find ϕ inside the domain.

Eq. (1) also arises in electrical applications. In electrostatics $\phi(x, y)$ is interpreted as the electric potential, or voltage, at the point (x, y), and its partial derivatives $\partial \phi / \partial x$ and $\partial \phi / \partial y$ are the components of the electric field intensity. Typically one specifies either the potential or the *normal* component of the intensity vector on the boundary of a domain and asks for the values of the potential inside (or outside) the domain. The curves defined by the equation $\phi = $ constant are called *equipotentials*.

Flow patterns of fluids can also be analyzed through Eq. (1). The fluid mechanical interpretation of ϕ that we shall adopt is the following: The curves given by

$$\phi(x, y) = \text{constant}$$

are the paths that the fluid particles follow. In other words, they are *streamlines*. Thus in studying flow around a nonporous obstacle, the perimeter of that obstacle must constitute part of a streamline, and we specify $\phi = $ constant there. Sometimes $\phi(x, y)$ is known as the *stream function*. For details of these physical interpretations of solutions of Eq. (1), the reader is directed to the references at the end of this chapter.

As we indicated in Sec. 7.1, the basic strategy in solving these problems is to map the given domain conformally onto a simpler domain, to determine the harmonic function that satisfies the "transplanted" boundary conditions, and to carry this function back via the conformal map.

Example 1

Find the function ϕ that is harmonic in the lens-shaped domain of Fig. 7.47(a) and takes the values 0 and 1 on the bounding circular arcs, as illustrated. Here ϕ can be interpreted as the steady-state temperature inside an infinitely long strip of material having this lens-shaped region as its cross section, with its sides maintained at the given temperatures.

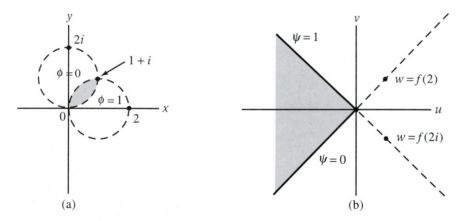

(a)   (b)

Figure 7.47 Lens region and its image.

Solution.   Because the domain is bounded by circular arcs, we are naturally inclined to see what we can do with Möbius transformations. If we choose the pole of the transformation to be at $z = 1 + i$, both circles will become straight lines—orthogonal lines, in fact, because conformality will preserve the right angle at $z = 0$. So let's consider

$$w = f(z) = \frac{z}{z - (1 + i)}, \tag{2}$$

which takes $z = 0$ to $w = 0$ and $z = 1 + i$ to $w = \infty$. To determine the image of the lens, we observe that since $z = 2$ goes to $w = 2/(1 - i) = 1 + i$ and $z = 2i$ goes to $w = 2i/(-1 + i) = 1 - i$, the lens is mapped onto the shaded wedge in Fig. 7.47(b), bounded by the rays $\text{Arg } w = 3\pi/4$ (the image of the arc where $\phi = 1$) and $\text{Arg } w = -3\pi/4$ (the image of the arc where $\phi = 0$). The corresponding harmonic function $\psi(w)$ in the w-plane is easily seen by the methods of Sec. 3.4 to be

$$\psi(w) = -\frac{2}{\pi} \arg w + \frac{5}{2},$$

taking the branch $0 < \arg w < 2\pi$. Carrying this back to the z-plane via Eq. (2), we find

$$\phi(x, y) = \frac{2}{\pi} \left(\frac{5\pi}{4} - \arg \frac{z}{z - (1 + i)} \right),$$

which can be expressed as

$$\phi(x, y) = \frac{2}{\pi} \left(\frac{\pi}{4} - \tan^{-1} \frac{x - y}{x(x - 1) + y(y - 1)} \right);$$

here $-\pi/2 < \tan^{-1}\theta < \pi/2$. ∎

Example 2

Find the function ϕ that is harmonic in the shaded domain depicted in Fig. 7.48(a) and takes the value 0 on the inner circle and 1 on the outer circle. One might interpret ϕ as the electrostatic potential inside a capacitor formed by two nested parallel (but nonconcentric) cylindrical conductors.

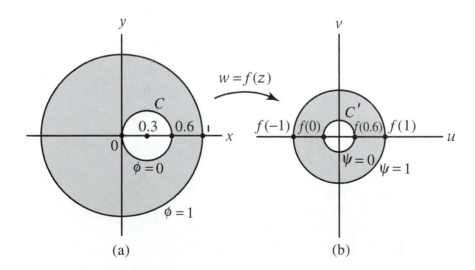

Figure 7.48 Cylindrical capacitor.

Solution.   This problem would be trivial if the inner circle were concentric with the outer circle; it would fall into the "washer" category of Sec. 3.4, and the solution would take the form $A \operatorname{Log} |z| + B$. Thus it seems we should try to map the given region onto a washer, as in Fig. 7.48(b).

The key to constructing such a map is to exploit the fact (demonstrated in Problem 11 in Exercises 7.5) that one can find a pair of (real) points $z = x_1$ and $z = x_2$ *that are symmetric with respect to both circles simultaneously.* To find them, note that the condition that x_1 and x_2 are symmetric with respect to the outer circle reads [Sec. 7.4, Eq. (11)]

$$x_2 = \frac{1}{x_1},$$

while symmetry with respect to the inner circle is expressed by

$$x_2 - 0.3 = \frac{(0.3)^2}{x_1 - 0.3}.$$

The solution to these equations is easily seen to be

$$x_1 = \frac{1}{3}, \qquad x_2 = 3.$$

Now what happens if we perform a Möbius transformation sending x_1 to 0 and x_2 to ∞ via

$$w = f(z) = \frac{z - \frac{1}{3}}{3 - z}? \tag{3}$$

Since Möbius mappings preserve symmetries of this nature, we end up with a pair of circles for which 0 and ∞ are symmetric points. But the point at infinity is symmetric to the *center* of a circle [see Sec. 7.4, Eq. (11)]. Therefore, the centers of the image circles coincide and the circles are concentric, as depicted in Fig. 7.48(b)!

The radius of the image of the inner circle can be calculated from

$$|w| = |f(0)| = \frac{1}{9},$$

and for the image of the outer circle

$$|w| = |f(1)| = \frac{1}{3}.$$

The solution to the problem in the w-plane is seen to be

$$\psi(w) = \frac{\operatorname{Log} |w|}{\operatorname{Log} 3} + 2 = \frac{\operatorname{Log} |9w|}{\operatorname{Log} 3}.$$

Transforming back, we find

$$\phi(z) = \psi\left(\frac{z - \frac{1}{3}}{z - 3}\right) = \frac{\operatorname{Log} \left| \dfrac{9z - 3}{z - 3} \right|}{\operatorname{Log} 3}$$

$$= \frac{1}{\operatorname{Log} 3} \left\{ \operatorname{Log} 3 + \frac{1}{2} \operatorname{Log}\left[(3x - 1)^2 + 9y^2 \right] - \frac{1}{2} \operatorname{Log}\left[(x - 3)^2 + y^2 \right] \right\}.$$

Example 3

Find a function $\phi(x, y)$ that is harmonic in the portion of the upper half-plane exterior to the circle $C : |z - 5i| = 4$ and that takes the value $+1$ on the circle and 0 on the real axis (Fig. 7.49). The solution can be interpreted as the electric potential due to a charged conducting cylinder lying above a conducting plane.

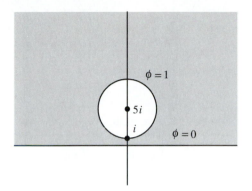

Figure 7.49 Charged cylinder over conducting plane.

Solution.   Actually, this configuration is similar to that of Example 2, if we interpret a straight line to be a circle with a center at infinity. Thus once again we look for a pair of points that are simultaneously symmetric with respect to both the line and the circle.

If z_p denotes a point in the upper half-plane, its symmetric point with respect to the x-axis is obviously $\overline{z_p}$. Now z_p and $\overline{z_p}$ will also be symmetric with respect to the circle if (recall Eq. (11), Sec. 7.4)

$$\overline{z_p} = \frac{4^2}{\left(\overline{z_p} + 5i\right)} + 5i.$$

Solving, we find $z_p = 3i$.

Reasoning as in Example 2, we argue that the Möbius transformation

$$w = f(z) = \frac{z - 3i}{z + 3i},$$

which carries $3i$ to the origin and $-3i$ to infinity, will map the two conductors onto concentric circles *centered at the origin*. The radius of the image of the real axis is

$$r_1 = |f(0)| = 1,$$

and the radius of the image of C is

$$r_2 = |f(i)| = \frac{1}{2}.$$

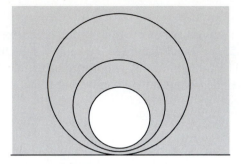

Figure 7.50 Equipotentials for Example 3.

The function that is harmonic in the annulus $r_2 < |w| < r_1$ and takes the proper boundary values is

$$\psi(w) = \frac{\text{Log}\,|w|}{\text{Log}\,\frac{1}{2}} = \frac{-\,\text{Log}\,|w|}{\text{Log}\,2},$$

so the solution to the problem is

$$\phi(x, y) = \frac{-1}{\text{Log}\,2}\,\text{Log}\left|\frac{z - 3i}{z + 3i}\right|.$$

Note that the equipotentials are circles in the z-plane (Fig. 7.50).   ■

Example 4

Find a *nonconstant* function ϕ harmonic inside the infinite domain depicted in Fig. 7.51 and taking the value 0 on the indicated polygonal path. As the sketched lines indicate, the curves $\phi = $ constant will be streamlines for the flow of a deep river over a discontinuous streambed.

Solution.   The Schwarz-Christoffel transformation is the tool for this geometry. Using the analysis surrounding Eq. (3) of Sec. 7.5, we map the x-axis onto the discontinuous streambed in the w-plane. (Notice that the Schwarz-Christoffel transformation maps the *simple* region onto the *complicated* one.) We assume that the corners $w_1 = i$ and $w_2 = 0$ are the images of $z = -1$ and $z = 1$, respectively. Since the streambed has a right-turn angle of $\pi/2$ at i and a left-turn angle of $\pi/2$ at 0, we must have

$$\frac{dw}{dz} = f'(z) = A(z + 1)^{+1/2}(z - 1)^{-1/2};$$

thus with the aid of an integral table we find

$$w = f(z) = A\left\{\left(z^2 - 1\right)^{1/2} + \log\left[z + \left(z^2 - 1\right)^{1/2}\right]\right\} + B. \tag{4}$$

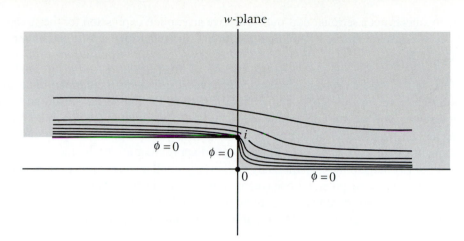

Figure 7.51 Discontinuous streambed.

We take branches of these functions that are real and positive for large real z and that are analytic in the upper half-plane.[†] Then the correspondences $f(-1) = i$, $f(1) = 0$ require

$$A \log(-1) + B = Ai\pi + B = i,$$
$$A \log(1) + B = B = 0,$$

with solution $A = 1/\pi$ and $B = 0$. Hence the mapping from the z-plane to the flow region is

$$w = f(z) = \frac{1}{\pi} \left\{ \left(z^2 - 1\right)^{1/2} + \log\left[z + \left(z^2 - 1\right)^{1/2}\right] \right\}. \tag{5}$$

Now we must find a nonconstant harmonic function of z in the upper half-plane that vanishes on the real axis. One answer is obvious:

$$\psi(x, y) = y = \text{Im}(z). \tag{6}$$

(This is similar to the "wall" geometry of Sec. 3.4.) To complete the problem we must carry $\psi(x, y)$ back to the flow region in the w-plane. But since we have constructed the map from the simple domain to the complicated domain, we need the inverse of the function (5) to complete the problem. Rather than going through the details of solving Eq. (5) for z in terms of w, let's simply abbreviate the answer by stating that the harmonic function $\phi(u, v)$ is given by

$$\phi(u, v) = \text{Im}\left(f^{-1}(w)\right).$$

[†]We are omitting many important details here. Branching is often a very subtle business when Schwarz-Christoffel transformations are used. A painstaking analysis would reveal that here $(z^2 - 1)^{1/2}$ is positive for large positive z and negative for large negative z and that the log function is handled by the restriction $-\pi/2 < \arg \zeta < 3\pi/2$.

Actually, this is not a serious "cop-out." We have an explicit expression for the stream-lines $\phi = $ constant simply by holding y constant and regarding x as a parameter in Eq. (5). ∎

One of the classic problems in elementary physics involves the parallel-plate capacitor. Here we have two oppositely charged flat conducting sheets separated by a fixed distance, and we must determine the electrostatic potential in the region between them. In the simple case of infinite (square) plates, ϕ is proportional to y and the equipotentials are as in Fig. 7.52. This is a good approximation to the more realistic problem of two large plates separated by a relatively small distance; however, near the edges of the plates the potential behaves in a more complicated manner, and this can be computed by conformal mapping. Example 5 should thus be interpreted as finding the potential in the region around two *semi*-infinite conducting plates holding opposite charges.

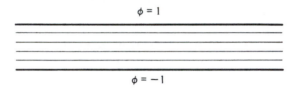

Figure 7.52 Infinite charged plates.

Example 5

Find a function ϕ that is harmonic in the doubly slit plane and takes the values -1 and $+1$ on the two slits indicated in Fig. 7.53.

Solution. We shall use the Schwarz-Christoffel transformation again, regarding the domain as the limiting form of the region sketched in Fig. 7.54, with the point w_0 going to $-\infty$. The limiting right-turn angles are π at $w = i$, $-\pi$ at $w = w_0$, and π at $w = -i$. If we select the preimages of $w = i$ and $w = w_0$ to be $z = -1$ and $z = 0$, respectively, the symmetry of the configuration clearly dictates that $z = +1$ will be the preimage of $w = -i$. Thus employing Eq. (3), Sec. 7.5 again, we have

$$\frac{dw}{dz} = A(z+1)z^{-1}(z-1) = A\left(z - \frac{1}{z}\right).$$

Consequently,

$$w = f(z) = A\left(\frac{z^2}{2} - \log z\right) + B.$$

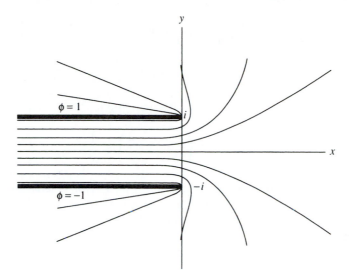

Figure 7.53 Semi-infinite charged plates.

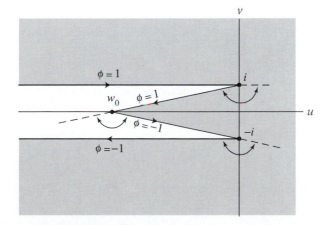

Figure 7.54 Approximate geometry for semi-infinite charged plates.

Enforcing $f(-1) = i$, $f(1) = -i$ and choosing a branch of $\log z$ that is positive for large positive z, we get

$$A\left(\frac{1}{2} - i\pi\right) + B = i,$$

$$A\left(\frac{1}{2} - 0\right) + B = -i,$$

yielding $A = -2/\pi$, $B = 1/\pi - i$. Hence the mapping is

$$w = f(z) = -\frac{2}{\pi}\left(\frac{z^2}{2} - \log z\right) + \frac{1}{\pi} - i. \tag{7}$$

Notice that we have not checked the condition at $z = 0$. This is unnecessary because of the symmetry of the situation. At any rate, it is easy to see that $|w| \to \infty$ as $z \to 0$.

The transformed problem is depicted in Fig. 7.55. The obvious solution is

$$\psi(z) = \frac{2}{\pi}\operatorname{Arg} z - 1.$$

Again the labor involved in inverting Eq. (7) is prohibitive, but the curves $\psi = $ constant are given parametrically by writing Eq. (7) as

$$w = -\frac{2}{\pi}\left(\frac{r^2}{2}e^{2i\theta} - \operatorname{Log} r - i\theta\right) + \frac{1}{\pi} - i$$

and holding θ constant while r varies from 0 to ∞.   ■

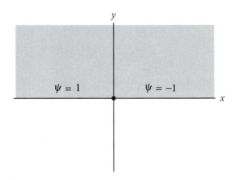

Figure 7.55 Transformed problem for semi-infinite charged plates.

Example 6
Find a nonconstant function ϕ that is harmonic in the slit upper half-plane of Fig. 7.56(a), taking the value $\phi = 0$ on the slit and the real axis. The lines $\phi = $ constant can be interpreted as the streamlines for fluid flow past a simple obstacle.

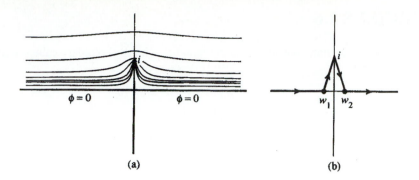

Figure 7.56 Flow past a simple obstacle.

Solution. Regarding the boundary as the limiting form of the polygonal path in Fig. 7.56(b), we construct a Schwarz-Christoffel transformation that maps -1 to w_1, 0 to i, and $+1$ to w_2, with limiting right-turn angles $-\pi/2$, π, and $-\pi/2$, respectively, as w_1 and w_2 approach 0. Hence

$$\frac{dw}{dz} = A(z+1)^{-1/2}z(z-1)^{-1/2} = \frac{Az}{\left(z^2-1\right)^{1/2}},$$

and

$$w = f(z) = A\left(z^2-1\right)^{1/2} + B. \tag{8}$$

Taking a branch that is positive for large positive z, we make the correspondences

$$f(-1) = B = 0,$$
$$f(1) = B = 0,$$
$$f(0) = Ai + B = i.$$

Hence $A = 1$, $B = 0$. (Again we are able to satisfy three conditions with only two constants in this case because of the symmetry of the region.)

In the z-plane the problem becomes, just as in Example 4, that of finding a nonconstant harmonic function vanishing on the real axis, so again we have $\psi(z) = \operatorname{Im} z = y$. In this case we can invert the map (8) to find

$$z = \left(w^2+1\right)^{1/2},$$

so that the required function is

$$\phi(u, v) = \operatorname{Im}\left\{\left(w^2+1\right)^{1/2}\right\},$$

taking a branch that is positive for large positive w. ∎

EXERCISES 7.6

1. Find the electrostatic potential ϕ in the semidisk with the boundary values as shown in Fig. 7.57.

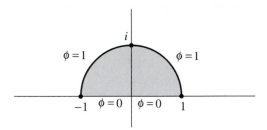

Figure 7.57 Region for Prob. 1.

2. Find the electrostatic potential in the upper half-plane exterior to the unit circle under the conditions shown in Fig. 7.58.

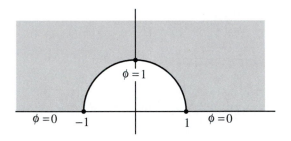

Figure 7.58 Region for Prob. 2.

3. Find the temperature distribution in Fig. 7.59.

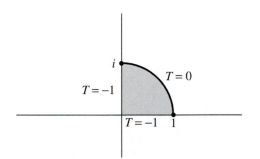

Figure 7.59 Region for Prob. 3.

4. Find the temperature distribution in the unit disk with boundary values as shown in Fig. 7.60. [HINT: Map to the upper half-plane; then use Prob. 3 of Exercises 7.1.]

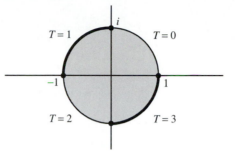

Figure 7.60 Region for Prob. 4.

5. Find the electrostatic potential in the slit upper half-plane with the boundary values as depicted in Fig. 7.61.

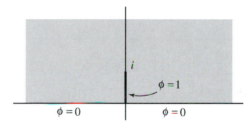

Figure 7.61 Region for Prob. 5.

6. Find the electrostatic potential ϕ in the region between two conducting cylinders under the conditions shown in Fig. 7.62.

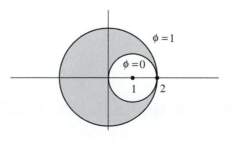

Figure 7.62 Region for Prob. 6.

7. Find the temperature inside the infinite regions depicted in Fig. 7.63.

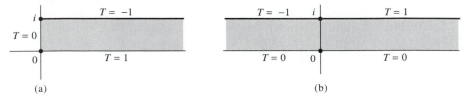

Figure 7.63 Region for Prob. 7.

8. Find the potential in the region exterior to two conducting cylinders charged as shown in Fig. 7.64. (This solution is used to predict the electromagnetic fields generated by transmission power lines.) [HINT: A Möbius transformation can be used to reduce this problem to the situation in Example 2.]

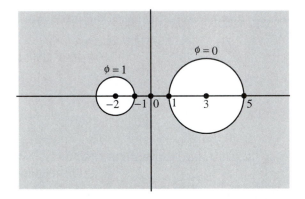

Figure 7.64 Region for Prob. 8.

9. Find the potential between two nested nonconcentric conducting cylinders charged as shown in Fig. 7.65.

10. Find the temperature distribution in the crescent-shaped region given in Fig. 7.66.

11. Find the streamlines for fluid flow in the region indicated in Prob. 7 in Exercises 7.5.

7.7   Further Physical Applications of Conformal Mapping

The examples in the previous section were rather straightforward. We were assigned the task of solving a boundary value problem in an irregular region, and we constructed a mapping to a simpler region by techniques that we had learned earlier. In the present

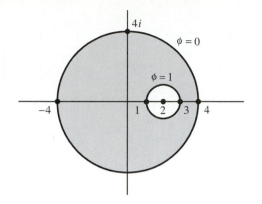

Figure 7.65 Region for Prob. 9.

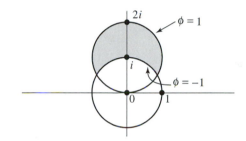

Figure 7.66 Region for Prob. 10.

section we shall study some problems wherein the mappings are not found by straight-forward methods but rather by techniques that may seem to arise from divine inspiration, dumb luck, or simply "experience."

The situation might be described as follows. When one is dealing with an area of mathematics that is so complicated that the direct solution of problems is not feasible, it is useful to try to gain some "feel" for the area by *postulating* a solution and then finding out what problem it solves. The mathematical community calls such procedures *inverse methods*, as opposed to the *direct methods* illustrated in Sec. 7.6. As our first experiment with an inverse method, let us consider what situations can be analyzed with the mapping $w = \sin z$ and its inverse $z = \sin^{-1} w$. This transformation turns out to be quite a versatile tool, and we shall utilize it to solve four very different physical problems.

We saw in Sec. 7.5 that $z = \sin^{-1} w$ maps the upper half-plane onto the semi-infinite strip depicted in Fig. 7.67, with the real axis mapping to the sides and bottom of the strip. We can thus use the harmonic function $\psi_1(u, v) = v = \text{Im}\, w$, which is zero on the real axis, to find the streamlines for flow in a blocked channel, as illustrated in the figure. They are given by the level curves

$$\phi_1(x, y) = \text{Im}(\sin z) = \text{constant}.$$

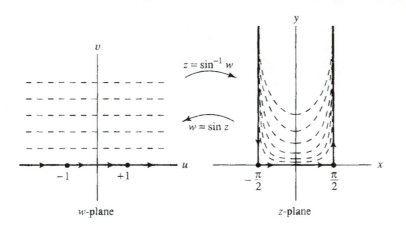

Figure 7.67 Blocked channel flow.

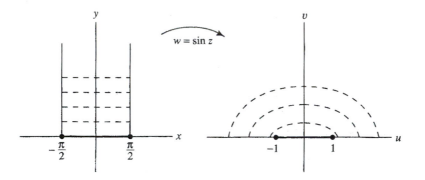

Figure 7.68 Charged conducting strip.

On the other hand, the harmonic function $\phi_2(x, y) = y = \text{Im}\, z$ in the z-plane is zero on the bottom of the strip, and the lines $\phi_2 = \text{constant}$ are perpendicular to the sides of the strip. Therefore, their images, the curves

$$\psi_2(u, v) = \text{Im}\left(\sin^{-1} w\right) = \text{constant}, \tag{1}$$

intersect the u-axis orthogonally for $|u| > 1$, while $\psi_2 = 0$ on the segment $-1 \leq u \leq 1$; see Fig. 7.68. Hence Eq. (1) can be interpreted as the equipotentials around an infinitely long charged conducting strip of width 2.

The harmonic function that is conjugate to this $\psi_2(u, v)$ is, obviously,

$$\eta(u, v) = \text{Re}\left(\sin^{-1} w\right) \tag{2}$$

(to be accurate, we should say that η is *one* of the harmonic conjugates of ψ_2; recall Sec. 2.5). The level curves of $\eta(u, v)$ intersect those of $\psi_2(u, v)$ orthogonally, so they

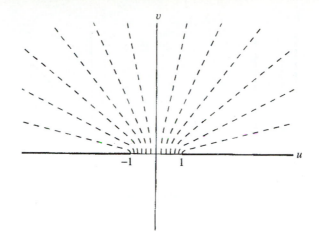

Figure 7.69 Coplaner charged plates.

look like the dashed curves in Fig. 7.69. In particular, $\eta(u, v) = -\pi/2$ on the u-axis to the left of -1, and $\eta(u, v) = \pi/2$ on the u-axis to the right of $+1$. Thus (2) can be interpreted as the potential due to two oppositely charged semi-infinite conducting plates lying side by side.

Finally, it is easy to verify that the mapping $w = \sin z$ as depicted in Fig. 7.67 is symmetric about the y-axis, so that we can restrict it to determine a one-to-one map of the strip in Fig. 7.70 onto the first quadrant. Consider the harmonic function $\phi_3(x, y) = 2x/\pi = (2\,\mathrm{Re}\,z)/\pi$. It is zero on the y-axis and $+1$ on the line $x = \pi/2$, and its level curves intersect the x-axis orthogonally. Hence the "inherited" function

$$\psi_3(u, v) = \frac{2}{\pi}\,\mathrm{Re}\left(\sin^{-1} w\right)$$

is zero on the v-axis and $+1$ on the u-axis for $u > 1$, while the curves $\psi_3 = $ constant intersect the segment $0 < u < 1$ orthogonally. The latter condition implies that the normal derivative of ψ_3 is zero on the segment. As a result, we can interpret ψ_3 as the steady-state temperature in the first quadrant when the v-axis is held at 0 degrees, the u-axis is held at 1 degree for $u > 1$, and the portion $0 < u < 1$ of the u-axis is thermally insulated.

Continuing in this vein, observe that we can solve an interesting variety of problems with the function $f(z) = z^\alpha$. For $\alpha > 0$, we have seen that a suitable branch of f maps the wedge $0 < \arg z < \pi/\alpha$ onto the upper half-plane, and thus $\mathrm{Im}\,z^\alpha$ is the stream function for fluid flow in a wedge. But, of course. *any* of the streamlines $\mathrm{Im}\,z^\alpha = $ constant could be considered as the perimeter of an obstacle placed in the flow. For example, if $\alpha = 2$,

$$\mathrm{Im}\,z^2 = 2xy,$$

and we have a stream function for flow inside, say, the rectilinear hyperbola $xy = 1$

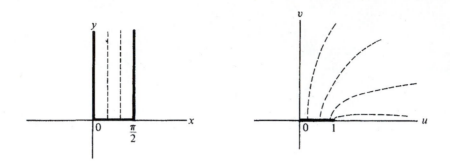

Figure 7.70 Thermal configuration with insulation.

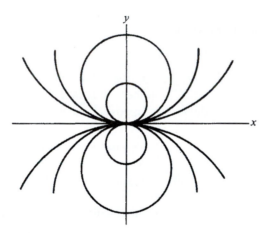

Figure 7.71 Dipole equipotentials.

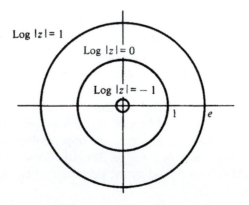

Figure 7.72 Vortex streamlines.

[cf. Fig. 2.6(b)]. For $\alpha = -1$, the level curves (see Figure 7.71)

$$\mathrm{Im}\, z^{-1} = \frac{-y}{x^2 + y^2} = \mathrm{constant}$$

are all circles through the origin. One can visualize these curves as equipotentials for a (two-dimensional) dipole at $z = 0$.

The level curves for the harmonic function $\mathrm{Log}\,|z|$ have an interesting interpretation. Recall that they are concentric circles (Fig. 7.72). Considered as streamlines, they represent the flow due to a *vortex*, or whirlpool, at the origin. On the other hand, if they are interpreted as isotherms or equipotentials, we infer the existence of a heat source or point charge[†] at $x = y = 0$. As we shall illustrate shortly, it is often instructive to study the effect of superimposing vortices or sources on a given pattern.

A very important application of the inverse method, which we shall describe presently, springs from considerations involving the following simple example.

Example 1

Find the streamlines for flow around a cylindrical obstacle as depicted in Fig. 7.73.

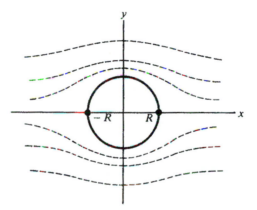

Figure 7.73 Flow around a cylinder.

Solution.   We assume for the moment that the flow is symmetric with respect to the x-axis. (We shall see later that other interpretations are possible.) Then we only have to deal with $y \geq 0$.

To map the flow region onto the upper-half w-plane we need a (nonconstant) analytic function $f(z)$ that is real on the x-axis (at least for $|x| > R$) and on the circle $|z| = R$. The first condition is satisfied by a large class of functions, for example, rational functions with real coefficients. To handle the second condition, we observe

---

[†]Remember that a point charge in two dimensions corresponds to a *line* charge in three dimensions.

that since the circle can be described by $z\bar{z} = R^2$, we have $\bar{z} = R^2/z$ there; hence the rational function $z + R^2/z$ is equal to $z + \bar{z} = 2\,\mathrm{Re}\,z$ on the circle. Consequently we are led to the mapping

$$w = f(z) = z + \frac{R^2}{z}. \tag{3}$$

It is easily verified that (3) produces a one-to-one map of the flow region (for $y > 0$) onto the upper half-plane. Thus the appropriate stream function corresponds to $\psi(u, v) = v = \mathrm{Im}\,w$, yielding the streamlines

$$\phi(x, y) = \mathrm{Im}\left(z + \frac{R^2}{z}\right) = \text{constant.} \quad \blacksquare$$

If we drop the symmetry assumption, we require only that the circle itself be a streamline. Hence we can add any constant multiple of $\mathrm{Log}\,|z|$ (since it has $|z| = R$ as a streamline) to the stream function and obtain "circulating" flow patterns as in Fig. 7.74.

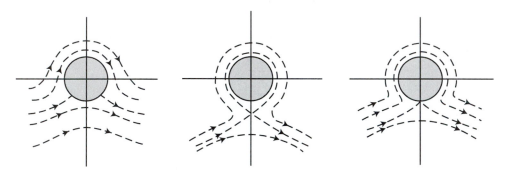

Figure 7.74 Circulating flows around a cylinder.

In 1908 the mathematician N. Joukowski had the inspiration to see what the mapping (3) would do, not to the circle $|z| = R$, but to an off-center circle such as C in Fig. 7.75(a). The result was an *airfoil*, as indicated in Fig. 7.75(b). By starting with different circles C we can generate a variety of these so-called Joukowski airfoils. Furthermore, since we have already found a wide class of flows around cylinders (Figs. 7.73 and 7.74), we can use the "Joukowski transformation" (3) to carry them over and compute flows around these airfoils! For instance, we can shape the airfoil to meet certain engineering specifications by suitably choosing C and introducing modifications into the mapping such as

$$w = f(z) = z + \frac{R^2}{z} + \frac{a}{z^2} + \frac{b}{z^3} + \cdots.$$

See Fig. 7.76.

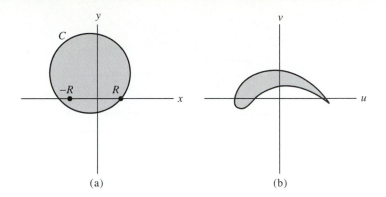

Figure 7.75 Joukowski airfoil.

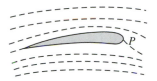

Figure 7.76 Point of attachment.

This technique has been extremely useful in aircraft design, and we direct the interested reader to the specialized literature for further study.

EXERCISES 7.7

1. Analyze the temperature distribution in the plate depicted in Fig. 7.77. [HINT: Use the solution to Prob. 7, Exercises 7.5, and the sine function.]

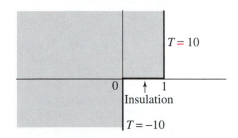

Figure 7.77 Region for Prob. 1.

2. Consider the problem of fluid flow around a straight obstacle inclined at an angle α, as in Fig. 7.78. The stream function ψ must be constant on the obstacle, and the streamlines ($\psi = $ constant) must tend to horizontal straight lines at large distances

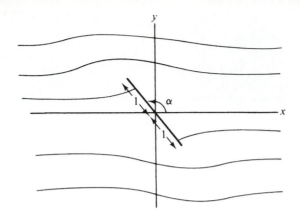

Figure 7.78 Region for Prob. 2.

from the origin. Show that

$$\psi(x, y) = \text{Im}\left[e^{-i\alpha}z\left(\cos\alpha + i\sin\alpha\sqrt{1 - \frac{e^{2i\alpha}}{z^2}}\right)\right]$$

satisfies these conditions.

3. Find the temperature distribution in the first quadrant under the boundary conditions indicated in Fig. 7.79.

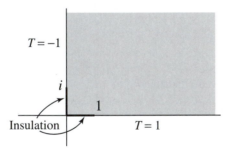

Figure 7.79 Region for Prob. 3.

4. Another feasible approach to Example 1 is to map the shaded region in Fig. 7.80 to the upper half-plane as follows: First use a Möbius transformation to map R to ∞ and $-R$ to 0. Argue that the shaded region then maps onto a 90° wedge, which can be rotated if necessary to coincide with the first quadrant. Squaring them maps onto the upper half-plane, and taking the imaginary part of the whole transformation should solve the problem of finding the stream function. Show that the implementation of this scheme leads to the mapping

$$w = -\left(\frac{z+R}{z-R}\right)^2$$

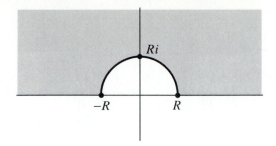

Figure 7.80 Region for Prob. 4.

and the stream function

$$\psi(z) = \operatorname{Im} w = \frac{4yR\left(x^2 + y^2 - R^2\right)}{\left[(x - R)^2 + y^2\right]^2}.$$

Although this function is harmonic in the shaded region and zero on the boundary, we reject it on the physical basis that the flow we seek must have nearly horizontal streamlines for large y. That is, the curves $\psi(x, y) = \text{constant}$ must approximate $y = \text{constant}$ far away from the obstacle. This is obviously not the case for the foregoing solution. (This problem illustrates an additional complication that we have ignored in our elementary treatment, namely, consideration of boundary conditions "at infinity" for unbounded domains.)

5. Analyze the temperature distribution in the slab $0 < y < 1$ under the conditions shown in Fig. 7.81.

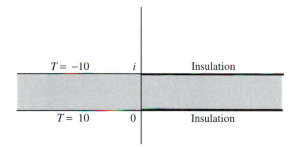

Figure 7.81 Region for Prob. 5.

6. Show that the mapping (3) takes two concentric circles, $|z| = R$ and $|z| = R' > R$, onto a line segment and an ellipse shown in Fig. 7.82. Use this to find the electrostatic potential between a conducting elliptic cylinder surrounding a conducting strip.

7. Using the mapping (3), find the streamlines for the flow indicated in Fig. 7.83.

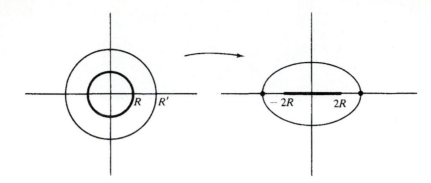

Figure 7.82 Region for Prob. 6.

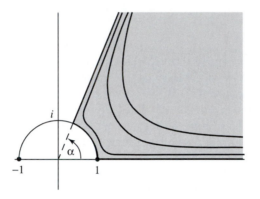

Figure 7.83 Region for Prob. 7.

8. In this problem we will verify a statement made in optional Sec. 2.7, that the orbit formed by iterating the function $f(z) = z^2 - 2$, starting from any point not lying in the interval $[-2,2]$, is unbounded.

    (a) Show that the Joukowski transformation $z = w + 1/w$ produces a one-to-one mapping between the exterior of the unit disk $|w| > 1$ and the complement of the interval $[-2,2]$ in the complex z-plane. Furthermore, if the sequence $\{w_n\}_{n=1}^{\infty}$ approaches infinity in the w-plane, then the corresponding sequence $\{z_n = w_n + 1/w_n\}_{n=1}^{\infty}$ approaches infinity in the z-plane.

    (b) Show that if $|w_1| > 1$ and $w_2 = w_1^2$, then the corresponding points $z_1 = w_1 + 1/w_1$ and $z_2 = w_2 + 1/w_2$ satisfy $z_2 = z_1^2 - 2$.

    (c) Argue that any orbit z_0, $z_1 = z_0^2 - 2$, $z_2 = z_1^2 - 2$, ... launched from a seed z_0 not in the interval $[-2,2]$ corresponds to an "orbit of squares" w_0, $w_1 = w_0^2$, $w_2 = w_1^2$, ... launched from a seed w_0 outside the unit disk, and conclude that both orbits approach infinity in their respective planes.

SUMMARY

One of the most important aspects of mappings generated by analytic functions is the persistence of solutions of Laplace's equation. That is, if $w = f(z)$ is a one-to-one analytic function mapping one domain onto another [which implies that $z = f^{-1}(w)$ is analytic also] and if $\phi(z)$ is harmonic in the first domain, then $\psi(w) := \phi\left(f^{-1}(w)\right)$ is harmonic in the second domain. This leads to a useful technique for solving boundary value problems for Laplace's equation; one performs a preliminary mapping to a domain such as a quadrant, half-plane, or annulus, where the corresponding problem is easy to solve, and then one carries the solution function back via the mapping.

An analytic mapping is conformal, that is, it preserves angles, at all points where the derivative is nonzero. Using this property and connectivity considerations, one can often determine the mapped image of a domain from the image of its boundary.

One important category of conformal mappings is the Möbius transformations. They are functions of the form $(az + b)/(cz + d)$, with $ad \neq bc$, and they can be expressed as compositions of translations, rotations, magnifications, and inversions. Because they preserve the class of straight lines and circles and their associated symmetries, they are usually the method of choice for solving problems in domains bounded by such figures.

On the other hand, Schwarz-Christoffel transformations map half-planes to polygons, and thus they are the appropriate tool for such geometries. These mappings are computed by determining the conditions imposed on the derivative f' at the corners of the polygon.

With these devices one can solve many two-dimensional problems in electrostatics, heat flow, and fluid dynamics. Sometimes insight into the nature of more complicated situations is provided by experimenting with inverse methods, where one first postulates solutions and then analyzes what problems they solve.

Suggested Reading

Most of the references listed previously treat conformal mapping, but the following are particularly useful:

Riemann Mapping Theorem and Geometric Considerations

[1] Goluzin, G. M. *Geometric Theory of Functions of a Complex Variable.* American Mathematical Society, 1969.

[2] Nehari, Z. *Conformal Mapping.* Dover Publications, Inc., New York, 1975.

Schwarz-Christoffel Transformation

[3] Levinson, N., and Redheffer, R. *Complex Variables.* Holden-Day, Inc., San Francisco, 1970.

[4] Henrici, P. *Applied and Computational Complex Analysis*, Vol. I. John Wiley & Sons, Inc., New York, 1974.

Compendium of Conformal Maps

[5] Kober, H. *Dictionary of Conformal Representations*, 2nd ed. Dover Publications, Inc., New York, 1957.

Applications

[6] Courant, R. *Dirichlet's Principle, Conformal Mapping, and Minimal Surfaces*. John Wiley & Sons, Inc. (Interscience Division), New York, 1950.

[7] Dettman, J. W. *Applied Complex Variables*. Dover Publications, Inc., New York, 1984.

[8] England, A. H. *Complex Variable Methods in Elasticity*. John Wiley & Sons, Inc. (Interscience Division), New York, 1971.

[9] Frederick, C., and Schwartz, E. L. "Conformal Image Warping," *IEEE Computer Graphics and Applications*, March 1990, pp. 54–61.

[10] Kyrala, A. *Applied Functions of a Complex Variable*. John Wiley & Sons, Inc., New York, 1972.

[11] Marsden, J. E., and Hoffman, M. J. *Basic Complex Analysis*, 2nd ed. W. H. Freeman and Company, Publishers, New York, 1987.

[12] Milne-Thomson, L. M. *Theoretical Hydrodynamics*, 2nd ed. The Macmillan Company, New York, 1968.

[13] Rothe, R., Ollendorff, F., and Pohlhausen, K. *Theory of Functions*. Dover Publications, Inc., New York, 1961.

[14] Smythe, W. R. *Static and Dynamic Electricity*, 3rd ed. Hemisphere Publications Corporation, New York, 1989.

Chapter 8

The Transforms of Applied Mathematics

In Sec. 3.6 we gave some indication of why, when analyzing linear time-invariant systems, it is particularly advantageous to deal with sinusoidal functions as inputs. Briefly, the virtues of employing an input of the form $Ae^{i\omega t}$ are as follows:

1. Compactness of notation—a real expression such as $\alpha \cos(\omega t + \phi) + \beta \sin(\omega t + \psi)$ can be represented simply by Re $\left(Ae^{i\omega t}\right)$.

2. The fact that differentiation amounts to multiplication by $i\omega$—thus, in a sense, replacing calculus by algebra.

3. The fact that the steady-state response of the system to this input will have the same form, a complex constant times $e^{i\omega t}$.

For these reasons it would be very helpful if a general input function $F(t)$ could be expressed as a sum of these sinusoids. One could then determine the output by finding the response to each sinusoidal component (which is an easier problem) and then adding these responses together (recall that superposition of solutions is permissible in a linear system).

Fourier analysis, as implemented through the Fourier series and the Fourier transform, is devoted to the decomposition of a function into these sinusoids. Other transforms—notably, the Mellin, the Laplace, and the z transforms—have been developed with the same objective: the decomposition of *arbitrary* functions into superpositions of elementary forms that are convenient for a particular analytical task at hand. Another transform, named for Hilbert, is intimately related to the others both theoretically and in applications, although it does not address the specific objective of functional decomposition.

The range of validity and applicability of these mathematical operations extends well beyond the domain of *analytic* functions, but the derivations of the key properties are much more transparent if we restrict ourselves. Thus we devote the final chapter of this book to a survey of the analytic-functional aspects of these transforms, and some proofs will have to be omitted because they go beyond these limitations.

8.1 Fourier Series (The Finite Fourier Transform)

As indicated in the introduction, the main goal of this chapter is to establish the possibility of expressing a (possibly complex-valued) function of a real variable, $F(t)$, as a sum of sinusoidal functions of the form $e^{i\omega t}$. The present section is devoted to the special case when $F(t)$ is periodic with period L; that is, $F(t) = F(t + L)$ for all t.

Naturally, we are inclined to seek a decomposition of F into sinusoids with the same period; that is, only those values of ω should occur such that $e^{i\omega(t+L)} = e^{i\omega t}$. This implies that $e^{i\omega L} = 1$, so that ω must be one of the numbers

$$\omega_n = \frac{2\pi n}{L} \qquad (n = 0, \pm 1, \pm 2, \ldots).$$

To be specific, we assume that $L = 2\pi$ (one can always rescale to achieve this condition). Our problem is thus to find (complex) numbers c_n such that

$$F(t) = \sum_{n=-\infty}^{\infty} c_n e^{int}. \tag{1}$$

Suppose, for the moment, that the series in Eq. (1) converges *uniformly* to $F(t)$ for $-\pi \le t \le \pi$ (and hence for all t). For any fixed integer m we can multiply by e^{-imt} to obtain

$$F(t)e^{-imt} = \sum_{n=-\infty}^{\infty} c_n e^{i(n-m)t}, \tag{2}$$

again converging uniformly, from which it follows that $F(t)e^{-imt}$ is a continuous function and that termwise integration of the series is valid [recall Theorem 8 of Sec. 5.3]. Integrating (2) over the interval $[-\pi, \pi]$ yields

$$\int_{-\pi}^{\pi} F(t)e^{-imt}\, dt = \sum_{n=-\infty}^{\infty} c_n \int_{-\pi}^{\pi} e^{i(n-m)t}\, dt; \tag{3}$$

however,

$$\int_{-\pi}^{\pi} e^{i(n-m)t}\, dt = \begin{cases} \dfrac{e^{i(n-m)t}}{i(n-m)} \bigg|_{-\pi}^{\pi} = 0 & \text{if } n \ne m, \\[2ex] t \big|_{-\pi}^{\pi} = 2\pi & \text{if } n = m. \end{cases}$$

Hence only the term $2\pi c_m$ survives on the right-hand side of Eq. (3). As a result we have the following formula for the coefficient c_m:

$$c_m = \frac{1}{2\pi} \int_{-\pi}^{\pi} F(t)e^{-imt}\, dt, \tag{4}$$

valid whenever the series in Eq. (1) is uniformly convergent.

Whether or not the series is convergent, we use the following terminology.

Definition 1. If F has period 2π and is integrable over $[-\pi, \pi]$, the (formal) series $\sum_{n=-\infty}^{\infty} c_n e^{int}$ with coefficients given by Eq. (4) is called the **Fourier series** for F; the numbers c_n are called the **Fourier coefficients** of F.

More generally, if $F(t)$ has period L, the Fourier[†] series looks like

$$\sum_{n=-\infty}^{\infty} c_n e^{in2\pi t/L},$$

and the Fourier coefficients become

$$c_n = \frac{1}{L} \int_{-L/2}^{L/2} F(t) e^{-in2\pi t/L} \, dt.$$

What we have shown is that under the assumption that $F(t)$ has a representation of the form (1) which is known to be uniformly convergent, then the series in question must be the Fourier series. Now we must investigate this assumption and try to determine *under what conditions the Fourier series will converge to F*.

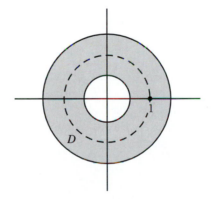

Figure 8.1 Annulus of analyticity.

A partial answer to this question can be derived from analytic function theory. Consider a function $f(z)$ analytic in some annulus, such as D in Fig. 8.1, which contains the unit circle. Then, of course, f can be represented by a Laurent series:

$$f(z) = \sum_{n=-\infty}^{\infty} a_n z^n \qquad (z \text{ in } D). \tag{5}$$

[†]Baron Jean Baptiste Joseph Fourier (1768–1830) was led to the discovery of this series while analyzing the manufacturing of cannons for Napoleon.

We shall be particularly concerned with the values of f on the unit circle where the series converges uniformly. Parametrizing this circle by $z = e^{it}$, $-\pi \leq t \leq \pi$, we introduce the notation $F(t) := f(e^{it})$ and rewrite Eq. (5) as

$$F(t) = \sum_{n=-\infty}^{\infty} a_n e^{int}. \tag{6}$$

Observe our good fortune; the function $F(t)$ has period 2π, and Eq. (6) is a decomposition of F into a series of sinusoids, converging uniformly! Thus Eq. (6) must be the Fourier series for $F(t)$.

In fact, we can even present an independent derivation of formula (4) for the Fourier coefficients in this case. According to Theorem 14 of Sec. 5.5 the coefficients in (5) are given by

$$a_n = \frac{1}{2\pi i} \oint_{|z|=1} \frac{f(\zeta)}{\zeta^{n+1}} \, d\zeta,$$

and inserting the parametrization we find

$$a_n = \frac{1}{2\pi i} \int_{-\pi}^{\pi} f(e^{it}) e^{-it(n+1)} i e^{it} \, dt$$

$$= \frac{1}{2\pi} \int_{-\pi}^{\pi} F(t) e^{-int} \, dt,$$

in agreement with Eq. (4).

Of course, we have not proved a great deal; we have only shown that *the Fourier series of a function F converges uniformly to F in those cases when the values of $F(t)$ coincide with the values of an analytic function $f(z)$ for $z = e^{it}$*. Furthermore, this technique of finding a Fourier series by way of a Laurent expansion is usually of more theoretical than practical value.

Example 1

Find the Fourier series for the periodic function

$$F(t) = e^{2\cos t}$$

using the preceding technique.

Solution.   First we must find an analytic function $f(z)$ that matches the values of $F(t)$ for z on the unit circle. This is easy; since

$$\cos t = \frac{e^{it} + e^{-it}}{2},$$

we see that

$$F(t) = e^{(z+1/z)} =: f(z)$$

when $z = e^{it}$. Hence the Fourier series for F can be obtained from the Laurent series for f. We have

$$e^{(z+1/z)} = e^z e^{1/z}$$

$$= \left(\sum_{m=0}^{\infty} \frac{z^m}{m!} \right) \left(\sum_{\ell=0}^{\infty} \frac{z^{-\ell}}{\ell!} \right),$$

and we can multiply these series termwise. (Termwise multiplication of Laurent series is valid, although we have not proved it in this text.) The term involving z^n in the result comes from the sum of products of terms $z^m/m!$ times $z^{-\ell}/\ell!$ with $m - \ell = n$; collecting these we have

$$e^{(z+1/z)} = \sum_{n=-\infty}^{\infty} z^n \left(\sum_{m=|n|}^{\infty} \frac{1}{m!} \cdot \frac{1}{(m - |n|)!} \right).$$

Hence the Fourier series for F is

$$F(t) = \sum_{n=-\infty}^{\infty} c_n e^{int}$$

with

$$c_n = \sum_{m=|n|}^{\infty} \frac{1}{m!(m - |n|)!}. \qquad \blacksquare$$

Example 2

Show that when $F(t) = f(e^{it})$ with f analytic, termwise differentiation of the Fourier series for F is valid.

Solution.   We know that the Laurent series (5) can be differentiated termwise:

$$\frac{df(z)}{dz} = \sum_{n=-\infty}^{\infty} n a_n z^{n-1}. \qquad (7)$$

For $z = e^{it}$, the chain rule yields

$$\frac{df}{dt} = \frac{df}{dz} \frac{dz}{dt} = \frac{df}{dz} i e^{it}.$$

Inserting Eq. (7) for df/dz and identifying $f(e^{it})$ as $F(t)$, we find

$$\frac{d}{dt} f(e^{it}) = \frac{dF(t)}{dt} = \sum_{n=-\infty}^{\infty} n a_n e^{i(n-1)t} i e^{it},$$

or

$$\frac{dF(t)}{dt} = \sum_{n=-\infty}^{\infty} i n a_n e^{int},$$

which agrees with termwise differentiation of Eq. (6). $\blacksquare$

As another illustration of the fertility of this approach, we shall present a heuristic derivation of *Poisson's formula* for harmonic functions on the unit disk. Since the validity of the formula has been stated in Sec. 4.7, we shall proceed formally and not worry about the rigorous justification of each detail.

We are given a continuous real-valued function $U(\theta)$ having period 2π, and we want to find a function $u(z)$ that is harmonic for $|z| < 1$ and approaches the value $U(\theta)$ as $z \rightarrow e^{i\theta}$; in other words, we want to solve the Dirichlet problem for the unit disk (see Sec. 4.7). First we assume that $U(\theta)$ has a Fourier expansion

$$U(\theta) = \sum_{n=-\infty}^{\infty} c_n e^{in\theta} = \sum_{n=-\infty}^{\infty} \left[\frac{1}{2\pi} \int_{-\pi}^{\pi} U(\phi) e^{-in\phi} \, d\phi \right] e^{in\theta},$$

where we have inserted the coefficient formula (4). If we combine the terms for n and $-n$, we derive (observe that $n = 0$ is exceptional)

$$U(\theta) = \frac{1}{2\pi} \int_{-\pi}^{\pi} U(\phi) \, d\phi + \sum_{n=1}^{\infty} \frac{1}{2\pi} \int_{-\pi}^{\pi} U(\phi) \left(e^{in(\theta-\phi)} + e^{-in(\theta-\phi)} \right) d\phi$$

$$= \frac{1}{2\pi} \int_{-\pi}^{\pi} U(\phi) \, d\phi + 2 \sum_{n=1}^{\infty} \frac{1}{2\pi} \int_{-\pi}^{\pi} U(\phi) \cos n(\theta - \phi) \, d\phi.$$

Now we use a device known to mathematicians as *Abel-Poisson summation* to sum the series. First we artificially introduce the variable r to obtain a function $g(r, \theta)$:

$$g(r, \theta) := \frac{1}{2\pi} \int_{-\pi}^{\pi} U(\phi) \, d\phi + \frac{2}{2\pi} \sum_{n=1}^{\infty} \int_{-\pi}^{\pi} U(\phi) r^n \cos n(\theta - \phi) \, d\phi. \qquad (8)$$

This yields three dividends; first, observe that the series

$$1 + 2 \sum_{n=1}^{\infty} r^n \cos n(\theta - \phi) \qquad (9)$$

converges uniformly in ϕ, if $0 \leq r < 1$. Hence it can be multiplied by $U(\phi)$ and integrated termwise. But this results in 2π times the right-hand side of Eq. (8). Thus we can rewrite Eq. (8) as

$$g(r, \theta) = \frac{1}{2\pi} \int_{-\pi}^{\pi} U(\phi) \left\{ 1 + 2 \sum_{n=1}^{\infty} r^n \cos n(\theta - \phi) \right\} d\phi. \qquad (10)$$

Second, observe that the series (9) is, in fact, the real part of the series

$$1 + 2 \sum_{n=1}^{\infty} r^n e^{in\theta} e^{-in\phi} = 1 + 2 \sum_{n=1}^{\infty} z^n e^{-in\phi}, \qquad (11)$$

a power series in $z = re^{i\theta}$. Since the latter series converges for $|z| < 1$, it defines an analytic function inside the unit disk, and consequently its real part, (9), is harmonic!

As a result, $g(r, \theta)$ is the real part of an analytic function [since $U(\phi)$ is real], and hence g is a harmonic function of $z = re^{i\theta}$ for $r < 1$. The formal substitution $r = 1$ in Eq. (10) yields the Fourier series for $U(\theta)$, so we are led to postulate that Eq. (10) solves the Dirichlet problem; that is, $u(z) = u\left(re^{i\theta}\right) = g(r, \theta)$ is a function which is harmonic for $|z| < 1$ and approaches $U(\theta)$ as $|z| \to 1$.

Finally, the third dividend of our labors follows from the equality

$$1 + 2 \sum_{n=1}^{\infty} r^n \cos n(\theta - \phi) = \frac{1 - r^2}{1 - 2r \cos(\theta - \phi) + r^2}, \tag{12}$$

which we invite the reader to prove as Prob. 4. Using this in Eq. (10), we arrive at the *Poisson formula*

$$u\left(re^{i\theta}\right) = \frac{1 - r^2}{2\pi} \int_{-\pi}^{\pi} \frac{U(\phi)}{1 - 2r \cos(\theta - \phi) + r^2} \, d\phi,$$

expressing a harmonic function inside the unit disk in terms of its "boundary values." As we indicated earlier, we refer the reader to Sec. 4.7 for a more precise statement of the validity of Poisson's formula.

At this point our achievements can be summarized as follows: Subject to some fairly restrictive analyticity assumptions, the equation

$$F(t) = \sum_{n=-\infty}^{\infty} c_n e^{int} \tag{13}$$

is valid when

$$c_n = \frac{1}{2\pi} \int_{-\pi}^{\pi} F(t) e^{-int} \, dt \qquad \text{(for all } n\text{)}. \tag{14}$$

Now notice that there is nothing in Eqs. (13) or (14) that would indicate the necessity of any analytic properties of F. Indeed, the coefficients (14) can be evaluated for any integrable F. So, we speculate, why should the validity of Eq. (13) hinge on analyticity? Shouldn't we expect that the Fourier series converges under weaker conditions? The answer is yes, but the proofs of the more general convergence theorems lie outside analytic function theory. We shall simply quote some of these results without proof.

The first theorem is more or less in line with our speculations. It postulates only the integrability of $|F|^2$, but it pays the price in that a much weaker type of convergence occurs.

Theorem 1. If the integral $\int_{-\pi}^{\pi} |F(t)|^2 \, dt$ exists, then the Fourier series defined by Eqs. (13) and (14) exists and converges to F in the mean square sense; that is,

$$\lim_{N \to \infty} \int_{-\pi}^{\pi} \left| F(t) - \sum_{n=-N}^{N} c_n e^{int} \right|^2 \, dt = 0.$$

Example 3

Prove *Parseval's identity* for the Fourier coefficients:

$$\int_{-\pi}^{\pi} |F(t)|^2 \, dt = \lim_{N \to \infty} 2\pi \sum_{n=-N}^{N} |c_n|^2, \qquad (15)$$

if $|F|^2$ is integrable over $[-\pi, \pi]$.

   Solution.   We have

$$\int_{-\pi}^{\pi} \left| F(t) - \sum_{n=-N}^{N} c_n e^{int} \right|^2 \, dt$$

$$= \int_{-\pi}^{\pi} \left[F(t) - \sum_{n=-N}^{N} c_n e^{int} \right] \left[\overline{F(t)} - \sum_{n=-N}^{N} \overline{c}_n e^{-int} \right] \, dt.$$

Since the conjugate of $F(t) \cdot \sum_{n=-N}^{N} \overline{c}_n e^{-int}$ is $\overline{F(t)} \cdot \sum_{n=-N}^{N} c_n e^{int}$, the right-hand side becomes

$$\int_{-\pi}^{\pi} |F(t)|^2 \, dt - 2 \operatorname{Re} \sum_{n=-N}^{N} \overline{c}_n \int_{-\pi}^{\pi} F(t) e^{-int} \, dt$$

$$+ \int_{-\pi}^{\pi} \left(\sum_{n=-N}^{N} c_n e^{int} \right) \left(\sum_{n=-N}^{N} \overline{c}_n e^{-int} \right) \, dt. \quad (16)$$

Recognizing the expression for the Fourier coefficient [Eq. (14)] in the preceding, we can write the second term as $-2(2\pi) \sum_{n=-N}^{N} |c_n|^2$. The third term can be expanded, but we must change one of the summation indices to avoid confusion; this term then becomes

$$\sum_{n=-N}^{N} c_n \sum_{m=-N}^{N} \overline{c}_m \int_{-\pi}^{\pi} e^{i(n-m)t} \, dt.$$

Recalling our previous evaluation of this integral, we see that this reduces to

$$2\pi \sum_{n=-N}^{N} c_n \overline{c}_n.$$

Thus we have shown

$$\int_{-\pi}^{\pi} \left| F(t) - \sum_{n=-N}^{N} c_n e^{int} \right|^2 \, dt = \int_{-\pi}^{\pi} |F(t)|^2 \, dt - 2\pi \sum_{n=-N}^{N} |c_n|^2.$$

According to Theorem 1, the left-hand side approaches zero as $N \to \infty$. Hence

$$\int_{-\pi}^{\pi} |F(t)|^2 \, dt - \lim_{N\to\infty} 2\pi \sum_{n=-N}^{N} |c_n|^2 = 0,$$

and Eq. (15) results. ■

The next Fourier convergence theorem is valuable in engineering applications, where switching circuits may produce (theoretically) discontinuous input functions, such as the periodic step function illustrated in Fig. 8.2.

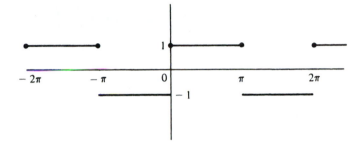

Figure 8.2 Periodic step function.

We restrict ourselves to periodic function F with a finite number of discontinuities in any period. Specifically, we assume that F has period 2π and that there is a finite subdivision of the interval $[-\pi, \pi]$ given by

$$-\pi = \tau_0 < \tau_1 < \tau_2 < \cdots < \tau_{n-1} < \tau_n = \pi$$

such that

1. $F(t)$ is continuously differentiable on each open subinterval (τ_j, τ_{j+1}) for $j = 0, 1, 2, \ldots, n-1$,

2. as t approaches any subdivision point τ_j from the left, $F(t)$ and $F'(t)$ approach limiting values denoted by $F(\tau_j-)$ and $F'(\tau_j-)$, respectively, and

3. as t approaches any τ_j from the right, $F(t)$ and $F'(t)$ again approach limiting values denoted $F(\tau_j+)$ and $F'(\tau_j+)$, respectively.

Such a function is said to be *piecewise smooth*.

Of course, if F is continuous at τ_j, then $F(\tau_j-) = F(\tau_j+) = F(\tau_j)$. For the step function in Fig. 8.2, $F(0-) = F(\pi+) = -1$, and $F(0+) = F(\pi-) = +1 = F(0) = F(\pi)$. Moreover, $F'(0-) = F'(0+) = 0$, but $F'(0)$ does not exist.

Theorem 2. Suppose that F is periodic and piecewise smooth. Then the Fourier series for F converges to $F(t)$ at all points t where F is continuous and converges to $\frac{1}{2}\left[F\left(\tau_j+\right)+F\left(\tau_j-\right)\right]$ at the points of discontinuity.

Example 4

Compute the Fourier series for the step function in Fig. 8.2, and state its convergence properties.

Solution.   The Fourier coefficients are given by

$$c_n = \frac{1}{2\pi}\int_{-\pi}^{\pi} F(t)e^{-int}\,dt = \frac{1}{2\pi}\int_{-\pi}^{0}(-1)e^{-int}\,dt + \frac{1}{2\pi}\int_{0}^{\pi}(1)e^{-int}\,dt$$

$$= \begin{cases} 0 & \text{if } n = 0, \\ \dfrac{i\left\{(-1)^n - 1\right\}}{\pi n} & \text{otherwise.} \end{cases}$$

Hence the Fourier series is

$$\frac{i}{\pi}\sum_{\substack{n=-\infty \\ (n\neq 0)}}^{\infty}\left[\frac{(-1)^n - 1}{n}\right]e^{int}. \tag{17}$$

According to Theorem 2, it converges to $+1$ for $0 < t < \pi$, to -1 for $-\pi < t < 0$, and to the average, 0, for $t = 0$ and $t = \pi$.   ∎

When Fourier analysis (or *frequency analysis*, as it is sometimes called) is used to solve linear systems governed by differential equations, the question naturally arises as to whether or not a Fourier series can legitimately be differentiated termwise. (Obviously, the result of Example 2 is much too restrictive.) The following argument seems to cover a great many cases of interest to engineers: Suppose that F has a convergent Fourier series expansion

$$F(t) = \sum_{n=-\infty}^{\infty} c_n e^{int}, \tag{18}$$

and suppose furthermore that the termwise-differentiated series

$$\sum_{n=-\infty}^{\infty} in c_n e^{int} \tag{19}$$

can be shown (by, say, the M-test) to be uniformly convergent on $[-\pi, \pi]$. Under such circumstances we know the "derived series" (19) can be legitimately *integrated* from $-\pi$ to t, termwise. But the result of this integration is the original series (18), up to a constant. Hence the sum function of (19) must be the *derivative* of $F(t)$, and we have proved the following.

Theorem 3. Suppose that the Fourier expansion (18) is valid and that the derived series (19) converges uniformly on $[-\pi, \pi]$. Then

$$\sum_{n=-\infty}^{\infty} inc_n e^{int} = \frac{d}{dt} \sum_{n=-\infty}^{\infty} c_n e^{int}.$$

Example 5

Find the Fourier series for the periodic function

$$F(t) = \left| \sin \frac{t}{2} \right|^5,$$

and state the convergence properties for the derived series.

Solution.   (Observe that F has period 2π.) The Fourier coefficients are given by

$$c_n = \frac{1}{2\pi} \int_{-\pi}^{\pi} \left| \sin \frac{t}{2} \right|^5 e^{-int} \, dt.$$

These integrals can be evaluated by standard techniques after application of the identity

$$\sin^5 \theta = \frac{5}{8} \sin \theta - \frac{5}{16} \sin 3\theta + \frac{1}{16} \sin 5\theta.$$

With some labor one finds that the Fourier series for $F(t)$ is given by

$$\sum_{n=-\infty}^{\infty} \frac{240/\pi}{225 - 1036n^2 + 560n^4 - 64n^6} e^{int}, \tag{20}$$

and, according to Theorem 2, it converges to $F(t)$. Differentiating termwise we derive

$$\sum_{n=-\infty}^{\infty} \frac{in(240/\pi)}{225 - 1036n^2 + 560n^4 - 64n^6} e^{int}. \tag{21}$$

This series converges uniformly, as can be seen by comparing its increasing and decreasing parts with the (convergent) series $\sum_{n=1}^{\infty} 2 \cdot 240 / \pi 64n^5$ [the factor 2 ensures that the terms of this series dominate those of (21) for large n]. Hence (21) represents $F'(t)$. Moreover, termwise differentiation of (21) can be justified by comparing the result with $\sum_{n=1}^{\infty} 2 \cdot 240 / \pi 64n^4$; thus

$$F''(t) = \sum_{n=-\infty}^{\infty} \frac{-n^2(240/\pi)}{225 - 1036n^2 + 560n^4 - 64n^6} e^{int}.$$

Clearly two more termwise differentiations are justified, leading to Fourier series for $F^{(3)}(t)$ and $F^{(4)}(t)$. In fact, the student should verify that the original function $F(t)$ is continuously differentiable exactly four times! (The fifth derivative jumps from $-\frac{15}{4}$ to $+\frac{15}{4}$ as t increases through $0, \pm 2\pi, \pm 4\pi$, etc.) ∎

The next example illustrates how Fourier series are used in practice to solve linear problems.

Example 6

Find a function f that satisfies the differential equation

$$\frac{d^2 f(t)}{dt^2} + 2\frac{df(t)}{dt} + 2f(t) = F(t), \tag{22}$$

where F is the periodic "sawtooth" function prescribed by

$$F(t) := \begin{cases} -1 - \dfrac{2t}{\pi}, & -\pi \le t \le 0, \\[2mm] -1 + \dfrac{2t}{\pi}, & 0 \le t \le \pi \end{cases}$$

(see Fig. 8.3).

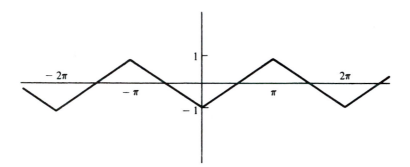

Figure 8.3 Sawtooth function.

Solution.   First we show how a solution can be found to the simpler equation

$$\frac{d^2 g(t)}{dt^2} + 2\frac{dg(t)}{dt} + 2g(t) = e^{i\omega t}, \tag{23}$$

where the "forcing function" on the right-hand side has been replaced by a simple sinusoid. The considerations outlined in the introduction to this chapter indicate that Eq. (23) has a solution of the form $g(t) = Ae^{i\omega t}$. To find A, we insert this expression into Eq. (23) and obtain

$$-\omega^2 Ae^{i\omega t} + 2i\omega Ae^{i\omega t} + 2Ae^{i\omega t} = e^{i\omega t},$$

or (dividing by $e^{i\omega t}$)

$$\left(-\omega^2 + 2i\omega + 2\right) A = 1.$$

Solving for A, we deduce that

$$g(t) = \frac{e^{i\omega t}}{-\omega^2 + 2i\omega + 2} \tag{24}$$

solves Eq. (23).

Next we expand the given function $F(t)$ into a Fourier series. Using formula (14) for the coefficients, we find

$$c_n = \frac{1}{2\pi} \int_{-\pi}^{\pi} F(t)e^{-int}\, dt$$

$$= \frac{1}{2\pi} \int_{-\pi}^{0} \left(-1 - \frac{2t}{\pi}\right) e^{-int}\, dt + \frac{1}{2\pi} \int_{0}^{\pi} \left(-1 + \frac{2t}{\pi}\right) e^{-int}\, dt$$

$$= \begin{cases} 0 & \text{if } n = 0, \\ \dfrac{2}{\pi^2 n^2} \{(-1)^n - 1\} & \text{if } n \neq 0. \end{cases}$$

Hence

$$F(t) = \sum_{\substack{n=-\infty \\ n\neq 0}}^{\infty} \frac{2\{(-1)^n - 1\}}{\pi^2 n^2} e^{int}, \tag{25}$$

which is valid by Theorem 2. Now we argue as follows: We have Eq. (25) expressing $F(t)$ as a linear combination, albeit infinite, of sinusoids, and we have Eq. (24) expressing a solution for a single sinusoid. By linearity, then, we are led to postulate that the same linear combination of these solutions ought to solve the given equation; that is,

$$f(t) = \sum_{\substack{n=-\infty \\ n\neq 0}}^{\infty} \frac{2\{(-1)^n - 1\}}{\pi^2 n^2} \frac{e^{int}}{-n^2 + 2in + 2} \tag{26}$$

should be valid. To complete the argument, we first make the observation that Eq. (26) certainly solves Eq. (22) termwise. Furthermore, by comparing the series in Eq. (26) with the convergent series $\sum (8/\pi^2 n^4)$ (having the same limits), we conclude that the former converges and that it can be legitimately differentiated termwise twice. Since Eq. (22) involves no derivatives higher than the second, we are done. ■

[The reader should observe that termwise differentiation of the sawtooth series (25) yields $2/\pi$ times the step function series (17), which is consistent with the fact that the derivative of the sawtooth is $2/\pi$ times the step function, except at the "break points" $0, \pm\pi, \pm 2\pi, \ldots$. This phenomenon, which is not predicted by Theorem 3, reflects the fact that there are more powerful convergence results for Fourier series. Some of these can be found in the references.]

Formula (14) for the coefficients in the Fourier series is sometimes known as the *finite Fourier transform* (the "infinite" Fourier transform is covered in Sec. 8.2). The efficient computation of this transform is of crucial importance in engineering applications. But practically speaking one usually has to evaluate the integral numerically, for the following reasons:

1. $F(t)$ may be known only through measured data—no formula is available.

2. Even if an analytic formula for $F(t)$ is given, there may be no closed-form expression for the indefinite integral.

Now, for n fixed, a Riemann sum approximating the integral

$$\int_{-\pi}^{\pi} \frac{F(t)}{2\pi} e^{-int} \, dt$$

takes the form (Sec. 4.2)

$$S_{n,N} = F(\tau_1)e^{-in\tau_1}\frac{t_1 - t_0}{2\pi} + F(\tau_2)e^{-in\tau_2}\frac{t_2 - t_1}{2\pi} + \cdots + F(\tau_N)e^{-in\tau_N}\frac{t_N - t_{N-1}}{2\pi},$$

where $-\pi = t_0 < t_1 < t_2 < \cdots < t_N = \pi$ and $t_{j-1} \leq \tau_j \leq t_j$. Let us choose the partition so that there are N equal intervals, giving $t_j = -\pi + 2\pi j/N$; and let us choose the "sample" points τ_j to be the left endpoints of each respective interval, $\tau_j = t_{j-1}$. Then the sum can be abbreviated

$$S_{n,N} = \sum_{j=0}^{N-1} F\left(-\pi + \frac{2\pi j}{N}\right)\frac{e^{-in(-\pi+2\pi j/N)}}{N} = \sum_{j=0}^{N-1} A_j e^{-i2\pi nj/N}, \qquad (27)$$

where $A_j := F(-\pi + 2\pi j/N)e^{in\pi} / N$. As N is increased, the sum $S_{n,N}$ converges to the coefficient c_n; thus error is controlled by choosing N large.

Of course, larger values of N also imply more computational effort. To estimate a single coefficient c_n by Eq. (27) requires N multiplications, and typically one evaluates the finite Fourier transform for N such coefficients—calling for a total of N^2 (complex) multiplications. In applications it is often desirable to take N to be several thousand, and the algorithm in this form is too computation-intensive. However, by judicious grouping of the terms in (27), the work can be reduced considerably.

Suppose, for instance, that $N = 16$, so that the computation of, say, $S_{1,16}$ takes the symbolic form

$$S_{1,16} = \sum_{j=0}^{15} A_j e^{-i2\pi j/16} = \sum_{j=0}^{15} A_j e^{-i\pi j/8}.$$

The numerical values of $\left\{e^{-ij\pi/8} : j = 0, 1, 2, \ldots, 15\right\}$ are quite redundant (see Table 8.1).

So the computation of $S_{1,16}$ can, in fact, be carried out with only three complex multiplications:

$$\begin{aligned}
S_{1,16} = \; & A_0 - A_8 - i\,(A_4 - A_{12}) \\
& + .924\,[A_1 - A_7 - A_9 + A_{15} - i\,(A_3 + A_5 - A_{11} - A_{13})] \\
& + .707\,[A_2 - A_6 - A_{10} + A_{14} - i\,(A_2 + A_6 - A_{10} - A_{14})] \\
& + .383\,[A_3 - A_5 - A_{11} + A_{13} - i\,(A_1 + A_7 - A_9 - A_{15})].
\end{aligned}$$

j	$e^{-ij\pi/8}$	
0	1.000	
1	.924	$-.383i$
2	.707	$-.707i$
3	.383	$-.924i$
4		$-1.000i$
5	$-.383$	$-.924i$
6	$-.707$	$-.707i$
7	$-.924$	$-.383i$
8	-1.000	
9	$-.924$	$+.383i$
10	$-.707$	$+.707i$
11	$-.383$	$+.924i$
12		$+1.000i$
13	.383	$+.924i$
14	.707	$+.707i$
15	.924	$+.383i$

Table 8.1: The numerical values of $\left\{e^{-ij\pi/8} : j = 0, 1, 2, \ldots, 15\right\}$

If we could achieve the same savings for 16 values of $S_{n,16}$, the number of multiplications would be reduced from $16^2 = 256$ to $16 \times 3 = 48$.

The *fast Fourier transform* (FFT) is an algorithm that systematically exploits these rearrangements of terms in the evaluation of (27). The emergence of the FFT in the late 1960s was a major milestone in modern system analysis and signal processing. For values of N of the form 2^m, the total number of multiplications required for N values of $S_{n,N}$ is reduced to roughly $Nm/2 = (N/2) \log_2 N$. Codes are readily available, and small computers can perform a 4096-point transform in seconds. An outline of the basic strategy of the FFT is given in Problem 12; applications and error analyses are discussed in the references.

EXERCISES 8.1

1. Compute the Fourier series for the following functions.

   (a) $F(t) = \sin^3 t$

   (b) $F(t) = \left|\cos^3 \dfrac{t}{3}\right|$

   (c) $F(t) = t^2 \qquad (-\pi < t < \pi)$

   (d) $F(t) = t|t| \qquad (-\pi < t < \pi)$

2. Verify the Fourier representation of the indicated function and state the convergence properties on the interval $[-\pi, \pi]$.

(a) $\displaystyle\sum_{\substack{n=-\infty \\ n \text{ even}}} \frac{-2}{\pi\left(n^2-1\right)} e^{int} = |\sin t|$

(b) $\displaystyle\sum_{n=-\infty}^{\infty} \frac{(-1)^n \sinh \pi}{(1-in)\pi} e^{int} = e^t$

(c) $\displaystyle\frac{1}{\pi} + \frac{1}{2}\sin t - \frac{2}{\pi}\sum_{n=1}^{\infty} \frac{\cos 2nt}{4n^2-1} = \begin{cases} 0, & -\pi \le t \le 0, \\ \sin t, & 0 \le t \le \pi \end{cases}$

(d) $\displaystyle\frac{\pi}{4} + \sum_{n=1}^{\infty} \frac{(-1)^n-1}{\pi n^2}\cos nt - \sum_{n=1}^{\infty} \frac{(-1)^n}{n}\sin nt = \begin{cases} 0, & -\pi \le t \le 0, \\ t, & 0 \le t < \pi \end{cases}$

3. Which of the series in Prob. 2 can be differentiated termwise?

4. Prove Eq. (12). [HINT: Use Eq. (11).]

5. Rewrite the series

$$\sum_{n=-\infty}^{\infty} c_n e^{int}$$

as a *trigonometric series* of the form

$$\sum_{n=0}^{\infty} \alpha_n \cos nt + \sum_{n=1}^{\infty} \beta_n \sin nt,$$

deriving the relations

$$\begin{aligned} \alpha_0 &= c_0, \\ \alpha_n &= c_n + c_{-n} & (n \ge 1), \\ \beta_n &= i\left(c_n - c_{-n}\right) & (n \ge 1). \end{aligned}$$

What are the conditions on the coefficients c_n such that the sum of the series is a real function?

6. **(a)** If $F(t)$ is defined only for $0 \le t \le \pi$, show that by defining $F(-t) := -F(t)$, $0 < t \le \pi$, and constructing the Fourier series for this function over the interval $[-\pi, \pi]$, one arrives at a *Fourier sine series*

$$\sum_{n=1}^{\infty} \beta_n \sin nt$$

for F, with coefficients given by

$$\beta_n = \frac{2}{\pi}\int_0^{\pi} F(t)\sin nt\, dt.$$

State conditions for the Fourier sine series to converge to $F(t)$ for $0 \le t \le \pi$.

(b) As in part (a), show that the definition $F(-t) := F(t), 0 < t \le \pi$, produces a *Fourier cosine series*

$$\sum_{n=0}^{\infty} \alpha_n \cos nt$$

with coefficients

$$\alpha_0 = \frac{1}{\pi} \int_0^{\pi} F(t)\, dt, \qquad \alpha_n = \frac{2}{\pi} \int_0^{\pi} F(t) \cos nt\, dt \quad (n \ge 1).$$

State the conditions for convergence on $[0, \pi]$.

7. Find the Fourier representation for the periodic solutions of the following equations.

(a) $\dfrac{d^2 f}{dt^2} + 3f = \sin^4 t$

(b) $\dfrac{d^2 f}{dt^2} + \dfrac{df}{dt} + f = t^2, -\pi \le t \le \pi$, continued with period 2π

(c) $\dfrac{d^2 f}{dt^2} + 4\dfrac{df}{dt} + 2f = $ (the step function in Fig. 8.2)

8. Show that the Fourier sine and cosine series for the function $F(t) = t, 0 \le t \le \pi$, are given by

$$\sum_{n=1}^{\infty} \frac{2(-1)^{n+1}}{n} \sin nt$$

and

$$\frac{\pi}{2} - \frac{4}{\pi} \sum_{\substack{n=1 \\ n \text{ odd}}}^{\infty} \frac{\cos nt}{n^2},$$

respectively. (See Prob. 6.)

9. Suppose that we wish to solve the Dirichlet problem for the unit disk and that the boundary values of the desired harmonic function for $z = e^{i\theta}$ are represented by the series

$$U(\theta) = \sum_{n=0}^{\infty} \alpha_n \cos n\theta + \sum_{n=1}^{\infty} \beta_n \sin n\theta.$$

Argue that the solution to the problem is given by

$$u\left(re^{i\theta}\right) = \sum_{n=0}^{\infty} \alpha_n r^n \cos n\theta + \sum_{n=1}^{\infty} \beta_n r^n \sin n\theta.$$

10. As an illustration of the power of Fourier methods in solving partial differential equations, consider the *nonstatic* problem of heat flow along a uniform rod of length π, whose ends are maintained at zero degrees temperature. The temperature T is now a function of position x along the rod ($0 \le x \le \pi$) and time t. If the initial

($t = 0$) temperature distribution is specified to be $f(x)$, the equations that T must satisfy are

$$\frac{\partial T(x, t)}{\partial t} = \frac{\partial^2 T(x, t)}{\partial x^2}$$

$$T(0, t) = T(\pi, t) = 0$$

$$T(x, 0) = f(x)$$

for $0 < x < \pi, t > 0$. Assuming the validity of termwise differentiation of Fourier expansions, show that

$$T(x, t) = \sum_{n=1}^{\infty} a_n \sin nx \, e^{-n^2 t}$$

solves the equations, where a_n is defined by

$$a_n = \frac{2}{\pi} \int_0^{\pi} f(\xi) \sin n\xi \, d\xi.$$

[HINT: You will need the Fourier sine series, Prob. 6.] What is the limiting value of $T(x, t)$ as $t \to \infty$? Interpret this.

11. Another illustration of the power of Fourier methods is provided by the vibrating string problem. A taut string fastened at $x = 0$ and $x = \pi$ is initially distorted into the shape $u = f(x)$, where u is the displacement of the string at the point x, and then the string is released. The equations governing the displacement $u(x, t)$ of the string are

$$\frac{\partial^2 u(x, t)}{\partial x^2} = \frac{\partial^2 u(x, t)}{\partial t^2}$$

$$u(0, t) = u(\pi, t) = 0$$

$$u(x, 0) = f(x)$$

$$\frac{\partial u(x, 0)}{\partial t} = 0$$

for $0 < x < \pi, t > 0$. Again assuming the validity of termwise differentiations of Fourier expansions, show that

$$u(x, t) = \sum_{n=1}^{\infty} b_n \sin nx \cos nt$$

solves the equations, where b_n is defined by

$$b_n = \frac{2}{\pi} \int_0^{\pi} f(\xi) \sin n\xi \, d\xi.$$

[HINT: Use the Fourier sine series again.] How would you modify this representation if the "initial conditions" were interchanged to read

$$u(x, 0) = 0$$

$$\frac{\partial u(x, 0)}{\partial t} = f(x)?$$

Combine these formulas to satisfy the more general set of initial conditions

$$u(x,0) = f_1(x)$$
$$\frac{\partial u(x,0)}{\partial t} = f_2(x).$$

12. (*Fast Fourier Transform*) Consider the evaluation of (27). As mentioned in the text, the computation of N values of $S_{n,N}$ apparently entails N^2 complex multiplications.

   (a) Suppose that N is even: $N = 2N_1$. Show that the formula for $S_{n,N}$ can be rewritten as

   $$S_{n,N} = \sum_{j=0}^{N_1-1} A_j e^{-i2\pi nj/N} + \sum_{j=0}^{N_1-1} A_{j+N_1} e^{-i2\pi n(j+N_1)/N}$$

   $$= \sum_{j=0}^{N_1-1} \left\{ A_j + (-1)^n A_{j+N_1} \right\} e^{-i2\pi nj/N} = \sum_{j=0}^{N_1-1} B_j e^{-i2\pi nj/N}.$$

   (b) Now how many complex multiplications will it take to compute N values of $S_{n,N}$? [ANSWER: N coefficients times $N_1 = N/2$ multiplications per coefficient $= N^2/2$; the multiplications by (-1), of course, are not counted.] We seek to iterate this process, halving the number of multiplications again (assuming N_1 is even). However the sum in (a) does not have the same form as that in (27)—the N in the exponent does not match the N_1 in the summation limits. So we have to back up.

   (c) Show that if n is even, $n = 2n_1$, then the sum formula in (a) takes the form

   $$S_{n,N} = \sum_{j=0}^{N_1-1} B_j e^{-i2\pi nj/N} = \sum_{j=0}^{N_1-1} B_j e^{-i2\pi n_1 j/N_1},$$

   whereas if n is odd, $n = 2n_1 + 1$, the formula can be written

   $$S_{n,N} = \sum_{j=0}^{N_1-1} \left\{ B_j e^{-i2\pi j/N} \right\} e^{-i2\pi n_1 j/N_1}$$

   $$= \sum_{j=0}^{N_1-1} C_j e^{-i2\pi n_1 j/N_1}.$$

   (d) Noting that in (c) the computation of the coefficients C_j requires a one-time "overhead" of $N_1 = N/2$ multiplications, how much work does it take to compute N values of $S_{n,N}$? [ANSWER: $N/2$ multiplications plus N coefficients times $N/2$ multiplications per coefficient $= N/2 + N^2/2$.]

   (e) At this point the sums in (c) have exactly the same form as the sum in (27), with N replaced by $N_1 = N/2$. If each of the two new sums is manipulated as before, it will be replaced by a sum of $N_2 = N_1/2 = N/4$ terms, with an overhead of N_2 multiplications per sum to form the new coefficients. Now how much work is required to compute N values $S_{n,N}$? [ANSWER: N_1 multiplications overhead to form the coefficients for (c) plus 2 times N_2 multiplications to perform the same overhead for each sum in (c) plus N coefficients times N_2 multiplications per coefficient $= N/2 + 2(N/4) + n^2/4 = 2(N/2) + N^2/4$.]

(f) If the trick in (c) is implemented yet again for the sums therein, how much work will be required to compute N values of $S_{n,N}$? [ANSWER: $3(N/2) + N^2/8$.]

(g) If N is a power of 2 and the trick in (c) is implemented to reduce the sums down to one term each, what is the net computational load to compute N Fourier coefficients? [ANSWER: $(\log_2 N)(N/2) + N$ multiplications.]

8.2   The Fourier Transform

We move on to the next stage in our program of decomposing arbitrary functions into sinusoids. We have seen how a periodic function can be expressed as a Fourier series, so now we seek a similar representation for nonperiodic functions.

To begin with, let's assume we are given a nonperiodic function $F(t)$, $-\infty < t < \infty$, which is, say, continuously differentiable. Then if we pick an interval of the form $(-L/2, L/2)$ we can represent $F(t)$ by a Fourier series *for t in this interval*:

$$F(t) = \sum_{n=-\infty}^{\infty} c_n e^{in2\pi t/L}, \qquad \frac{-L}{2} < t < \frac{L}{2}, \tag{1}$$

with coefficients given by

$$c_n = \frac{1}{L} \int_{-L/2}^{L/2} F(t) e^{-in2\pi t/L} \, dt \qquad (n = 0, \pm 1, \pm 2, \ldots). \tag{2}$$

Actually the series in Eq. (1) defines a *periodic function* $F_L(t)$, $-\infty < t < \infty$, which coincides with $F(t)$ on $(-L/2, L/2)$; see Fig. 8.4. [Notice that $F_L(t)$ may be discontinuous even though $F(t)$ is smooth.]

Thus we have a sinusoidal representation of $F(t)$ over an interval of length L. If we now let $L \to \infty$, it seems reasonable to conjecture that this might evolve into a sinusoidal representation of $F(t)$ valid for *all t*. Let's explore this possibility.

We are going to rewrite these equations in what will seem at first like a rather bizarre form, but it will aid in interpreting them as $L \to \infty$. We define g_n to be $c_n L/2\pi$, and introduce the factor $[(n+1) - n] \equiv 1$ into the series in Eq. (1). Then we have

$$F_L(t) = \sum_{n=-\infty}^{\infty} g_n e^{in2\pi t/L} \frac{[(n+1) - n]2\pi}{L} \tag{3}$$

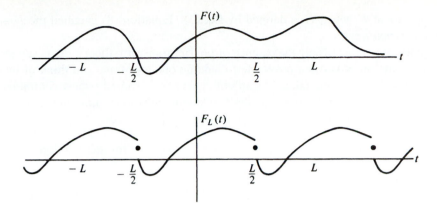

Figure 8.4 Periodic replica of $F(t)$.

and

$$g_n = \frac{1}{2\pi} \int_{-L/2}^{L/2} F(t)e^{-in2\pi t/L} \, dt. \tag{4}$$

Now write $\omega_n = n2\pi/L$, producing

$$F_L(t) = \sum_{n=-\infty}^{\infty} G_L(\omega_n) \, e^{i\omega_n t} \, (\omega_{n+1} - \omega_n), \tag{5}$$

where the function $G_L(\omega)$ is defined for *any* real ω by

$$G_L(\omega) := \frac{1}{2\pi} \int_{-L/2}^{L/2} F(t)e^{-i\omega t} \, dt. \tag{6}$$

As L goes to infinity, $G_L(\omega)$ evolves rather naturally into a function $G(\omega)$ which is known as the *Fourier transform* of F:

$$\boxed{G(\omega) := \frac{1}{2\pi} \int_{-\infty}^{\infty} F(t)e^{-i\omega t} \, dt.} \tag{7}$$

Moreover, since $\Delta\omega_n := \omega_{n+1} - \omega_n$ goes to zero as $L \to \infty$ and since ω_n ranges from $-\infty$ to $+\infty$, Eq. (5) begins to look very much like a Riemann sum for the integral

$$\int_{-\infty}^{\infty} G(\omega)e^{i\omega t} \, d\omega.$$

Thus we are led to propose the equality

$$\boxed{F(t) = \int_{-\infty}^{\infty} G(\omega)e^{i\omega t} \, d\omega} \tag{8}$$

for nonperiodic F, when G is defined by Eq. (7). Equation (8) is called the *Fourier inversion formula*.

Equations (7) and (8) are the essence of Fourier transform theory. As is suggested by this discussion, it is often profitable to indulge one's whimsy and think of the integral in Eq. (8) as a generalized "sum" of sinusoids, summed over a *continuum* of frequencies ω. Equation (7) then dictates the "coefficients," $G(\omega)d\omega$, in the sum.

Example 1
Find the Fourier transform and verify the inversion formula for the function

$$F(t) = \frac{1}{t^2 + 4}.$$

Solution.   Observe that

$$F(t) = \frac{1}{t^2 + 4} = \frac{1}{(t - 2i)(t + 2i)}$$

is analytic except for simple poles at $t = \pm 2i$. We shall use residue theory to evaluate the Fourier transform, interpreting the integral as a principal value:

$$G(\omega) = \frac{1}{2\pi}\text{p.v.}\int_{-\infty}^{\infty} \frac{e^{-i\omega t}}{t^2 + 4}\, dt.$$

If $\omega \geq 0$, we close the contour with expanding semicircles in the lower half-plane; by the techniques of Chapter 6 we find

$$G(\omega) = \frac{1}{2\pi}(-2\pi i)\, \text{Res}\left(\frac{e^{-i\omega t}}{t^2 + 4}; -2i\right)$$

$$= -i \cdot \lim_{t \to -2i} \frac{e^{-i\omega t}}{t - 2i} = \frac{e^{-2\omega}}{4} \qquad (\omega \geq 0).$$

Similarly, for $\omega < 0$ we close in the upper half-plane and find

$$G(\omega) = \frac{1}{2\pi}(2\pi i)\, \text{Res}\left(\frac{e^{-i\omega t}}{t^2 + 4}; 2i\right) = \frac{e^{2\omega}}{4} \qquad (\omega < 0).$$

In short,

$$G(\omega) = \frac{e^{-2|\omega|}}{4}.$$

To verify the Fourier inversion formula we compute

$$\int_{-\infty}^{\infty} G(\omega)e^{i\omega t}\, d\omega = \int_{-\infty}^{\infty} \frac{e^{-2|\omega|}}{4} \cdot e^{i\omega t}\, d\omega.$$

By symmetry, the imaginary part vanishes, and this integral equals

$$\text{Re}\int_{-\infty}^{\infty} \frac{e^{-2|\omega|}}{4}e^{i\omega t}\, d\omega = 2 \cdot \text{Re}\int_{0}^{\infty} \frac{e^{-2\omega}}{4}e^{i\omega t}\, d\omega$$

$$= \frac{1}{2}\text{Re}\, \frac{e^{(-2+it)\omega}}{-2 + it}\bigg|_{\omega=0}^{\infty} = \frac{1}{t^2 + 4}.$$

Hence

$$\frac{1}{t^2 + 4} = \int_{-\infty}^{\infty} \frac{e^{-2|\omega|}}{4} \cdot e^{i\omega t}\, d\omega. \quad \blacksquare \tag{9}$$

As is the case of Fourier series, a wealth of theorems has been discovered stating conditions under which the Fourier integral representations (7) and (8) are valid. A very useful one for applications deals with piecewise smooth functions $F(t)$ like those in Theorem 2 of the previous section; that is, on every bounded interval $F(t)$ is continuously differentiable for all but the finite number of values $t = \tau_1, \tau_2, \ldots, \tau_n$, and at each τ_j the "one-sided limits" of $F(t)$ and $F'(t)$ exist. Note the *principal value* interpretation of the integral is called for in the theorem; this ensures that the inverse transform converges at the points of discontinuity.

> **Theorem 4.** Suppose that $F(t)$ is piecewise smooth on every bounded interval and that $\int_{-\infty}^{\infty} |F(t)|\, dt$ exists. Then the Fourier transform, $G(\omega)$, of F exists and
>
> $$\text{p.v.} \int_{-\infty}^{\infty} G(\omega)e^{i\omega t}\, d\omega = \begin{cases} F(t) & \text{where } F \text{ is continuous,} \\ \dfrac{F(t+) + F(t-)}{2} & \text{otherwise.} \end{cases}$$

Example 2

Find the Fourier transform of the function

$$F(t) = \begin{cases} 1, & -\pi \le t \le \pi, \\ 0, & \text{otherwise} \end{cases}$$

(Fig. 8.5), and confirm the inversion formula.

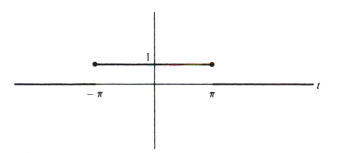

Figure 8.5 "Boxcar" function.

Solution.   We have

$$G(\omega) = \frac{1}{2\pi} \int_{-\pi}^{\pi} (1)e^{-i\omega t}\, dt = \frac{\sin \omega\pi}{\omega\pi}.$$

Hence Theorem 4 tells us that

$$\text{p.v.} \int_{-\infty}^{\infty} \frac{\sin \omega\pi}{\omega\pi} e^{i\omega t}\, d\omega = \begin{cases} 1, & |t| < \pi, \\ 0, & |t| > \pi, \\ \dfrac{1}{2}, & t = \pm\pi. \end{cases} \tag{10}$$

To confirm this, rewrite the left-hand side of (10) as

$$\text{p.v.} \frac{1}{2\pi i} \int_{-\infty}^{\infty} \frac{e^{i\omega(\pi+t)} - e^{i\omega(-\pi+t)}}{\omega}\, d\omega. \tag{11}$$

Now recall from Example 1, Sec. 6.5 that

$$\text{p.v.} \int_{-\infty}^{\infty} \frac{e^{ix}}{x}\, dx = i\pi$$

which, with the changes of variables $x = C\omega$, generalizes to

$$\text{p.v.} \int_{-\infty}^{\infty} \frac{e^{iC\omega}}{\omega}\, d\omega = \begin{cases} i\pi & \text{if } C > 0, \\ -i\pi & \text{if } C < 0. \end{cases}$$

Of course,

$$\text{p.v.} \int_{-\infty}^{\infty} \frac{1}{x}\, dx = 0.$$

Therefore we derive

if $t < -\pi$, then (11) becomes $\dfrac{1}{2\pi i}[-i\pi - (-i\pi)] = 0$;

if $t = -\pi$, then (11) becomes $\dfrac{1}{2\pi i}[0 - (-i\pi)] = \dfrac{1}{2}$;

if $-\pi < t < \pi$, then (11) becomes $\dfrac{1}{2\pi i}[i\pi - (-i\pi)] = 1$;

if $t = \pi$, then (11) becomes $\dfrac{1}{2\pi i}[i\pi - 0] = \dfrac{1}{2}$;

if $\pi < t$, then (11) becomes $\dfrac{1}{2\pi i}[i\pi - i\pi] = 0$. ∎

Example 3
Find the Fourier transform of the function

$$F(t) = \begin{cases} \sin t, & |t| \le 6\pi, \\ 0, & \text{otherwise} \end{cases}$$

(Fig. 8.6), and confirm the inversion formula. (Physicists call this function a *finite wave train*.)

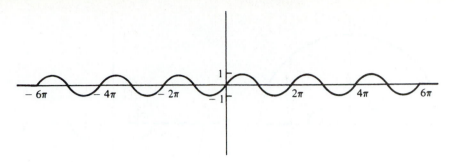

Figure 8.6 Finite wave train.

Solution.   We have

$$G(\omega) = \frac{1}{2\pi} \int_{-6\pi}^{6\pi} (\sin t)e^{-i\omega t}\, dt = \frac{i \sin 6\pi \omega}{\pi\left(1 - \omega^2\right)}.$$

Since $F(t)$ is continuous everywhere, the inversion formula implies

$$\int_{-\infty}^{\infty} \frac{i \sin 6\pi \omega}{\pi\left(1 - \omega^2\right)} e^{i\omega t}\, d\omega = F(t). \tag{12}$$

To confirm this rewrite the left-hand side as

$$\frac{i}{2i\pi} \int_{-\infty}^{\infty} \frac{e^{i\omega(6\pi + t)} - e^{i\omega(-6\pi + t)}}{\left(1 - \omega^2\right)}\, d\omega = \text{p.v.} \frac{-1}{2\pi} \int_{-\infty}^{\infty} \frac{e^{i\omega(6\pi + t)} - e^{i\omega(-6\pi + t)}}{(\omega - 1)(\omega + 1)}\, d\omega$$

(because of the removable singularities at $\omega = \pm 1$).

Now the integral

$$\text{p.v.} \frac{-1}{2\pi} \int_{-\infty}^{\infty} \frac{e^{i\omega(6\pi + t)}}{(\omega - 1)(\omega + 1)}\, d\omega$$

can be evaluated using the indented-contour techniques of Sec. 6.5. For $t \geq -6\pi$ we employ the contour shown in Fig. 8.7(a) and invoke Lemmas 3 and 4 of Chapter 6 to obtain

$$\text{p.v.} \frac{-1}{2\pi} \int_{-\infty}^{\infty} \frac{e^{i\omega(6\pi + t)}}{(\omega - 1)(\omega + 1)}\, d\omega = \frac{-1}{2\pi}(\pi i)\{\text{Res}(-1) + \text{Res}(1)\}$$

$$= \frac{-i}{2} \left[\frac{e^{-i(6\pi + t)}}{-2} + \frac{e^{i(6\pi + t)}}{2} \right]$$

$$= \frac{\sin(6\pi + t)}{2} = \frac{\sin t}{2}.$$

Similarly, for $t \leq -6\pi$ we use the contour of Fig. 8.7(b) and find

$$\text{p.v.} \frac{-1}{2\pi} \int_{-\infty}^{\infty} \frac{e^{i\omega(6\pi + t)}}{(\omega - 1)(\omega + 1)}\, d\omega = \frac{-1}{2\pi}(-\pi i)\{\text{Res}(-1) + \text{Res}(1)\}$$

$$= -\frac{\sin t}{2}.$$

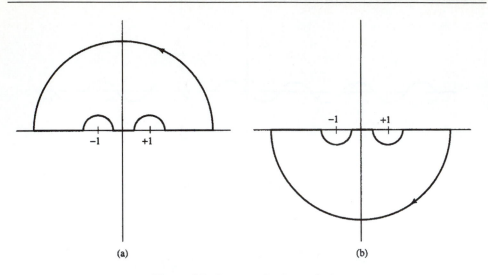

Figure 8.7 Contours for Example 3.

By the same reasoning one obtains

$$\text{p.v.}\frac{-1}{2\pi}\int_{-\infty}^{\infty}\frac{e^{i\omega(-6\pi+t)}}{(\omega-1)(\omega+1)}\,d\omega = \begin{cases} \dfrac{\sin t}{2} & \text{if } t \geq 6\pi, \\[2mm] -\dfrac{\sin t}{2} & \text{if } t \leq 6\pi. \end{cases}$$

Piecing this together we validate (12). ■

The Fourier transform equations are used just like Fourier series in solving linear systems; as an illustration, consider the following.

Example 4

Find a function that satisfies the differential equation

$$\frac{d^2 f(t)}{dt^2} + 2\frac{df(t)}{dt} + 2f(t) = \begin{cases} \sin t, & |t| \leq 6\pi, \\ 0, & \text{otherwise.} \end{cases} \tag{13}$$

Solution.   In Example 6, Sec. 8.1, we learned that a solution to $f'' + 2f' + 2f = e^{i\omega t}$ is

$$\frac{e^{i\omega t}}{-\omega^2 + 2i\omega + 2}.$$

Now the right-hand side of Eq. (13) is the function $F(t)$ in the previous example, and Eq. (12) can be interpreted as expressing F as a "superposition" of sinusoids of the form $e^{i\omega t}$. Hence we propose that the corresponding superposition of solutions

$$f(t) = \int_{-\infty}^{\infty} \frac{i\sin 6\pi\omega}{\pi\left(1-\omega^2\right)}\left(\frac{e^{i\omega t}}{-\omega^2+2i\omega+2}\right)d\omega \tag{14}$$

solves the given equation. As before, we should establish that this expression converges and can be differentiated twice under the integral sign, but, instead, we invite the student to verify that residue theory yields the expression (see Prob. 2)

$$
f(t) = \begin{cases}
0, & t \le -6\pi, \\
\frac{2}{5}\left(e^{-6\pi-t}-1\right)\cos t + \frac{1}{5}\left(e^{-6\pi-t}+1\right)\sin t, & -6\pi \le t \le 6\pi, \\
-\left(e^{6\pi}-e^{-6\pi}\right)e^{-t}\left(\frac{2}{5}\cos t + \frac{1}{5}\sin t\right), & t \ge 6\pi,
\end{cases}
$$

which, as direct computation shows, solves the differential equation (13). ∎

As an amusing exercise in the manipulation of contour integrals we now present an informal derivation of an identity that can be considered as a Fourier expansion theorem, if we are lenient in interpreting relations (7) and (8) between the function and its transform.

Example 5

Suppose that the function $F(t)$ is analytic and bounded by a constant M in an open strip $|\operatorname{Im} t| < \delta$, and define $G_L(\omega)$ as in Eq. (6). Argue that, as $L \to \infty$,

$$
\text{p.v.} \int_{-\infty}^{\infty} G_L(\omega)e^{i\omega t}\, d\omega \to F(t) \tag{15}
$$

for each real t.

Solution. Notice, first of all, that if $F(t)$ has a Fourier transform, it will be given by the limit of $G_L(\omega)$ as $L \to \infty$. Hence (15) looks very much like a Fourier inversion formula. In fact, we shall argue that the members of (15) are equal whenever $L > 2|t|$.

For this purpose we define I_r via

$$
\text{p.v.} \int_{-\infty}^{\infty} G_L(\omega)e^{i\omega t}\, d\omega
$$

$$
= \lim_{r \to \infty} \frac{1}{2\pi} \int_{-r}^{r} \left[\int_{-L/2}^{L/2} F(\tau)e^{-i\omega\tau}\, d\tau \right] e^{i\omega t}\, d\omega =: \lim_{r \to \infty} I_r. \tag{16}
$$

We state without proof that the order of integration can legitimately be reversed under these circumstances, producing

$$
I_r = \frac{1}{2\pi} \int_{-r}^{r} \left[\int_{-L/2}^{L/2} F(\tau)e^{-i\omega\tau}\, d\tau \right] e^{i\omega t}\, d\omega
$$

$$
= \frac{1}{2\pi} \int_{-L/2}^{L/2} F(\tau) \left[\int_{-r}^{r} e^{i\omega(t-\tau)}\, d\omega \right] d\tau
$$

$$
= \frac{1}{2\pi} \int_{-L/2}^{L/2} \frac{F(\tau)}{i(t-\tau)} \left(e^{ir(t-\tau)} - e^{-ir(t-\tau)} \right) d\tau.
$$

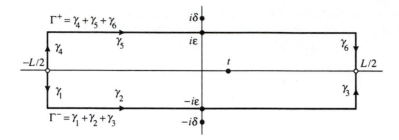

Figure 8.8 Contour for Example 5.

We write this as

$$I_r = \frac{1}{2\pi i} \int_{-L/2}^{L/2} \frac{F(\tau)}{t - \tau} \left(e^{ir(t-\tau)} - 1 \right) d\tau$$

$$+ \frac{1}{2\pi i} \int_{-L/2}^{L/2} \frac{F(\tau)}{t - \tau} \left(1 - e^{-ir(t-\tau)} \right) d\tau, \tag{17}$$

for reasons that will become apparent shortly. Observe that each integrand is analytic in τ as long as we stay in the strip $|\operatorname{Im} \tau| < \delta$, because the singularity at $\tau = t$ is removable. Hence the integrals are independent of path. We choose to evaluate the first integral in Eq. (17) along the contour Γ^- in Fig. 8.8 and the second along Γ^+:

$$I_r = \frac{1}{2\pi i} \int_{\Gamma^-} \frac{F(\tau)}{t - \tau} \left(e^{ir(t-\tau)} - 1 \right) d\tau + \frac{1}{2\pi i} \int_{\Gamma^+} \frac{F(\tau)}{t - \tau} \left(1 - e^{-ir(t-\tau)} \right) d\tau.$$

If we rewrite this as

$$I_r = \frac{1}{2\pi i} \int_{\Gamma^-} \frac{F(\tau)}{\tau - t} d\tau - \frac{1}{2\pi i} \int_{\Gamma^+} \frac{F(\tau)}{\tau - t} d\tau$$

$$+ \frac{1}{2\pi i} \int_{\Gamma^-} \frac{F(\tau)}{t - \tau} e^{ir(t-\tau)} d\tau - \frac{1}{2\pi i} \int_{\Gamma^+} \frac{F(\tau)}{t - \tau} e^{-ir(t-\tau)} d\tau, \tag{18}$$

we recognize that the first two integrals combine to give the integral of $F(\tau)/2\pi i(\tau - t)$ around the simple closed positively oriented contour $(\Gamma^-, -\Gamma^+)$, and according to Cauchy's integral formula, this is precisely $F(t)$. Thus (15) will follow if we show that the last two integrals go to zero as $r \to \infty$. This is most encouraging, since $\exp[ir(t - \tau)] \to 0$ for τ in the lower half-plane and $\exp[-ir(t - \tau)] \to 0$ for τ in the upper half-plane.

To fill in the details, consider first the part of the integration along $\gamma_1 : \tau = -L/2 - iy, 0 \le y \le \varepsilon$ (see Fig. 8.8). We have

$$\left| \frac{1}{2\pi i} \int_{\gamma_1} \frac{F(\tau)}{t - \tau} e^{ir(t-\tau)} d\tau \right| \le \frac{1}{2\pi} \int_0^\varepsilon \frac{M}{L/2 - |t|} \left| e^{irt} e^{irL/2} e^{-ry} \right| dy$$

$$= \frac{1}{\pi} \frac{M}{L - 2|t|} \int_0^\varepsilon e^{-ry} dy = \frac{1}{\pi} \frac{M}{L - 2|t|} \frac{1 - e^{-r\varepsilon}}{r},$$

where M is the bound for $|F|$. This certainly approaches zero as $r \to \infty$. For the integration along $\gamma_2 : \tau = x - i\varepsilon$, $-L/2 \leq x \leq L/2$ (Fig. 8.8 again),

$$\left| \frac{1}{2\pi i} \int_{\gamma_2} \frac{F(\tau)}{t - \tau} e^{ir(t-\tau)} \, d\tau \right| \leq \frac{1}{2\pi} \int_{-L/2}^{L/2} \frac{M}{\varepsilon} \left| e^{ir(t-x+i\varepsilon)} \right| dx$$

$$= \frac{1}{2\pi} \frac{M e^{-r\varepsilon}}{\varepsilon} L,$$

again vanishing as $r \to \infty$. The integrals over γ_3, γ_4, and γ_6 are similar to that over γ_1, and the integral over γ_5 is handled like that over γ_2. ■

EXERCISES 8.2

1. Find the Fourier transform of the following functions and express the inversion formula.

   (a) $F(t) = e^{-|t|}$

   (b) $F(t) = e^{-t^2}$ [HINT: Complete the square and use the fact that $\int_{-\infty}^{\infty} e^{-x^2} \, dx = \sqrt{\pi}$.]

   (c) $F(t) = te^{-t^2}$ [HINT: Integrate by parts and use part (b).]

   (d) $F(t) = \dfrac{\sin t}{t}$ [HINT: Use Eq. (10).]

   (e) $F(t) = \dfrac{\sin \pi t}{1 - t^2}$ [HINT: Exploit Example 3.]

2. Use residue theory to evaluate the integral in Eq. (14).

3. Find the Fourier transform and confirm the inversion formula (by residue theory or some other method) for the following functions.

   (a) $F(t) = \dfrac{1}{t^4 + 1}$ (b) $F(t) = \dfrac{t}{t^4 + 1}$

   (c) $F(t) = e^{-t^2}$ [HINT: See Prob. 1(b).]

4. Show that if $F(t)$ is a real function, then the real part of its Fourier transform is even, and the imaginary part of its Fourier transform is odd.

5. Show that if $G(\omega)$ is the Fourier transform of $F(t)$, then $G(\omega - \Omega)$ is the Fourier transform of $F(t)e^{i\Omega t}$, for any real constant Ω.

6. Find the Fourier integral representations for solutions to the following differential equations.

   (a) $\dfrac{d^2 f}{dt^2} + \dfrac{df}{dt} + f = e^{-t^2}$

(b) $\dfrac{d^2 f}{dt^2} + 4\dfrac{df}{dt} + f = \begin{cases} 0, & t < 0 \\ e^{-t}, & t \geq 0 \end{cases}$

(c) $\dfrac{d^2 f}{dt^2} + 2\dfrac{df}{dt} + 3f = \begin{cases} 1, & |t| < 1 \\ 0, & \text{otherwise} \end{cases}$

7. If $f(x)$ is a given function whose Fourier transform is $G(\omega)$, show that the function

$$T(x, t) = \int_{-\infty}^{\infty} G(\omega) e^{i\omega x} e^{-\omega^2 t} \, d\omega$$

solves the partial differential equation

$$\frac{\partial T(x, t)}{\partial t} = \frac{\partial^2 T(x, t)}{\partial x^2}$$

for $t > 0$ and $-\infty < x < \infty$, and the initial condition

$$T(x, 0) = f(x).$$

[These equations describe the flow of heat in an infinite rod heated initially to the temperature $T = f(x)$. Recall Prob. 10 in Exercises 8.1.] Assume the validity of the Fourier representations and differentiation under the integral sign.

Insert the expression for the transform $G(\omega)$, interchange the order of integration, and use the hint accompanying Prob. 1(b) to derive the formula

$$T(x, t) = \frac{1}{2\sqrt{\pi t}} \int_{-\infty}^{\infty} f(\xi) e^{-(x - \xi)^2/4t} \, d\xi.$$

8. If $f(x)$ is a given function whose Fourier transform is $G(\omega)$, show that the function

$$u(x, t) = \int_{-\infty}^{\infty} G(\omega) e^{i\omega x} \cos \omega t \, d\omega$$

solves the partial differential equation

$$\frac{\partial^2 u(x, t)}{\partial x^2} = \frac{\partial^2 u(x, t)}{\partial t^2}$$

and the initial conditions

$$u(x, 0) = f(x)$$
$$\frac{\partial u}{\partial t}(x, 0) = 0$$

for $t > 0$ and $-\infty < x < \infty$. [These equations govern the motion of an infinite taut string initially displaced to the configuration $u = f(x)$ and released at $t = 0$. Recall Prob. 11 in Exercises 8.1.] Assume the validity of the Fourier representations and of differentiation under the integral sign.

How would you modify this representation if the initial conditions were

$$u(x, 0) = 0$$

$$\frac{\partial u(x, 0)}{\partial t} = f(x)?$$

Combine these results to handle the general set of initial conditions

$$u(x, 0) = f_1(x)$$

$$\frac{\partial u}{\partial t}(x, 0) = f_2(x).$$

9. The *Mellin transform*[†] can be obtained from the Fourier transform by a change of variables. Suppose $f(r)$ is defined for $0 < r < \infty$. Let $x = -\operatorname{Log} r$ (so $r = e^{-x}$) and set $F(x) := f(e^{-x})$; then x runs from ∞ down to $-\infty$.

   (a) Write the Fourier transform equations for $F(x)$ and recast them in terms of r and f to obtain

   $$f(r) = \int_{-\infty}^{\infty} g(\omega)e^{-i\omega \operatorname{Log} r}\, d\omega, \quad g(\omega) = \frac{1}{2\pi}\int_{0}^{\infty} f(r)e^{i\omega \operatorname{Log} r} r^{-1}\, dr.$$

   (b) The Mellin transform of f is formally defined by

   $$M[s; f] := \int_{0}^{\infty} f(r)r^{s-1}\, dr.$$

   Show that the inverse transform can be expressed as

   $$f(r) = \frac{1}{2\pi}\int_{-\infty}^{\infty} M[i\omega; f]r^{-i\omega}\, d\omega.$$

10. The two-dimensional Laplace equation for a harmonic function ϕ, expressed in polar coordinates as $\phi = \phi(r, \theta)$, is given

   $$\frac{1}{r}\frac{\partial}{\partial r}\left(r\frac{\partial \phi}{\partial r}\right) + \frac{1}{r^2}\frac{\partial^2 \phi}{\partial \theta^2} = 0.$$

   (a) Presuming that the order of differentiation and integration can be interchanged freely, use the Mellin transform to show that

   $$\phi(r, \theta) = \frac{1}{2\pi}\int_{-\infty}^{\infty} \frac{M[i\omega; f]}{\sinh \omega\theta_0} r^{-i\omega} \sinh \omega\theta\, d\omega$$

   solves Laplace's equation in the wedge $0 < \theta < \theta_0$ and takes the boundary values $\phi(r, 0) = 0$, $\phi(r, \theta_0) = f(r)$.

   (b) Construct a solution to Laplace's equation in the wedge, taking the boundary values $\phi(r, 0) = f(r)$, $\phi(r, \theta_0) = 0$.

[†]Hjalmar Mellin (1854-1933) was one of the founders of the Finnish Academy of Sciences.

(c) Construct a solution to Laplace's equation in the wedge, taking the boundary values $\phi(r, 0) = f_1(r)$, $\phi(r, \theta_0) = f_2(r)$.

(d) Construct a solution to Laplace's equation in the wedge satisfying the boundary conditions $\partial\phi(r, 0)/\partial\theta = 0$, $\phi(r, \theta_0) = f(r)$.

8.3   The Laplace Transform

In the two previous sections we were motivated by the desire to solve linear systems by means of frequency analysis. The strategy we were employing can be stated as follows: If a linear system is forced by a sinusoidal input function, $e^{i\omega t}$, then we expect that there ought to be a solution that is a sinusoid having the same frequency.

Now this is probably not the *only* solution; for example, consider the problem of finding a function $f(t)$ that satisfies the differential equation

$$\frac{d^2 f(t)}{dt^2} + 2\frac{df(t)}{dt} + f(t) = e^{i2t}. \tag{1}$$

It has a solution of the form Ae^{i2t} with $A = 1/(4i - 3)$. But if $g(t)$ is a solution of the so-called *associated homogeneous equation*

$$\frac{d^2 g(t)}{dt^2} + 2\frac{dg(t)}{dt} + g(t) = 0,$$

then the function g may be added to a solution f of Eq. (1) to produce another solution of Eq. (1). For example, the function

$$\frac{1}{4i - 3} e^{i2t} + 7e^{-t} \tag{2}$$

also solves Eq. (1), since $7e^{-t}$ is a "homogeneous solution." The reader should verify the solution (2) by direct computation to see exactly what's going on.

Now for most *physical* systems, these homogeneous solutions are transient in nature; that is, they die out as time increases [like e^{-t} in (2)]. This is evidenced by the fact that most physical systems, if not forced, eventually come to rest due to dissipative phenomena such as resistance, damping, radiation loss, etc. Such systems are called *asymptotically stable*. In these cases, we argue that the analysis of the preceding sections provides the *unique* solutions for the types of problems formulated there, because for both the periodic functions and the functions integrable over the whole real line the inputs have been driving the system "since $t = -\infty$" and hence the transients must have died out by any (finite) time.

Now it is time to become more flexible and to develop some mathematical machinery that will handle the transients. That is, we must take into account two considerations. The input is "turned on" at $t = 0$ and has *not* been driving the system for all time, and the system starts in some "initial configuration" at $t = 0$ that probably does *not* coincide with the steady-state solution. The *Laplace transform*, as we shall see, handles both these effects. It also accommodates nondissipative systems.

Let us begin by dealing with the input function. We have $F(t)$ defined for all $t \geq 0$. For this discussion it is convenient to extend the domain to the whole line, so we set $F(t) = 0$ for $t < 0$ (such a function is commonly called "causal") and then consider the Fourier transform of F:

$$G(\omega) = \frac{1}{2\pi} \int_{-\infty}^{\infty} F(t)e^{-i\omega t}\, dt.$$

In our case,

$$G(\omega) = \frac{1}{2\pi} \int_{0}^{\infty} F(t)e^{-i\omega t}\, dt. \tag{3}$$

Now if F is sufficiently well behaved near infinity (we shall not be precise here), one can show that Eq. (3) defines a function of ω that is analytic in the lower half-plane $\operatorname{Im}\omega < 0$. Indeed, the derivative is given, as expected, by

$$\frac{dG(\omega)}{d\omega} = \frac{-i}{2\pi} \int_{0}^{\infty} t F(t)e^{-i\omega t}\, dt;$$

the *lower* half-plane is appropriate because

$$\left| e^{-i\omega t} \right| = e^{(\operatorname{Im}\omega)t}$$

is bounded there. If we let ω be pure imaginary, say $\omega = -is$ with s nonnegative, we create

$$g(s) := 2\pi G(-is), \tag{4}$$

a function called the *Laplace transform* of $F(t)$:

$$g(s) = \int_{0}^{\infty} F(t)e^{-st}\, dt. \tag{5}$$

It is often useful to indicate the relation between $g(s)$ and $F(t)$ by employing the notation

$$g(s) = \mathcal{L}\{F\}(s).$$

As an example, consider $F(t) = e^{-t}$; its Laplace transform is

$$g(s) = \mathcal{L}\left\{e^{-t}\right\}(s) = \int_{0}^{\infty} e^{-t}e^{-st}\, dt = -\left.\frac{e^{-(s+1)t}}{s+1}\right|_{0}^{\infty} = \frac{1}{s+1}.$$

We remark that the integral in Eq. (5) may converge even if F does not approach zero as $t \to \infty$, provided s is sufficiently large. Indeed, for the function $F(t) = e^{7t}$, which might characterize a nondissipative "runaway" physical system, we have

$$\int_{0}^{b} e^{7t}e^{-st}\, dt = \frac{e^{(7-s)b} - 1}{7 - s},$$

and if $s > 7$, this approaches $(s - 7)^{-1}$ as $b \to \infty$. In fact, whenever there exist two positive numbers M and α such that

$$|F(t)| \le M e^{\alpha t}, \qquad \text{for all } t \ge 0,$$

one can show that the integral in Eq. (5) converges for any *complex* s satisfying $\text{Re}(s) > \alpha$. Accordingly, we shall say that Eq. (5) defines the Laplace transform $\mathcal{L}\{F\}(s)$ for any (complex) value of s for which the integral converges. In essence the Laplace transform is able to encompass more functions than the Fourier transform, by allowing the frequency variable ω to be complex.

As a simple extension of the preceding computation shows, the Laplace transform of the function e^{at} is $1/(s - a)$ for $\text{Re}(s) > \text{Re}(a)$. By interpreting this statement with various choices of the constant a, we are able to derive the first eight entries in the Laplace transform table. Entry (ix) is obtained by integration by parts, and it leads immediately to entries (x), (xi), and (xii).

The derivation of (xiii) proceeds as follows:

$$\mathcal{L}\left\{F(t)e^{-at}\right\}(s) = \int_0^\infty F(t)e^{-at}e^{-st}\,dt$$

$$= \int_0^\infty F(t)e^{-(s+a)t}\,dt = \mathcal{L}\{F\}(s+a).$$

Entry (xiv) says, of course, that the Laplace transform is linear.

By looking at the transform of the derivative $F'(t)$, we can see how the Laplace transform takes initial configurations into account; we have

$$\mathcal{L}\left\{F'\right\}(s) = \int_0^\infty e^{-st}F'(t)\,dt.$$

Now if F' is sufficiently well behaved so that integration by parts is permitted, this becomes

$$\mathcal{L}\left\{F'\right\}(s) = -\int_0^\infty (-s)e^{-st}F(t)\,dt + e^{-st}F(t)\Big|_0^\infty,$$

and assuming that $e^{-st}F(t) \to 0$ as $t \to \infty$, we find

$$\mathcal{L}\left\{F'\right\}(s) = s\mathcal{L}\{F\}(s) - F(0). \tag{6}$$

Iterating this equation results in

$$\mathcal{L}\left\{F''\right\}(s) = s\mathcal{L}\left\{F'\right\}(s) - F'(0)$$
$$= s^2\mathcal{L}\{F\}(s) - sF(0) - F'(0), \tag{7}$$

and, in general,

$$\mathcal{L}\left\{F^{(k)}\right\}(s) = s^k\mathcal{L}\{F\}(s) - s^{k-1}F(0) - s^{k-2}F'(0) - \cdots - F^{(k-1)}(0). \tag{8}$$

Sufficient conditions for the validity of these equations are given in the following theorem.

TABLE OF LAPLACE TRANSFORMS

(i) $\mathcal{L}\left\{e^{at}\right\} = \dfrac{1}{s-a}$ $[\mathrm{Re}(s) > \mathrm{Re}(a)]$

(ii) $\mathcal{L}\{1\} = \mathcal{L}\left\{e^{0t}\right\} = \dfrac{1}{s}$ $[\mathrm{Re}(s) > 0]$

(iii) $\mathcal{L}\{\cos \omega t\} = \mathrm{Re}\,\mathcal{L}\left\{e^{i\omega t}\right\} = \dfrac{s}{s^2 + \omega^2}$ $[\omega \text{ real, } \mathrm{Re}(s) > 0]$

(iv) $\mathcal{L}\{\sin \omega t\} = \mathrm{Im}\,\mathcal{L}\left\{e^{i\omega t}\right\} = \dfrac{\omega}{s^2 + \omega^2}$ $[\omega \text{ real, } \mathrm{Re}(s) > 0]$

(v) $\mathcal{L}\{\cosh \omega t\} = \mathcal{L}\{\cos i\omega t\} = \dfrac{s}{s^2 - \omega^2}$ $[\omega \text{ real, } \mathrm{Re}(s) > |\omega|]$

(vi) $\mathcal{L}\{\sinh \omega t\} = \mathcal{L}\{-i \sin i\omega t\} = \dfrac{\omega}{s^2 - \omega^2}$ $[\omega \text{ real, } \mathrm{Re}(s) > |\omega|]$

(vii) $\mathcal{L}\left\{e^{-\lambda t} \cos \omega t\right\} = \mathrm{Re}\,\mathcal{L}\left\{e^{(-\lambda + i\omega)t}\right\} = \dfrac{s + \lambda}{(s + \lambda)^2 + \omega^2}$
    $[\omega, \lambda \text{ real, } \mathrm{Re}(s) > -\lambda]$

(viii) $\mathcal{L}\left\{e^{-\lambda t} \sin \omega t\right\} = \mathrm{Im}\,\mathcal{L}\left\{e^{(-\lambda + i\omega)t}\right\} = \dfrac{\omega}{(s + \lambda)^2 + \omega^2}$
    $[\omega, \lambda, \text{ real, } \mathrm{Re}(s) > -\lambda]$

(ix) $\mathcal{L}\left\{t^n e^{at}\right\} = \dfrac{n!}{(s-a)^{n+1}}$ $[\mathrm{Re}(s) > \mathrm{Re}(a)]$

(x) $\mathcal{L}\{t^n\} = \dfrac{n!}{s^{n+1}}$ $[\mathrm{Re}(s) > 0]$

(xi) $\mathcal{L}\{t \cos \omega t\} = \mathrm{Re}\,\mathcal{L}\left\{te^{i\omega t}\right\} = \dfrac{s^2 - \omega^2}{\left(s^2 + \omega^2\right)^2}$ $[\omega \text{ real, } \mathrm{Re}(s) > 0]$

(xii) $\mathcal{L}\{t \sin \omega t\} = \mathrm{Im}\,\mathcal{L}\left\{te^{i\omega t}\right\} = \dfrac{2s\omega}{\left(s^2 + \omega^2\right)^2}$ $[\omega \text{ real, } \mathrm{Re}(s) > 0]$

(xiii) $\mathcal{L}\left\{F(t)e^{-at}\right\}(s) = \mathcal{L}\{F\}(s + a)$

(xiv) $\mathcal{L}\{aF(t) + bH(t)\} = a\mathcal{L}\{F(t)\} + b\mathcal{L}\{H(t)\}$

> **Theorem 5.** Suppose that the function $F(t)$ and its first $n - 1$ derivatives are continuous for $t \geq 0$ and that $F^{(n)}(t)$ is piecewise smooth on every finite interval $[0, b]$. Also, suppose that there are positive constants M, α such that for $k = 0, 1, \ldots, n - 1$
>
> $$\left| F^{(k)}(t) \right| \leq M e^{\alpha t} \qquad (t \geq 0).$$
>
> Then the Laplace transforms of F, F', F'', $\ldots$, $F^{(n)}$ exist for $\mathrm{Re}(s) > \alpha$, and Eq. (8) is valid for $k = 1, 2, \ldots, n$.

The reader is invited to prove this theorem in Prob. 2.

To illustrate how the Laplace transform is used in solving the so-called initial-value problems, we consider an example.

Example 1

Find the function $f(t)$ that satisfies

$$\frac{d^2 f(t)}{dt^2} + 2 \frac{df(t)}{dt} + f(t) = \sin t \tag{9}$$

for $t \geq 0$ and which at $t = 0$ has the properties

$$f(0) = 1, \qquad f'(0) = 0. \tag{10}$$

Solution.   We begin by taking the transform of the Eq. (9). Thanks to the linearity property (xiv) we have

$$\mathcal{L}\left\{ f''(t) \right\} + 2\mathcal{L}\left\{ f'(t) \right\} + \mathcal{L}\{f(t)\} = \mathcal{L}\{\sin t\}.$$

Using Eq. (8) and the initial conditions (10) we find

$$\mathcal{L}\left\{ f'(t) \right\} = s\mathcal{L}\{f(t)\} - 1, \tag{11}$$
$$\mathcal{L}\left\{ f''(t) \right\} = s^2\mathcal{L}\{f(t)\} - s \cdot 1 - 0. \tag{12}$$

Thus our equation is transformed to

$$\left(s^2 + 2s + 1 \right) \mathcal{L}\{f(t)\} - s - 2 = \mathcal{L}\{\sin t\},$$

or, from entry (iv) of the table,

$$\mathcal{L}\{f(t)\} = \frac{s + 2}{s^2 + 2s + 1} + \frac{1}{\left(s^2 + 2s + 1 \right) \left(s^2 + 1 \right)}.$$

Writing the first term on the right as

$$\frac{s + 2}{s^2 + 2s + 1} = \frac{s + 1}{(s + 1)^2} + \frac{1}{(s + 1)^2} = \frac{1}{s + 1} + \frac{1}{(s + 1)^2},$$

we find that, according to entries (i) and (ix), it is the Laplace transform of the function

$$e^{-t} + te^{-t}.$$

To analyze the second term, we use partial fractions to express

$$\frac{1}{\left(s^2 + 2s + 1\right)\left(s^2 + 1\right)} = \frac{1}{2}\frac{1}{s+1} + \frac{1}{2}\frac{1}{(s+1)^2} - \frac{1}{2}\frac{s}{s^2+1},$$

which is the Laplace transform of

$$\frac{1}{2}e^{-t} + \frac{1}{2}te^{-t} - \frac{1}{2}\cos t$$

[see entries (i), (ix), and (iii)].

Hence the solution is

$$f(t) = e^{-t} + te^{-t} + \frac{1}{2}e^{-t} + \frac{1}{2}te^{-t} - \frac{1}{2}\cos t$$

$$= \frac{3}{2}e^{-t} + \frac{3}{2}te^{-t} - \frac{1}{2}\cos t,$$

which can be directly verified.   ∎

In the example we found it fairly easy to solve for the Laplace transform of the solution; to find the solution itself we had to invert the transform. Often, as illustrated, this can be done by referring to a table of Laplace transforms. However, since this transform was derived from the Fourier transform, which has an inversion formula, we suspect that a formula also exists for the inverse Laplace transform. To see this, we recall that by Theorem 4, the Fourier inversion formula for a continuously differentiable, integrable F is

$$F(t) = \int_{-\infty}^{\infty} G(\omega)e^{i\omega t}\, d\omega.$$

Recall, also, that the Laplace transform $\mathcal{L}\{F\}$ was expressed in terms of the Fourier transform by formula (4), or, equivalently,

$$G(\omega) = \frac{1}{2\pi}\mathcal{L}\{F\}(i\omega).$$

Hence we have immediately

$$F(t) = \frac{1}{2\pi}\int_{-\infty}^{\infty} \mathcal{L}\{F\}(i\omega)e^{i\omega t}\, d\omega$$

for such functions. This formula is often written (substituting $-is$ for ω) as

$$F(t) = \frac{1}{2\pi i}\int_{-i\infty}^{i\infty} \mathcal{L}\{F\}(s)e^{st}\, ds, \tag{13}$$

with the obvious interpretation of these imaginary limits of integration.

We would like to generalize formula (13) to cover *nonintegrable* functions whose Laplace transforms are defined only for Re(s) sufficiently large. This is easy to achieve if we can find a positive number a sufficiently large so that $F(t)e^{-at}$ is integrable. Then we write the inversion formula (13) for the function $F(t)e^{-at}$:

$$F(t)e^{-at} = \frac{1}{2\pi i} \int_{-i\infty}^{i\infty} \mathcal{L}\left\{F(t)e^{-at}\right\}(s)e^{st}\,ds. \tag{14}$$

Inserting entry (xiii) of the table in Eq. (14) we multiply by e^{at} to derive

$$F(t) = \frac{1}{2\pi i} \int_{-i\infty}^{i\infty} \mathcal{L}\{F\}(s+a)e^{(s+a)t}\,ds, \tag{15}$$

which we interpret as the so-called *Bromwich integral*[†]

$$F(t) = \frac{1}{2\pi i} \int_{a-i\infty}^{a+i\infty} \mathcal{L}\{F\}(s)e^{st}\,ds.$$

A rigorous analysis produces the following generalization:

Theorem 6. Suppose that $F(t)$ is piecewise smooth on every finite interval $[0, b]$ and that $|F(t)|$ is bounded by $Me^{\alpha t}$ for $t \geq 0$. Then $\mathcal{L}\{F\}(s)$ exists for Re(s) > α, and for all $t > 0$ and any $a > \alpha$,

$$\frac{F(t+) + F(t-)}{2} = \frac{1}{2\pi i}\,\text{p.v.} \int_{a-i\infty}^{a+i\infty} \mathcal{L}\{F\}(s)e^{st}\,ds.$$

Example 2

Find the piecewise smooth function with Laplace transform $1/(s^4 - 1)$.

Solution.   It is possible to employ partial fractions and the transform table to solve this problem, but we shall illustrate the use of the inversion formula. Observe that this function is certainly analytic for Re(s) > 1. To get the inverse transform, let us evaluate the integral

$$I := \text{p.v.} \int_{a-i\infty}^{a+i\infty} \frac{1}{s^4 - 1} e^{st}\,ds \qquad (t \geq 0)$$

with, say, $a = 2$. This can be done by residue theory. I is the limit, as $\rho \to \infty$, of the contour integral

$$I_\rho := \int_{\gamma_\rho} \frac{e^{zt}}{z^4 - 1}\,dz,$$

[†]Thomas John l'Anson Bromwich (1875-1929).

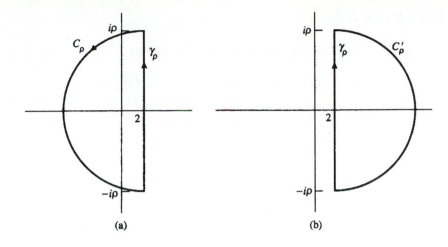

Figure 8.9 Contours for Example 2.

where γ_ρ is the vertical segment from $2 - i\rho$ to $2 + i\rho$. For $t \geq 0$ we close the contour with the half-circle $C_\rho : z = 2 + \rho e^{i\theta}$, $\pi/2 \leq \theta \leq 3\pi/2$; see Fig. 8.9(a). The integral over C_ρ is bounded by

$$\max_{\pi/2 \leq \theta \leq 3\pi/2} \frac{\left| e^{(2+\rho\cos\theta)t} e^{i\rho t \sin\theta} \right|}{(\rho - 2)^4 - 1} \pi\rho = \frac{e^{2t}\pi\rho}{(\rho - 2)^4 - 1},$$

which goes to zero as $\rho \to \infty$.

Now the integrand has four simple poles at ± 1, $\pm i$, all of which eventually lie inside the semicircular contour of Fig. 8.9(a); in fact,

$$\frac{e^{zt}}{z^4 - 1} = \frac{e^{zt}}{(z - 1)(z + 1)(z - i)(z + i)}.$$

Hence

$$I = \lim_{\rho \to \infty} I_\rho = 2\pi i [\mathrm{Res}(1) + \mathrm{Res}(-1) + \mathrm{Res}(i) + \mathrm{Res}(-i)]$$

$$= 2\pi i \left[\frac{e^t}{2(1 - i)(1 + i)} + \frac{e^{-t}}{(-2)(-1 - i)(-1 + i)} \right.$$

$$\left. + \frac{e^{it}}{(i - 1)(i + 1)(2i)} + \frac{e^{-it}}{(-i - 1)(-i + 1)(-2i)} \right]$$

$$= \pi i \sinh t - \pi i \sin t.$$

So the inverse transform is

$$F(t) = \frac{1}{2\pi i} I = \frac{\sinh t - \sin t}{2}, \qquad t \geq 0.$$

For $t < 0$ we close the contour as in Fig. 8.9(b); the integral over C'_ρ is characterized by $-\pi/2 \leq \theta \leq \pi/2$, and for negative t it goes to zero. Since this contour encloses no singularities, we confirm $F(t) = 0$ for $t < 0$. ■

EXERCISES 8.3

1. Compute the Laplace transforms of the following functions.

   (a) $F(t) = 3 \cos 2t - 8e^{-2t}$ **(b)** $F(t) = 2 - e^{4t} \sin \pi t$

   (c) $F(t) = \begin{cases} 1, & t < 1 \\ 0, & t \geq 1 \end{cases}$ **(d)** $F(t) = \begin{cases} 0, & t < 1 \\ 1, & 1 \leq t \leq 2 \\ 0, & t > 2 \end{cases}$

   (e) $F(t) = \sin^2 t$ **(f)** $F(t) = 1/\sqrt{t}$

   [HINT: In (f) let $\chi = \sqrt{st}$ and use the fact that $\int_0^\infty e^{-\chi^2} d\chi = \frac{1}{2}\sqrt{\pi}$.]

2. Prove Theorem 5.

3. Find the inverse transform of the following functions.

   (a) $\dfrac{1}{s^2 + 4}$ **(b)** $\dfrac{4}{(s-1)^2}$ **(c)** $\dfrac{s+1}{s^2 + 4s + 4}$

   (d) $\dfrac{1}{s^3 + 3s^2 + 2s}$ **(e)** $\dfrac{s+3}{s^2 + 4s + 7}$

4. The effect of a time delay in a physical system is described mathematically by re-placing a function $f(t)$ by the *delayed function*

$$f_\tau(t) := \begin{cases} 0, & 0 \leq t < \tau, \\ f(t-\tau), & \tau \leq t < \infty. \end{cases}$$

   Show that $\mathcal{L}\{f_\tau(t)\} = e^{-\tau s}\mathcal{L}\{f(t)\}$. [Compare Prob. 1(c) and (d).]

5. Use the Laplace transform to solve the following initial-value problems.

   (a) $\dfrac{df}{dt} - f = e^{3t}$, $f(0) = 3$

   (b) $\dfrac{d^2 f}{dt^2} - 5\dfrac{df}{dt} + 6f = 0$, $f(0) = 1, \; f'(0) = -1$

   (c) $\dfrac{d^2 f}{dt^2} - \dfrac{df}{dt} - 2f = e^{-t} \sin 2t$, $f(0) = 0, \; f'(0) = 2$

   (d) $\dfrac{d^2 f}{dt^2} - 3\dfrac{df}{dt} + 2f = \begin{cases} 0, & 0 \leq t < 3 \\ 1, & 3 \leq t \leq 6 \\ 0, & t > 6 \end{cases}$, $f(0) = 0, \; f'(0) = 0$

   [HINT: See Prob. 4.]

6. Verify the inversion formula for the following functions.

   (a) $F(t) = e^{-t}$ **(b)** $F(t) \equiv 1$

7. In control theory the differential equation

$$a_n \frac{d^n f}{dt^n} + a_{n-1} \frac{d^{n-1} f}{dt^{n-1}} + \cdots + a_1 \frac{df}{dt} + a_0 f = u$$

is interpreted as a relation between the (known) input $u(t)$ and the (unknown) output $f(t)$.

(a) Show that the Laplace transforms $U(s)$ and $F(s)$ of the input and output are related by

$$F(s) = \frac{U(s)}{a_n s^n + \cdots + a_1 s + a_0} + \frac{P(s)}{a_n s^n + \cdots + a_1 s + a_0},$$

where $P(s)$ is a polynomial in s whose coefficients depend on the a_i and the initial values of $f(t)$ and its derivatives. The fraction multiplying $U(s)$ in this expression is called the *Laplace domain transfer function* of the system.

(b) Show that every solution of the equation with $u(t) \equiv 0$ will go to zero as $t \to +\infty$ if all the poles of the transfer function are simple and lie in the left half-plane. (In other words, the system is *stable*.)

8. Identify the transfer functions (Prob. 7) and determine the stability of the differential equations in Prob. 5.

9. Consider the mass-spring system shown in Fig. 8.10. Each spring has the same natural (unstretched) length L, but when it is compressed or elongated it exerts a force proportional to the amount of compression or elongation (Hooke's law); the constant of proportionality is denoted by K. If we let x and y be the respective displacements of the masses m_1 and m_2 from equilibrium, we have the situation depicted in Fig. 8.11 (for x and y both positive).

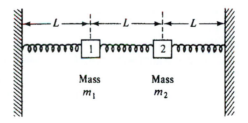

Figure 8.10 Mass-spring system unstretched.

(a) By writing Newton's law (force equals mass times acceleration) for each mass, derive the equations of motion

$$m_1 \frac{d^2 x}{dt^2} = -Kx - K(x - y),$$

$$m_2 \frac{d^2 y}{dt^2} = -Ky + K(x - y).$$

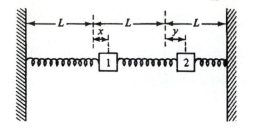

Figure 8.11 Mass-spring system stretched.

(b) Setting $K = 1$, $m_1 = m_2 = 1$, use Laplace transforms to find the solutions $x(t)$, $y(t)$ of these equations if the masses are released from rest (zero velocity) with each of the following initial displacements:

    i. $x(0) = 1$, $y(0) = -1$.

    ii. $x(0) = 1$, $y(0) = 1$.

    iii. $x(0) = 1$, $y(0) = 0$.

(c) For which initial conditions in part (b) is the *system's* response periodic [note that this requires *both* $x(t + T) = x(t)$ *and* $y(t + T) = y(t)$]? Can you visualize these responses? (They are called *normal modes* for the system.)

8.4   The z-Transform

As we have seen, the application of transforms to various families of functions can often bring about certain computational advantages—for example, the replacement of differentiation by multiplication. For continuous functions on a finite interval, or periodic functions, we use the Fourier series; if the interval is semi-infinite, we use the Laplace transform; and the Fourier transform is employed when the interval comprises the whole real line.

Often in practice we encounter "functions" that are *discrete* data structures. For example, when a continuous function $f(t)$ is measured in a laboratory it is sampled at a discrete set of points $\{t_j\}$. The transform tool that facilitates the mathematical manipulation of such discrete data streams is known as the *z-transform*.

Let us denote a discrete sequence of numbers by $a(n)$; we assume n to take integer values from $-\infty$ to ∞, and we allow the possibility that $a(n)$ is complex. As examples consider

$$a(0)$$
$$\downarrow \qquad\qquad\qquad\qquad\qquad\qquad (1)$$
$$\ldots, \frac{1}{16}, \frac{1}{8}, \frac{1}{4}, \frac{1}{2}, 1, \frac{1}{2}, \frac{1}{4}, \frac{1}{8}, \frac{1}{16}, \ldots$$

(sampled values of $f(x) = 2^{-|x|}$ at integers $x = n$);

$$a(0)$$
$$\downarrow$$
$$\ldots, -1, 1, -1, 1, -1, 1, -1, 1, \ldots \qquad (2)$$

(sampled values of $\cos \pi x$);

$$a(0)$$
$$\downarrow$$
$$\ldots, 0, 0, 0, 3, 1, 4, 1, 5, 9, \ldots \qquad (3)$$

(digits in decimal representation of π);

$$a(1990) \qquad\qquad\qquad a(2000)$$
$$\downarrow \qquad\qquad\qquad\qquad \downarrow$$
$$\ldots, 0, 0, 0, 2, 20K, 30M, 20M, 2M, 1K, 1, 0, 0, 0, 0, \ldots \qquad (4)$$

(number of requests for "the Macarena" at wedding receptions).

The z-transform of the sequence $a(n)$ is defined as the sum of the series

$$A(z) := \sum_{n=-\infty}^{\infty} a(n) z^{-n} \qquad (5)$$
$$= \cdots + a(-2)z^2 + a(-1)z + a(0) + a(1)z^{-1} + a(2)z^{-2} + \cdots$$

at all points where (5) converges. Note the exponent for z is *minus n*.

For the sequence $a(n) = 2^{-|n|}$ in (1) the z-transform can be reorganized as follows:

$$\sum_{n=-\infty}^{\infty} 2^{-|n|} z^{-n} = \sum_{n=-\infty}^{0} 2^n z^{-n} + \sum_{n=1}^{\infty} 2^{-n} z^{-n}$$
$$= \sum_{n=0}^{\infty} \left(\frac{z}{2}\right)^n + \sum_{n=1}^{\infty} \left(\frac{1}{2z}\right)^n . \qquad (6)$$

Both "sums" are geometric series. The first converges to $1/(1 - z/2)$ for $|z| < 2$, whereas the second converges to

$$\sum_{n=1}^{\infty} \left(\frac{1}{2z}\right)^n = \left(\frac{1}{2z}\right) \sum_{n=0}^{\infty} \left(\frac{1}{2z}\right)^n = \frac{\frac{1}{2z}}{1 - \frac{1}{2z}} = \frac{1}{2z - 1}$$

for $|z| > \frac{1}{2}$. Thus, in the common annulus of convergence $\dfrac{1}{2} < |z| < 2$, the z-transform of the sequence $2^{-|n|}$ is the analytic function

$$A(z) = \frac{1}{1 - z/2} + \frac{1}{2z - 1} = \frac{-3z}{2(z - 2)\left(z - \frac{1}{2}\right)}. \qquad (7)$$

Clearly the z-transform is the Laurent series of $A(z)$ in this annulus.

The z-transform for the oscillating sequence (2) has the form $\sum_{n=-\infty}^{\infty}(-1)^n z^{-n}$; the sum for $n \geq 0$ converges for $|z| > 1$, whereas the sum for $n < 0$ converges only for $|z| < 1$. These regions are disjoint, so the z-transform converges nowhere.

Since the negatively indexed elements of the "pi" sequence (3) are all zero, the corresponding portion of the series converges for all z. The terms of the positively indexed portion of the series are bounded by $9|z|^{-n}$ and thus by the M-test (Theorem 13, Sec. 5.4) this subseries converges for $|z| > 1$, which is then the common region of convergence and the domain of definition of the z-transform.

The sequence (4) has only a finite number of nonzero terms, so its z-transform is simply a polynomial in $1/z$ converging for all values of $z \neq 0$.

From these examples we can see the general nature of the z-transform of a sequence; it is an analytic function defined by the Laurent series whose coefficients are the terms of the sequence taken in reverse order (since $a(n)$ multiplies z^{-n}). From the convergence theory surveyed in Sec. 5.4 we deduce that the positively indexed portion of the series converges for

$$|z| > \limsup \sqrt[n]{|a(n)|},$$

and the negatively indexed portion converges for

$$|z| < \frac{1}{\limsup \sqrt[n]{|a(-n)|}}.$$

The z-transform is thus well defined for z in the annulus

$$\limsup \sqrt[n]{|a(n)|} < |z| < \frac{1}{\limsup \sqrt[n]{|a(-n)|}}, \tag{8}$$

if this set is nonempty. The transform is analytic in this annulus, and the Laurent series enjoys the usual properties of termwise differentiation, integration, and multiplication.

If $a(n) = 0$ for $n < 0$ [as in (3)], the sequence is said to be *causal*. From (8) we see that the z-transform of a causal sequence converges *outside a circle* (that is, the outer radius of the annulus is infinite).

A given analytic function can be the z-transform of more than one sequence, since its Laurent series representation is not unique (it depends on the region of convergence). The computations in Example 2 of Sec. 5.5 show that the function $1/[(z - 1)(z - 2)]$ is the transform of each of the following sequences:

$$a(n) = \begin{cases} 1 - 2^{n-1}, & n \leq 0 \\ 0, & n > 0 \end{cases} \qquad \text{for } |z| < 1;$$

$$a(n) = \begin{cases} -2^{n-1}, & n \leq 0 \\ -1, & n > 0 \end{cases} \qquad \text{for } 1 < |z| < 2;$$

$$a(n) = \begin{cases} 0, & n < 1 \\ 2^{n-1} - 1, & n \geq 1 \end{cases} \qquad \text{for } |z| > 2.$$

The third of these sequences is causal; in general any Laurent series converging in the exterior of a circle is the *z*-transform of a causal sequence.

When the *z*-transform of a sequence can be written in closed form, it provides a very compact representation of the sequence. Also, as we shall see, it facilitates the solution of recursion relations, or "difference equations," involving sequences. The tools for recovering a sequence from its *z*-transform are precisely the tools for constructing Laurent series, which we explored in Sec. 5.5; one employs Maclaurin series such as the geometric series, partial fractions, etc. In this regard, the following version of Theorem 14 of that section can be interpreted as an inverse *z*-transform formula:

Theorem 7. Let $A(z)$ be the *z*-transform of the sequence $\{a(n) : -\infty < n < \infty\}$ in the annulus $a < |z| < b$. Then

$$a(n) = \frac{1}{2\pi i} \oint_\Gamma A(\zeta)\zeta^{n-1}\, d\zeta \qquad (n = 0, \pm 1, \pm 2, \ldots)$$

where Γ is any positively oriented simple closed contour lying in the annulus and encircling the origin.

The key to most applications of the *z*-transform is the following property.

Theorem 8. Let $A(z)$ be the *z*-transform of the sequence $\{a(n) : -\infty < n < \infty\}$ in the annulus $a < |z| < b$. Then the corresponding *z*-transform of the shifted sequence $\{b(n) = a(n+1) : -\infty < n < \infty\}$ is given by $zA(z)$. More generally, the *z*-transform of the sequence $\{c(n) = a(n+N) : -\infty < n < \infty\}$ equals $z^N A(z)$ for any N (positive or negative).

The proof is transparent: the *z*-transform of $b(n)$ is

$$\sum_{n=-\infty}^{\infty} b(n)z^{-n} = \sum_{n=-\infty}^{\infty} a(n+1)z^{-n} = z \sum_{n=-\infty}^{\infty} a(n+1)z^{-(n+1)} = zA(z),$$

and the generalization follows easily.

This property is most commonly used to express the transform of the *delayed* sequence $c(n) = a(n-1)$ as $z^{-1}A(z)$. Some examples will demonstrate how the *z*-transform enables the solution of difference equations.

Example 1

Let $a(n)$ represent the balance in a savings account at the end of month n. Starting from month 1 a monthly deposit of t dollars is made and compound interest is paid on the previous month's principal at the rate of $r \cdot 100$ percent (per month). If the account holds P dollars at the beginning of month 0 and no monies are withdrawn, develop the formula for $a(n)$.

Solution.   The balance in the account on successive months has the pattern

$$\vdots$$

$$a(-2) = 0$$
$$a(-1) = 0$$
$$a(0) = P$$
$$a(1) = a(0) + ra(0) + t$$
$$a(2) = a(1) + ra(1) + t$$

$$\vdots$$

From this display we can formulate the difference equation relating the balance on successive months:

$$a(n) = a(n - 1) + ra(n - 1) + D(n), \tag{9}$$

where

$$D(n) := \begin{cases} 0, & n < 0 \\ P, & n = 0 \\ t, & n \geq 1. \end{cases}$$

Clearly $a(n) = 0$ for $n < 0$, so $a(n)$ is causal. Thus we assume that $a(n)$ has a z-transform $A(z)$ for z sufficiently large. Taking the z-transform of both sides of Eq. (9), we use Theorem 8 and the fact that

$$\sum_{n=1}^{\infty} z^{-n} = z^{-1} \sum_{n=0}^{\infty} z^{-n} = z^{-1} \frac{1}{1 - 1/z} = \frac{1}{z - 1} \qquad \text{for } |z| > 1$$

to obtain

$$A(z) = [1 + r]z^{-1}A(z) + P + \frac{t}{z - 1},$$

or

$$A(z) = \frac{Pz^2 + (t - P)z}{(z - 1)(z - 1 - r)} = z\frac{Pz + (t - P)}{(z - 1)(z - 1 - r)} \qquad (|z| \text{ "large"}).$$

With a partial fraction decomposition this becomes

$$A(z) = z\left(\frac{P + t/r}{z - 1 - r} - \frac{t/r}{z - 1}\right),$$

which has the following Laurent series expansion for large $|z|$ (for $|z| > 1 + r$, in fact):

$$A(z) = \left(P + \frac{t}{r}\right)\frac{1}{1 - (1 + r)/z} - \frac{t}{r}\frac{1}{1 - 1/z}$$

$$= \left(P + \frac{t}{r}\right)\sum_{n=0}^{\infty}(1 + r)^n z^{-n} - \frac{t}{r}\sum_{n=0}^{\infty} z^{-n}.$$

The corresponding sequence is therefore given by

$$a(n) = \begin{cases} \left(P + \dfrac{t}{r}\right)(1+r)^n - \dfrac{t}{r}, & n \geq 0, \\ 0, & n < 0. \end{cases} \qquad \blacksquare$$

Consider the linear constant-coefficient difference equation given by

$$y(n) = \beta(1)y(n-1) + \beta(2)y(n-2) + \beta(3)y(n-3)$$
$$+ b(0)x(n) + b(1)x(n-1) + b(2)x(n-2) .$$

In statistics such an equation is called an *autoregressive-moving-average* relationship between the input $x(n)$ and the output $y(n)$, of order (3,2). More generally, the "ARMA(p,q)" equation, characterized by

$$y(n) = \sum_{j=1}^{p} \beta(j)y(n-j) + \sum_{k=0}^{q} b(k)x(n-k) , \qquad (10)$$

is frequently used to model random sequences $y(n)$ driven by "white noise" $x(n)$. (The first sum in (10) is the autoregression and the second is the moving average.)

Example 2

For the ARMA(1,1) model express the z-transform of the output in terms of the z-transform of the input and the constants $\beta(0)$, $b(0)$, and $b(1)$.

Solution. Taking the z-transform of the equation

$$y(n) = \beta(1)y(n-1) + b(0)x(n) + b(1)x(n-1) \qquad (11)$$

we find

$$Y(z) = \beta(1)z^{-1}Y(z) + b(0)X(z) + b(1)z^{-1}X(z)$$

or

$$Y(z) = \frac{b(0) + b(1)z^{-1}}{1 - \beta(1)z^{-1}}X(z) = \frac{b(0)z + b(1)}{z - \beta(1)}X(z) . \qquad \blacksquare \qquad (12)$$

We have discussed how engineers often characterize a system by its frequency response. For discrete systems modeled by difference equations like (10), this means using samples of the complex sinusoid $e^{i\omega t}$ as the input

$$x(n) = \{e^{i\omega t}\}_{t=n} = e^{i\omega n} , \qquad (13)$$

solving for outputs having the same form

$$y(n) = H(\omega)x(n) = H(\omega)e^{i\omega n} , \qquad (14)$$

and observing the dependence of the *frequency domain transfer function* $H(\omega)$ on the frequency ω in the interval $[-\pi, \pi]$.[†] One says that $H(\omega)$ *filters* the input.

Example 3

Find necessary and sufficient conditions on the coefficients $\beta(1)$, $b(0)$, and $b(1)$ in the ARMA(1,1) model for the modulus of the transfer function $H(\omega)$ to be unity for all frequencies ω. Such a transfer function is known as an *all-pass filter*.

Solution. Rigorously speaking, it would be incorrect to take the z-transform of $x(n)$, since the series $\sum_{n=-\infty}^{\infty} e^{i\omega n} z^{-n}$ converges nowhere. So we directly substitute the forms (13) and (14) into (11)

$$H(\omega)e^{i\omega n} = \beta(1)H(\omega)e^{i\omega(n-1)} + b(0)e^{i\omega n} + b(1)e^{i\omega(n-1)}$$

and solve:

$$H(\omega) = \frac{b(0) + b(1)e^{-i\omega}}{1 - \beta(1)e^{-i\omega}} = \frac{b(0)e^{i\omega} + b(1)}{e^{i\omega} - \beta(1)} \quad . \tag{15}$$

This has the same form as the factor occurring in Eq. (12), with the substitution $z = e^{i\omega}$. In other words, $H(\omega)$ coincides with the Möbius transformation

$$\tilde{H}(z) = \frac{b(0)z + b(1)}{z - \beta(1)} \equiv \frac{z + b(1)/b(0)}{z/b(0) - \beta(1)/b(0)} \tag{16}$$

for z on the unit circle. The all-pass requirement $|H(\omega)| = 1$, then, is the requirement that this Möbius transformation maps the unit circle to the unit circle. From what we know about Möbius transformations, $\tilde{H}(z)$ will also map the interior of the circle to itself, or to the exterior.

In Example 3 of Sec. 7.4 and Prob. 16 of Exercises 7.4, we proved that all Möbius transformations mapping the unit circle to itself have the form

$$\tilde{H}(z) = e^{i\theta} \frac{z - \alpha}{\bar{\alpha}z - 1}.$$

Matching this with Eq. (16) we conclude

$$\beta(1) = b(0) = \frac{1}{\bar{\alpha}}, \quad b(1) = -\frac{\alpha}{\bar{\alpha}}, \quad |\alpha| \neq 1 . \quad \blacksquare$$

The solution of difference equations with initial conditions is most conveniently accomplished with the *unilateral z-transform* $A^+(z)$, which omits the negatively indexed terms in the series (5):

$$A^+(z) := \sum_{n=0}^{\infty} a(n)z^{-n}. \tag{17}$$

[†] Since $e^{i(\omega+2\pi)n} \equiv e^{i\omega n}$, only the frequencies in $(-\pi, \pi]$ need to be considered in the present context.

Clearly if $a(n)$ is causal, then $A^+(z) = A(z)$, and the region of convergence of $A^+(z)$ is the exterior of a circle ($|z| > \limsup \sqrt[n]{|a(n)|}$). The shifting property for the unilateral *z*-transform is similar to that described in Theorem 8, modified along the lines of the Laplace transform formula for derivatives [Eq. (8), Sec. 8.3].

Theorem 9. Let $A^+(z)$ be the unilateral *z*-transform of the sequence $\{a(n) : -\infty < n < \infty\}$ in the region $a < |z|$. Then the corresponding unilateral *z*-transform of the shifted sequence $\{b(n) = a(n+1) : -\infty < n < \infty\}$ is given by $z[A^+(z) - a(0)]$. More generally, the unilateral *z*-transform of the sequence $\{c(n) = a(n+N) : -\infty < n < \infty\}$ equals

$$z^N\left[A^+(z) - a(0) - a(1)z^{-1} - a(2)z^{-2} - \cdots - a(N-1)z^{-(N-1)}\right]$$

for any positive N.

Again the proof is easy. The unilateral *z*-transform of $b(n)$ is

$$\sum_{n=0}^{\infty} b(n)z^{-n} = \sum_{n=0}^{\infty} a(n+1)z^{-n} = z\sum_{n=0}^{\infty} a(n+1)z^{-(n+1)}$$

$$= z\sum_{m=0}^{\infty} a(m)z^{-m} - za(0)$$

$$= z[A^+(z) - a(0)].$$

The generalization is left to the reader.

Example 4

Suppose the sequence $a(n)$ satisfies the difference equation

$$a(n+2) - 3a(n+1) + 2a(n) = 0 \tag{18}$$

for $n \geq 0$ and that $a(0) = 1$ and $a(1) = -1$. Find a formula for $a(n)$ for $n \geq 0$.

Solution. Since (18) holds for all $n \geq 0$ we are justified in applying the unilateral *z*-transform. Employing Theorem 9 we derive

$$z^2\left[A^+(z) - 1 - (-1)z^{-1}\right] - 3z\left[A^+(z) - 1\right] + 2A^+(z) = 0$$

or

$$A^+(z) = \frac{z^2 - 4z}{z^2 - 3z + 2} = z\frac{z - 4}{(z-1)(z-2)} = z\left(\frac{3}{z-1} + \frac{-2}{z-2}\right)$$

$$= 3\frac{1}{1 - 1/z} - 2\frac{1}{1 - 2/z}$$

$$= 3\sum_{n=0}^{\infty} z^{-n} - 2\sum_{n=0}^{\infty} 2^n z^{-n}.$$

Thus

$$a(n) = 3 - 2^{n+1}, \qquad \text{for } n \geq 0. \quad \blacksquare$$

Further properties of the z-transform and its engineering applications are developed in the references.

EXERCISES 8.4

1. Show that the z-transform is a linear operator; that is, if $A(z)$ and $B(z)$ denote the z-transforms of $\{a(n)\}$ and $\{b(n)\}$, respectively, then the transform of $\{\alpha a(n) + \beta b(n)\}$ is $\alpha A(z) + \beta B(z)$, in the common region of convergence.

2. If $A(z)$ is the z-transform of $\{a(n)\}$, show that the z-transform of the "linearly weighted" sequence $\{na(n)\}$ is $-zA'(z)$. Show that the annulus of convergence is unchanged. [HINT: $\lim_{n\to\infty} \sqrt[n]{n} = 1$.]

3. If $A(z)$ is the z-transform of $\{a(n)\}$, show that the z-transform of the "exponentially weighted" sequence $\{\alpha^n a(n)\}$ is $A(z/\alpha)$. How is the new annulus of convergence related to the old?

4. Verify the entries in the following table of *causal* z-transforms: for $n \geq 0$,

   (a) $a(n) = \begin{cases} 1, & n = 0 \\ 0, & n > 0 \end{cases}$ $A(z) = 1$

   (b) $a(n) = 1$ $A(z) = \dfrac{z}{z-1}$

   (c) $a(n) = n$ $A(z) = \dfrac{z}{(z-1)^2}$

   (d) $a(n) = \alpha^n$ $A(z) = \dfrac{z}{z-\alpha}$

   (e) $a(n) = \sin n\omega$ $A(z) = \dfrac{z\sin\omega}{z^2 - 2z\cos\omega + 1}$

   (f) $a(n) = \cos n\omega$ $A(z) = \dfrac{z(z-\cos\omega)}{z^2 - 2z\cos\omega + 1}$

5. Find inverse z-transforms for the following functions in the indicated annuli:

   (a) $A(z) = \dfrac{1}{1 + 1/(3z)}, \qquad |z| < \dfrac{1}{3}$

   (b) $A(z) = \dfrac{1}{1 + 1/(3z)}, \qquad |z| > \dfrac{1}{3}$

   (c) $A(z) = \dfrac{z^4}{z+2}, \qquad |z| < 2$

   (d) $A(z) = \dfrac{z^4}{z+2}, \qquad |z| > 2$

(e) $A(z) = \dfrac{z+2}{2z^2 - 7z + 3}, \qquad \dfrac{1}{2} < |z| < 3$

(f) $A(z) = \dfrac{1 - 1/(2z)}{1 + 3/(4z) + 1/(8z^2)}, \qquad |z| > \dfrac{1}{2}$

(g) $A(z) = \dfrac{z}{(z - \frac{1}{2})(z - 1)^2}, \qquad |z| > 1$

(h) $A(z) = \dfrac{1 - \alpha/z}{\alpha - 1/z}, \qquad |z| > \dfrac{1}{|\alpha|}$

6. If $A(z)$ is the z-transform of a causal sequence $\{a(n)\}$, show that $a(0) = \lim_{z \to \infty} A(z)$.

7. Use unilateral z-transforms to solve the following difference equations.

    (a) $a(n+1) = (0.5)a(n), \qquad a(0) = 2$

    (b) $a(n+1) + 2a(n) = 1, \qquad a(0) = 1$

    (c) $a(n+2) - 5a(n+1) + 6a(n) = 1, \qquad a(0) = 2, a(1) = 3$

8. Derive the formula for the unilateral z-transform of the backward-shifted sequence $a(n - N), N > 0$.

8.5   Cauchy Integrals and the Hilbert Transform

An integral of the form

$$\int_\Gamma \frac{f(\zeta)}{\zeta - z} \, d\zeta$$

is known as a *Cauchy integral*. The study of Cauchy integrals is quite provocative and rewarding, and in this section we shall explore some of the theoretical and practical aspects of such forms.

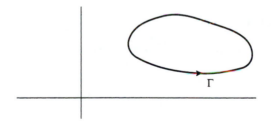

Figure 8.12 Contour for Cauchy integral.

If Γ is a simple smooth closed curve as in Fig. 8.12, and f is analytic inside and on Γ, then Cauchy's formula and theorem tell us that

$$\oint_\Gamma \frac{f(\zeta)}{\zeta - z} \, d\zeta = \begin{cases} 2\pi i f(z) & \text{if } z \text{ lies inside } \Gamma, \\ 0 & \text{if } z \text{ lies outside } \Gamma. \end{cases} \tag{1}$$

Figure 8.13 Indented contour.

The question naturally arises as to how the values of the integral evolve as the point z *crosses* Γ. To explore this, consider the indented contour Γ'_ε in Fig. 8.13. The point z_0 lies on the original contour Γ, but since it falls inside Γ'_ε, Cauchy's formula tells us that

$$\oint_{\Gamma'_\varepsilon} \frac{f(\zeta)}{\zeta - z_0}\, d\zeta = 2\pi i f(z_0). \tag{2}$$

Here we have presumed that the radius ε of the semicircular indentation S'_ε is sufficiently small that Γ'_ε remains within the domain of analyticity for f. Now as ε goes to zero, the contribution from the semicircle S'_ε to the integral along Γ'_ε approaches $\pi i f(z_0)$ (Lemma 4, Sec. 6.5). This is half the value shown in Eq. (2); therefore, the remaining $\pi i f(z_0)$ must come from the rest of the contour, which is a facsimile of Γ snipped "symmetrically" around z_0 in a manner generalizing the *principal value* concept (Sec. 6.5). To summarize: On the basis of Fig. 8.13 we express

$$\underset{\substack{\| \\ 2\pi i f(z_0) \\ \text{(Cauchy's formula)}}}{\oint_{\Gamma'_\varepsilon} \frac{f(\zeta)}{\zeta - z_0}\, d\zeta} = \text{p.v.} \int_\Gamma \frac{f(\zeta)}{\zeta - z_0}\, d\zeta + \underset{\substack{\| \\ \pi i f(z_0) \\ \text{(Lemma 4)}}}{\lim_{\varepsilon \to 0} \int_{S'_\varepsilon} \frac{f(\zeta)}{\zeta - z_0}\, d\zeta}, \tag{3}$$

from which we conclude

$$\text{p.v.} \int_\Gamma \frac{f(\zeta)}{\zeta - z_0}\, d\zeta = \pi i f(z_0). \tag{4}$$

Of course, our Fig. 8.13 benignly sidesteps any topological complications; see Prob. 3, for example. We direct the reader to the references for a more rigorous statement and derivation of (4).

Example 1

Confirm Eq. (4) for the case when Γ is the positively oriented unit circle centered around $z = 1$, $f(z) \equiv 1$, and $z_0 = 0$.

   Solution.   Obviously $\int_\Gamma 1/(\zeta - z)\, d\zeta$ equals $2\pi i$ for $|z - 1| < 1$ and zero for $|z - 1| > 1$. For $z = z_0 = 0$ we refer to Fig. 8.14 to derive

$$\text{p.v.} \int_\Gamma \frac{1}{\zeta - 0}\, d\zeta = \lim_{z_1, z_2 \to 0} \left[\text{Log}\, z_1 - \text{Log}\, z_2 \right]$$

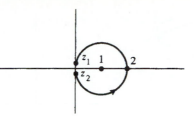

Figure 8.14 Contour for Example 1.

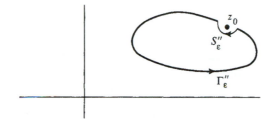

Figure 8.15 Interior indentation.

with $|z_1| = |z_2|$ as shown. Thus we have

$$\text{p.v.} \int_\Gamma \frac{1}{\zeta - 0} \, d\zeta = \lim_{z_1, z_2 \to 0} \left[i \, \text{Arg} \, z_1 - i \, \text{Arg} \, z_2 \right] = i\frac{\pi}{2} + i\frac{\pi}{2} = \pi i. \quad \blacksquare$$

What happens if we construct an indentation penetrating the *interior of* Γ, as in Fig. 8.15? Then the (clockwise-oriented) semicircle S_ε'' contributes *minus* $\pi i f(z_0)$ (Lemma 4, Sec. 6.5) and the decomposition of the integral (3) takes the form

$$\oint_{\Gamma_\varepsilon''} \frac{f(\zeta)}{\zeta - z_0} \, d\zeta \quad = \text{p.v.} \int_\Gamma \frac{f(\zeta)}{\zeta - z_0} \, d\zeta + \lim_{\varepsilon \to 0} \int_{S_\varepsilon''} \frac{f(\zeta)}{\zeta - z_0} \, d\zeta.$$

$$\begin{array}{cccc}
\| & & \| & \| \\
0 & & \pi i f(z_0) & -\pi i f(z_0) \\
\text{(Cauchy's theorem)} & & [\text{Eq. (4)}] & \text{(Lemma 4)}
\end{array} \qquad (5)$$

From these considerations we can visualize what happens to the Cauchy integral (1) as the point z crosses the contour Γ. In Fig. 8.16 the points $\{z_n^+ : n = 1, 2, 3, \ldots\}$ approach z_0 from the inside, and the Cauchy integrals equal $2\pi i f(z_n^+)$. Furthermore, if the contour is indented as in Fig. 8.13, these integrals are unchanged (by the deformation invariance theorem, Sec. 4.4a). They approach the limit $2\pi i f(z_0)$, and invoking Eqs. (3) and (4) we can attribute half of this limit to the principle value and the other half to the exterior indentation S_ε'.

Now for the sequence $\{z_n^-\}$ approaching z_0 from *outside* Γ (Fig. 8.17), we use the interior indentation S_ε'' to argue that the contributions to the limit (zero) of the Cauchy

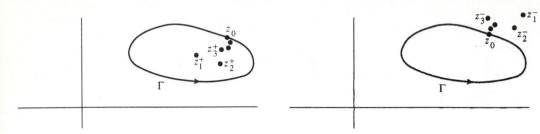

Figure 8.16 Approaching z_0 from inside Γ.

Figure 8.17 Approaching z_0 from outside Γ.

integrals are $\pi i f(z_0)$ from the principal value and $-\pi i f(z_0)$ from the semicircle S''_ε. Thus as the point z crosses the contour Γ, we can ascribe the jump in the Cauchy integral to the substitution of one indentation for the other; S'_ε "opens the gate" and lets z through; then S''_ε closes the gate behind it. The difference between the interior and exterior limits is due to the opposing contributions of the semicircles:

$$\lim_{z_n^+ \to z_0} \oint_\Gamma \frac{f(\zeta)}{\zeta - z_n^+} \, d\zeta - \lim_{z_n^- \to z_0} \oint_\Gamma \frac{f(\zeta)}{\zeta - z_n^-} \, d\zeta = \lim_{\varepsilon \to 0} \int_{S'_\varepsilon} \frac{f(\zeta)}{\zeta - z_0} \, d\zeta - \lim_{\varepsilon \to 0} \int_{S''_\varepsilon} \frac{f(\zeta)}{\zeta - z_0} \, d\zeta$$

$$= \pi i f(z_0) - [-\pi i f(z_0)]$$

$$= 2\pi i f(z_0). \tag{6}$$

Note that by similar accounting the *average* of the interior and exterior limits yields the principal value of the integral:

$$\lim_{z_n^+ \to z_0} \oint_\Gamma \frac{f(\zeta)}{\zeta - z_n^+} \, d\zeta + \lim_{z_n^- \to z_0} \oint_\Gamma \frac{f(\zeta)}{\zeta - z_n^-} \, d\zeta$$

$$= 2 \left(\text{p.v.} \int_\Gamma \frac{f(\zeta)}{\zeta - z_0} \, d\zeta \right) + \lim_{\varepsilon \to 0} \int_{S'_\varepsilon} \frac{f(\zeta)}{\zeta - z_0} \, d\zeta + \lim_{\varepsilon \to 0} \int_{S''_\varepsilon} \frac{f(\zeta)}{\zeta - z_0} \, d\zeta$$

$$= 2 \left(\text{p.v.} \int_\Gamma \frac{f(\zeta)}{\zeta - z_0} \, d\zeta \right). \tag{7}$$

If the contour Γ is *not* closed we have no general theorem to tell us the values of $\int_\Gamma f(\zeta)/(\zeta - z) \, d\zeta$, but the argumentation motivated by Figs. 8.13 through 8.17 still validates Eqs. (6) and (7) as long as we interpret the sequence $\{z_n^+\}$ as approaching z_0 from the *left* of Γ (as determined by its orientation) and $\{z_n^-\}$ as approaching from the right. The *Sokhotskyi-Plemelj formulas* (proved in the references) extend these considerations to more general (not necessarily analytic) functions and contours; they state that the difference between the limiting values of the Cauchy integral $\int_\Gamma f(\zeta)/(\zeta - z) \, d\zeta$ as z approaches z_0 (on Γ) from the left and from the right is always equal to $2\pi i f(z_0)$, whereas their average equals the principal value.

A particularly useful identity results when Γ "encloses" a half-plane, as in Fig. 8.18. If $f(z)$ is analytic in, say, the upper half-plane and goes to zero at infinity so rapidly

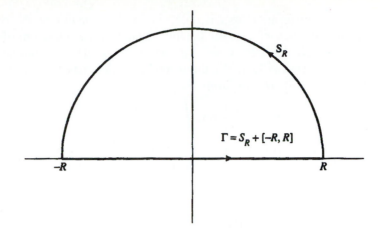

Figure 8.18 Contour enclosing half-plane (as $R \to \infty$).

that the contribution of the semicircle S_R vanishes as $R \to \infty$, then Eq. (4) takes the form

$$\text{p.v.} \int_{-\infty}^{\infty} \frac{f(\xi)}{\xi - x} \, d\xi = \pi i f(x). \tag{8}$$

The integral over S_R will disappear, for instance, if $|f(z)| \leq K/|z|$ (by the usual estimates) or if $|f(z)| \leq K|e^{imz}|$ for positive m (by Jordan's lemma).

Expressing $f(x + iy) = u(x, y) + iv(x, y)$ and separating (8) into its real and imaginary parts, we obtain

$$v(x, 0) = -\frac{1}{\pi} \, \text{p.v.} \int_{-\infty}^{\infty} \frac{u(\xi, 0)}{\xi - x} \, d\xi, \quad u(x, 0) = \frac{1}{\pi} \, \text{p.v.} \int_{-\infty}^{\infty} \frac{v(\xi, 0)}{\xi - x} \, d\xi. \tag{9}$$

The first of these formulas motivates the following definition.

Definition 2. The **Hilbert transform** of an arbitrary real-valued function $\phi(x)$ $(-\infty < x < \infty)$ is defined by

$$\psi(x) := -\frac{1}{\pi} \, \text{p.v.} \int_{-\infty}^{\infty} \frac{\phi(\xi)}{\xi - x} \, d\xi \tag{10}$$

(when the integral exists).

Our derivation shows that whenever $\phi(x)$ and $\psi(x)$ are a pair of functions such that the combination $\phi + i\psi$ can be extended as an analytic function in the upper half-plane, suitably "dying off" at infinity therein, then ψ is the Hilbert[†] transform of

---

[†] David Hilbert (1862-1943) proposed the "23 Paris problems," many of which still challenge and guide mathematics research today.

ϕ. Thus the Hilbert transform of $\cos x$ is $\sin x$, since $\cos x + i \sin x = e^{iz}$ for $z = x$ and Jordan's lemma implies e^{iz} dies off suitably. Note that the transform of $\sin x$ is *minus* $\cos x$ (because $\sin x - i \cos x = -ie^{iz}$ for $z = x$). This is consistent with the appearance of the signs in Eqs. (9), the second of which may be regarded as the formula for the *inverse Hilbert transform*:

$$\phi(x) := \frac{1}{\pi} \text{ p.v.} \int_{-\infty}^{\infty} \frac{\psi(\xi)}{\xi - x} d\xi \tag{11}$$

Clearly, a collection of Hilbert transforms can be generated by writing down analytic functions with the requisite properties in the upper half-plane and separating their real and imaginary parts on the x-axis. In this manner one derives (see Prob. 1) the accompanying table.

TABLE OF HILBERT TRANSFORMS

Function $\phi(x)$		Hilbert transform $\psi(x)$
$\cos \omega x$	$(\omega > 0)$	$\sin \omega x$
$\cos \omega x$	$(\omega < 0)$	$-\sin \omega x$
$\sin \omega x$	$(\omega > 0)$	$-\cos \omega x$
$\sin \omega x$	$(\omega < 0)$	$\cos \omega x$
$\dfrac{a}{a^2 + x^2}$	$(a > 0)$	$\dfrac{x}{a^2 + x^2}$
$\dfrac{\sin ax}{x}$	$(a > 0)$	$\dfrac{1 - \cos ax}{x}$

Example 2
Verify Eq. (10) for the first entry in the transform table.

Solution.   We break up the Cauchy integral into two parts:

$$\text{p.v.} \int_{-\infty}^{\infty} \frac{\cos \omega \xi}{\xi - x} d\xi = \text{p.v.} \int_{-\infty}^{\infty} \frac{e^{i\omega\xi}}{2(\xi - x)} d\xi + \text{p.v.} \int_{-\infty}^{\infty} \frac{e^{-i\omega\xi}}{2(\xi - x)} d\xi.$$

To evaluate the first integral we close the contour with a semicircle and an indentation in the upper half ξ-plane, as shown in Fig. 8.19(a). Using Jordan's lemma, (Lemma 3 from Sec. 6.4) and Cauchy's theorem, we obtain

$$\text{p.v.} \int_{-\infty}^{\infty} \frac{e^{i\omega\xi}}{2(\xi - x)} d\xi = \pi i \text{ Res}(\xi = x) = \frac{\pi i e^{i\omega x}}{2}.$$

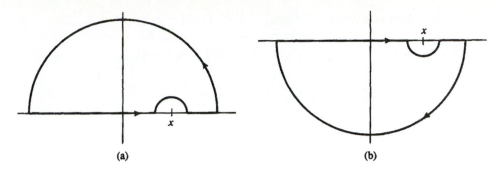

Figure 8.19 Contours for Example 2.

The contour for the second integral is closed as in Fig. 8.19(b), and we have

$$\text{p.v.} \int_{-\infty}^{\infty} \frac{e^{-i\omega\xi}}{2(\xi - x)}\, d\xi = -\pi i \operatorname{Res}(\xi = x) = -\frac{\pi i e^{-i\omega x}}{2}.$$

The sum of these equals

$$\text{p.v.} \int_{-\infty}^{\infty} \frac{\cos \omega\xi}{\xi - x}\, d\xi = \pi i \frac{e^{i\omega x} - e^{-i\omega x}}{2} = -\pi \sin \omega x$$

and dividing by $-\pi$, we confirm the entry.   ∎

The reader may have observed that a methodology for calculating the Hilbert transform is already at hand: starting from the function $\phi(x)$, one could use Poisson's formula for the half-plane (Prob. 14, Exercises 4.7) to extend it as a suitable harmonic function $u(x, y)$ in the upper half-plane; then Theorem 25 of that section establishes the harmonic conjugate $v(x, y)$ whose restriction to the x-axis becomes the transform $\psi(x)$. Eq. (10), then, accomplishes all this directly. It also affords the extension of the transform to a wider class of functions. These generalizations can be found in the references.

Many applications of the Hilbert transform are based upon the way it interacts with the Fourier transform. Let us write the latter and its inversion formula as

$$\Phi(\omega) = \frac{1}{2\pi} \int_{-\infty}^{\infty} \phi(x)e^{-i\omega x}\, dx, \tag{12}$$

$$\phi(x) = \int_{-\infty}^{\infty} \Phi(\omega)e^{i\omega x}\, d\omega. \tag{13}$$

Now the Hilbert transform is a linear operation whose effect on cosines and sines has been determined. Therefore, if we adopt the point of view (espoused in Sec. 8.2) that Eq. (13) expresses ϕ as a superposition of sinusoids, then we rewrite (13) as

$$\begin{aligned}
\phi(x) = &\int_{-\infty}^{0} \Phi(\omega)[\cos \omega x + i \sin \omega x]\, d\omega \\
&+ \int_{0}^{\infty} \Phi(\omega)[\cos \omega x + i \sin \omega x]\, d\omega
\end{aligned} \tag{14}$$

and take its Hilbert transform as follows:

$$\psi(x) = \int_{-\infty}^{0} \Phi(\omega)[-\sin \omega x + i \cos \omega x]\, d\omega + \int_{0}^{\infty} \Phi(\omega)[\sin \omega x - i \cos \omega x]\, d\omega$$

$$= i \int_{-\infty}^{0} \Phi(\omega)e^{i\omega x}\, d\omega + (-i) \int_{0}^{\infty} \Phi(\omega)e^{i\omega x}\, d\omega$$

$$= \int_{-\infty}^{0} \Phi(\omega)e^{i(\omega x + \pi/2)}\, d\omega + \int_{0}^{\infty} \Phi(\omega)e^{i(\omega x - \pi/2)}\, d\omega. \tag{15}$$

Comparing (15) with (14) we observe that the Hilbert transform multiplies the positive-frequency portion of $\Phi(\omega)$ by $e^{-i\pi/2} = -i$, and the negative-frequency portion by $e^{i\pi/2} = i$; positive frequencies are "phase-delayed" by $\pi/2$ radians, and negative frequencies are phase-advanced by the same amount. The Hilbert transform is thus a 90° phase-shift operation.

This suggests an efficient numerical algorithm for the implementation of the Hilbert transform. First we use the fast Fourier transform algorithm to approximate $\Phi(\omega)$. Then we reverse the real and imaginary parts and change the sign of the former for $\omega < 0$ and the latter for $\omega > 0$. Finally, we take the inverse Fourier transform, again employing the FFT. This procedure is commonly used in signal-processing applications.

The following example shows how the Hilbert transform arises in radio communications.

Example 3

The human ear can detect sound waves with frequencies up to a certain level Ω_0; higher frequencies go unheard. (The frequency Ω_0 is about $25,000$ radians per second, or 4 kiloHertz.) Thus in practice when we perform a Fourier transform on a sound message $\phi(t)$, we need to retain only the contributions for $|\omega|$ below Ω_0:

$$\phi(t) = \int_{-\infty}^{\infty} \Phi(\omega)e^{i\omega t}\, d\omega \approx \int_{-\Omega_0}^{\Omega_0} \Phi(\omega)e^{i\omega t}\, d\omega.$$

Effectively, then, the real and imaginary parts of $\Phi(\omega)$ can be visualized as in Fig. 8.20. (Recall the symmetry properties of Φ for a real $\phi(t)$, Prob. 4 of Exercises 8.2.)

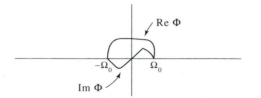

Figure 8.20 Original signal (Fourier transform).

In radio broadcasting it is difficult to transmit at such low frequencies, so often the *amplitude-modulation* scheme is employed. The signal is electronically multiplied by a *carrier wave* $\cos \Omega_1 t = [e^{i\Omega_1 t} + e^{-i\Omega_1 t}]/2$ at a high frequency Ω_1 (in the megahertz range); according to Prob. 5, Exercises 8.2, this shifts the transform to the left and right as indicated in Fig. 8.21, resulting in a high-frequency signal $\phi(t) \cos \Omega_1 t$ that is easily transmitted and received.

Figure 8.21 Lower Single Sideband.

Because of the symmetries in Φ, one can save power by first zeroing (filtering) out the shaded portion of $\Phi(\omega \pm \Omega_1)$ in Fig. 8.21, without loss of information. The receiver then extracts ("detects") the original message by multiplying the "Lower Single SideBand" signal (LSS) by $\cos \Omega_1 t$ (again), resulting in the transform indicated in Fig. 8.22, filtering out the high-frequency portions, and multiplying by 4 (to account for the filtering losses). This procedure is known as *synchronous detection*.

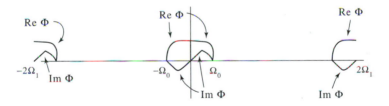

Figure 8.22 Original signal recovered.

It is important for detection that the LSS is multiplied by $\cos \Omega_1 t$, and not by an *unsynchronized* version

$$\cos(\Omega_1 t - \beta) = \cos \Omega_1 t \, \cos \beta + \sin \Omega_1 t \, \sin \beta.$$

Indeed, unsynchronized detection would result in a superposition of the original message plus a distorted message, resulting from multiplying by $\sin \Omega_1 t$.

Show that the distorted message is the Hilbert transform of the original message (times .25).

Solution.   Of course, $\sin \Omega_1 t = [e^{i\Omega_1 t} - e^{-i\Omega_1 t}]/2i$, so multiplication by this sinusoid not only shifts the transform, but it also mixes the real and imaginary parts.

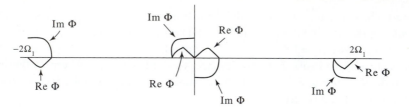

Figure 8.23 Hilbert transform recovered.

A little thought reveals that the resulting transform looks like Fig. 8.23. After dropping the high-frequency portions, comparison with Fig. 8.20 shows that the positive-frequency portion of $\Phi(\omega)$ has been multiplied by $(-i)$ and the negative-frequency portion by i. "Synchronous detection" using the sine, rather than the cosine, has produced the Hilbert transform $\psi(t)$ of the message. ∎

In communications applications such as this the combination $\phi(t) + i\psi(t)$, where ϕ is the original signal and ψ is its Hilbert transform, is called the "analytic signal" associated with $\phi(t)$ (resulting in untold headaches for mathematicians who consult in the industrial sector!).

The Hilbert transform is also useful in other areas. Recall that in Sec. 3.6 we argued that RLC (resistor-inductor-capacitor) electrical circuits, when subjected to a sinusoidal external voltage, will eventually reach a steady state wherein all the internal voltages and currents oscillate sinusoidally at the same frequency as the driving voltage. In particular, we showed that for the circuit of Fig. 3.20, the (complex) current I_s resulting from the voltage $V_s(t) = e^{i\omega t}$ was given by

$$I_s = \frac{e^{i\omega t}}{R_{\text{eff}}},$$

where

$$R_{\text{eff}} = \frac{R/i\omega C}{R + 1/i\omega C} + i\omega L. \tag{16}$$

This "synchronous" behavior is characteristic of most closed (autonomous) physical systems, in that the steady-state response $y(t)$ to an input $e^{i\omega t}$ takes the form $k(\omega)e^{i\omega t}$; the *frequency domain transfer function* $k(\omega)$ (in this case $1/R_{\text{eff}}$) almost always depends on the applied frequency.

An alternative description of the driver-response relationship for such physical systems is available from differential equation theory (see the references). Using a technique known traditionally as the *variation of parameters*, one can express the response $y(t)$ of the system as a *weighted sum* of the functional values of the input $u(t)$. The identity takes the form

$$y(t) = \int_{-\infty}^{\infty} G(t - \tau)u(\tau)\,d\tau. \tag{17}$$

$G(t - \tau)$, the *Green's function* for the system, thus measures the extent to which the values of u at time τ affect the output at time t. G is also called the *impulse response*.

Let's put these two observations together. If the input $u(t)$ is a sinusoid $e^{i\omega t}$, the transfer function description and the Green's function description of the output must agree. Therefore,

$$k(\omega)e^{i\omega t} = \int_{-\infty}^{\infty} G(t - \tau)e^{i\omega\tau}\, d\tau,$$

or

$$k(\omega) = \int_{-\infty}^{\infty} G(t - \tau)e^{-i\omega(t-\tau)}\, d\tau = \int_{-\infty}^{\infty} G(T)e^{-i\omega T}\, dT, \qquad (18)$$

where $T = t - \tau$. Equation (18) has the form of a Fourier transform, except for a factor of 2π. Thus we can invert it and express the Green's function in terms of the transfer function:

$$G(T) = \frac{1}{2\pi}\int_{-\infty}^{\infty} k(\omega)e^{i\omega T}\, d\omega. \qquad (19)$$

If the system is a true physical model (such as the RLC circuit), it must obey the *causality* principle: the value of the response y at a given time t cannot depend on the values of the input u at *later* times τ. Thus from Eq. (17) we see that for causal systems the Green's function $G(t - \tau)$ equals zero for $\tau > t$, and as a result Eq. (19) yields the equality

$$\int_{-\infty}^{\infty} k(\omega)e^{i\omega T}\, d\omega = 0 \qquad \text{for all } T < 0. \qquad (20)$$

As a criterion for causality, condition (20) is intriguing. For $T < 0$ the exponential factor, as a function of the complex variable ω, decays in the *lower* half-plane. Thus Jordan's lemma (Sec. 6.4) would guarantee (20) if the analytic continuation of $k(\omega)$ into the lower half-plane turned out to be analytic there and bounded by (constant$/|\omega|$). In practice $k(\omega)$ is often a rational function (as in the case of the RLC circuit), and subject to a few assumptions about the nature of the physical system, condition (20) can be shown to be *necessary*, as well as sufficient, for causality.

Example 4

Verify the causality condition (20) for the transfer function of the circuit in Fig. 3.20.

Solution. From Eq. (16) we find, with some algebra,

$$k(\omega) = \frac{1}{R_{\text{eff}}} = \frac{i\omega C R + 1}{-\omega^2 C R L + i\omega L + R},$$

whose poles are

$$\omega_{\pm} = \frac{i \pm \sqrt{-1 + 4C R^2 / L}}{2C R}. \qquad (21)$$

If $4C R^2 / L > 1$, the radical is real and both poles lie in the upper half-plane. Otherwise, we have $0 < 4C R^2 / L < 1$ and the radical is an imaginary number between 0 and i; thus the poles still stay in the upper half-plane. Therefore, $k(\omega)$ is analytic in the lower half-plane, and since it falls off like $|\omega|^{-1}$, the system is causal. ∎

How does the Hilbert transform come into play in this context? The causal transfer function $k(\omega)$ has all the properties that are necessary to validate identity (8); it goes to zero at infinity like $|\omega|^{-1}$ and it is analytic in a half-plane, *albeit the wrong one*. But if we modify the derivation of (8) by using a lower half-plane contour instead of Fig. 8.18, we conclude that the real and imaginary parts of $k(\omega)$ are related by the Hilbert transform equations, with a sign change to account for the clockwise orientation of the contour. Thus we have the *Kramers-Kronig* relations

$$\operatorname{Im} k(\omega) = \frac{1}{\pi} \text{p.v.} \int_{-\infty}^{\infty} \frac{\operatorname{Re} k(\eta)}{\eta - \omega} \, d\eta,$$

$$\operatorname{Re} k(\omega) = -\frac{1}{\pi} \text{p.v.} \int_{-\infty}^{\infty} \frac{\operatorname{Im} k(\eta)}{\eta - \omega} \, d\eta. \tag{22}$$

The first identity is very useful in circuit theory and atomic and electromagnetic scattering applications, because the real part of the transfer function can often be determined efficiently and accurately using power-loss measurements. Then (22) enables the calculation of the experimentally more elusive imaginary part. In optical applications the dielectric constant plays the role of the transfer function k, and Eqs. (22) are known as *dispersion relations* in this context.

Example 5
Verify the relations (22) for the transfer function of the circuit shown in Fig. 8.24.

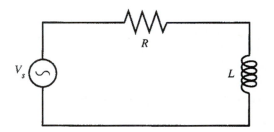

Figure 8.24 Circuit for Example 5.

Solution.   The effective impedance is seen by the methods of Sec. 3.6 to be $R_{\text{eff}} = R + i\omega L$. Therefore,

$$I_s = \frac{V_s}{R_{\text{eff}}} = \frac{V_s}{R + i\omega L}$$

and $k(\omega) = 1/[R + i\omega L]$. This function is analytic in the lower half-plane and goes to zero like $|\omega|^{-1}$. We have

$$\operatorname{Re} k(\omega) = \frac{R}{R^2 + \omega^2 L^2}, \qquad \operatorname{Im} k(\omega) = \frac{-\omega L}{R^2 + \omega^2 L^2}.$$

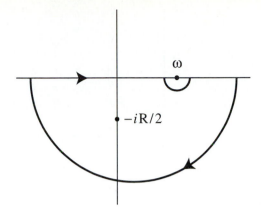

Figure 8.25 Contour for Example 5.

From the contour Γ in Fig. 8.25 we derive

$$\text{p.v.} \int_{-\infty}^{\infty} \frac{\text{Re}\, k(\eta)}{\eta - \omega}\, d\eta = \text{p.v.} \int_{-\infty}^{\infty} \frac{R}{R^2 + \eta^2 L^2}\, \frac{1}{\eta - \omega}\, d\eta$$

$$= -2\pi i \,\text{Res}(-iR/L) - \pi i \,\text{Res}(\omega)$$

and some algebra reveals this to be

$$\frac{-\pi\omega L}{R^2 + \omega^2 L^2} = \pi \,\text{Im}\, k(\omega),$$

in accordance with the first of Eqs. (22). The verification of the second equation is left as an exercise. ■

EXERCISES 8.5

1. Verify the entries in the Hilbert transform table. [HINT: For the last two entries consider the analytic functions $i/(z + ai)$ and $(e^{iaz} - 1)/iz$.]

2. Confirm Eq. (10) for some of the entries in the Hilbert transform table.

3. Argue that for the contour shown in Fig. 8.26, the contributions of the semicircles S'_ε and S''_ε to the integrals in Eqs. (3) and (5) will be $3\pi i f(z_0)/2$ and $-\pi i f(z_0)/2$, respectively. (Note that the Sokhotskyi-Plemelj prediction remains valid, however.)

4. Use Euler's formula and the Hilbert transform table to show directly that if $|\omega| < \Omega_1$, the Hilbert transform of $\cos \omega x \, \cos \Omega_1 x$ is given by $\cos \omega x \, \sin \Omega_1 x$.

5. Suppose $f(z)$ is analytic in a domain that encloses the unit disk $|z| \leq 1$. Give an argument that the Hilbert transform of $g(x) = f(e^{i\omega x}) + f(e^{-i\omega x})$ is given by $-i[f(e^{i\omega x}) - f(e^{-i\omega x})]$, for any constant ω. [HINT: Examine the Maclaurin series.]

6. Verify the second dispersion relation (22) for the circuit in Fig. 8.24.

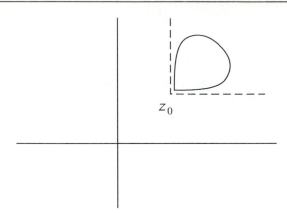

z_0

Figure 8.26 Contour for Prob. 3.

7. Verify that the value of the jump in the following integrals, as z crosses the contour at any point z_0, is $2\pi i f(z_0)$, in accordance with the Sokhotskyi-Plemelj formula:

(a) $\displaystyle\oint_{|\zeta|=1} \frac{1/\zeta}{\zeta - z}\, d\zeta$ (b) $\displaystyle\int_{-\infty}^{\infty} \frac{1}{\xi - z}\, d\xi$

(c) $\displaystyle\int_{-\infty}^{\infty} \frac{\cos \xi}{\xi - z}\, d\xi$ [HINT: Consult Prob. 10, Exercises 6.4.]

(d) $\displaystyle\int_{-\infty}^{\infty} \frac{\xi/(\xi^2 + 1)}{\xi - z}\, d\xi$ [HINT: Consult Prob. 18, Exercises 4.7.]

8. The identity

$$\lim_{\varepsilon \downarrow 0} \frac{1}{x - x_0 - i\varepsilon} = \text{p.v.}\frac{1}{x - x_0} + i\pi \delta(x - x_0) \qquad (23)$$

is frequently used by theoretical physicists. Here $\delta(x - x_0)$ is the *Dirac delta function*, an "idealized function" postulated to have the property that

$$\int_{-\infty}^{\infty} f(x)\delta(x - x_0)\, dx = f(x_0)$$

for any continuous function $f(x)$. Equation (23) is, strictly speaking, a crude abbreviation for the identity resulting when it is multiplied by $f(x)$ and integrated over $(-\infty, \infty)$, with the limits reversed:

$$\lim_{\varepsilon \downarrow 0} \int_{-\infty}^{\infty} \frac{f(x)}{x - x_0 - i\varepsilon}\, dx = \text{p.v.} \int_{-\infty}^{\infty} \frac{f(x)}{x - x_0}\, dx + i\pi f(x_0). \qquad (24)$$

(a) Derive Eq. (24), assuming that $f(z)$ is analytic for $\text{Im } z \geq 0$ and approaches zero at infinity sufficiently rapidly that one can close the contour with a semicircle C_ρ^+ over which the integral goes to zero. [HINT: Let your analysis be guided by the sketches in Fig. 8.27.]

(b) What is $\displaystyle\lim_{\varepsilon \downarrow 0} \frac{1}{x - x_0 + i\varepsilon}$ [in the spirit of (23)]?

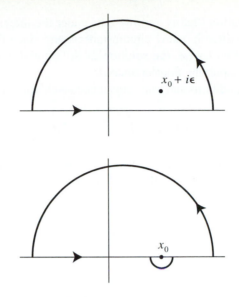

Figure 8.27 Contours for Prob. 8.

SUMMARY

The chief value of sinusoidal analysis in applied mathematics lies in the fact that when a linear system is driven by a function $Ae^{i\omega t}$ its response takes the same form, under the proper circumstances. To take advantage of this we have to be able to express a given function of a real variable as a superposition of sinusoids.

When the function F is periodic, say of period L, and satisfies certain continuity conditions, the appropriate decomposition is given by the Fourier series

$$F(t) = \sum_{n=-\infty}^{\infty} c_n e^{in2\pi t/L},$$

with coefficients defined by

$$c_n = \frac{1}{L} \int_{-L/2}^{L/2} F(t)e^{-in2\pi t/L} \, dt.$$

The Fast Fourier Transform is an efficient algorithm for approximating these coefficients.

If F is not periodic but $|F|$ is integrable (and the continuity conditions still hold), then the decomposition takes the form

$$F(t) = \int_{-\infty}^{\infty} G(\omega)e^{i\omega t} \, d\omega,$$

where G is the Fourier transform of F, defined by

$$G(\omega) = \frac{1}{2\pi} \int_{-\infty}^{\infty} F(t)e^{-i\omega t} \, dt.$$

"Direct" differentiation (that is, termwise or under the integral sign) of both representations can be justified in many circumstances, and since this operation merely amounts to multiplication by $i\omega$, the solution of differential equations can often be greatly simplified by employing Fourier analysis.

To handle initial conditions and transients it is convenient to use the Laplace transform of F,

$$\mathcal{L}\{F\}(s) = \int_0^\infty F(t)e^{-st}\,dt.$$

The Laplace transform can be identified as the Fourier transform of a function that is "turned on" at $t = 0$ [that is, $F(t) = 0$ for $t < 0$], with $i\omega$ replaced by s. The effect of the initial conditions is displayed in the formulas for the derivatives $F^{(n)}(t)$:

$$\mathcal{L}\left\{F^{(n)}\right\}(s) = s^n \mathcal{L}\{F\}(s) - s^{n-1}F(0) - s^{n-2}F'(0) - \cdots - F^{(n-1)}(0).$$

Hence the Laplace transform is the appropriate tool for solving initial-value problems.

Another advantage of the Laplace transform is its ability to handle certain nonintegrable functions. There is an inversion formula for recovering $F(t)$ from $\mathcal{L}\{F\}(s)$, but the use of tables is frequently more convenient.

The tool that plays the role of the Fourier/Laplace transform in the cases where the data set is discrete is the z-transform. It can be related to the theory of Laurent series, from which many of its properties are derived.

The behavior of Cauchy integrals as their singularities cross the contours is discontinuous, with jumps predicted by the Sokhotskyi-Plemelj formulas. When the contour is the real axis the integrals yield the Hilbert transform formulas, which relate the real and imaginary parts of the integrand. This transform finds applications in the analysis of causal autonomous systems such as modulation schemes and electrical networks. It is the theoretical underpinning of the dispersion and Kramers-Kroenig relations.

Suggested Reading

The references mentioned at the end of Chapter 3 under the heading "Sinusoidal Analysis" may also be useful for a further study of the material in this chapter. In addition, the following texts deal with the subjects from a more rigorous viewpoint.

Fourier Series

[1] Bachman, G., and Narici, L. *Functional Analysis*, 2nd ed., Dover Pubns., New York, 2000.

[2] Edwards, R. E. *Fourier Series*, 2nd ed., Vols. I and II. Springer-Verlag, New York, 1979.

[3] Friedman, A. *Advanced Calculus*. Holt, Rinehart and Winston, Inc., New York, 1971.

Fourier and Laplace Transforms

[4] Churchill, R. V. *Operational Mathematics*, 3rd ed. McGraw-Hill Book Company, New York, 1972.

[5] Kammler, D.A., *A First Course in Fourier Analysis*, Prentice Hall, Upper Saddle River, N. J., 2000.

[6] Snider, A.D., *Partial Differential Equations: Sources and Solutions*, Prentice-Hall, Inc., Upper Saddle River, N.J., 1999.

z-Transforms

[7] Oppenheim, A. V., and Schafer, R. W. *Digital Signal Processing*. Prentice-Hall, Inc., Upper Saddle River, N.J., 1975.

[8] Papoulis, A. *Signal Analysis*. McGraw-Hill Book Company, New York, 1977.

Cauchy Integrals and Hilbert Transforms

[9] Bendat, J. S. *The Hilbert Transform and Applications to Correlation Measurements*. Bruel and Kjaer, Naerum, Denmark.

[10] Carlson, A.B., Crilly, P.B., and Rutledge, J.C., *Communication Systems*, 4th ed., McGraw-Hill, 20001.

[11] Henrici, P. *Applied and Computational Complex Analysis*, Vol. 3. John Wiley & Sons, Inc., New York, 1986.

[12] Levinson, N., "Simplified Treatment of Integrals of Cauchy Type, the Hilbert Problem, and Singular Integral Equations," *SIAM Review*, 7, October 1965.

[13] Panofsky, W. K. H., and Phillips, M. *Classical Electricity and Magnetism*, 2nd ed. Addison-Wesley Publishing Company, Inc., Boston, Mass., 1962.

Appendix A

Numerical Construction of Conformal Maps

In Chapter 7 we looked at conformal mapping and some of its applications, and we developed a formula to represent these maps in the case of arbitrary polygonal domains, the Schwarz-Christoffel transformation. It might seem that once we have this formula, applying it to construct a conformal map onto a given polygon should be just a matter of paperwork. But as we indicated in Chapter 7, this is not so. In fact, for nearly a century after Schwarz and Christoffel discovered their formula in the 1860s, the construction of conformal maps for all but the simplest polygons remained in practice impossible. The arrival of computers changed this situation.

The purpose of this appendix is to describe some of the mathematical techniques that can be applied on the computer to make the Schwarz-Christoffel transformation a practical reality. In doing this we can only scratch the surface of the issues of numerical analysis involved; this is a lively field in its own right, which the student is encouraged to explore at his or her leisure. But we hope at least to convey the flavor of how conformal mapping, and complex analysis in general, can be used in serious scientific problems. At the end we will also give an indication of how the Schwarz-Christoffel ideas are related to the broader problem of constructing conformal maps onto arbitrary domains.

A.1 The Schwarz-Christoffel Parameter Problem

Recall from Section 7.5 that in the Schwarz-Christoffel problem our aim is to find an analytic function f that maps the upper half z-plane onto the interior of a polygon P in the w-plane defined by n vertices $w_1, \ldots, w_n$. In Theorem 7 and Eq. (8) of Sec. 7.5, we showed that such a map can always be represented by the formula

$$
\begin{aligned}
f(z) &= A g(z) + B \\
&= A \int_0^z (\zeta - x_1)^{\theta_1/\pi} (\zeta - x_2)^{\theta_2/\pi} \cdots (\zeta - x_{n-1})^{\theta_{n-1}/\pi} \, d\zeta + B
\end{aligned} \qquad (1)
$$

for some suitable choice of complex numbers A and B and real numbers $x_1 < x_2 < \cdots < x_{n-1}$. Here θ_i is the right-turn angle of the polygon at the vertex w_i, as indicated in Fig. 7.36. The numbers x_i, the "prevertices," are the points along the real axis that are mapped to the vertices w_i by f.

As we mentioned in Section 7.5, there are two reasons why applying Eq. (1) is not a trivial task. First, there is the problem of *numerical integration*. Even the simple triangular domain of Figure 7.39 led to an integral (1) that could not be evaluated in closed form. For more complicated polygons, the evaluation of Schwarz-Christoffel integrals is almost invariably beyond the reach of exact formulas. Therefore one is forced to look for efficient numerical approximations. We will consider this problem in Sec. A.3. But until then, let us pretend that all integrals (1) can be evaluated effortlessly.

The second and more serious difficulty arises as soon as n, the number of vertices of P, is four or more. As we pointed out, only three of the prevertices x_i can be specified arbitrarily, and the rest are then implicitly determined by the geometry of P. If P is symmetrical, as was the rectangle of Fig. 7.45 for example, then the correct values of these missing parameters are often evident. But for general polygons they are emphatically unknown. What use then is Eq. (1)? Obviously the prevertices must first be determined somehow, and the problem of finding them is known as the Schwarz-Christoffel *parameter problem*.

To be precise, let us fix the following three values: $x_1 = -1$, $x_2 = 0$, and (implicitly) $x_n = \infty$. What remains are then $n - 3$ prevertices whose values are sought:

$$n - 3 \text{ unknowns}: \quad x_3, x_4, \ldots, x_{n-1} \tag{2}$$

subject to the constraints

$$0 < x_3 < x_4 < \cdots < x_{n-1} < \infty. \tag{3}$$

For any choice of these quantities, Eq. (1) defines a function g that maps the real axis onto a polygon P'. Furthermore, P' has the same corner angles as P; our problem is to find a set of values x_i for which P' is geometrically similar to P. Once we have accomplished this, as mentioned in Sec. 7.5, the constants A and B in Eq. (1) can be adjusted so that the two polygons actually coincide.

For example, suppose the right-turn angles in Fig. 7.36 are

$$\theta_1 = -146°, \quad \theta_2 = -82°, \quad \theta_3 = 160°, \quad \theta_4 = -150°, \quad \theta_5 = -142°,$$

and suppose we try the prevertex values

$$x_3 = 1, \quad x_4 = 2.$$

Fig. A.1(a) shows the resulting polygon P', computed numerically and scaled and rotated so that the side $\left[w_1', w_2'\right]$ is the same as $[w_1, w_2]$. As expected, the angles are correct but the side lengths are far off. To lengthen the side $\left[w_2', w_3'\right]$, let us try the new guess

$$x_3 = 3, \quad x_4 = 4.$$

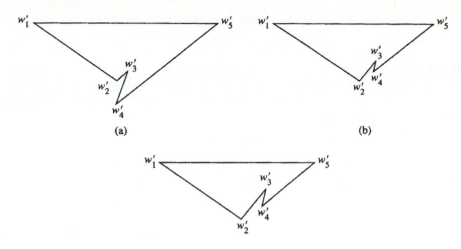

Figure A.1 Schwarz-Christoffel approximations to the polygon P of Fig. 7.36 based on different choices of prevertices $0 < x_3 < x_4 < \infty$.

The result shown in Fig. A.1(b) looks better, but $[w'_2, w'_3]$ is still too short, and $[w'_3, w'_4]$ has shrunk ominously. If we now try

$$x_3 = 10, \qquad x_4 = 15,$$

we get the polygon of Fig. A.1(c), which is beginning to look about right. But it is obvious that it is going to take a long time to make P' precisely similar to P—say, to six digits of accuracy—by trial and error. If n were 10 or 20 instead of 5, the process would be impossibly slow.

To make speedier progress, we need to formulate the condition of similarity algebraically. Consider the following set of $n - 3$ equations:

$$\frac{|g(x_i) - g(x_{i-1})|}{|g(x_2) - g(x_1)|} - \frac{|w_i - w_{i-1}|}{|w_2 - w_1|} = 0, \quad i = 3, 4, \dots, n - 1. \tag{4}$$

Notice that each absolute value in the right-hand quotient of this formula represents a side length in P while, by Eq. (1), each absolute value in the left-hand quotient represents a side length in P'. Thus these identities assert that the sides $[w'_1, w'_2]$ through $[w'_{n-2}, w'_{n-1}]$ of P' have the same lengths as the corresponding sides of P, except for a uniform scale factor $C = |g(x_2) - g(x_1)| / |w_2 - w_1|$. It may appear that this leaves the two sides $[w'_{n-1}, w'_n]$ and $[w'_n, w'_1]$ unaccounted for, but suppose P' has all but these two side lengths equal to C times the corresponding side lengths in P. Then, since the angles are fixed by the exponents in Eq. (1), and since any two nonparallel lines have a unique point of intersection, the remaining two side lengths of P and P' automatically have to be related in the same way too. In other words, Eq. (4) fully expresses the desired condition that P' and P are *similar*.

Equations (2) to (4) constitute a *constrained system of $n - 3$ nonlinear equations in $n - 3$ unknowns*. The reason we have put both quotients on the left-hand side in

Eq. (4) is so that the system can be written symbolically in the form

$$F_i(\mathbf{x}) = 0, \quad i = 3, 4, \ldots, n - 1,$$

where $\mathbf{x}$ denotes the vector $(x_3, \ldots, x_{n-1})$ of unknowns, of length $n - 3$, and each F_i is a real-valued function of $\mathbf{x}$. If we go further and let $\mathbf{F}$ denote the vector-valued function $\mathbf{F}(\mathbf{x}) = (F_3(\mathbf{x}), \ldots, F_{n-1}(\mathbf{x}))$, and $\mathbf{0}$ the vector $(0, \ldots, 0)$, then the system takes the even more compact form

$$\mathbf{F}(\mathbf{x}) = \mathbf{0}. \tag{5}$$

We have now translated the Schwarz-Christoffel parameter problem into the following algebraic task: Find a root $\mathbf{x}^*$ of Eq. (5), subject to (3). It can be proved that there is a unique solution to this problem, and that this solution is precisely the vector of prevertices for the Schwarz-Christoffel mapping (1) onto P.

Of course, in getting to (5) we have not really simplified the mathematics, just the notation. Written out in full, for example, the component $i = 3$ of Eq. (5) looks like this for our polygon with $n = 5$:

$$\frac{\left| \int_{x_2}^{x_3} (\zeta - x_1)^{-146°/\pi} (\zeta - x_2)^{-82°/\pi} (\zeta - x_3)^{+160°/\pi} (\zeta - x_4)^{-150°/\pi} \, d\zeta \right|}{\left| \int_{x_1}^{x_2} (\zeta - x_1)^{-146°/\pi} (\zeta - x_2)^{-82°/\pi} (\zeta - x_3)^{+160°/\pi} (\zeta - x_4)^{-150°/\pi} \, d\zeta \right|}$$
$$- \frac{|w_3 - w_2|}{|w_2 - w_1|} = 0. \tag{6}$$

(The degree symbol represents $\pi/180$, and thus $146°/\pi$ is equivalent to $146/180$, and so on.) Nevertheless, a simplification of notation is often highly useful in formulating a problem for numerical computation, because it may reveal that the problem has a standard form for which numerical methods are already in existence.

Such is exactly the case here. In fact, solving a nonlinear system of equations of the form (5) is one of the fundamental, extensively studied problems in numerical analysis. In general, there is no hope of finding the root $\mathbf{x}^*$ exactly, so one has to approximate it by iteration. First, some initial guess vector $\mathbf{x}^{(0)}$ is chosen. Then a sequence of new guesses $\mathbf{x}^{(1)}, \mathbf{x}^{(2)}, \ldots$ is successively computed, which it is hoped will converge to $\mathbf{x}^*$. Most of the commonly used algorithms for generating these guesses are variants of *Newton's method*, with which the reader may be familiar for the case of a single scalar nonlinear equation. Many computer programs have been written to implement this kind of iteration, and they can be found in the software libraries of most large computer systems and also in interactive numerical problem-solving environments such as MATLAB. Under favorable circumstances, such a computer program will often converge astonishingly fast. It is not unusual that after a few dozen iterations, $\mathbf{x}^{(k)}$ will be so near $\mathbf{x}^*$ that $\mathbf{F}\left(\mathbf{x}^{(k)}\right)$ has magnitude 10^{-16} or smaller; in other words, $\mathbf{x}^{(k)}$ has converged to "machine accuracy" in double precision on a typical computer with a 32-bit word length.

If we blindly apply a program for solving systems of equations to our Schwarz-Christoffel problem, however, the correct solution will usually not be found. The reason is that we have ignored the constraints (3). Even if $\mathbf{x}^{(0)}$ satisfies these constraints,

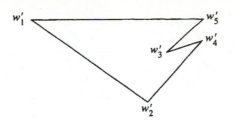

Figure A.2 An approximation to P obtained with prevertices out of order: $x_3 > x_4$.

it is highly likely that a later iterate $\mathbf{x}^{(k)}$ will violate them, unless we take pains to prevent this. In such a situation Eq. (1) will define a polygon P' in which the vertices are out of order. For example, suppose we take

$$x_3 = 2, \qquad x_4 = 1$$

in the pentagonal example considered above. Figure A.2 shows the result. It is a perfectly good polygon, but it belongs to a different family from the shape for which we are looking, and adjusting its side lengths to satisfy Eq. (4) will be a waste of time.

A simple but effective way to get around the problem of constraints is by introducing a change of variables. If $\mathbf{x}$ is a vector of positive numbers satisfying (3), and $x_2 = 0$ as usual, then the formula

$$\hat{x}_i = \log (x_i - x_{i-1}) \in \mathbf{R}, \quad i = 3, 4, \ldots, n-1 \tag{7}$$

defines a new vector $\hat{\mathbf{x}} = (\hat{x}_3, \ldots, \hat{x}_{n-1})$ of unconstrained real numbers. Conversely, given an arbitrary vector $\hat{\mathbf{x}}$, the inverse formula

$$x_i = x_{i-1} + e^{\hat{x}_i}, \quad i = 3, 4, \ldots, n-1 \tag{8}$$

defines a vector $\mathbf{x}$ that satisfies (3). To solve the parameter problem (2), (3), and (5), we can rewrite Eq. (5) as a nonlinear system in the new unconstrained variables,

$$\widehat{\mathbf{F}} (\hat{\mathbf{x}}) = \mathbf{0}. \tag{9}$$

The parameter problem becomes the unconstrained nonlinear system of Eqs. (7) and (9). Now we can apply a numerical algorithm based on Newton's method without worrying about constraints.

A.2 Examples

Let us look at a few examples to see how these ideas perform in practice. First, we should finish off the polygon of Fig. 7.36, but before we can do this, we have to specify its dimensions. Let us assume

$$|w_2 - w_1| = 8, \qquad |w_3 - w_2| = 3, \qquad |w_4 - w_3| = 1;$$

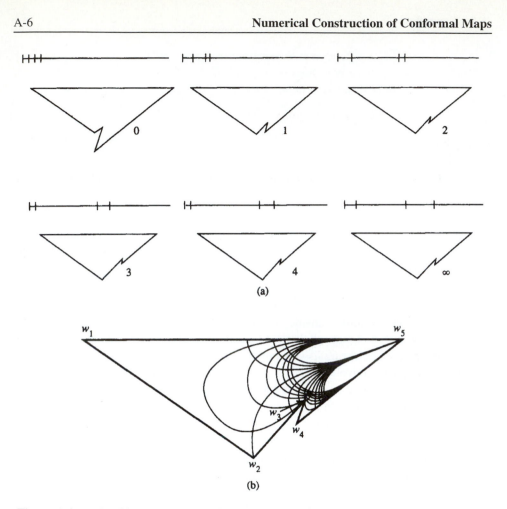

Figure A.3 (a) Rapid convergence to the correct prevertices and the correct P by a variant of Newton's method. (b) The polygon P successfully mapped onto the upper half-plane. The curves inside P are the conformal images of vertical and horizontal lines in the half-plane.

the other two side lengths are then determined implicitly, as discussed earlier. Beginning with the initial guess $\hat{x}_3^{(0)} = \hat{x}_4^{(0)} = 0$ in Eq. (7), that is, $x_3^{(0)} = 1$ and $x_4^{(0)} = 2$ by Eq. (8), we now iterate numerically with our computer program to solve the parameter problem of Eq. (9). Figure A.3(a) shows the polygon P' at some of the first few steps (ignoring $n - 3$ uninteresting initial steps required to estimate a *Jacobian matrix* needed for the approximate Newton method). Figure A.3(b) shows the final polygon, obtained to 15-digit accuracy in 14 iterations. The shapes of the intermediate polygons, as well as the exact number of iterations, would differ depending on the details of the programming.

The line above each polygon in Figure A.3(a) represents the segment $[-1, 15]$ of the real axis, and the vertical cross-segments indicate the points $x_1, \ldots, x_{n-1}$ for the given step. At the beginning, the four prevertices are evenly spaced. The points

$x_1 = -1$ and $x_2 = 0$ remain fixed throughout the iteration, but x_3 and x_4 change. One can see the rapid convergence of these parameters as the final polygon is approached. The converged values turn out to be

$$x_3 = 8.22544, \qquad x_4 = 11.07029.$$

Notice that these numbers are surprisingly far from the guess $x_3 = 10$, $x_4 = 15$ that led to Fig. A.1(c).

In Fig. A.3(b), the final polygon P is plotted with additional curves drawn on the inside. These curves are the images of the horizontal and vertical lines

$$\text{Im } z = 0.2, 1, 2, 3, 4, 5, 6 \quad \text{and} \quad \text{Re } z = -6, -4, -2, 0, \ldots, 18$$

in the z-plane. All 20 curves meet at w_5, which corresponds to $z = \infty$. But inside the polygon, as they must, the curves meet everywhere at right angles. In fact, they form a grid in which each individual cell looks like a distorted rectangle of length-to-width ratio 2. This confirms what we know about conformal maps in general: locally, they change the scale but not the shape.

Figures A.4 and A.5 show analogous plots for two other polygons, again beginning with the initial guess $x_j^{(0)} = j - 2$ for $1 \leq j \leq n - 1$. In Fig. A.4, P is a six-sided L shape. Since this polygon happens to be symmetrical about the line through the corners w_3 and w_6, the converged prevertices end up symmetrically located around x_3 (and around $x_6 = \infty$). In Fig. A.5, P is a seven-sided polygon shaped like a claw. In this latter case, the *preliminary* steps shown represent polygons that fold over themselves. This means that the intermediate functions g are not one-to-one.

Notice in Fig. A.5(a) that although it is easy to distinguish x_4, x_5, and x_6 early in the iteration, their final values lie nearly on top of each other. Here are the converged values:

$$x_1 = -1, \qquad x_2 = 0, \qquad x_3 = 2.48883,$$
$$x_4 = 3.89729, \quad x_5 = 3.91233, \quad x_6 = 3.96037.$$

Thus $|x_4 - x_5| \approx .015$, for example. Such uneven distribution of prevertices is not the exception but the rule in conformal mapping, and indeed, the unevenness is often far more pronounced. The root of the phenomenon is that the two ends of our "claw" have very little influence on each other. For example, if we think of the claw as a heat conductor, it is intuitively clear that a change in the temperature applied along the boundary near one end will have almost no effect near the other.

To carry these observations further, perhaps the most valuable insight to be gained from looking at pictures of numerical conformal maps is an appreciation of the great distortions they often introduce. Consider Figs. A.3(b), A.4(b), and A.5(b). One kind of distortion that appears in all three cases is near w_n, where the curves bunch up tightly. Since w_n corresponds to ∞ under the conformal map, this is not surprising. Another kind of distortion appears near outward-pointing or "salient" corners, which the interior image curves tend to avoid. This is especially pronounced in Fig. A.5(b), where the lower claw is entirely bare. Salient corners are "deadwater" regions, little affected by what goes on in the rest of the polygon. In contrast, at an inward-pointing

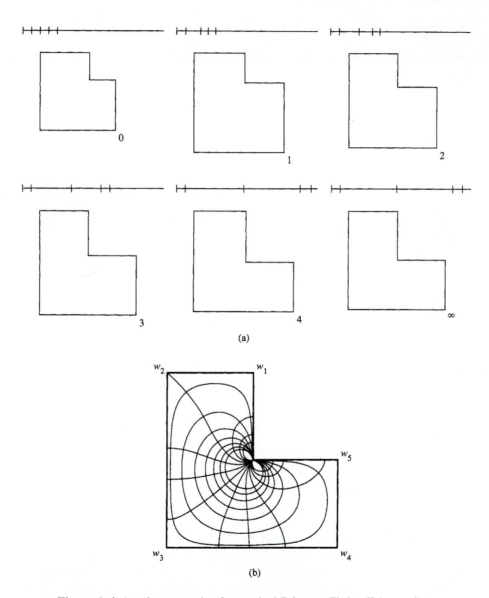

Figure A.4 Another example of numerical Schwarz-Christoffel mapping.

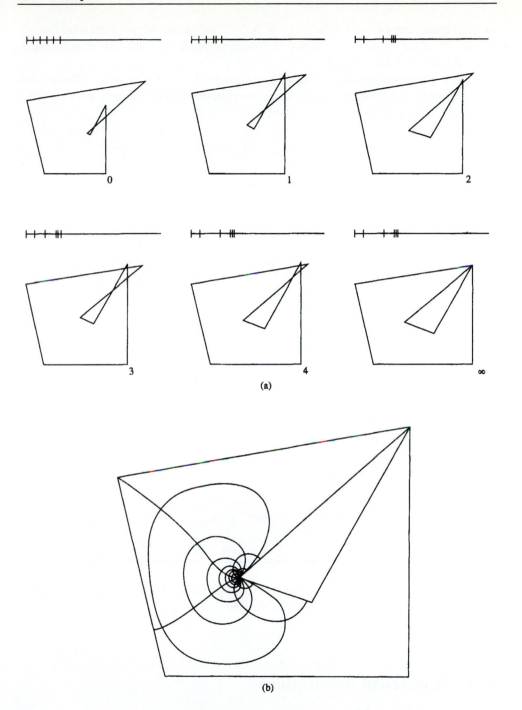

(a)

(b)

Figure A.5 A third example.

or "reentrant" corner, such as w_3 in Fig. A.3(b), the image curves bunch up closely. Physically, at such a corner the speed of flow, the electric field strength, or the temperature gradient will be infinite.

The explanation of this is revealed in Examples 4 and 6 of Sec. 7.6. The streamlines (or isotherms or equipotentials) are expressed in terms of the imaginary part of z, which is $f^{-1}(w)$; if, as in Eq. (3) of Sec. 7.5,

$$\frac{dw}{dz} = f'(z) = A(z - x_1)^{\alpha_1}(z - x_2)^{\alpha_2} \cdots (z - x_n)^{\alpha_n} \tag{1}$$

with $\alpha_j = \theta_j/\pi$ for a right turn θ_j, then

$$\frac{dz}{dw} = A^{-1}(z - x_1)^{-\theta_1/\pi}(z - x_2)^{-\theta_j/\pi} \cdots (z - x_n)^{-\theta_j/\pi},$$

so the gradients become unbounded at the reentrant ($\theta_j > 0$) corners.

For flows in regions *exterior* to polygons, the point of view is reversed; the flow is stagnant in the coves or fiords, while swift currents prevail at the ends of promontories or peninsulas. Indeed, the streamlines for the flow over the simple polygonal streambed in Example 4 of Sec. 7.6 are displayed in Fig. 7.51. For any more complicated streambeds, we have to resort to the numerical techniques described in this appendix. Figure A.6 shows a plot generated by the computer for the flow over an idealized square boulder. The vertex x_n lies at infinity, but this is no problem since x_n does not appear in Eq. (1). Schwarz and Christoffel would have liked this picture!

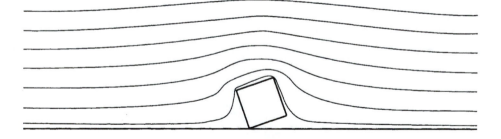

Figure A.6 "Flow over a square boulder," the idealized potential flow streamlines computed by a Schwarz-Christoffel map to the upper half-plane.

A.3   Numerical Integration

Having shown these successful results, we must now return to the first question mentioned at the start of this appendix. How were these Schwarz-Christoffel integrals computed? Notice that every time the vector-valued function $\mathbf{F}(\mathbf{x})$ of Eqs. (4) and (5) is evaluated, the computer program must compute $n - 2$ integrals of the form

$$g(x_i) - g(x_{i-1}) = \int_{x_{i-1}}^{x_i} (\zeta - x_1)^{\theta_1/\pi} \cdots (\zeta - x_{n-1})^{\theta_{n-1}/\pi} \, d\zeta. \tag{1}$$

Since $F(x)$ will be evaluated many times during the course of the iterative solution of the parameter problem, this may add up to hundreds or thousands of numerical integrations. Moreover, once the parameter problem is solved, further integrations will be necessary to produce plots like Figs. A.3 to A.6. Clearly, it is imperative to be able to compute Eq. (1) efficiently.

   Numerical integration, also called *quadrature*, is another old and well-understood problem of numerical analysis. Suppose we want to compute the integral from a to b of a function $\phi(z)$. (It does not matter in what we say here whether a, b, and ϕ are real or complex.) For a first attempt, an obvious idea is to calculate a Riemann sum based on a division of the interval $[a, b]$ into subintervals. Let N be a positive integer, and set

$$\Delta z = \frac{b - a}{N}, \qquad z_i = a + \left(i - \frac{1}{2} \right) \Delta z, \quad i = 1, \ldots, N.$$

Then the integral can be approximated by

$$\int_a^b \phi(z) \, dz \approx \sum_{i=1}^{N} \phi(z_i) \, \Delta z. \tag{2}$$

The method of approximate integration is called the *midpoint rule*, and it is not hard to show that if ϕ is smooth, then the difference between the numerical estimate and the exact integral will decrease at least in proportion to N^{-2} as $N \to \infty$. We write this as

$$\int_a^b \phi(z) \, dz - \sum_{i=1}^{N} \phi(z_i) \, \Delta z = O\left(N^{-2}\right).$$

Thus the midpoint rule is a usable numerical method in principle; but its performance can at best be described as so-so. To get eight significant digits, we will typically need $N \approx 10,000$.

   For greater efficiency, we can generalize (2) to a formula

$$\int_a^b \phi(z) \, dz \approx \sum_{i=1}^{N} \omega_i \phi(z_i), \tag{3}$$

where the numbers $\{\omega_i\}$ are a set of real *weights*, not necessarily equal, and the numbers $\{z_i\}$ are a set of *points* or *nodes* in $[a, b]$, not necessarily equally spaced. Of course, everything will depend on how the weights and nodes are chosen. In *Simpson's rule*, with which the reader may be familiar, the points are equally spaced but the weights alternate in magnitude in a pattern $1 - 4 - 2 - 4 - 2 - \cdots - 4 - 2 - 4 - 1$, and this leads to an error estimate $O\left(N^{-4}\right)$. Now only roughly 100 points (rather than 10,000) are needed to get eight significant digits. There are also formulas of higher order based on equally spaced points, called *Newton-Cotes formulas*. But the king of them all is the method known as *Gaussian quadrature*. In this method, the points are unevenly distributed in an "optimal" way (they depend on the zeros of the so-called *Legendre polynomial* of order N), and the weights are irrational numbers. The result is a formula under which the errors decrease faster than $O\left(N^{-K}\right)$ for any fixed K. In fact, if ϕ is analytic in an open region of the complex plane containing the interval $[a, b]$, then the error in Gaussian quadrature decreases at a geometric rate $O\left(\rho^N\right)$ for some $\rho < 1$, as $N \to \infty$.

Now what if we go ahead and use Gaussian quadrature to integrate Eq. (1) for our Schwarz-Christoffel problem? The results will be virtually useless! The reason is that the integrand of Eq. (1), far from being analytic on $\left[x_{i-1}, x_i\right]$, has singularities at both endpoints. For example, there are singularities with exponents $-82°/\pi$ and $+160°/\pi$ in the numerator on the left in Eq. (6). As a consequence, it turns out that both Gaussian quadrature and the midpoint rule reduce to thoroughly unacceptable rates of convergence for this integral, sometimes worse than $O\left(N^{-1}\right)$. Simpson's rule is in general not even defined, since it insists on evaluating the integrand at the endpoints, where the function may be infinite. Certainly the presence of endpoint singularities is a key issue in Schwarz-Christoffel integration.

Once this difficulty has been recognized, there are many ways to cope with it. One of these is to use a modified procedure called *Gauss-Jacobi quadrature*. A Gauss-Jacobi formula has the usual form (3), but the weights and nodes are specially chosen on the basis of the assumption that ϕ has singularities of type $(z - a)^\alpha$ at one endpoint and of type $(z - b)^\beta$ at the other, for some numbers $\alpha, \beta > -1$. (The nodes now depend on the zeros of the so-called *Jacobi polynomials*.) In fact, if ϕ can be written as a product

$$\phi(z) = (z - a)^\alpha (z - b)^\beta \psi(z),$$

where ψ is analytic in an open region containing $[a, b]$, then the Gauss-Jacobi formula will also have errors decreasing geometrically with N, as the Gaussian formula did in the case where ϕ was analytic. See the Suggested Reading for references on how these nodes and weights can be computed. Obviously, Gauss-Jacobi quadrature is made to order for our Schwarz-Christoffel application, where we have singularities of type $\alpha = \theta_{i-1}/\pi, \beta = \theta_i/\pi$.

For an illustration of the power of Gauss-Jacobi quadrature, consider again the integral that came up in Example 1 of Sec. 7.5, for the conformal map onto a triangle. The integral reduces to $(1 + i)\left/\sqrt{2}\right.$ times

$$\int_0^1 \left(1 - x^2\right)^{-3/4} dx \approx 2.622058. \tag{4}$$

Here we have $a = 0$, $\alpha = 0$, $b = 1$, and $\beta = -3/4$. With $N = 4$, for example, the

Gauss-Jacobi nodes and weights turn out to be approximately

$$z_1 = .02867, \quad z_2 = .38609, \quad z_3 = .75428, \quad z_4 = .98366,$$
$$\omega_1 = .13109, \quad \omega_2 = .31759, \quad \omega_3 = .56220, \quad \omega_4 = 1.36754.$$

Inserting these values in (2), we get the approximate integral 2.622057, which is accurate to five places! The following table shows more systematically how fast the approximation converges as N increases.

N	Approx. integral
1	2.57
2	2.6208
3	2.62202
4	2.622057

Simpson's rule and its relatives, even if we took care of the singularity somehow, could not come close to this performance.

In applying the Gauss-Jacobi quadrature to more complicated Schwarz-Christoffel maps, there is one more difficulty we have to address before we can count on accurate results at low cost. Recall that in mapping the "claw" of Fig. A.5, we found that the prevertices ended up quite unevenly distributed. Now consider in that problem, for example, the integral in Eq. (1) from x_3 to x_4, which we will have to compute repeatedly during the iteration to determine the length of the corresponding side $[w'_3, w'_4]$. By using the appropriate Gauss-Jacobi formula, we can avoid any ill effects due to the singularity at x_4. But x_5, with its own singularity, lies only a distance .015 away. It turns out that, as a result, Gauss-Jacobi quadrature will not begin to give accurate answers until N is at least as large as $1/.015 \approx 10^2$—quite unacceptable! This problem of crowded singularities comes up in almost every practical Schwarz-Christoffel mapping problem.

Fortunately, this difficulty too can be circumvented once it is recognized. One way to do so is to use *compound Gauss-Jacobi quadrature*, in which troublesome intervals of integration are automatically subdivided near the endpoints into small subintervals whose length is comparable to the distance to the nearest singularity. With this improvement, the remarkable behavior of geometric convergence—errors $O\left(\rho^N\right)$—can be maintained for the conformal mapping of arbitrary polygons.

A.4   Conformal Mapping of Smooth Domains

The Schwarz-Christoffel integral of Eq. (1) contains a product of terms $(\zeta - x_1)^{\theta_1/\pi}$, one for each corner w_i of the polygon P. Now what could we do if, instead of a polygon, we wanted to map the upper half-plane conformally onto a more general domain D bounded by a smooth curve C? There is an obvious way to try to go about this. We attempt to treat C as a polygon with infinitely many corners w_i, at each of which the right turn angle θ_i is infinitesimal.

Making this idea precise is easier than one might expect. Recall the form of Eq. (1)

$$f'(z) = A \, (z - x_1)^{\theta_1/\pi} \, (z - x_2)^{\theta_2/\pi} \cdots (z - x_{n-1})^{\theta_{n-1}/\pi}$$

for a conformal map onto a polygon. Let us rewrite this as

$$f'(z) = A \exp \left[\frac{1}{\pi} \sum_{i=1}^{n-1} \theta_i \log (z - x_i) \right]. \tag{1}$$

Each term $\log (z - x_i)$ here cannot be defined as a single-valued function in a neighborhood of x_i, but fortunately, we only need to evaluate $f'(z)$ in the upper half-plane. Thus let $\log (z - x_i)$ be defined to be single-valued for $\text{Im}\, z \geq 0$, $z \neq x_i$, and then Eq. (1) becomes a well-defined function where we need it. And now it should be obvious what formula to use for curved boundaries—just replace the sum in Eq. (1) above by an integral!

$$f'(z) = A \exp \left[\frac{1}{\pi} \int_{-\infty}^{\infty} \theta(x) \log(z - x) \, dx \right]. \tag{2}$$

In this equation, $\theta(x)$ is a function that can be identified as the amount of turning of *C per unit length* along the real x-axis. Integrating, we get the following analog of Eq. (1) for smooth domains:

$$f(z) = A \int_0^z \exp \left[\frac{1}{\pi} \int_{-\infty}^{\infty} \theta(x) \log(\zeta - x) \, dx \right] d\zeta + B. \tag{3}$$

This is a *continuous Schwarz-Christoffel formula.*

The reader is entitled at this point to be skeptical. If applying the usual Schwarz-Christoffel integral, with its finite number of vertices, turned out to be such a major undertaking, what hope can there be of solving Eq. (3)? In particular, we now face a seemingly much harder *continuous parameter problem*. Before, we had to determine the point x_i corresponding to each corner w_i. Now, we need to determine *for every x* the "amount of corner" $\theta(x)$ present there.

Nevertheless, it turns out that Eq. (3) can be solved numerically with a reasonable computational effort. It is a (somewhat irregular) example of an *integral equation*, a term for an equation in which an unknown function—$\theta(x)$—appears inside an integral. Several methods are known for solving integral equations numerically. This particular equation has been applied by a number of computational aerodynamicists for determining maps onto complicated domains (see reference [5]). In addition, there are a dozen or more other well-known integral equations that can also be used to compute conformal maps. Determining how these equations can be efficiently implemented, and which of them is best in practice, is an area of active research today, more than 150 years after Riemann first proposed his mapping theorem.

A.5   Conformal Mapping Software

Fortunately, one does not need to be an expert numerical analyst in order to use computational conformal maps, because there are several high-quality software packages available for free. The most user friendly is the *Schwarz–Christoffel Toolbox* for MAT-LAB (see the following list on how to obtain this and the other packages described in this section). The SC Toolbox operates in the interactive, graphical environment of MATLAB and requires very little expertise. For instance, an L shape like Figure A.4 can be created by entering

```
>> p = polygon( [i -1+i -1-i 1-i 1 0] );
>> f = hplmap(p);
>> plot(f)
```

There is also a graphical interface, complete with a powerful "polygon editor," that can be used to avoid typing commands altogether. In addition to the standard Schwarz–Christoffel transformation, the SC Toolbox is also capable of computing well-known variations of the transformation for maps from the unit disk, an infinite strip, or a rectangle, or from the unit disk to the exterior of a polygon. Some examples of these variations and their applications are illustrated in Figure A.7.

The other software packages to be discussed are distributed as source libraries in Fortran or C and may require a modest amount of compiling and additional programming by the user. One of these is DSCPACK, which computes a variation on the Schwarz–Christoffel transformation for regions that are polygonal but doubly connected—that is, they have one hole (recall Sec. 4.4). Two such maps computed using data from DSCPACK are shown in Figure A.8. Note that the Riemann Mapping Theorem does not apply to such a region. In fact, each must be mapped to a disk with a hole whose size is determined uniquely by the original geometry, and finding the size of the hole is part of the parameter problem.

We briefly mention the packages *CONFPACK*, which implements an alternative integral formula called *Symm's equation* and is appropriate for simply connected regions interior or exterior to a boundary consisting of one or more piecewise smooth curves; *zipper*, a fast algorithm for general simply connected regions; and *CirclePack*, which approximates conformal maps by packing circles into a given region. These packages can be obtained at the following Web sites.

SC Toolbox: www.math.udel.edu/~driscoll/SC

DSCPACK, CONFPACK: www.netlib.org/conformal

zipper: www.math.washington.edu/~marshall/zipper.html

CirclePack: www.math.utk.edu/~kens

Suggested Reading

Most textbooks of numerical analysis describe the methods of quadrature and nonlinear systems we have discussed. Good examples are the following:

Figure A.7 Examples of variations on the Schwarz–Christoffel transformation, computed using the SC Toolbox for MATLAB. The graphs are the images of orthogonal circles and radii (top row) or horizontal and vertical lines.

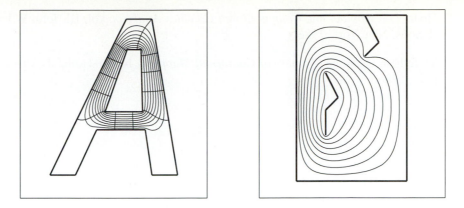

Figure A.8 Schwarz–Christoffel maps to two doubly connected regions, computed using DSCPACK.

Numerical Analysis

[1] Atkinson, K. E., *An Introduction to Numerical Analysis*, 2nd ed., John Wiley and Sons, Inc., New York, 1989.

[2] Gautschi, W., *Numerical Analysis: An Introduction*, Birkhauser, Boston, 1997.

Many such textbooks that use and teach MATLAB are listed at the Web site of The MathWorks, Inc., www.mathworks.com. Among the best is

[3] Mathews, J. H. and Fink, K. D. *Numerical Methods Using MATLAB*. Prentice Hall, Upper Saddle River, N. J., 1999.

The original presentation of the numerical implementation of Schwartz-Christoffel algorithm discussed herein appears in

[4] Trefethen, L. N. "Numerical Computation of the Schwarz-Christoffel Transformation," *SIAM Journal on Scientific and Statistical Computing* 1, 1980, 82–102.

Applications to aerodynamics are given in

[5] Davis, R. T. "Numerical Methods for Coordinate Generation Based on Schwarz-Christoffel Transformations," 4*th AIAA Comp. Fluid Dynamics Conf. Proc.*, 1978.

Among the most comprehensive references on numerical methods for conformal mapping are

[6] Driscoll, T. A. and Trefethen, L. N. *Schwarz–Christoffel Mapping.* Cambridge University Press, Cambridge, 2001.

[7] Henrici, P. *Applied and Computational Complex Analysis*, vol. III. John Wiley & Sons, New York, 1986.

[8] Trefethen, L. N., ed. *Numerical Conformal Mapping*. North–Holland, Amsterdam, 1986.

Appendix B

Table of Conformal Mappings

The following pages list some conformal mappings that arise in applications. Notice that the corresponding points are labled by the same letter; thus the point a is mapped to the point A, etc. A more extensive tabulation appears in Ref. [5] at the end of Chapter 7.

B.1 Möbius Transformations

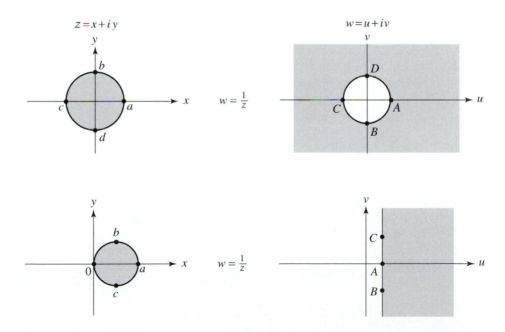

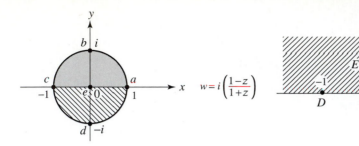

$$w = i\left(\frac{1-z}{1+z}\right)$$

$z = x + iy$ $w = u + iv$

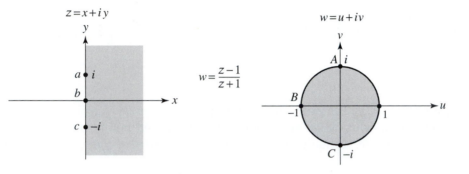

$$w = \frac{z-1}{z+1}$$

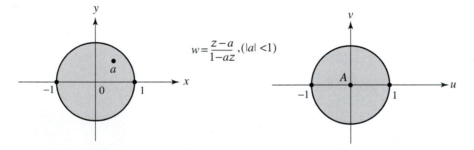

$$w = \frac{z-a}{1-az}, \ (|a| < 1)$$

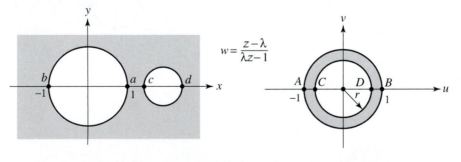

$$w = \frac{z-\lambda}{\lambda z - 1}$$

$$\lambda = \frac{1 + cd + \sqrt{(c^2-1)(d^2-1)}}{c+d}, \ (1 < c < d)$$

$$r = \frac{cd - 1 - \sqrt{(c^2-1)(d^2-1)}}{d-c}, \ (0 < r < 1)$$

$$w = \frac{z - \lambda}{\lambda z - 1}$$

$$\lambda = \frac{c + d}{1 + cd + \sqrt{(1 - c^2)(1 - d^2)}}$$

$$\text{inner radius} = \frac{d - c}{1 - cd + \sqrt{(1 - c^2)(1 - d^2)}}$$

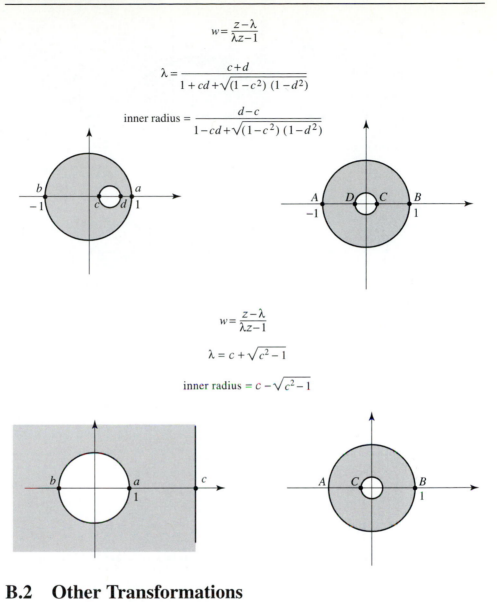

$$w = \frac{z - \lambda}{\lambda z - 1}$$

$$\lambda = c + \sqrt{c^2 - 1}$$

$$\text{inner radius} = c - \sqrt{c^2 - 1}$$

B.2   Other Transformations

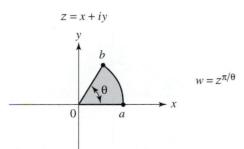

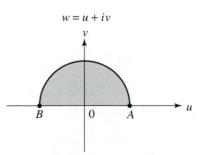

$z = x + iy$

$w = u + iv$

$w = z^{\pi/\theta}$

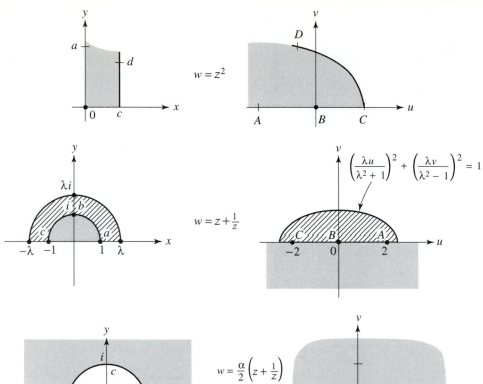

$$w = z^2$$

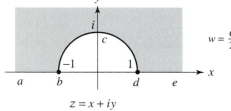

$$w = z + \frac{1}{z}$$

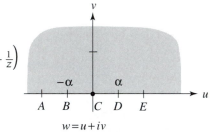

$$\left(\frac{\lambda u}{\lambda^2 + 1}\right)^2 + \left(\frac{\lambda v}{\lambda^2 - 1}\right)^2 = 1$$

$$w = \frac{\alpha}{2}\left(z + \frac{1}{z}\right)$$

$$z = x + iy$$

$$w = u + iv$$

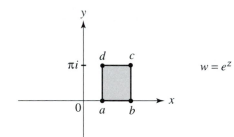

$$w = e^z$$

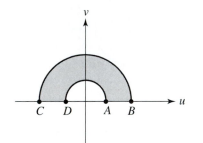

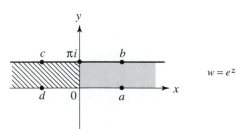

$$w = e^z$$

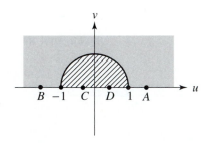

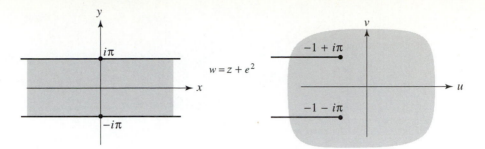

$$w = z + e^2$$

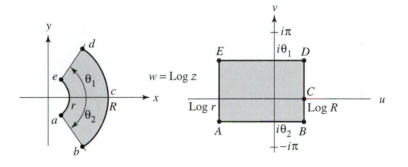

$$w = \text{Log } z$$

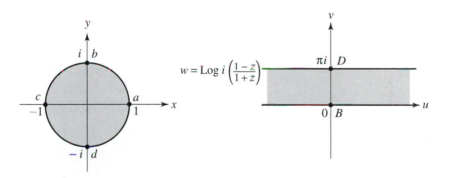

$$w = \text{Log } i \left(\frac{1-z}{1+z} \right)$$

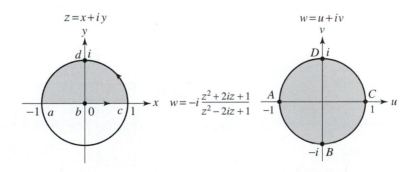

$$z = x + iy \qquad w = u + iv$$

$$w = -i \frac{z^2 + 2iz + 1}{z^2 - 2iz + 1}$$

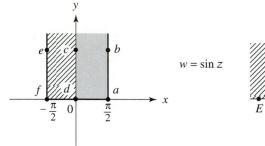

$$w = \sin z$$

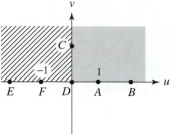

$$z = x + iy \qquad\qquad w = u + iv$$

$$w = 2\sqrt{z+1} + \mathrm{Log}\,\frac{\sqrt{z+1}-1}{\sqrt{z+1}+1}$$

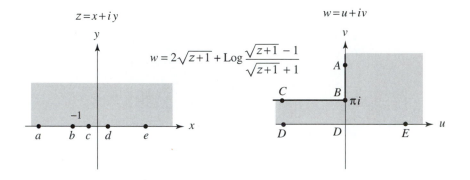

$$w = \frac{i\sigma_0}{\pi}\left\{\mathrm{Log}\!\left(\frac{1+\zeta}{1-\zeta}\right) - i\alpha\,\mathrm{Log}\!\left(\frac{1+i\alpha\,\zeta}{1-i\alpha\,\zeta}\right)\right\}$$

$$\alpha = \frac{\tau_0}{\sigma_0}\,;\quad \zeta^2 = \frac{z+1}{z-\alpha^2}$$

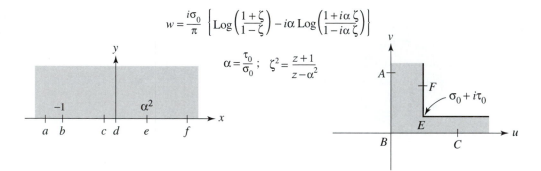

$$w = \tanh^{-1}z - \tan^{-1}z$$

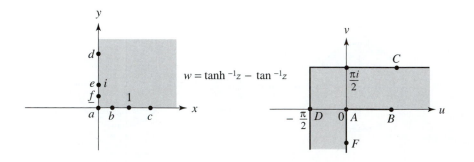

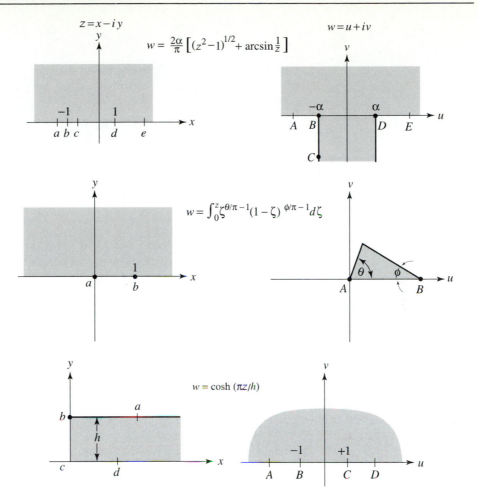

Answers to Odd-Numbered Problems

CHAPTER 1

Exercises 1.1, Page 4

5. (a) $0 + (-3/2)i = -3i/2$ (b) $3 + 0i = 3$ (c) $0 + (-2)i = -2i$

7. (a) $8 + i$ (b) $1 + i$ (c) $0 + (-8/3)i = -8i/3$

9. $\dfrac{61}{185} - \dfrac{107}{185}i$

11. $2 + 0i = 2$

13. $6 + 5i$

17. $8 - 10i$

19. $a^3 - 3ab^2 + 5a^2 - 5b^2 = a$, $3a^2b - b^3 + 10ab = b + 3$, where $z = a + bi$

21. $z_1 = 1 + i$, $z_2 = -i$

Exercises 1.2, Page 12

3. -3

7. (a) horizontal line $y = -2$
 (b) circle, center $= 1 - i$, rad $= 3$
 (c) circle, center $= i/2$, rad $= 2$
 (d) perpendicular bisector of segment joining $z = 1$ and $z = -i$
 (e) parabola $y^2 = 4(x + 1)$ with vertex -1, focus 0
 (f) ellipse with foci at ± 1
 (g) circle, center $= \frac{9}{8}$, rad $= \frac{3}{8}$
 (h) half-plane consisting of all points on or to the right of the vertical line $x = 4$
 (i) all points inside the circle with center i and rad $= 2$ (open disk)
 (j) all points outside the circle $|z| = 6$

Exercises 1.3, Page 22

3. $|z_1 + z_2|^2 + |z_1 - z_2|^2 = 2|z_1|^2 + 2|z_2|^2$

5. (a) 1   **(b)** $5\sqrt{26}$ **(c)** $5\sqrt{5}/2$ **(d)** 1

7. (a) $\arg = \pi + 2k\pi$, polar form: $\frac{1}{2}\text{cis}(\pi)$

  (b) $\arg = 3\pi/4 + 2k\pi$, polar form: $3\sqrt{2}\,\text{cis}(3\pi/4)$

  (c) $\arg = -\dfrac{\pi}{2} + 2k\pi$, polar form: $\pi\,\text{cis}(-\pi/2)$

  (d) $\arg = 7\pi/6 + 2k\pi$, polar form: $4\text{cis}(7\pi/6)$

  (e) $\arg = 7\pi/12 + 2k\pi$, polar form: $2\sqrt{2}\,\text{cis}(7\pi/12)$

  (f) $\arg = 5\pi/3 + 2k\pi$, polar form: $4\text{cis}(5\pi/3)$

  (g) $\arg = 5\pi/12 + 2k\pi$, polar form: $\frac{\sqrt{2}}{2}\,\text{cis}(5\pi/12)$

  (h) $\arg = 13\pi/12 + 2k\pi$, polar form: $\frac{\sqrt{14}}{2}\,\text{cis}(13\pi/12)$

9. rotation of vector z about origin through an angle ϕ in the counterclockwise direction

13. (b), (d)

21. $r = \sqrt{r_1^2 + r_2^2 + 2r_1r_2\cos(\theta_1 - \theta_2)}$, and θ is determined by the pair of equations: $\cos\theta = (r_1\cos\theta_1 + r_2\cos\theta_2)/r$, $\sin\theta = (r_1\sin\theta_1 + r_2\sin\theta_2)/r$

23. The center of mass of three particles that lie inside or on the unit circle also lies inside or on the circle.

29. $l = .0732$ m, $dl/dt = -.1155$ m/sec

Exercises 1.4, Page 31

1. (a) $\dfrac{\sqrt{2}}{2} - \dfrac{\sqrt{2}}{2}i$ **(b)** $e^2 i$ **(c)** $e^{\cos 1}\cos(\sin 1) + ie^{\cos 1}\sin(\sin 1)$

3. (a) $\dfrac{\sqrt{2}}{3}e^{-i\pi/4}$ **(b)** $16\pi e^{i4\pi/3}$ **(c)** $8e^{i3\pi/2}$

11. (a), (c), (d)

17. (a) circle $|z| = 3$ traversed counterclockwise

  (b) circle $|z - i| = 2$ traversed counterclockwise

  (c) upper half of circle $|z| = 2$ traversed counterclockwise

  (d) circle $|z - (2 - i)| = 3$ traversed clockwise

21. $\left|\dfrac{1 - z^n}{1 - z}\right| = |1 + z + \cdots + z^{n-1}| \leq 1 + 1 + \cdots + 1 = n$ for $z = e^{i\theta} \neq 1$

23. (a) $\frac{35\pi}{64}$ **(b)** $\frac{5\pi}{8}$

Exercises 1.5, Page 37

5. **(a)** $2(\cos \pi/4 + i \sin \pi/4)$, $2(\cos 3\pi/4 + i \sin 3\pi/4)$, $2(\cos 5\pi/4 + i \sin 5\pi/4)$, $2(\cos 7\pi/4 + i \sin 7\pi/4)$

(b) 1, $\cos 2\pi/5 + i \sin 2\pi/5$, $\cos 4\pi/5 + i \sin 4\pi/5$, $\cos 6\pi/5 + i \sin 6\pi/5$, $\cos 8\pi/5 + i \sin 8\pi/5$

(c) $\cos \pi/8 + i \sin \pi/8$, $\cos 5\pi/8 + i \sin 5\pi/8$, $\cos 9\pi/8 + i \sin 9\pi/8$, $\cos 13\pi/8 + i \sin 13\pi/8$

(d) $\sqrt[3]{2}[\cos(-\pi/9) + i \sin(-\pi/9)]$, $\sqrt[3]{2}(\cos 5\pi/9 + i \sin 5\pi/9)$, $\sqrt[3]{2}(\cos 11\pi/9 + i \sin 11\pi/9)$

(e) $\sqrt[4]{2}(\cos 3\pi/8 + i \sin 3\pi/8)$, $\sqrt[4]{2}(\cos 11\pi/8 + i \sin 11\pi/8)$

(f) $\sqrt[12]{2}\exp[i\pi(1 + 8k)/24]$, $k = 0, 1, 2, 3, 4, 5$

7. **(a)** $-\dfrac{1}{4} \pm i\dfrac{\sqrt{23}}{4}$ **(b)** $2 - i$, $1 - i$ **(c)** $1 \pm \sqrt[4]{2}e^{-i\pi/8}$

9. 1, $1 + i\sqrt{3}$, $1 - i\sqrt{3}$

11. $z = 1/(w - 1)$, $w = e^{i2k\pi/5}$, $k = 1, 2, 3, 4$

15. $\sqrt[4]{8}(\cos 5\pi/8 + i \sin 5\pi/8)$, $\sqrt[4]{8}(\cos 13\pi/8 + i \sin 13\pi/8)$

19. **(b)** $\dfrac{\pm\sqrt{2} + 2i}{3}$

21. **(a)** $\pm(3 + i)$ **(b)** $\pm(3 + 2i)$ **(c)** $\pm(5 + i)$
(d) $\pm(2 - i)$ **(e)** $\pm(1 + 3i)$ **(f)** $\pm(3 - i)$

Exercises 1.6, Page 42

3. (b), (c), (f)

5. (a), (c)

7. (a), (b), (c), (d), (e)

11. all points of S and 0

17. No, $S \cap T$ might not be connected.

19. D is not connected.

21. $\ln(x^2 + y^2) + C$

23. (a) and (d)

Exercises 1.7, Page 50

1. (a) $(0, 1, 0)$ **(b)** $(\frac{12}{101}, \frac{-16}{101}, \frac{99}{101})$ **(c)** $(\frac{-12}{25}, \frac{16}{25}, \frac{-3}{5})$

5. (a) The hemisphere $x_1 > 0$ **(b)** The bowl $x_3 < -\frac{3}{5}$

    (c) The zone $0 < x_3 < \frac{3}{5}$ **(d)** The dome $x_3 > 0.8$

    (e) The great circle $x_1 = x_2,\ -1 \leq x_3 \leq 1$ (45 and 225 degrees longitude)

CHAPTER 2

Exercises 2.1, Page 56

1. (a) $3x^2 - 3y^2 + 5x + 1 + i(6xy + 5y + 1)$ **(b)** $x/(x^2 + y^2) - iy/(x^2 + y^2)$

    (c) $x/[x^2 + (y-1)^2] + i(1-y)/[x^2 + (y-1)^2]$

    (d) $(2x^2 - 2y^2 + 3)/\sqrt{(x-1)^2 + y^2} + i4xy/\sqrt{(x-1)^2 + y^2}$

    (e) $e^{3x}\cos 3y + ie^{3x}\sin 3y$ **(f)** $2\cos y \cosh x + i2\sin y \sinh x$

3. (a) half-plane Re $w > 5$ **(b)** upper half-plane Im $w \geq 0$ **(c)** $|w| \geq 1$

    (d) circular sector $|w| < 2,\ -\pi < $ Arg $w < \pi/2$

5. (a) domain of definition $= \mathbf{C}$, range $= \mathbf{C} \smallsetminus \{0\}$ **(c)** circle $|w| = e$

    (d) ray (half-line) Arg $w = \pi/4$ **(e)** infinite sector $0 \leq $ Arg $w \leq \pi/4$

Exercises 2.2, Page 63

1. spirals to 0

7. (a) converges to 0   **(b)** does not converge   **(c)** converges to π

    (d) converges to $2 + i$ **(e)** converges to 0   **(f)** does not converge

11. (a) $-8i$ **(b)** $-7i/2$ **(c)** $6i$ **(d)** $-\dfrac{1}{2}$ **(e)** $2z_0$ **(f)** $4\sqrt{2}$

13. limits exist except at $z = -1$; continuous except at $z = -1$ and $z = 0$; removable discontinuity at $z = 0$

17. no

19. $-\dfrac{1}{2} - i$

21. (a) 1   **(b)** 0   **(c)** $-\dfrac{\pi}{2} + i$ **(d)** 1

Exercises 2.3, Page 70

7. (a) $18z^2 + 16z + i$

(b) $-12z\left(z^2 - 3i\right)^{-7}$

(c) $\dfrac{\left(iz^3 + 2z + \pi\right)2z - \left(z^2 - 9\right)\left(3iz^2 + 2\right)}{\left(iz^3 + 2z + \pi\right)^2}$

(d) $(z + 2)^2[-5z^2 + (-16 - i)z + 3 - 8i]/(z^2 + iz + 1)^5$

(e) $24i\left(z^3 - 1\right)^3\left(z^2 + iz\right)^{99}\left(53z^4 + 28iz^3 - 50z - 25i\right)$

9. (a) $2 - 3i$ (b) $\pm i$ (c) $\tfrac{1}{2}(-1 \pm i\sqrt{15})$ (d) $\dfrac{1}{2}, 1$

11. (a) nowhere analytic   (b) nowhere analytic   (c) analytic except at $z = 5$
(d) everywhere analytic   (e) nowhere analytic   (f) analytic except at $z = 0$
(g) nowhere analytic   (h) nowhere analytic

13. (a), (b), (d), (f), (g)

15. $\dfrac{3}{5}$

Exercises 2.4, Page 77

3. $g(z) = 3z^2 + 2z - 1$

5. $f'(z) = 2e^{x^2 - y^2}(x + iy)[\cos(2xy) + i\sin(2xy)]$

7. Hint: If $f' = g'$ in D, consider $h := f - g$.

9. If F were analytic, then $\operatorname{Im} F(z) \equiv 0$ implies $F(z)$ is constant, which is a contradiction.

11. Hint: Consider $f(z) + \overline{f(z)}$.

13. Hint: $|f(z)| \equiv$ constant by condition (8). Now use the result of Prob. 12.

Exercises 2.5, Page 84

3. (a) $v = -x + a$ (b) $v = -e^x \cos y + a$ (c) $v = y^2/2 - y - x^2/2 - x + a$
(d) $v = \cos x \sinh y + a$ (e) $v = \operatorname{Tan}^{-1}(y/x) + a = \operatorname{Arg} z + a$
(f) $v = -\operatorname{Re}(e^{z^2}) + a = -e^{x^2 - y^2}\cos(2xy) + a$

7. $\phi(x, y) = x + 1$

9. $\phi(x, y) = xy - 1$

13. (one example) $\phi(r, \theta) = r^4 \sin 4\theta = 4x^3 y - 4xy^3$

15. $\operatorname{Re} \frac{40}{63}(z^3 - z^{-3})$

17. (a) $\phi(x, y) = \operatorname{Re}(z^2 + 5z + 1)$ (b) $\phi(x, y) = 2\operatorname{Re}\left(\dfrac{z^2}{z + 2i}\right)$

19. $\operatorname{Re}\left(\dfrac{1}{2z^2}\right) + \dfrac{1}{2}$

Exercises 2.6, Page 90

3. See Fig. 2.6(a).

Exercises 2.7, Page 95

1. $\zeta_\pm = \frac{1 \pm \sqrt{1-4c}}{2}$ are fixed, ζ_- attracts for $-\frac{3}{4} < c < \frac{1}{4}$

3. (a) i and $-i$ are fixed and repellors

(b) $\frac{1}{2}$ and -1 are fixed and repellors, $-\frac{1}{2}$ is fixed and an attractor

5. $(-1 + \sqrt{5})/2$ is fixed and attractor, $(-1 - \sqrt{5})/2$ is fixed and repellor

9. If $|\alpha| \le 1$ the whole complex plane is the filled Julia set; if $|\alpha| > 1$ the origin is the filled Julia set.

CHAPTER 3

Exercises 3.1, Page 108

1. $2(z + 1)^2(z^2 + 9)$

3. (a) $z^3(z + 1 + i)^2$ (b) $(z - 2)(z + 2)(z + 2i)(z - 2i)$

(c) $(z - \omega_7)(z - \omega_7^2) \cdots (z - \omega_7^6)$, where $\omega_7 = e^{i2\pi/7}$

5. (a) $42 + 83(z - 2) + 80(z - 2)^2 + 40(z - 2)^3 + 10(z - 2)^4 + (z - 2)^5$

(b) $\sum_{k=0}^{10} \binom{10}{k} 2^{10-k}(z - 2)^k$ (binomial expansion of $[(z - 2) + 2]^{10} = z^{10}$)

(c) $(z - 2)^3 + (z - 2)^4$

11. (a) Pole at 0 of order 3, at $(-1 - \sqrt{2})i$ of order 1, and at $(-1 + \sqrt{2})i$ of order 1

(b) Pole at 2 of order 1 and at 3 of order 2

(c) Pole at -2 of order 6

(d) Pole at -2 of order 1

13. (a) $\dfrac{(3+i)/2}{z} - \dfrac{3+i}{z+1} + \dfrac{(3+i)/2}{z+2}$

(b) $\dfrac{i}{z} + \dfrac{i/2}{z+i} - \dfrac{3i/2}{z-i}$

(c) $\dfrac{1/6+i\sqrt{3}/6}{(z+1/2+i\sqrt{3}/2)^2} - \dfrac{i\sqrt{3}/9}{z+1/2+i\sqrt{3}/2} + \dfrac{1/6-i\sqrt{3}/6}{(z+1/2-i\sqrt{3}/2)^2} + \dfrac{i\sqrt{3}/9}{z+1/2-i\sqrt{3}/2}$

(d) $\dfrac{5}{2}z^2 - \dfrac{15}{4}z + \dfrac{47}{8} + \dfrac{33/16}{z+1/2} - \dfrac{9}{z+1}$

15. (a) $\dfrac{3-2i}{4}$ (b) $-\dfrac{1}{512}$ (c) 6 (d) 0 (e) 3

Exercises 3.2, Page 115

3. sum $= \dfrac{1 - e^{101z}}{1 - e^z}$ for $z \neq 2k\pi i$; sum $= 101$ for $z = 2k\pi i$

5. (a) $e^2 \dfrac{\sqrt{2}}{2} + i e^2 \dfrac{\sqrt{2}}{2}$ **(b)** $i e^2$ **(c)** $i \sinh 2$
(d) $\cos(1)\cosh(1) + i \sin(1)\sinh(1)$
(e) $-\sinh(1)$ **(f)** 0

9. (a) $2\pi z \exp\left(\pi z^2\right)$ **(b)** $-2\sin(2z) - \left(i/z^2\right)\cos(1/z)$
(c) $2\cos(2z)\exp[\sin(2z)]$ **(d)** $3\tan^2 z \sec^2 z$
(e) $2(\sinh z + 1)\cosh z$ **(f)** $1 - \tanh^2 z = \operatorname{sech}^2 z$

17. (a) $z = ik\pi/2,\ k = 0, \pm 1, \pm 2, \ldots$ **(b)** $z = 2k\pi - i\ln 3,\ k = 0, \pm 1, \pm 2, \ldots$
(c) no solution

19. Hint: If $z_1 \neq z_2$ and $e^{z_1} = e^{z_2}$, then $|z_1 - z_2| = |2k\pi i| \geq 2\pi$

21. (a) The plane cut along the negative imaginary axis and along the interval $[-1, 1]$.
(b) Upper half plane

25. $e^{1/z} = w$ when $z = \dfrac{1}{\operatorname{Log} w + i2\pi k}$, so it is easy to choose an integer k that makes $|z| \leq 0.001$ for any of these w.

Exercises 3.3, Page 123

1. (a) $i(\pi/2 + 2k\pi),\ k = 0, \pm 1, \pm 2, \ldots$
(b) $\frac{1}{2}\operatorname{Log} 2 + i(7\pi/4 + 2k\pi),\ k = 0, \pm 1, \pm 2, \ldots$
(c) $-i\pi/2$ **(d)** $\operatorname{Log} 2 + i(\pi/6)$

5. (a) $\operatorname{Log} 2 + i(\pi/2 + 2k\pi),\ k = 0, \pm 1, \pm 2, \ldots$ **(b)** $\pm\sqrt[4]{2}\exp(i\pi/8)$
(c) $i(2\pi/3 + 2k\pi),\ i(4\pi/3 + 2k\pi),\ k = 0, \pm 1, \pm 2, \ldots$

9. cut plane: $\mathbf{C} \setminus \{z = x + i : x \geq 4\}$, $f'(z) = -1/(4 + i - z)$

11. (one example) $f(z) = \operatorname{Log}\left(z^2 + 2z + 3\right)$, $f'(-1) = 0$

13. **(a)** $\operatorname{Log}(2z - 1)$
(b) $\mathcal{L}_0(2z - 1)$, where $\mathcal{L}_0\left(re^{i\theta}\right) = \operatorname{Log} r + i\theta,\ 0 < \theta < 2\pi$
(c) $\mathcal{L}_{\pi/2}(2z - 1)$, where $\mathcal{L}_{\pi/2}\left(re^{i\theta}\right) = \operatorname{Log} r + i\theta,\ \pi/2 < \theta < 5\pi/2$

15. $w = (1/\pi)\operatorname{Log} z$

19. Choose a branch of $\arg z$ that is continuous on the complement of the half-parabola; say, $\arg(re^{i\theta}) = \theta$, where $g(r) < \theta < g(r) + 2\pi$,
$g(r) = \operatorname{Tan}^{-1}\sqrt{2/(\sqrt{1 + 4r^2} - 1)}$.

Exercises 3.4, Page 129

1. 0.5

3. 0

5. $\dfrac{B-A}{\text{Log}\,r_1 - \text{Log}\,r_2}\,\text{Log}\,|z| + \dfrac{A\,\text{Log}\,r_1 - B\,\text{Log}\,r_2}{\text{Log}\,r_1 - \text{Log}\,r_2}$

7. $\text{Im}[(\log z)^2/2]$

Exercises 3.5, Page 136

1.   **(a)** $\exp(-\pi/2 - 2k\pi)$, $k = 0, \pm1, \pm2, \ldots$

    (b) $+1, -\dfrac{1}{2} + i\dfrac{\sqrt{3}}{2}, -\dfrac{1}{2} - i\dfrac{\sqrt{3}}{2}$

    (c) $\exp\left[-2k\pi^2 + \pi i\,\text{Log}\,2\right]$, $k = 0, \pm1, \pm2, \ldots$

    (d) $(1+i)\exp[2k\pi + \pi/4 - (i/2)\,\text{Log}\,2]$, $k = 0, \pm1, \pm2, \ldots$

    (e) $-2 + 2i$

3. **(a)** 2   **(b)** $e^{-\pi}$ **(c)** $(1+i)\exp[(i/2)\,\text{Log}\,2 - \pi/4]$

5. Take, for example, $z_1 = -1 + i$, $z_2 = i$, $\alpha = 1/2$.

7. $(1+i)\exp(-\pi/2)$

11. $z = \pi/4 + k\pi$, $k = 0, \pm1, \pm2, \ldots$

15. **(a)** $i\exp\left[\frac{1}{2}\text{Log}\left(1-z^2\right)\right]$ **(b)** $z\exp\left[\frac{1}{2}\text{Log}\left(4/z^2 + 1\right)\right]$

    (c) $z^2\exp\left[\frac{1}{2}\text{Log}\left(1 - 1/z^4\right)\right]$ **(d)** $z\exp\left[\frac{1}{3}\text{Log}\left(1 - 1/z^3\right)\right]$

17. $0 < \text{Sec}^{-1}x < \pi/2$ for $x > 1$; $\pi/2 < \text{Sec}^{-1}x < \pi$ for $x < -1$

19. $z = \log(-1 \pm \sqrt{1+w})$; $z = \text{Log}\,3 + i\pi(2k+1)$ or $z = i2\pi k$, k any integer

Exercises 3.6, Page 143

1. $I_s = \dfrac{\sin(\omega t - \phi_0)}{R}$

5. $I_s \to \dfrac{\cos(\omega t)}{R}$

7. **(b)** Any number of symmetrically placed inductors will work, other than for two. (Two will work if they are placed at right angles.)

CHAPTER 4

Exercises 4.1, Page 159

1. **(a)** $z(t) = (1 + i) + t(-3 - 4i), 0 \le t \le 1$
   (b) $z(t) = 2i + 4e^{-it}, 0 \le t \le 2\pi$
   (c) $z(t) = Re^{it}, \pi/2 \le t \le \pi$
   (d) $z(t) = t + it^2, 1 \le t \le 3$

3. $z(t) = a \cos t + ib \sin t, 0 \le t \le 2\pi$

5. yes

7. $z(t) = \begin{cases} -1 - i + 8t & 0 \le t \le \frac{1}{4} \\ 1 - i + 8i\left(t - \frac{1}{4}\right) & \frac{1}{4} \le t \le \frac{1}{2} \\ 1 + i - 8\left(t - \frac{1}{2}\right) & \frac{1}{2} \le t \le \frac{3}{4} \\ -1 + i - 8i\left(t - \frac{3}{4}\right) & \frac{3}{4} \le t \le 1 \end{cases}$

   length $= 8$

9. $z(t) = \begin{cases} -2 + \exp(-6\pi it) & 0 \le t \le \frac{1}{3} \\ -1 + 6\left(t - \frac{1}{3}\right) & \frac{1}{3} \le t \le \frac{2}{3} \\ 2 - \exp\left[6\pi i\left(t - \frac{2}{3}\right)\right] & \frac{2}{3} \le t \le 1 \end{cases}$

11. 15π

13. **(a)** instantaneous velocity at time t
    (b) instantaneous speed at time t
    (c) infinitesimal (differential) distance traveled during time interval dt
    (d) distance traveled from time $t = a$ to time $t = b$

Exercises 4.2, Page 170

1. yes

3. **(a)** $1 + i/3$ **(b)** $(1 + i) \sinh 2$
   (c) $\dfrac{i}{12}\left[1 - (1 + 2i)^6\right] = \dfrac{11}{3} - \dfrac{29i}{3}$ **(d)** $\dfrac{1}{2i} - \dfrac{1}{8 + 2i} = -\dfrac{2}{17} - \dfrac{8i}{17}$

5. $4\pi i$

7. $\dfrac{1}{2} + i$

9. $\dfrac{13}{10} + \dfrac{i}{6}$

11. **(a)** $3 + i$ **(b)** $3 + i$ **(c)** $3 + i$

13. $-2i$

Exercises 4.3, Page 178

1. (a) $-3 + 2i$ **(b)** $-2\sinh 1$ **(c)** $i\pi$ **(d)** 0 **(e)** $-\dfrac{i}{3}\sinh^3 1$

(f) $\dfrac{e^\pi}{2} + \dfrac{e^i}{2}(\cosh 1 + i\sinh 1)$ **(g)** $-\dfrac{\sqrt{2}}{3} - \dfrac{2}{3}\left(\sqrt{\pi}\right)^3 + \dfrac{i\sqrt{2}}{3}$

(h) $\pi - 2 + i\left(2 - \dfrac{\pi^2}{4}\right)$ **(i)** $\dfrac{\pi}{4} - \dfrac{1}{2}\arctan 2 + \dfrac{i}{4}\operatorname{Log} 5$

5. Hint: Consider Theorem 7.

7. Hint: Consider a branch of $\log(z - z_0)$ whose branch cut does not intersect C.

Exercises 4.4, Page 199

1. (a), (c)

3. (a), (b), (d), (e)

5. $z(s, t) = (2 - s)\cos 2\pi t + i(3 - 2s)\sin 2\pi t,\ 0 \le s, t \le 1$

9. (a), (c), (d), (f)

11. Since the whole plane $\mathbf{C}$ is simply connected, Theorem 10 (or Theorem 13) applies.

13. (a) π **(b)** 0 **(c)** $-\pi$

15. $-4\pi i$

17. 0

Exercises 4.5, Page 212

1. 0

3. (a) $-2\pi i$ **(b)** $\dfrac{3\pi i e^{3/2}}{2}$ **(c)** $\dfrac{2\pi i}{9}$ **(d)** $10\pi i$ **(e)** $-2e\pi i$ **(f)** $-i\pi/2$

5. $G(1) = 4\pi i,\ G'(i) = -2\pi(2 + i),\ G''(-i) = 4\pi i$

7. $\dfrac{-2\pi i}{9}$

9. $\left| f^{(n)}(0) \right| \le Mn!$

11. $\dfrac{\partial^2 u}{\partial x^2} = \operatorname{Re} f''$

13. $g(z)$ is not analytic inside Γ; note $G(z) \equiv 0$.

Exercises 4.6, Page 219

3. Hint: Apply the Cauchy estimates to the disk $\{\zeta : |\zeta - z| \leq r - |z|\}$.

5. $\left|e^{f(z)}\right| \leq e^M$, so $e^{f(z)} \equiv$ constant, which implies $f'(z)e^{f(z)} \equiv 0$. Thus $f'(z) \equiv 0$.

7. If f is entire and $|f(z)| \leq M|z|^n$ for $|z| > r_0$, where n is a nonnegative integer, then f must be a polynomial of degree at most n.

15. Hint: Suppose f does not vanish and apply the maximum modulus principle as well as the minimum modulus principle (Prob. 14).

17. $\dfrac{9\sqrt{2}}{8}$

Exercises 4.7, Page 225

1. $\phi(z) \equiv -5$

5. Consider $\phi_1(x, y) = y$ and $\phi_2(x, y) \equiv 0$ in the upper half-plane.

11. 3

15. $\dfrac{1}{\pi}\left[\tan^{-1}\left(\dfrac{1-x}{y}\right) - \tan^{-1}\left(\dfrac{-1-x}{y}\right)\right]$ for $y > 0$.

CHAPTER 5

Exercises 5.1, Page 239

1. **(a)** $\dfrac{9}{10} + i\dfrac{3}{10}$ **(b)** $3(1-i)$ **(c)** $\dfrac{3}{5}$ **(d)** $\dfrac{-2+i}{5 \cdot 2^{13}}$ **(e)** $\dfrac{9}{8}$ **(f)** -1

3. Hint: If $z_n, z_{n+1}, z_{n+2}, \ldots$ are within ε of their limit L, how far apart can any two z_j be?

5. Apply Prob. 3.

7. **(a)** diverges   **(b)** converges   **(c)** diverges   **(d)** diverges   **(e)** converges   **(f)** diverges

9. Hint: How does $|z|$ compare with $|x| + |y|$?

11. **(a)** $|z| < 1$ **(b)** $|z - i| < 2$ **(c)** all z **(d)** $|z + 5i| < 1$

17. Hint: Apply Log to the inequality $x^n < \dfrac{1}{2}$.

Exercises 5.2, Page 249

3. Hint: Use the chain rule to find the derivatives of the composite functions.

5. (a) $\displaystyle\sum_{j=0}^{\infty}(-z)^j$, $|z| < 1$ **(b)** $\displaystyle\sum_{j=0}^{\infty}\frac{(-1)^j z^{2j}}{j!}$, all z

(c) $\displaystyle\sum_{j=0}^{\infty}\frac{(-1)^j 3^{2j+1} z^{2j+4}}{(2j+1)!}$, all z

(d) $\displaystyle\sum_{j=0}^{\infty}\frac{i^j\left[1+(-1)^j\right]-i}{j!}z^j$, all z

(Note that $\left[1+(-1)^j\right]$ vanishes for odd j.)

(e) $i + \displaystyle\sum_{j=1}^{\infty}\frac{2}{(1-i)^{j+1}}(z-i)^j$, $|z-i| < \sqrt{2}$

(f) $\dfrac{1}{\sqrt{2}}\left\{1-\left(z-\dfrac{\pi}{4}\right)-\dfrac{1}{2!}\left(z-\dfrac{\pi}{4}\right)^2+\dfrac{1}{3!}\left(z-\dfrac{\pi}{4}\right)^3+\dfrac{1}{4!}\left(z-\dfrac{\pi}{4}\right)^4-\cdots\right\}$,
all z

(g) $\displaystyle\sum_{j=1}^{\infty}jz^j\left(=z\frac{d}{dz}\frac{1}{1-z}\right)$, $|z| < 1$

7. $2\displaystyle\sum_{j=0}^{\infty}\frac{z^{2j+1}}{2j+1}$, $|z| < 1$

11. (a) $1+z-\dfrac{z^3}{3}+\cdots$ **(b)** $-1-2z-\dfrac{5z^2}{2}-\cdots$

(c) $1+\dfrac{z^2}{2}+\dfrac{5z^4}{24}+\cdots$ **(d)** $z-\dfrac{z^3}{3}+\dfrac{2z^5}{15}+\cdots$

13. $\dfrac{z(1+z)}{(1-z)^3}$

17. $f(z)=(1-z)^{-1}$ is not analytic at $z=1$.

19. nine terms ($n = 0$ to 8)

Exercises 5.3, Page 258

3. (a) $|z| = 1$ **(b)** $|z-1| = \dfrac{1}{2}$ **(c)** $|z| = 0$ **(d)** $|z-i| = 3$

(e) $|z+2| = \dfrac{1}{\sqrt{10}}$ **(f)** $|z| = 2$

5. (a) $\dfrac{6!6^3}{3^6}$ **(b)** $2\pi i$ **(c)** 0 **(d)** 0

7. $z + \dfrac{z^3}{3} + \dfrac{z^5}{10}$

9. Hint: The polynomials are analytic inside and on C.

11. (b) Hint: Argue that two analytic functions that agree on a real interval must have identical derivatives.

13. **(a)** $\displaystyle\sum_{k=0}^{\infty} \dfrac{z^{2k}}{2^k k!} = e^{z^2/2}$

(b) $1 + z - \dfrac{4}{2!}z^2 - \dfrac{4}{3!}z^3 + \dfrac{4^2}{4!}z^4 + \dfrac{4^2}{5!}z^5 + \cdots$

$$= \left[1 - \dfrac{(2z)^2}{2!} + \dfrac{(2z)^4}{4!} - \cdots\right] + \dfrac{1}{2}\left[(2z) - \dfrac{(2z)^3}{3!} + \dfrac{(2z)^5}{5!} - \cdots\right]$$

$$= \cos 2z + \dfrac{1}{2}\sin 2z$$

(c) $\displaystyle\sum_{k=0}^{\infty}(k+1)z^{2k} = \dfrac{1}{\left(1 - z^2\right)^2}$

15. **(a)** $\displaystyle\sum_{k=0}^{\infty} \dfrac{(-1)^k}{(2k+1)!}\left[\int_{-1}^{2} t^{2k+1} g(t)\,dt\right] z^{2k+1}$

(b) Hint: Differentiate (a) termwise.

Exercises 5.4, Page 266

1. (a) 2 **(b)** ∞ **(c)** 0 **(d)** ∞

3. (a) 2 **(b)** 1 **(c)** $\dfrac{1}{3}$

(d) e (Hint: Use the ratio test.) **(e)** 1 **(f)** 1

5. (a) R **(b)** R^4 **(c)** $\sqrt{R}$ **(d)** R

(e) ∞ (if $R > 0$)

9. $f(z) = a_0 + \displaystyle\sum_{j=0}^{\infty} z^{2^j}$, $|z| < 1$

11. $1, \zeta, \dfrac{(3\zeta^2 - 1)}{2}, \dfrac{(5\zeta^3 - 3\zeta)}{2}$

Exercises 5.5, Page 276

1. (a) $\displaystyle\sum_{j=-1}^{\infty}(-1)^{j+1}z^j$ **(b)** $\displaystyle\sum_{j=2}^{\infty}(-1)^j z^{-j}$

(c) $-\displaystyle\sum_{j=-1}^{\infty}(z+1)^j$ **(d)** $\displaystyle\sum_{j=2}^{\infty}(z+1)^{-j}$

3. (a) $\dfrac{1}{3} \sum\limits_{j=0}^{\infty} \left[(-1)^j - \left(\dfrac{1}{2}\right)^j \right] z^j$

    (b) $\dfrac{1}{3} \sum\limits_{j=1}^{\infty} (-1)^{j-1} z^{-j} - \dfrac{1}{3} \sum\limits_{j=0}^{\infty} \left(\dfrac{1}{2}\right)^j z^j$

    (c) $\dfrac{1}{3} \sum\limits_{j=1}^{\infty} \left[(-1)^{j-1} + 2^j \right] z^{-j}$

5. $\dfrac{\frac{5}{4}}{(z-4)^3} + \sum\limits_{j=-2}^{\infty} (-1)^{j+1} \left(\dfrac{1}{4}\right)^{j+4} (z-4)^j$

7. (a) $\dfrac{1}{z^2} + \dfrac{1}{z^3} + \dfrac{3}{2z^4} + \cdots$ (b) $\dfrac{1}{z} - \dfrac{1}{2} + \dfrac{z}{12} + \cdots$

    (c) $\dfrac{1}{z} + \dfrac{z}{6} + \dfrac{7z^3}{360} + \cdots$ (d) $\dfrac{1}{e} \left[1 + z + \dfrac{z^2}{2!} + \cdots \right]$

9. $\dfrac{1}{2} < |z| < 2$

11. $\sum\limits_{j=n}^{\infty} \alpha^{j-n} \dfrac{(j-1)!}{(j-n)!(n-1)!} z^{-j}$

13. Hint: Use Eq. (1).

Exercises 5.6, Page 285

1. (a) pole of order 2 at 0, removable singularity at -1

    (b) essential singularity at 0   (c) simple poles at $\pm i$

    (d) simple poles at $2n\pi i$ $(n = 0, \pm 1, \pm 2, \ldots)$

    (e) simple poles at $\dfrac{2n+1}{2}\pi$ $(n = 0, \pm 1, \pm 2, \ldots)$

    (f) essential singularity at 0

    (g) removable singularity at 0

    (h) essential singularity at 0, simple poles at $\dfrac{1}{n\pi}$ $(n = \pm 1, \pm 2, \ldots)$

3. Possible answers are:

    (a) $\dfrac{(z-i)^2}{(z-2+3i)^5}$ (b) $ze^{1/(z-1)}$ (c) $\dfrac{(\sin z)e^{1/(z-i)}}{z(z-1)^6}$ (d) $\dfrac{e^{1/[z(z-1)]}}{(z-1-i)^2}$

5. (a) false   (b) true   (c) true   (d) false   (e) true

7. essential

9. Yes; $e^{1/z}$ is bounded on the negative real axis.

11. Hint: If, say, Re $f \le M$ then $e^{f(z)}$ would have a removable singularity. Now take the log.

13. Hint: Choose a tiny contour in the formula for a_{-j}.

19. (b) $0 + (\frac{1}{6} - \frac{2}{\pi^2})z + (0)z^2 + (\frac{7}{360} - \frac{2}{\pi^4})z^3 + (0)z^4 + \cdots$, radius $= 2\pi$

(c) $\cdots + \frac{-2\pi^2}{z^3} - \frac{1}{z} + (\frac{1}{6} - \frac{2}{\pi^2})z + (\frac{7}{360} - \frac{2}{\pi^4})z^3 + \cdots$

Exercises 5.7, Page 290

1. (a) essential singularity   **(b)** essential singularity   **(c)** analytic
   (d) zero of order 2   **(e)** pole of order 2   **(f)** essential singularity
   (g) essential singularity   **(h)** not isolated   **(i)** analytic

3. (a) $1 + 2 \sum_{j=1}^{\infty} (-1)^j / z^j$, $|z| > 1$ **(b)** $\sum_{j=0}^{\infty} (-1)^j / z^{2j}$, $|z| > 1$

(c) $\sum_{j=0}^{\infty} i^j / z^{3j+3}$, $|z| > 1$.

5. $(\deg Q) - (\deg P)$

7. Observe $\oint_{|z|=1} \frac{dz}{z} = 2\pi i$

11. $\deg(P) - \deg(Q) = 1, 0,$ or -1

Exercises 5.8, Page 301

1. z^2

3. $\sin \frac{1}{1-z}$ vanishes at $z = 1 - \frac{1}{n\pi}$, $n = 1, 2, \ldots$

5. all values except $0 \le \alpha < 1$

7. They both sum to $\frac{1}{1-z}$.

9. (a) no   **(b)** yes   **(c)** yes   **(d)** no   **(e)** yes   **(f)** yes

11. $f(z) = zg'(z)$

15. If $\phi(x, y) \to 0$, it can be harmonically extended as an odd function of y. If $\partial\phi/\partial y \to 0$, ϕ can be harmonically extended as an even function of y.

CHAPTER 6

Exercises 6.1, Page 313

1. (a) $\text{Res}(2) = e^6$ **(b)** $\text{Res}(1) = -2$, $\text{Res}(2) = 3$ **(c)** $\text{Res}(0) = 0$

   (d) $\text{Res}(-1) = -6$ **(e)** $\text{Res}(0) = 1$, $\text{Res}(-1) = -5/2e$

   (f) $\text{Res}(0) = \dfrac{1}{3}$

   (g) $\text{Res}\left[\pm\dfrac{(2n+1)\pi}{2}\right] = -1$, $n = 0, 1, 2, \ldots$

   (h) $\text{Res}(n\pi) = (-1)^n(n\pi - 1)$, $n = 0, \pm 1, \pm 2, \ldots$

   (i) $\text{Res}(1) = -2$

3. (a) $\pi i \sin 2$ **(b)** $\dfrac{\pi i \left(e^2 - 1\right)}{4}$ **(c)** $-8\pi i$

   (d) $\pi i \left[\dfrac{(2-5i)e^{2i}}{58} - \dfrac{12-5i}{50}\right]$ **(e)** $\dfrac{\pi i}{3}$ **(f)** 0 **(g)** 0

5. no; yes $(1/z^2$, for example)

7. $2\pi i$

Exercises 6.4, Page 336

5. $\dfrac{\pi \sin 3}{2e^3}$

7. $\dfrac{\pi}{3e}\left(1 - \dfrac{1}{2e}\right)$

9. $\dfrac{i\pi}{e^6}$

11. $m > 0$, $\deg P < \deg Q$

Exercises 6.5, Page 344

1. (a) $\dfrac{i\pi}{2}$ **(b)** $\dfrac{3\pi i e^{3i}}{8}$ **(c)** 0 **(d)** $-\pi i$

9. $\dfrac{3\pi}{4}$

11. $-\pi \cot(a\pi)$

Exercises 6.6, Page 354

9. (a) $\pi/4 - \text{Log}\,\sqrt{2}$ **(b)** $2\pi\sqrt{3}/9$

13. $2\pi\sqrt{3}/3$

Exercises 6.7, Page 364

1. (a), (c), (e), and (f)

3. 1

7. Hint: Compare with $f(z) = 27$.

9. 4

13. Easy: let $h(z) = -f(z)$

15. Hint: f and $f + h$ have the same poles; now apply Prob. 14.

21. A "zero" for $F(z)$ is a "minus 1" for $P(z)$; apply the argument principle.

CHAPTER 7

Exercises 7.1, Page 374

1. $\text{Log}\,|w| + \text{Arg}\,w$; $e^x \cos y + e^x \sin y$

3. $a_4 + \dfrac{1}{\pi} \displaystyle\sum_{k=1}^{3} (a_k - a_{k+1})\,\text{Arg}\,(z - x_k)$

5. $\dfrac{1}{2} - \dfrac{1}{2}\dfrac{x^2 + y^2 - 1}{(1+x)^2 + y^2}$

7. These are the Cauchy-Riemann equations for $f^{-1}(w)$.

Exercises 7.2, Page 382

1. (a) 1; $f(-1 + \zeta) = f(-1 - \zeta)$ **(b)** 1; $f(n\pi + \zeta) = f(n\pi - \zeta)$
 (c) 2; $f(r) = f\left(re^{i2\pi/3}\right) = f\left(re^{i4\pi/3}\right)$

3. Angles increase (decrease) for $\alpha > 1$ ($\alpha < 1$).

5. pure imaginary constants

11. **(a)** the whole upper half-plane: $\text{Im}\,w > 0$
 (b) the whole plane minus the logarithmic spiral $\rho = e^\phi$, $-\infty \le \phi < \infty$
 (c) $\{w : |w| < 1, \text{Im}\,w > 0\}$
 (d) $\{w : |w| > 1, \text{Im}\,w > 0\}$

    (e) the upper half-annulus $\{w : e < |w| < e^2, \operatorname{Im} w > 0\}$

    (f) $\{w : |w| > 1\}$ and $\{w : |w| < 1\}$

13. (a) the upper half-plane: $\operatorname{Im} w > 0$ **(b)** fourth quadrant

    (c) the whole plane minus the real intervals $(-\infty, -1], [1, \infty)$

    (d) the interior of the ellipse $(u^2/\cosh^2 1) + (v^2/\sinh^2 1) = 1$ excluding the real segments $[-\cosh 1, -1], [1, \cosh 1]$

Exercises 7.3, Page 392

1. $w = 3iz + 5$

3. (a) $\{w : |w - 2 + 2i| \le 1\}$ **(b)** $\{w : |w - 6i| \le 3\}$

    (c) $\{w : \operatorname{Re}(w) \le \frac{1}{2}\}$

    (d) $\{w : \operatorname{Re}(w) \ge \frac{3}{2}\}$ **(e)** $\{w : |w - \frac{2}{3}| \le \frac{1}{3}\}$

5. $w = e^{3\pi i/4} \left(\dfrac{z+i}{z-1} \right)$

7. (a) $w = iz$ **(b)** $w = \dfrac{2z}{z+1}$ **(c)** $w = \dfrac{z+i}{z-1}$ **(d)** $w = \dfrac{z+1}{z-1}$

9. the region exterior to both of the circles $C_1 : |w - (1-i)/2| = 1/\sqrt{2}$ and $C_2 : |w - (1+i)/2| = 1/\sqrt{2}$

11. $w = \exp\left(4\pi \left[-i \left(\dfrac{1}{z-2} + \dfrac{1}{4} \right) \right] \right)$

Exercises 7.4, Page 403

1. $z - 2$

3. (a) $\dfrac{4-3i}{25}$ **(b)** $\dfrac{7-i}{6}$ **(c)** $\dfrac{5-2i}{3}$

5. $\lambda > 0$

7. No. (This would violate the symmetry principle.)

9. 0

17. $w = \dfrac{i-2z}{2+iz} e^{i\theta}$ (any real θ)

19. $w = \dfrac{az+b}{cz+d}$ with a, b, c, d real and $ad - bc > 0$

21. (b) No. (A shift, for example, does not commute with an inversion.)

Exercises 7.5, Page 416

1. $w = A (z - x_1)^2 - 1$, where $A < 0$

3. $w = \dfrac{i}{2} - \dfrac{i}{\pi} \sin^{-1} z$

5. $w = -\dfrac{2}{\pi} \left(\sin^{-1} z + z\sqrt{1 - z^2} \right)$

7. $w = \dfrac{i}{\pi} \sqrt{z^2 - 1} + \dfrac{\sin^{-1} z}{\pi} + \dfrac{1}{2}$

9. $w = \dfrac{1}{\pi} \mathrm{Log} \left(\dfrac{z - x_2}{x_2 - x_1} \right)$, where $x_1 < x_2$

11. The analytic continuation of the S-C transformation across the interval (x_{j-1}, x_j) maps the lower half-plane onto the figure obtained by reflecting the polygon P through the mirror containing $f(x_{j-1})$ and $f(x_j)$.

Exercises 7.6, Page 430

1. $\phi(x, y) = \dfrac{2}{\pi} \mathrm{Arg} \left(\dfrac{1 + z}{1 - z} \right)$

3. $T(x, y) = \dfrac{2}{\pi} \mathrm{Arg} \left(\dfrac{1 + z^2}{1 - z^2} \right) - 1$

5. $\phi(x, y) = \dfrac{1}{\pi} \mathrm{Arg} \left(\dfrac{\sqrt{z^2 + 1} - 1}{\sqrt{z^2 + 1} + 1} \right)$

7. **(a)** $T(x, y) = 1 - \dfrac{1}{\pi}[\mathrm{Arg}(\cos(\pi i z) + 1) + \mathrm{Arg}(\cos(\pi i z) - 1)]$

(b) $T(x, y) = \dfrac{2}{\pi} \mathrm{Arg} \left(e^{\pi z} + 1 \right) - \dfrac{1}{\pi} \mathrm{Arg} \left(e^{\pi z} \right)$

9. $\phi(x, y) = (\mathrm{Log}\, s)^{-1} \mathrm{Log} \left| \dfrac{4 - \lambda z}{4\lambda - z} \right|$, where $\lambda = \dfrac{19 + \sqrt{105}}{16}$ and $s = \dfrac{13 - \sqrt{105}}{8}$.

11. A parametric representation of the streamlines is obtained by holding y constant in the S-C mapping equation $w = f(x + iy)$.

Exercises 7.7, Page 439

1. isotherms: $z(t) = g(a + it)$, $t \geq 0$, $-\dfrac{\pi}{2} < a < \dfrac{\pi}{2}$, where $g(w) = \dfrac{1}{2} + \dfrac{w - \cos w}{\pi}$ maps the half-strip $-\dfrac{\pi}{2} < u < \dfrac{\pi}{2}$, $v > 0$ onto the given region.

3. $T(z) = \dfrac{2}{\pi} \mathrm{Re} \left[\sin^{-1} \left(z^2 \right) \right]$

5. $T(z) = -\dfrac{20}{\pi} \operatorname{Re}\left[\sin^{-1}\left(-e^{-\pi z}\right)\right]$

7. streamlines: $\operatorname{Im}\left(z^{\pi/\alpha} + z^{-\pi/\alpha}\right) = \text{constant}$

CHAPTER 8

Exercises 8.1, Page 459

1. **(a)** $\dfrac{i}{8}\left[-3e^{it} + 3e^{-it} + e^{3it} - e^{-3it}\right]$

   (b) $\displaystyle\sum_{n=-\infty}^{\infty} \dfrac{12(-1)^n}{\pi\left(9 - 4n^2\right)\left(1 - 4n^2\right)} e^{i2nt/3}$

   (c) $c_n = \dfrac{2(-1)^n}{n^2}$ for $n \neq 0$; $c_0 = \dfrac{\pi^2}{3}$

   (d) $c_n = (-1)^n\left[\dfrac{i\pi}{n} - \dfrac{2i}{\pi n^3}\right] + \dfrac{2i}{\pi n^3}$ for $n \neq 0$; $c_0 = 0$

3. (a), (b)

5. $\bar{c}_n = c_{-n}$

7. **(a)** $\dfrac{1}{8} + \dfrac{\cos 2t}{2} - \dfrac{\cos 4t}{104}$

   (b) $\dfrac{\pi^2}{3} + \displaystyle\sum_{\substack{n=-\infty \\ n\neq 0}}^{\infty} \dfrac{c_n e^{int}}{1 + in - n^2}$, c_n as in Prob. 1(c).

   (c) $\displaystyle\sum_{\substack{n=-\infty \\ n\neq 0}}^{\infty} \dfrac{c_n e^{int}}{2 + 4in - n^2}$, c_n as in Example 4.

11. $u(x,t) = \displaystyle\sum_{n=1}^{\infty} b_n \sin nx \cos nt + \sum_{n=1}^{\infty} c_n \sin nx \sin nt$

$$b_n = \dfrac{2}{\pi}\int_0^{\pi} f_1(\xi)\sin n\xi\, d\xi$$

$$c_n = \dfrac{2}{\pi n}\int_0^{\pi} f_2(\xi)\sin n\xi\, d\xi$$

Exercises 8.2, Page 473

1. **(a)** $G(\omega) = \dfrac{1}{\pi}\left(\dfrac{1}{1 + \omega^2}\right)$ **(b)** $G(\omega) = \dfrac{1}{2\sqrt{\pi}} e^{-\omega^2/4}$ **(c)** $G(\omega) = \dfrac{-i\omega e^{-\omega^2/4}}{4\sqrt{\pi}}$

   (d) $G(\omega) = \begin{cases} \frac{1}{2} & \text{if } |\omega| < 1 \\ 0 & \text{if } |\omega| > 1 \\ \frac{1}{4} & \text{if } \omega = \pm 1 \end{cases}$ **(e)** $G(\omega) = \begin{cases} -\frac{i}{2}\sin\omega & \text{if } |\omega| \leq \pi \\ 0 & \text{if } |\omega| \geq \pi \end{cases}$

3. (a) $G(\omega) = \dfrac{e^{-|\omega|/\sqrt{2}}}{2\sqrt{2}} \left(\cos \dfrac{|\omega|}{\sqrt{2}} + \sin \dfrac{|\omega|}{\sqrt{2}} \right)$

(b) $G(\omega) = -\dfrac{ie^{-|\omega|/\sqrt{2}}}{2} \sin \dfrac{\omega}{\sqrt{2}}$

(c) $G(\omega) = \dfrac{1}{2\sqrt{\pi}} e^{-\omega^2/4}$

Exercises 8.3, Page 484

1. (a) $\dfrac{3s}{s^2 + 4} - \dfrac{8}{s + 2}$ (b) $\dfrac{2}{s} - \dfrac{\pi}{(4 - s)^2 + \pi^2}$ (c) $\dfrac{1}{s} \left(1 - e^{-s} \right)$

(d) $\dfrac{1}{s} \left(e^{-s} - e^{-2s} \right)$ (e) $\dfrac{2}{s \left(s^2 + 4 \right)}$ (f) $\sqrt{\dfrac{\pi}{s}}$

3. (a) $\dfrac{1}{2} \sin(2t)$ (b) $4te^t$ (c) $e^{-2t} - te^{-2t}$

(d) $\dfrac{1}{2} \left[1 - 2e^{-t} + e^{-2t} \right]$ (e) $e^{-2t} \left(\cos \sqrt{3}t + \dfrac{1}{\sqrt{3}} \sin \sqrt{3}t \right)$

5. (a) $f(t) = \dfrac{1}{2} e^{3t} + \dfrac{5}{2} e^t$

(b) $f(t) = 4e^{2t} - 3e^{3t}$

(c) $f(t) = \dfrac{28}{39} e^{2t} - \dfrac{5}{6} e^{-t} - \dfrac{1}{13} e^{-t} \sin 2t + \dfrac{3}{26} e^{-t} \cos 2t$

(d) $f(t) = \begin{cases} 0 & \text{if } 0 \le t \le 3 \\ \frac{1}{2} + \frac{1}{2} e^{2t-6} - e^{t-3} & \text{if } 3 \le t \le 6 \\ \frac{1}{2} e^{2t-6} - \frac{1}{2} e^{2t-12} + e^{t-6} - e^{t-3} & \text{if } t \ge 6 \end{cases}$

7. (b) Hint: Consider the partial fraction expansion of $F(s)$, and compare with the table entries.

9. (b) i. $x(t) = \cos \sqrt{3}t, \ y(t) = -\cos \sqrt{3}t$;

ii. $x(t) = y(t) = \cos t$;

iii. $x(t) = \dfrac{1}{2} \left(\cos t + \cos \sqrt{3}t \right), \ y(t) = \dfrac{1}{2} \left(\cos t - \cos \sqrt{3}t \right).$

(c) (i), (ii)

Exercises 8.4, Page 494

3. Multiply the radii by $|\alpha|$.

5. (a) $a(n) = -(-3)^{-n} (n \le -1), 0$ otherwise

(b) $a(n) = \left(-\dfrac{1}{3} \right)^n (n \ge 0), 0$ otherwise

(c) $a(n) = (-1)^n 2^{n+3} (n \leq -4)$, 0 otherwise

(d) $a(n) = (-2)^{n+3} (n \geq -3)$, 0 otherwise

(e) $a(n) = -\left(\dfrac{1}{2}\right)^n (n \geq 1),\ -3^{n-1} (n \leq -1),\ -\dfrac{1}{3} (n = 0)$

(f) $a(n) = 4\left(-\dfrac{1}{2}\right)^n - 3\left(-\dfrac{1}{4}\right)^n (n \geq 0)$, 0 otherwise

(g) $a(n) = 4\left(\dfrac{1}{2}\right)^n + 2n - 4\ (n \geq 0)$, 0 otherwise

(h) $a(n) = \dfrac{1}{\alpha^n}\left[\alpha^{-1} - \alpha\right] (n \geq 1),\ \dfrac{1}{\alpha} (n = 0)$, 0 otherwise

7. (a) 2^{-n+1} (b) $\dfrac{1}{3} + \dfrac{2}{3}(-2)^n$ (c) $\dfrac{1}{2} + 2^{n+1} - \dfrac{3^n}{2}$

Exercises 8.5, Page 507

3. Hint: Use Lemma 4, Section 6.5.

7. (b) p.v. $\dfrac{1}{x - x_0} - i\pi\delta(x - x_0)$

Index